MICROBIAL ECOLOGY OF FOODS

VOLUME II

Food Commodities

MICROORGANISMS IN FOODS 3

Sponsored by the
INTERNATIONAL COMMISSION ON MICROBIOLOGICAL
SPECIFICATIONS FOR FOODS
of the
INTERNATIONAL ASSOCIATION OF MICROBIOLOGICAL SOCIETIES

CONTRIBUTORS

A. Alaoui
A. C. Baird-Parker
M. H. Brown
F. L. Bryan
N. S. de Caruso
J. H. B. Christian
D. S. Clark
C. Cominazzini
D. A. Corlett, Jr.
A. N. Al-Dulaimi
R. P. Elliott
O. Emberger
J. M. Goepfert
J. Gomez-Ruiz
R. Hablas
B. C. Hobbs
A. Hurst
S. T. Iaria
M. Ingram
M. Kalember-Radosavljević
I. Kashoulis

W. J. Kooiman
K. H. Lewis
J. Liston
H. Lundbeck
S. Mendoza
G. Mocquot
S. El-Mossalami
Y. K. Motawa
C. F. Niven, Jr.
P. M. Nottingham
J. C. Olson, Jr.
N. Othman
H. Pivnick
F. Quevedo
T. A. Roberts
H. Sidahmed
J. H. Silliker
B. Simonsen
H. J. Sinell
J. Takács
M. Van Schothorst

MICROBIAL ECOLOGY OF FOODS

VOLUME II

Food Commodities

by
The International Commission on Microbiological
Specifications for Foods

Editorial Committee
J. H. Silliker (Chairman), R. P. Elliott (Editorial Coordinator)
A. C. Baird-Parker, F. L. Bryan, J. H. B. Christian
D. S. Clark, J. C. Olson, Jr., T. A. Roberts

1980

ACADEMIC PRESS

New York   London   Toronto   Sydney   San Francisco

ACADEMIC PRESS, INC.
111 Fifth Avenue, New York, New York 10003

United Kingdom Edition published by
ACADEMIC PRESS, INC. (LONDON) LTD.
24/28 Oval Road, London NW1 7DX

Library of Congress Cataloging in Publication Data

International Commission on Microbiological
    Specifications for Foods.
    Food commodities.

    (Microbial ecology of foods ; v. 2)
    Bibliography: p.
    Includes index.
    1. Food——Microbiology. 2. Food industry
and trade. I. Silliker, John H. II. Elliot,
Paul R. III. Title. IV. Series.
QR115.M45 vol. 2 576'.163s [664] 80–14888
ISBN 0–12–363502–0 (v. 2)

PRINTED IN THE UNITED STATES OF AMERICA

80 81 82 83   9 8 7 6 5 4 3 2 1

Dedicated to

MAURICE INGRAM
1912–1977

Founder of the ICMSF

DISTINGUISHED SCIENTIST, WARM FRIEND

Contents*

*Persons listed in Table of Contents are the chairpersons of their respective committees. Their affiliations are listed in Appendix III

Preface

Microbial Ecology of Foods was written by a group of over 30 scientists chosen from 22 countries for their expert knowledge in special fields of food microbiology. The work is not, however, a collection of separately authored chapters, but each chapter fits into an overall scheme to provide complete coverage of all important aspects of food microbiology. The book is intended as a source of information for those who must interpret the results of microbiological analyses conducted on foods.

Volume I, "Factors Affecting Growth and Death of Microorganisms," encompasses the environmental factors that affect microorganisms in foods, namely, temperature, irradiation, water activity, pH, E_h, organic acids, curing salts, antibiotics, gases, packaging, and cleaning systems. Special chapters treat the effects of cell injury on survival and recovery of microorganisms in food and the metabolic aspects of mixed populations. The treatment accorded each of these factors includes methods of measurement, effects on spoilage organisms and pathogens, control measures, and interrelationships with the other factors.

Volume I can stand alone for those who want a basic theoretical background in food microbiology. In this respect it is particularly suitable as an undergraduate or postgraduate textbook for students who have had at least one course in general microbiology. Volume I also gives valuable background information in the theoretical aspects for those whose interest is primarily the applied microbiology of Volume II.

Volume II, "Food Commodities," is a comprehensive treatise on the microbiology of specific commodity groups. Each chapter covers (1) the important properties of the food commodity that affects the microbial content, (2) the initial microbial flora on flesh foods at slaughter or on vegetable foods at harvest, (3) the effects of harvest, transport, processing, and storage on the microbial content, and (4) the means of controlling the process and the microbial content. Each chapter is an up-to-date review of applied microbiology, compiled by leading authorities selected solely for their expert knowledge.

Volume II is meant to be used by those interested primarily in applied aspects of food microbiology, namely, food processors, food microbiologists, food technologists, veterinarians, public health workers, and regulatory officials. Although some will use Volume II alone, most will want Volume I as well, to establish a background of understanding of the theoretical aspects of foods as substrates for microbial development and destruction.

The subject of each chapter in this book could justify a separate volume, and accordingly this text does not include all the material that has been written on a given subject. In each chapter, the reader has been directed to appropriate key publications for further study. The ICMSF is concerned specifically with foods which move in international commerce. Thus discussion of products of local or regional importance is necessarily minimal.

This book is a logical sequel to the two earlier works published by the Commission (see Appendix I).

Acknowledgments

Microbial Ecology of Foods is the result of four years of study by ICMSF members and consultants, involving workshops held in Alexandria (1976) and Cairo (1977, 1978), Egypt. The content was planned by Dr. J. H. Silliker and Dr. J. H. B. Christian and debated and approved by the Commission in plenary session. Chapters were assigned to subcommittees, whose chairmen, listed in the Table of Contents, were responsible for writing the texts. The Editorial Committee was under the Chairmanship of Dr. J. H. Silliker. R. Paul Elliott coordinated and edited the various parts of the book and proofread copy.

The Commission is most grateful for the generous financial sponsorship from the U.S. Department of Health, Education and Welfare, Public Health Service, Center for Disease Control (CDC); the Ministry of Health, Arab Republic of Egypt; the World Health Organization; the Ministry of Health of Kuwait; and the various companies within the food industry (see Appendix II). This assistance, of course, does not constitute endorsement of the findings and views expressed herein. Special thanks are given to Dr. D. J. Sencer, Dr. A. E. Najjar, Dr. F. L. Bryan, and others at the Center for Disease Control, U.S. Department of Health, Education, and Welfare, and to Dr. Fouad Mohy El-Din, Dr. Ibrahim Badran, Dr. Ahmed El-Akkad, Dr. Hekmat El-Sayed Aly, Dr. Mohammed Fahmi Saddick Ahmed, and others at the Ministry of Health, Arab Republic of Egypt for developing and coordinating the project that supported workshops at which most of the work on this book was done. Appreciation is extended to Dr. J. C. Olson, Jr., and Dr. A. C. Baird-Parker for organizing meetings of the Editorial Committee and to the Instructional Media Division, Bureau of Training, CDC, for drawing the figures. Finally, thanks are expressed to the respective national governments, universities, and private companies for supporting the participation of their staff and the work of the Commission, of which the present text is but one result.

Introduction:
The Evolution of Processing Techniques

The raw materials of the food industry are living plants and animals, whose external surfaces are contaminated with a heterogenous microflora from the farm and whose internal tissues are usually sterile. From the time of slaughter or harvest, the raw materials are subjected to the many processes that destroy or remove the initial flora, or that permit further contamination or multiplication.

The microbiology of foods is a study of selective environments. Each processing step or manipulation influences qualitatively and quantitatively both the surviving flora and the flora that will develop during subsequent processing, handling or storage. The selective nature of these environments in turn influences the spoilage patterns which, when characteristic, help to safeguard the consumer. At the same time, this selectivity may pose specific health hazards if it favors the survival or growth of pathogenic organisms. A single preservation technique is seldom utilized in the production of a particular food. Generally, we deal with a combination of physical and chemical agents, which by their interactions, establish stability.

The evolution of processing techniques has been largely by trial and error. This is not surprising if one considers that microorganisms were first discovered only within the past three centuries. For almost all of history man has sought means to preserve food from destruction by agents about which he had little knowledge. Food processing and handling procedures evolved empirically, and the science necessary to understand successful food preservation systems was born, in many cases, centuries later.

One of man's earliest discoveries was that low temperatures extend the keeping quality of meats. He probably noted that carcasses of slaughtered animals kept longer when stored in caves. At the same time, he unknowingly used factors other than low temperature to extend the edible life of his kill. As the warm carcass cooled, the surface dried (reduced water activity), which delayed the development of microorganisms. Furthermore, as the carcass cooled, postmortem glycolysis formed acid, making the interior of the meat more resistant to spoilage. Finally, muscle metabolism caused a drop in oxidation-reduction potential, which restricted the growth of microorganisms in interior tissues to the anaerobes. Thus prehistoric man, capitalizing on the observation that chilled meat kept longer, inadvertently utilized three other factors which limited spoilage.

We do not know when man discovered that partial removal of water increased the shelf life of foods, but the Bible mentions dried grapes and dried figs. No doubt the sun's heat was the first means of drying. However, there is also a Bible reference to "parched corn" which suggests the use of drying by means other than the sun (Labuza, 1976). Early in history, man discovered that foods could be protected from spoilage during drying if they were treated with salt. The addition of salt reduced the water activity in the meat, and this, combined with the further loss of water during sun drying, modified its flora, discouraging the growth of pseudomonads thousands of years before Leeuwenhoek took the first rudimentary steps that eventually led to their identification. The salt drying of fish was described in Egyptian hieroglyphics, and by 1000 B.C. salted and smoked meats were generally available. Even before the dawn of recorded history man probably used a combination of methods to reduce water activity.

Man first used sodium nitrate on meats because it was present as a contaminant in salt from some sources. It is not unrealistic to postulate that early man discovered that such meat had a more desirable color and better keeping qualities that did normal salted meat. Only toward the end of the nineteenth century was it recognized that in meat curing brines it was *nitrite* from bacterial reduction of nitrate that fixed the color and preserved the meat. We are still gaining understanding of the interaction of factors which early man first imposed on meat as the result of his empirical observations. Only within the past century has man applied thermal processing to cured meats to yield completely shelf-stable canned products, and only in the past few decades have we begun to understand why cured meats have greater stability than uncured meats.

A similar evolution of processing techniques has occurred with dried foods. More than 3000 years ago the Incas in Peru produced dried potatoes and vegetables by applying the elements of freeze drying, a process which we now consider sophisticated modern technology. Vegetables were allowed to freeze overnight and then stomped upon to squeeze out the juices. This was repeated to produce a thin layer which was dried in the sun. The process was carried out on mountains at heights above 3000 meters, and hence the drying process was effected under reduced atmospheric pressures (Labuza, 1976). Similarly the Sioux and Cree Indians of North America made a product called *pemmican* by cutting and pounding buffalo meat into thin strips, drying it in the sun, mixing it with acid berries and nuts, and then embedding it in fat. This semi-dry product stored in fat was a forerunner of the rapidly expanding field of semimoist foods. By the use of acidic berries, yet another preservation technique (acidification) was used in combination with drying (Labuza, 1976).

Fermentation was without doubt accidentally discovered early in man's history when certain fruits, vegetables, and milk became acid or alcoholic when held at ambient temperatures. The changed product had a longer storage life than the original material, and indeed in many cases it had a desirable new flavor and texture.

Other foods evolved from fermented products—e.g., cheese from naturally soured milk. Cheese was described by Greek and Roman writers several centuries before the birth of Christ, and certainly was produced long before that. Primitive peoples learned that the curd of sour cream could be concentrated by pressing or partial drying and that the product could be preserved longer than could sour milk itself. The milk was curdled in several ways—by natural microbial souring, by rennet from the stomachs of nursing animals, by vinegar, or by the juices of certain plants. Spices and salt were often added to the curd, and some varieties of cheese were smoked (Foster *et al.*, 1957). Thus fermentation was combined with reduction in water activity, and upon this were imposed chemical additives, yielding new foods with increased stability.

Man's first directed attempt to create a new method of food conservation can be traced to the work of Nicholas Appert, who invented canning more than a century and a half ago (Goldblith, 1976). Appert's book clearly and concisely described his canning process (Appert, 1810), and no other major book on thermal processing of any consequence appeared for over 70 years. Furthermore science was not applied to the study of the process for 85 years after the first publication of his book, even though millions of cases of food had been packed and consumed in the interim (Bitting, 1937).

The evolution of thermal processing depended greatly on the development of equipment and quite secondarily on an understanding of the biological mechanisms involved. These advances have been reviewed (Goldblith, 1971, 1972). In about 1860 Isaac Solomon, a Baltimore canner, modified the technology by adding calcium chloride to the cooking water. This increased its boiling point to 115°C which greatly speeded production and, as we now know, also increased the microbicidal effectiveness of the process. In 1874 A. K. Shriver was issued a patent on his invention of the pressure cooker, or retort, which wrought a revolution in the canning industry. In 1910, Peter Durand received a United Kingdom patent which followed Appert's procedure except that it used containers of tin and other metals instead of glass.

During the century following Shriver's invention, the evolution of thermal processing has largely involved aseptic canning, high temperature–short time processing and hydrostatic sterilizing. These and other advances were primarily engineering improvements directed toward heating the canned food more effectively or efficiently.

Concomitant with engineering advances, processes evolved which combined thermal processing with other preservation techniques—e.g., concentration, or water removal. In 1856, Gail Borden obtained a patent on condensing milk, a process that combined a substerilizing thermal treatment with reduced water activity to yield a stable product (Borden, 1856). Such relatively mild heat treatments are also used to produce a variety of high acid canned foods, the growth of surviving organisms being inhibited by low pH.

The stability of canned cured meats can be traced to the effects of curing salts as well as thermal processing to a degree that surviving spore-forming bacteria are injured and therefore incapable of out-growth in the cured meat environment. This combination of factors permits the production of shelf-stable canned cured meats using a heat treatment equivalent to approximately 1/25th of the so-called botulinum cook. An even milder heat process, termed pasteurization, will yield a canned cured meat product which has an indefinite shelf life when held under refrigeration. Hence the stability is the result of proper sequential application of curing salts, heat and refrigeration.

In summary, throughout man's history, food handling practices have evolved empirically. Almost without exception, the procedures involved combinations of preservation techniques, most of which were applied long before man had the basis for understanding the mechanisms of his success. Interestingly, man's first directed attempt to create a new method of food preservation, canning, has perhaps involved less innovation than the variety of other processes which developed by trial and error during man's early history.

Contents of Volume I

FACTORS AFFECTING LIFE AND DEATH OF MICROORGANISMS

15

Meats and Meat Products

I. INTRODUCTION

In all parts of the world, meat has long been regarded as a nutritious, highly desirable food, but recently, in more affluent countries, the role of meat as a basic foodstuff has changed. Technological developments in food processing, preservation, and handling have given consumers a much greater choice over the foods they can buy. Consequently, meat eaters have become more selective, more conscious of quality, and more concerned about value for money. They are more aware of and sensitive to spoilage, to off odors, strong flavors, discolorations, and other indications of lack of freshness. As a result, quality of product has become a more significant factor in marketing meat products.

Along with concern about freshness, there is increased emphasis on health aspects. By its very nature and origin, meat is not only highly susceptible to spoilage but is also frequently implicated in the spread of foodborne disease. Despite many improvements in meat-processing hygiene over the past 100 years, concern about the role of meat products as a cause of food poisoning is increasing rather than diminishing. To some extent, this happens because the virtual elimination in many parts of the world of serious diseases such as tuberculosis, typhoid fever, cholera, and smallpox has focused attention on the foodborne diseases. The widespread distribution of meat products makes consequences of contamination with food-poisoning bacteria more serious.

A. Definitions

Meat is primarily the muscular tissue of "red meat" mammals—in usual practice a restricted number of species of "meat animals." The muscle is made up of contractile myofibrillar elements and soluble sarcoplasmic proteins. Up to one-quarter by weight of muscle is connective tissue, and, depending upon the particular muscle, one-third may be fat. Although the raw connective tissue is relatively resistant to attack by microbes, its presence has not yet been clearly demonstrated to have any significant microbiological consequences. The fatty tissue, on the other hand, has properties significantly different from those of the muscle.

Meat, as legally defined, commonly includes various organs ("offals") besides muscular tissue, the line being drawn at somewhat different points in different countries. Some of these (e.g., heart) closely resemble muscular tissue; some are rather different in properties and composition (e.g., liver), others very different (e.g., lungs and tripe).

In this chapter, meat is regarded as any edible portion of the carcass of any cattle, sheep, swine, goats, or horses, exclusive of lips, snouts, ears, caul fat, leaf fat, kidney fat, and other visceral fat, and exclusive of all organs, except the heart, tongue and esophagus (U.S. Department of Agriculture, 1976).

There is, however, little information about the microbiology of meat other than muscular (sometimes plus fatty) tissue, and almost none for specific organs, excepting liver and kidney.

Meat products are unusually diverse, ranging from those consisting wholly of meat (e.g., pure ground beef, salami) through those with admixtures of nonmeat protein (e.g., casein or soya protein) or of carbohydrate (sausage rusk) to those in which the meat is only a minor element (e.g., certain kinds of meat pie or vegetable and meat stews). There are widely different views about what should or should not be regarded as meat products. For present purposes, the important point is that the presence of significant proportions of alien material introduces quite different elements into the microbiology; hence, this chapter is confined to meat products consisting almost entirely of meat.

B. Important Properties

First among properties of microbiological importance, meat has a high water content (Table 15.1), corresponding to an a_w approximately 0.99, which is suitable for growth of most microorganisms. To change this, a large proportion of the water must be removed (at least from the surface) as explained in this chapter, Section VII.

Muscle contains about 75% water, in which is dissolved a variety of major growth substrates and supporting nutrilites (Table 15.1). Consequently, muscle is a very good medium for growing a wide variety of microorganisms, particularly bacteria, which are favored by wet conditions.

TABLE 15.1

Approximate Composition of Adult Mammalian Muscle after Rigor Mortis [a]

Component		% Wet weight
Water		75
Protein		
Structural connective	2.0 ⎫	
Myofibrillar	11.5 ⎬	19
Sarcoplasmic	5.5 ⎭	
Fat		2.5
Carbohydrate		
Glycogen	0.1 ⎫	
Glucose + phosphates	0.2 ⎬	1.2 [b]
Lactic acid	0.9 ⎭	
Miscellaneous soluble		
Nitrogenous: amino acids	0.35 ⎫	
creatine	0.55 ⎬	1.65
minor ingredients	0.75 ⎭	
Inorganic: K	0.35 ⎫	
P	0.2 ⎬	0.65
others	0.1 ⎭	
Vitamins: most B-vitamins present in useful amounts		

[a] Based on Lawrie (1975).
[b] Variable to almost nil—see text.

It is characteristic of muscle that the proportion of carbohydrates to nitrogenous compounds is relatively low, and may sometimes be very low. In the latter case, it is not possible to develop a significant degree of acidity even in the presence of a lactic flora. There are minor differences among species (e.g., pork contains relatively high concentrations of thiamin) that might presumably correlate with differences in microflora but this has not yet been demonstrated.

The growth of microorganisms on meat is primarily at the expense of the soluble constituents—carbohydrates, lactic acid, and amino acids. Significant breakdown of the mass of protein cannot be demonstrated in normal spoilage; softening and digestion of structural proteins become evident only at a very late stage (Dainty et al., 1975). Present evidence suggests that the breakdown products of spoilage are almost entirely harmless, the "ptomaines" of the older literature probably being undetected microbial toxins developed by microorganisms other than typical spoilage organisms.

The redox properties of meat have a major microbiological influence.

The central factor is a tissue respiration that continues to consume oxygen (if present) and produce CO_2. In life, the demand is more than compensated for by oxygen transported in the blood, and the oxygen tension and the redox potential in living muscle are high (except in some muscles in periods of violent exercise). With the cessation of blood supply at death, oxygen content and redox potential in the muscle gradually fall, leading to anaerobic production and accumulation of lactic acid. The acidity developed may suffice greatly to diminish tissue metabolism, which nevertheless continues for several days at a rate—even at low temperature—exceeding that at which oxygen can diffuse into meat for more than a few millimeters. Therefore, the bulk of meat becomes anaerobic within a few hours *post mortem* and remains so except for an aerated surface layer a few millimeters thick that is indicated by its brighter red color and that does not develop acid. When frozen, cooked, or salted, the tissue respiration is inhibited; nevertheless, sufficient reducing activity normally remains to maintain anaerobic conditions within any piece of meat more than approximately 10 mm thick. Consequently, although an aerobic flora develops on the surface, only anaerobes or facultative anaerobes can grow within the meat. (Some of the latter grow relatively slowly.) Because few of these anaerobic organisms grow readily at low temperatures, they scarcely develop within the bulk of chilled meat, even after long periods. When meat is minced, it is reaerated throughout, but if it is then packed together again, anaerobic conditions are gradually reestablished within its mass. If numerous ($> 10^6$/g) microorganisms are present, their respiration augments the tissue respiration.

The pH of meat may range naturally from about 7.0, which is nearly optimal for many pathogenic and spoilage bacteria, to levels approaching 5.0. Values approaching 5.5 are unfavorable in themselves to the growth of many of the important bacteria, and in combination with other unfavorable factors such as low temperature may almost prevent bacterial growth. However, some isolates of *Pseudomonas* (fluorescent and nonfluorescent), *Enterobacter,* and *Microbacterium thermosphactum** grow at their maximum rate within the pH range 5.5–7.0 (Gill and Newton, 1977). Low pH in combination with curing salts is especially effective against bacteria commonly found in cured meats. The pH of meat is inversely proportional to the amount of lactic acid developed by muscular glycolysis following death: pH 7.0 corresponding to almost none; pH 5.5 to approximately 1%. The amounts of lactic acid depend, in turn, on the amount of glycogen in the muscles at death. This may be low if the muscle has been exercised before slaughter, in which event the ultimate pH will be

---

* *Incertae sedis* (Buchanan and Gibbons, 1974).

relatively high and the muscle *dry* and *firm* in texture and *dark* in color (DFD condition). In a muscle not exercised at slaughter, all or most of the glycogen will be converted gradually after death to give a low ultimate pH near 5.5 with normal appearance and texture. If however, such a non-exercised muscle is stimulated just before death, as readily occurs in stress-susceptible pigs, the largely unused reserve of glycogen is rapidly converted to lactic acid and the low ultimate pH is approached before the tissues have had time to cool; this denatures sarcoplasmic protein, aggravating an apparent whitening of the tissue and a loss of water-holding power, called the *pale soft exudative* (PSE) condition.

The widely held view that the pH throughout the muscle is necessarily the same as at the surface of meat is questioned by Carse and Locker (1974) with obvious implications for the rate of microbial growth on the meat surface. The situation at fat and connective tissue surfaces may also differ from that in muscle.

The pH value of meat is also important for technological reasons. A low pH favors prompt and effective curing; a high pH favors retention of water and a closed texture. Hence pH may be modified artificially, e.g., with glucono-δ-lactone to lower it or with alkaline polyphosphates to raise pH.

C. Methods of Preservation

Meat is a highly perishable commodity, i.e., spoilage microorganisms grow quickly on it. Thus, trade in meat, even at the local retail level, depends on some degree of preservation that controls the spoilage flora. The larger the processing operation and the more distant the market, the greater the need for preservation.

The most important means of preservation are chilling or freezing, cooking (includes canning), curing (often with smoking), and drying. In modern technology several procedures are often combined; for example, a cured smoked ham may be cooked in the can and kept under refrigeration where it may be safely stored for long periods.

Low-temperature storage alone—i.e., chilling or freezing—enables meat to be held for some time without appreciable change in properties. On the other hand, other methods of preservation—cooking, canning, drying, curing, and smoking—all affect the characteristic properties of meat in some way and produce a product that is clearly different from fresh meat.

D. Classification of Meats into Types

The effects of heating and freezing have been described in Chapter 1, those of curing in Chapter 8, and those of packaging in Chapter 11 (Vol I).

Heating destroys vegetative spoilage organisms and most pathogens, the range of survivors dwindling as the heat process increases. Adequate curing controls the normal spoilage flora and pathogens, besides changing the appearance and flavor. Freezing, drying, or packaging may be applied to any kind of meat—raw, cured, or cooked.

Consequently, the microbiological problems of raw meats are quite different from those of dried, cured, or cooked meats. In raw meats, concern is with organisms adapted to high a_w levels (mostly bacteria). In dried meats, it is with organisms adapted to relatively low a_w levels (mostly molds); in cured meats it is with salt-tolerant organisms, especially *Micrococcus* and *Lactobacillus* and some fungi. In cooked meats, cured or not, the effect of cooking is to bring into increased prominence the heat resistant flora that normally is minor and unimportant. Selection becomes more strict with increased cooking, and likewise, the stability conferred increases. Pasteurization at temperatures below 85°C may leave some organisms capable of growing in the product. These products have only limited stability ("semiconserves"). Higher temperatures result in products that are microbiologically stable ("conserves"), the few surviving spores being unable to grow. Vacuum-packaging (which favors facultatively anaerobic or anaerobic species), chilling (which limits growth to psychrotrophic species), or freezing (which arrests all growth until thawed) modify these basic relations.

For consideration of the microbiology of meats and processed meats, therefore, it is advantageous to distinguish meat foods according to their method of preservation. Primary divisions are based on whether the meat is untreated, cured, or heated to various degrees, with secondary consideration of packaging or freezing. This leads to the following categories, which form the basis of the rest of this chapter:

Raw chilled meats
    1. Carcass meat
    2. Vacuum-packed meat
    3. Retail cuts
    4. Comminuted meat
Frozen meats
Low-temperature-rendered meats
Dried meats
Raw, cured meats
    1. High a_w
    2. Low a_w

Cooked, uncured meats
   1. Pasteurized meats
   2. Botulinum-cooked meats
Cooked, cured meats
   1. Pasteurized meats
   2. Hermetically sealed, shelf-stable meats
   3. Fully sterilized meats.

II. INITIAL MICROFLORA

With the exception of the external surface and the gastrointestinal and respiratory tracts, the tissues of normal healthy animals contain few microorganisms. The animals' defense mechanisms effectively control infectious agents in the healthy living body. However, this defense begins to fail after slaughter.

There is evidence that the number of microorganisms within tissues is raised under stress, and falls again after rest (Narayan and Takács, 1966). Viable counts at 37°C of the order of $10^3/g$ of tissue may indicate invasion under unusual stress, but there are indications of differences among animal species, and the subject is still debated.

The cleanliness of livestock at slaughter depends on a number of factors that will vary with the location of the farm, the method of transport, and the holding conditions at the slaughterhouse. Contamination on animals grazed on pasture under wet, muddy conditions will differ from that on livestock from dry, dusty areas. The prevailing temperature of the soil will influence the proportion of psychrotrophic organisms in contamination from this source. Tropical soils contain fewer psychrotrophic bacteria than soils from temperate zones, and the organisms on the skin of cattle and on meat follow a similar trend (Empey and Scott, 1939). Cattle from feed lots may carry fewer soil organisms and more fecal bacteria than those from pasture. The nature and amount of fecal contamination will be affected by dietary and other factors (Grau and Smith, 1974).

III. PRIMARY PROCESSING

A. Effects of Processing on Microorganisms

Various diseases of animals may also affect man, hence meat should be prepared only from healthy animals. Veterinary inspection before slaughter ensures that only healthy animals are accepted for meat production. Meat inspection also removes material that can be detected as unsatisfactory only after slaughter. The diseases (e.g., anthrax, brucellosis, tuberculosis, taeniasis, trichinosis) and the organisms that cause them are thereby excluded and fall outside the scope of the present chapter.

Meat is contaminated by contact with the hide, skin, or feet; stomach and intestinal contents; milk from udders; plant and equipment; hands and clothing of personnel; water used for washing carcasses and equipment; and even air in the processing and storage areas. Contamination may take place during almost every operation of the slaughtering, cutting, processing, storage, and distribution of meat. The extent to which this occurs reflects the standards of hygiene and sanitation in the slaughterhouse and processing plant. The composition of this flora reflects the various sources of contamination and the effectiveness of hygiene measures aimed at preventing the spread of infection (Ayres, 1955).

The primary processing of meat consists of slaughtering, dressing, chilling, and cutting. The initial flora is mainly determined by surface contamination received during these operations.

1. Slaughter

There is still disagreement in the literature over the presence of microorganisms within the tissues of animals at slaughter. Generally, it has been assumed that unless the viscera of slaughtered cattle, pigs, and sheep are removed soon after death, the meat will become contaminated by bacteria from the intestinal tract. Bacterial spoilage deep within the muscles ("bone-taint") has been attributed to the multiplication of intrinsic bacteria originating from the intestine. However, if aseptic precautions are observed a high proportion of sterile tissue samples can be obtained from normal healthy animals (Gardner and Carson, 1967; Gill et al., 1976), which indicates that the number of organisms within the tissue is normally very small, approximately 1/10–100 g. Evidence from bacteriological examination in connection with meat inspection also has shown this to be the case.

Bacteria can enter the tissues through the bloodstream from contamination on the sticking knife, but the ease with which sterile tissue can be obtained from healthy animals slaughtered in the normal way suggests that contamination through the blood circulatory system is usually not great. Failure in ham curing has been ascribed to this cause (Jensen, 1954), although this does not seem to have been proved definitely.

2. Dressing

The first major source of contamination affecting meat is from the skin or hide of the particular animal being dressed and others in close proximity. Microbial contamination from this source includes the normal flora

TABLE 15.2

Effects of Washing with Warm, Chlorinated Water on Bacterial Numbers on Lamb Carcasses [a]

| Water temp. (°C) | Residual count $\log_{10}/cm^2$ [b] | | |
| | Chlorine | | |
	None	30 ppm	95 ppm
50	3.7	3.5	3.3
65	3.5	2.6	2.3
80	3.1	2.7	2.6

[a] Kelly *et al.* (1974).

[b] Three carcasses in each treatment. Average before washing: 4.22. Counts underlined are significantly lower than initial count.

of the skin (staphylococci, micrococci, pseudomonads, yeasts, and molds) as well as organisms of fecal and soil origin. The total number of organisms on the skin may exceed $10^9/cm^2$ (Ayres, 1955).

Further sources of contamination during dressing include the knives, hands, and clothing of the workers and the water used to wash carcasses, hands, and equipment. Some microorganisms of human origin reach the meat in this way, but if a reasonable standard of hygiene is maintained, contamination from these sources is likely to be negligible compared with that from the animal itself. Attempts are often made to clean carcasses, but these procedures have little effect on microbiological contamination. Under practical conditions cloths and brushes become so contaminated that they add rather than remove bacteria, hence their use is prohibited in some countries. Suitable sprays may diminish surface contamination approximately 10-fold (see results of an experiment in Table 15.2). Water pressure, temperature, and nozzle design are also important (Bailey, 1972). Another experiment on washing beef quarters in a meat packaging plant is summarized in Table 15.3.

At the completion of hygienic dressing, beef carcasses are likely to carry a surface contamination of 10^3 to 10^5 viable aerobic bacteria/cm^2, mostly less than 10^2 psychrotrophs/cm^2 and from 10^1 to 10^2 coliforms/cm^2. Sheep carcasses usually have a slightly higher level of contamination than beef with 10^3 to 10^6 viable aerobes/cm^2 (cf. Fig. 15.1). Although psychrotrophic bacteria will often be less than $10^2/cm^2$, about 20% of samples

TABLE 15.3

Reductions (log$_{10}$) in Total Counts on Marketed Beef Carcasses Sprayed with Water Containing 200 ppm Chlorine [a]

Spray pressure at nozzle	Temperature	
(kg/cm²)	13 °C	52 °C
4	1.4	2.9
25	3.1	3.5

[a] Simplified from Kotula *et al.* (1974).

will have up to 10^3 or more psychrotrophs/cm². Coliforms also tend to be more numerous on mutton than on beef carcasses. Numbers of bacteria, especially those of enteric species, tend to be higher on pig carcasses, apparently because the skin is usually not removed, and the practices used to remove the hair are hygienically unsatisfactory.

The level of contamination is usually lower in the body cavity than on the external surface (Ingram and Roberts, 1976). Besides *Escherichia coli,* skin, hide, and meat surfaces carry various genera of the Enterobacteriaceae (Newton *et al.,* 1977b).

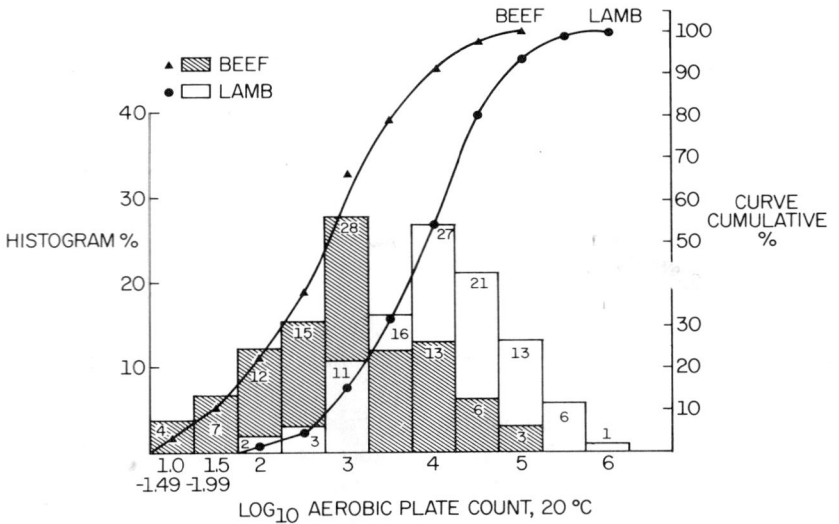

Fig. 15.1. Range of bacterial counts on 240 commercial carcasses. (From Ingram and Roberts, 1976.)

The practice of removing bones from the carcass before chilling ("hot boning") permits more rapid chilling with the implication of less bacterial multiplication. In practice, however, it may be difficult to achieve a cooling rate in a bulk pack of hot boned meat that is as fast as that achieved on the surface of a normally chilled carcass. There is a risk of greater contamination from an environment that may be warmer than in normal operations. This hazard is minimized if the recommendations of the Codex Alimentarius Commission are observed (Food and Agriculture Organization/World Health Organization, 1976a,b).

3. Chilling

The effect of chilling on the microbial flora depends on several conditions. Fast chilling at low temperatures with high air speeds and low humidities may reduce bacterial numbers. Under less rigorous conditions, growth of psychrotrophic organisms can occur thus altering the proportion of psychrotrophs to mesophiles. The time the carcass is held in the chiller may have more effect on the microbial population than the temperature of chilling (Nottingham and Wyborn, 1975). Provided the chilling facilities are properly maintained, aerial contamination should not exceed 10^2 organisms/m^2/min, which would contribute about 14 organisms/cm^2 of carcass surface/day. If carcasses are allowed to cool at ambient temperatures of 15° to 20°C or more, growth of mesophiles, including pathogens, may occur. Edible offals often are more heavily contaminated than is carcass meat and therefore should be immediately chilled. Some countries, e.g., in the European Community, require cooling to below 3°C.

4. Cutting and Boning

Contamination received by meat during cutting, boning, and packaging depends on local conditions. The meat receives extensive handling during these operations, and fresh surfaces are exposed, thus making the meat more susceptible to the effects of contamination. Factors such as the temperature of the boning room, the time meat is held there, and the cleanliness of the cutting tables, conveyor belts, saws, knives, and other equipment all affect the microbial flora.

In a properly managed boning or cutting room maintained at 10°C or below, aerial contamination and microbial growth should not make any significant contribution to the microbial load on the meat. The major source of contamination is likely to be the surface of the incoming carcasses, and the extent to which this contamination is spread on freshly cut surfaces by contact with equipment can have an important effect on the keeping quality of the meat. Contamination that builds up on the equipment through inadequate sanitation is likely to contain a high proportion of psychrotrophic bacteria.

Products that undergo extensive handling are likely to become contaminated with bacteria of human origin. For example, there is evidence that the proportion of human derived strains of staphylococci present on meat increases during processing. Enteric viruses that are sometimes found in meat products can be of human origin but most often are associated with animals.

B. Spoilage

In hygienically prepared meat, the number of the pathogens is very small, and the microflora consists substantially of saprophytic species. The most numerous are gram-negative rods and micrococci (cf. Table 15.4, which gives an example of the composition of the microbial flora on beef); among the saprophytes are: *Acinetobacter, Aeromonas, Alcaligenes, Flavobacterium, Moraxella,* coryneforms, and *Pseudomonas,* as well as various Enterobacteriaceae. The micrococci readily detected on the fresh carcass are mainly *Micrococcus* spp. and *Staphylococcus* spp. Fecal streptococci are present in very low numbers. Lactic acid bacteria, *Microbacterium thermosphactum* (incertae sedis), and various *Bacillus* species initially are present in low numbers. Yeasts and molds are also present. The origin of many of these organisms is still not clear.

TABLE 15.4

Approximate Percentage Composition of Microbial Flora on Fresh Beef Carcasses and on Beef at the Retail Store [a]

Microorganisms	After slaughter	After chill	Before shipment	Carcasses in store	Loins	Steaks
Pseudomonas						
fluorescens	—	—	2	5	—	9
P. fragi- [b]	29	20	23	54	62	65
P. geniculata [b]	9	1	22	31	12	17
P. rugosa [b]	2	8	4	—	—	—
Acinetobacter–						
Moraxella	—	—	2	9	27	10
Micrococcus	45	65	38	—	—	—
Bacillus	12	13	3	—	—	—
Other	2	2	6	—	—	—

[a] After Stringer *et al.* (1969).
[b] The Bergey's Manual, 8th ed. (Buchanan and Gibbons, 1974) considers these as *species incertae sedis.*

The importance of the various types of microorganisms initially present on carcass meat depends on the future use of the product. *Pseudomonas, Acinetobacter,* and *Moraxella* dominate the flora of unprocessed meat exposed to air at chill temperatures. If the surface of the meat becomes dry, yeasts (e.g., *Trichosporon scottii*) and molds (e.g., *Cladosporium, Sporotrichum* and *Thamnidium* spp) replace the bacteria. If the meat is packed to exclude air, *Microbacterium thermosphactum* (incertae sedis) or atypical lactobacilli prevail. None of these are important in frozen meat unless there is an opportunity for microbial growth during thawing. The psychrotrophs are heat sensitive and are therefore of no importance in heated products.

There is a direct relation between the initial number of bacteria on meat and the time it takes to spoil (see Fig. 15.2). Further, if the initial number is too high, the durability of color and flavor—even in frozen storage—may be compromised.

C. Pathogens

Pathogens that may be present include *Salmonella, Staphylococcus aureus, Yersinia enterocolitica, Clostridium perfringens,* and occasionally *C. botulinum.*

Enteric organisms such as fecal coliforms and fecal streptococci are frequently found on meat, indicating that the gut is a common source of contamination. Certain enteric viruses also may be present. The food-poisoning organisms of most concern with meat—*C. perfringens, Salmonella, S. aureus* and probably enteropathogenic *Escherichia coli*—are associated with enteric contamination.

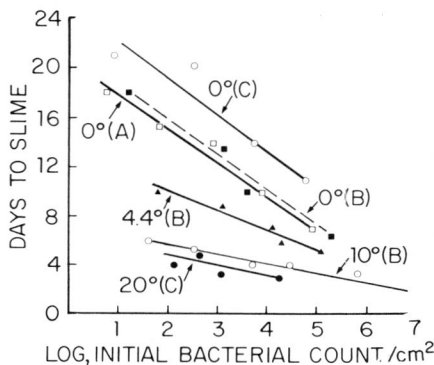

Fig. 15.2. Shelf-life of beef as affected by initial bacterial count and storage temperature (°C). [Elliott and Michener (1965), reviewing the work of (A) Haines (1937), (B) Ayres (1951, 1955), (C) Ayres (1960a).]

Given appropriate temperatures, most enteric organisms can multiply readily in meat and on material in slaughterhouses so that, unlike in water, their numbers, in general, do not give a reliable indication of the degree of fecal pollution. Moreover, the presence of psychrotrophic Enterobacteriaceae has no significance as an indication of fecal contamination.

Salmonellae are probably the most troublesome pathogens. Outbreaks caused by salmonellae in meat or meat products are prominent in the food-poisoning records of various countries, and alien serotypes from one country continue to appear in others. Where adequate surveillance systems exist, it has been demonstrated several times that an alien serotype, first recorded in an imported feedstuff, may subsequently appear among domestic animals and later in meatborne outbreaks among the human population. For this reason, some countries require that feedstuffs of animal origin be free from salmonellae before import.

Salmonellae entering the rumen of cattle and sheep die out over a period of several days when the animals are fed; but if they are starved so that the rumen becomes less acid, salmonellae multiply to levels of $10^3/g$ of rumen contents. If feeding is then resumed, salmonellae populations rise to $10^4-10^5/g$ in feces and disappear only gradually over a period of several weeks. Feed intake in the period before slaughter thus could have an important influence on the status of *Salmonella* at slaughter (Grau *et al.*, 1969). Salmonellae have been isolated from nasal swabs of calves just after slaughter (Nottingham *et al.*, 1972).

The general situation is clear from Table 15.5: A small proportion of

TABLE 15.5

Animals Carrying Salmonellae at Stages of Meat Production [a]

		Percentage infected		
Animal	Country	On the farm	Awaiting slaughter	Finished carcass
Calf	United Kingdom	0.5	20.1	35.6
	Netherlands	3.1	6	22.7
Cow	United States	1.7	—	12
Pig	United Kingdom	2.9	13.5	18
	United States	7	25	50
		5	25	30
	Netherlands	8	13	38

[a] Ingram (1972a).

live animals carry salmonellae; the status is maintained by feeding contaminated animal by-products (cf. Chapter 17), by the use of "carrier" animals in breeding (Hobbs, 1974), and by environmental contamination. The proportion of animals with salmonellae may have risen greatly by the time the live animals reach slaughter, indicating that cross-infection during transport and lairage is a major problem. General levels of salmonellae contamination are clearly much higher in pigs and calves than in adult cattle.

The effects of stress during transport have been confirmed experimentally. Demonstration of an increasing proportion of carriers with time in lairage led the United Kingdom to prohibit holding animals in lairage for more than 72 hr. Because infected animals at this stage are symptomless carriers, the problem cannot be controlled by inspection. The proportion may more than double during slaughter and dressing, indicating that cross-contamination during these operations is important. The serotypes found on the carcasses show that contamination which enters with the animals is much more important than foci of infection in the abattoir.

Although some of the data from Table 15.5 are based on extensive surveys, the usual percentage of *Salmonella*-contaminated carcasses in abattoirs is considerably smaller. In a survey of the incidence of *Salmonella* in carcass meat (with whole carcasses sampled), Roberts (1976) found 0.4% were positive for beef, none were positive for lamb, and 2.6% were positive for pigs. This probably reflects the usual situation in modern abattoirs, but the incidence of *Salmonella* varies markedly in different countries and at different times. (See also the review in Meara, 1973.)

There are regulations that require rooms for cutting or boning to be held at temperatures below 10°C in the belief, not yet supported by evidence, that this will diminish the incidence of *Salmonella*. Following demonstrations that some salmonellae grow at 10°C and occasional reports that some grow at lower temperatures (to 5.3°C, Matches and Liston, 1968b), there recently have been requirements that before cutting meat be cooled to 7°C, or if that is not possible, that cleaning and disinfection be done every 4 hr.

Improvement of modern slaughter hygiene is incapable of removing, or even greatly diminishing the cross-contamination by salmonellae. Neglect of slaughter hygiene, however, can make the situation much worse. Given present methods of producing and distributing live animals, salmonellae inevitably occur from time to time in or on carcass meat (National Academy of Sciences/National Research Council, 1969), and in certain situations the greatest improvement would apparently come from attention to transport and lairage (Williams and Spencer, 1973). Solutions lie in treatments that destroy salmonellae or in producing *Salmonella*-free animals.

Low numbers of staphylococci occur regularly on meat, but only become important when selected by appropriate holding conditions and repression by competing organisms. By means of phage typing and biochemical characteristics, it is possible to distinguish "animal" from "human" strains. On this basis, "animal" types prevail on carcasses immediately after slaughter, but "human" types predominate in meat and meat products at time of sale (Sinell, 1973). The primary hazard arising from staphylococci in meat is, apparently, due to contamination by human handlers.

For information on *Yersinia enterocolitica* see Section IV,B,3.

D. Control

Because contamination of the meat is often very uneven, it is desirable to take a number of samples from different parts and to examine as many samples as possible. Conclusions on levels of contamination in a particular lot of meat should never be based on examination of a small number of samples. Generally, spread plates are preferable to pour plates; replication of dilutions or plates gives little increase in accuracy and should not be employed at the cost of reducing the number of samples examined (Nottingham *et al.,* 1975).

The temperature used for incubation of plates should be chosen in relation to the type of sample being examined. To determine the likelihood of contamination with mesophilic pathogens on freshly dressed carcasses, it is advisable to incubate the plates at 35° or 37°C. With meat or equipment that has been held under chilled conditions, plates should be incubated at 25°C or below to include psychrotrophs and thus to give an indication of the microbial growth that has occurred.

The microorganisms on meat depend on the care with which it was prepared. For example, carcasses become contaminated from tools and machinery used by slaughtermen; retail cuts are further contaminated by fingers, knives, and cutting blocks; ground meat by unclean mincing machines. An equilibrium is reached by continual exchange of contamination between these objects and the meat itself, which is the main cause of the increased contamination shown in Table 15.5. Such contamination can readily be reduced by effective and regular cleaning of hands and equipment. The reader is referred to existing publications: for slaughter, *Meat Hygiene* (Food and Agriculture Organization, 1957); for meat handling, *Code of Hygienic Practice for Fresh Meat* (Food and Agriculture Organization/World Health Organization, 1976a); for meat processing, *Code of Hygienic Practice for Processed Meat Products* (Food and Agriculture Organization/World Health Organization, 1976b); and for cleaning and disinfection, Chapter 14.

It is most important to identify the critical areas in which sanitary measures should be applied, and such decisions are frequently misguided. In this connection, the concepts of Hazard Analysis Critical Control Points (HACCP) (Bauman, 1974) and suggestions of the Food Protection Committee (National Academy of Sciences/National Research Council, 1975) may be helpful.

IV. RAW CHILLED MEAT

A. Carcass Meat

1. Effects of Processing on Microorganisms

If carcass meat is exposed to temperatures above approximately 20°C, many mesophilic pathogenic bacteria are able to grow. For example, clostridia can multiply within the tissues and cause obnoxious anaerobic putrefaction, which is an indication that pathogenic species (*Clostridium perfringens, C. botulinum*) might have grown to dangerous numbers. There may be an appreciable development of fecal streptococci before the clostridia grow. Members of the Enterobacteriaceae can also multiply, most rapidly on the surface. For these reasons it is normal practice in tropical countries to organize slaughter, distribution, and sale, so that the meat is consumed within 6–8 hr after slaughter.

"Bone-taints" may develop in deep-seated joints, when carcasses fail to cool soon enough. Various genera have been reported, predominantly *Clostridium* and *Bacillus,* but also non-spore-formers. Some of the predominant species appear to be different in cattle and pigs, but there is insufficient information for reliable comparisons.

Much carcass meat has traditionally been handled in temperate countries at air temperatures in the range 10°–20°C without the catastrophic occurrences mentioned above, primarily because turnover is rapid and the development of mesophilic pathogens (especially clostridia) is greatly restricted in this temperature range. Perhaps the greatest risk, not appreciated in the past, may arise from the multiplication of invasive pathogens like *Salmonella* spp.

An outbreak of salmonellosis in Sweden in 1953, involving approximately 9000 cases was caused by improperly cooled carcass meat from one slaughterhouse. The refrigeration facilities were inadequate to cool an un-usually large number of carcasses, and cross-contamination and growth of salmonellae occurred (Lundbeck *et al.,* 1955).

The flora of meat stored at 10°–20°C is likely to include Enterobac-teriaceae, micrococci, and staphylococci, as well as *Pseudomonas, Acine-tobacter, Moraxella,* and *Aeromonas.* Many of these strains may be similar to those growing on meat chilled below 10°C (though this has not been established), but the proportion of mesophilic strains seems likely to be higher. The bacteriology of meat handled in this way has not been in-vestigated systematically, and there is no statistical information. Hence, it is not possible to suggest what might be regarded as normal.

It is desirable to place meat carcasses in chill rooms at temperatures below 10°C, preferably near 0°C, as soon as possible after slaughter. Chilling to below 10°C within about 10 hr *post mortem* is, however, liable to cause toughening. Electrical treatment can prevent this. The aim is to cool not only the surface of the meat, which rapidly attains a temperature similar to that of the ambient air, but also to bring the temperature of the deep tissues below about 20°C before internal development of mesophiles has time to take place. A falling oxygen tension and redox potential and an increasing acidity are probably important factors in the control of these bacteria in the first few hours after death before the internal temperature falls (Ingram, 1972b). Drying of the surface can reduce the a_w sufficiently to stop growth. Figure 15.3 shows the effect of different relative humidi-ties and Figs. 1.6 (Vol. I) and 15.2 show the effect of temperature on bacterial growth on beef.

If chilling is carried out correctly, the interior of the meat will contain few bacteria, and only those kinds that multiply slowly or not at all at

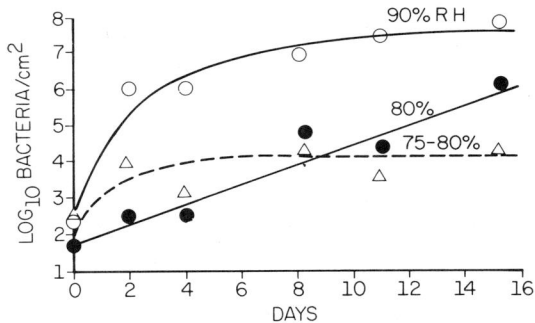

Fig. 15.3. Effect of relative humidity (RH) on bacterial growth on beef car-casses at 2°C. (From Buttiaux and Catsaras, 1966.)

chilled temperatures. Hence, during storage of chilled meat, the important microbiological changes take place at the surface.

2. Spoilage

When carcass meat is kept at temperatures below 10°C the aerobic conditions, high a_w, and low temperatures select psychrotrophic bacteria (Ayres, 1955). A flora consisting of *Pseudomonas, Acinetobacter,* and *Moraxella* increasingly predominates as storage progresses and the total number of bacteria rises (cf. Tables 15.4 and 15.6). This was originally called the *Pseudomonas–Achromobacter* association (Mossel and Ingram, 1955), but the old genus *Achromobacter* has since been divided among *Acinetobacter, Alcaligenes, Moraxella* and perhaps other genera. Many of the relevant strains are intermediate between *Acinetobacter* and *Moraxella* and still of doubtful identity (Buchanan and Gibbons, 1974, p. 436). The genus *Alteromonas* has recently been created. It includes *A. putrefaciens,* previously called *Pseudomonas putrefaciens.* This organism is important in the spoilage of fish, less so for poultry, and is not implicated clearly in the spoilage of red meat. Other gram-negative rods occur less commonly.

As a consequence of the increasing proportion of psychrotrophs during chilled storage, aerobic plate counts incubated at 37°C remain stationary or fall as those incubated at 0° to 5°C increase. When plates are incubated between 20° and 30°C, the counts are highest because both psychrotrophs and mesophiles grow in this range. Counts at about 1°C rarely equal those at 20°C, indicating that strict psychrophiles (incapable of growth at 20°C) are of little account in refrigerated meat (Ingram, 1966) and that both mesophiles (incapable of growth at 1°C) and psychrotrophs grow at 20°C. With meat that has been refrigerated, the highest viable counts are obtained at 25°C, and this is probably the most convenient temperature for fol-

TABLE 15.6

Aerobic Plate Counts (21°C, 2 Days) on Beef Carcasses at Different Times after Slaughter and on Receipt at Retail Stores [a]

Days after slaughter	No. of carcasses	Range of averages for five sites ($\log_{10}$/in.²)	Range of individual samples ($\log_{10}$/in.²)
1	88	3.80–4.59	1.60–6.08
2	72	4.18–4.99	1.60–6.70
3	27	4.19–4.79	1.60–6.13
4	11	4.22–6.21	1.78–7.24
At retail	39	4.98–6.35	1.30–7.45

[a] Stringer *et al.* (1969).

lowing the flora in general. Significantly lower counts may be obtained at
30°C, a temperature widely favored for many other foods. To count the
psychrotrophic organisms separately, a temperature of 20°–22°C is often
used; it has recently been suggested that 17°C is better because plates are
less likely to include mesophiles. In either case, the time of incubation (2–
4 days) represents a compromise between maximizing the psychrotrophic
count and excluding mesophiles, for in the temperature range where both
grow, psychrotrophs have a shorter lag period (Fig. 1.3, Vol. I).

The effect of cool temperatures on the development of the *Pseudomonas–Acinetobacter–Moraxella* association has been thoroughly investigated in water-saturated atmospheres. Together with the initial level of
contamination (ignoring the mesophiles, which become relatively negligible), the rate of multiplication has been shown to bear a direct relation
to the time of spoilage (Ingram, 1972b).

In general terms, multiplication and spoilage are delayed in atmospheres
of lower humidities, especially at low temperatures. The relations are complex and have yet to be established satisfactorily because of the difficulty
of controlling the surface dryness of pieces of meat. For example, it has
long been known that, in controlled artificial systems, *Pseudomonas–
Acinetobacter–Moraxella* bacteria cannot grow at a relative humidity
(RH) less than 95%, but in practice they can grow on meat in atmospheres
with RH lower than this, though less rapidly than in saturated atmospheres. This growth depends on the continued diffusion of water to the
surface from below, and where this diffusion is hindered (e.g., on fatty
tissue) or where moving air creates too high a rate of evaporation, such
growth is less likely.

If the RH of the atmosphere is < 80%, and especially if the air moves
rapidly over the surface of the meat, evaporation takes place faster than
diffusion of water from within the meat, and the surface layers lose the
greater part of their water, forming a tough skin. On this, yeasts (e.g.,
Trichosporon) or molds (*Geotrichum, Thamnidium, Cladosporium*) may
grow in place of bacteria. The situation resembles that in dried meat (cf.
Section VII). Such changes once were important—when meat was cooled
in ambient air of relatively high temperature and low RH. But modern
technology minimizes evaporation to prevent weight loss, keeping the meat
surface moist and thereby encouraging bacteria. Because of these influences, the most perishable parts of the carcass are those protected from
desiccation, e.g., under the diaphragm, the axillae, or on the cut muscles
of the neck.

The multiplication of many spoilage bacteria is influenced also by
natural variation in the pH of the meat, i.e., between about 5.5 and 7.0.

The effect of pH appears on the whole much smaller with fresh meats than with cured meats.

The so-called DFD and PSE meats (see Section I,B) are distinguished basically by their differences in pH, that of DFD meat being above pH 6.2 and that of PSE falling below pH 6.2 within 1 hr *post mortem*. Consistent with this difference, bacterial growth is somewhat earlier on DFD than on PSE meat—i.e., the former is more perishable. But the difference is small.

The use of about 15% CO_2 in the atmosphere approximately doubles the time for the *Pseudomonas–Acinetobacter–Moraxella* association to develop, at temperatures near 0°C; CO_2 is less effective at higher temperatures where CO_2-insensitive bacteria grow (Chapter 10). Meat may be held in stores with carefully controlled ventilation; or, more crudely, packed in boxes with CO_2 pellets that also reduce the temperature (and doubtless cause local freezing).

The general effect of handling under chill refrigeration without other special treatment is illustrated in Table 15.6. On some of the meat the bacterial numbers increase to high levels; on some the numbers remain low. The latter represent meats that are accidentally cooler or drier. But the low counts may also be due to variation as a result of sampling. The increased preponderance of the *Pseudomonas–Acinetobacter–Moraxella* association by the time of retail sale is illustrated by Table 15.4. These organisms replace the micrococci that were numerous on the freshly slaughtered carcasses.

After a lapse of time, the metabolic activities of the microbes become detectable to human senses. The meat surface becomes discolored and odorous. Then slime appears, and obscures the "sheen" of the meat surface. There is general agreement that spoilage usually begins to be evident when the number of bacteria reaches $10^7/cm^2$ of meat surface. The point at which meat should be called spoiled is a matter of subjective opinion, hence there is some disagreement as to the precise number of bacteria at the point of spoilage. Differences in techniques also lead to differing estimates of bacterial numbers (reviewed in Elliott and Michener, 1961).

Information on biochemical changes in carcass meat is nonexistent (excepting taints). The relevant changes are discussed under chilled retail cuts (Section IV,C,2).

Orthodox chemical tests for spoilage remain negative or weak until the meat is organoleptically spoiled. Various tests have been proposed (Jay, 1972): e.g., NH_3 production; high pH (but the pH may be naturally high, cf. Section I,B); water-holding capacity or extract release volume, which are directly related to pH; and dye reduction tests, especially with resazurin (but meat enzymes may reduce the dyes). For the reasons just given, none of these have much useful predictive value. The precise nature of the organoleptic spoilage is still uncertain. The changes have not yet been associated with any readily identifiable group of the bacteria (Ingram and Dainty, 1971). Consequently, the most useful predictive index of potential spoilage at temperatures below 25°C is the number and activity of psychrotrophic bacteria.

Lipases occur in some species of bacteria, yeasts, and molds, but the rates of attack may be very different among species of the same genus and may be affected by the nature of the substituent fatty acids. Lipolysis is usually slow compared with putrefactive changes, hence it is important chiefly in meat that has been preserved (e.g., by curing). The longer shelf-life of such products permits microbial activity over longer periods. Whether lipolysis is regarded as spoilage is a matter of circumstances. Specifications for animal fat (which might have been prepared from decomposed carcasses), at one time included low limits in free-fatty-acid content. On the other hand, the desired ripe flavors of fermented sausages may depend on microbial breakdown of the fats (cf. Section VIII).

3. Pathogens

Official records indicate that the most frequent causes of food-poisoning associated with meat are *Clostridium perfringens, Salmonella,* and *Staphylococcus aureus. Clostridium botulinum* is important if the meat is heat processed (see Section IX).

a. Clostridia. Early surveys, which were primarily based on a search for spores, found clostridia only infrequently in carcass meat. However, *C. perfringens* occurs mainly as vegetative cells, which until recently were overlooked.

It has been difficult to develop suitable selective media for this organism. Present indications are that *C. perfringens* occurs normally at a level of one or fewer per 100 g deep within the muscle tissues immediately after slaughter, mostly as vegetative cells. The level may rise to approximately 1 or more per 10 g if the animals are slaughtered in a fatigued condition. The level is higher in organs like the liver. Contamination on the surface seems to be of a similar order immediately after slaughter and is composed in like manner predominantly of vegetative cells (Tables 15.7 and 15.8). Presumably further contamination occurs, for numbers are frequently greater by the time the meat reaches commercial channels; spore counts of 1/10 g are not uncommon.

Clostridium botulinum has been regarded as extremely rare, but recent observations with better methods suggest that the frequency may occa-

TABLE 15.7

Frequency of Occurrence of *Clostridium perfringens* on the Surface of Carcass Meat before Chilling [a, b]

Product	Percent positive after enrichment
Beef	29
Lamb	85
Pork	66

[a] Smart *et al.* (1979).
[b] Method note: 2 swabs of 100 cm^2, unheated.

sionally be disturbingly high, at least in pork carcasses (Roberts and Smart, 1976).

b. Salmonellae. Salmonellae occur on red meat carcasses but with variable frequency. Many are likely to be missed unless large samples (100 g or more) are examined by sensitive methods. Liquid enrichment procedures are essential, direct plating being wholly inadequate because numbers are usually small. However, highly contaminated areas may be found. Recovery depends greatly on methods; some observed rates of contamination are illustrated in Table 15.5.

Most of the quantitative data relate to feces, mesenteric lymph nodes, and other organs; there is surprisingly little useful numerical data about

TABLE 15.8

Proportions of Vegetative Cells and Spores of *Clostridium perfringens* on Surface of Carcass Meat [a, b]

	Number/100 cm^2					
Meat	<20	20–50	50–100	100–150	150–200	>200
Pork						
vegetative cells	4	7	1	4	3	1
spores	10	7	1	2	0	0
Beef						
vegetative cells	18	1	0	0	0	1
spores	20	0	0	0	0	0

[a] J. L. Smart and T. A. Roberts (personal communication).
[b] Half each sample heated 30 min at 75°C for spores; 20 samples each of pork and beef.

the meat itself. The surveys suggest that salmonellae may occur more frequently in organs like liver in pigs and spleen in calves. In addition, salmonellae may occur more frequently in lymph nodes rather than in muscular tissue. With carcass meat, however, it appears that the majority of salmonellae occur on the surface, suggesting they come from fecal contamination. Compared with the proportion of animals carrying salmonellae just before slaughter, the proportions of carcasses carrying salmonellae just after slaughter are usually 2-fold or greater, which implies that half or more of the finally contaminated carcasses suffer cross-contamination from initially contaminated ones. Slaughterhouse hygiene cannot prevent the contamination of meat by this organism (see Table 15.5). Commercial experience shows that the incidence of salmonellae in fresh meat is much reduced, however, if carcasses are dipped in hot water and protected from subsequent recontamination.

The location of the salmonellae has not been investigated directly; there are obvious difficulties because the organisms are so few. A presumed location on the surface is the only justification for the use of techniques involving rinsing or swabbing to detect salmonellae. These techniques would clearly miss any organisms within the carcass.

4. Interrelations

Because the members of the *Pseudomonas–Acinetobacter–Moraxella* association are not pathogenic or toxicogenic, and because the circumstances in which it develops are unfavorable to recognized food-poisoning species (excepting perhaps the psychrotrophic *Yersinia enterocolitica*), the numbers of that association give no indication of hazards to public health. That is to say, a plate count at or below 25°C has no significance as an "indicator" with meat held under refrigeration at temperatures too low for growth of pathogens (approximately 5°C). It is significant only as regards potential for spoilage.

This is borne out by the lack of correlation among plate counts, indicator counts, and the occurrence of salmonellae, with chilled carcass meat. Indeed, direct experimentation shows that, at temperatures below about 10°C, pathogens introduced into the surface of meat gradually die out while the psychrotrophic flora continues to multiply. This inability to compete results mainly from an inability of the pathogens to grow under somewhat adverse competitive conditions at these low temperatures. They can grow, however, in presence of the meat flora at higher temperatures if initial numbers are large enough.

It will be clear that the type of spoilage produced at chill temperatures (unlike that at high temperatures) has little public health significance, and that pathogens must be sought separately. Nevertheless, the utility of such meat is significantly impaired.

5. Control

In the virtual absence of studies relating health risks to surface counts on carcass meat, it is inappropriate to reject consignments for public health reasons solely on the basis of high plate counts. It is a different matter, however, to use microbiological criteria for quality control purposes, aimed at producing an acceptable product (Nottingham, 1974). Microbiological quality control involves development and use of processing designed to keep microbial numbers low by reducing contamination and reducing microbial proliferation. The effectiveness of the method is monitored by microbiological examination of plant, equipment, materials, and products. Microbial counts are used to assess the effectiveness of hygiene measures in controlling initial contamination and of environmental factors in limiting microbial growth. Such counts do not relate to the safety of the product.

The value of quality control is determined by the use made of the results of the microbiological examinations. Rapid feedback of information is required to ensure that remedial action is taken before there is any need to reject product, e.g., high counts on chilled carcass meat indicate the necessity to adjust chilling conditions. Before the introduction of the concept of microbiological quality control, safety and quality were assured by taking care to follow good manufacturing practice. Now microbiological control is used to decide whether good practice is being observed.

Clostridium perfringens occurs within carcass meat as vegetative cells in low numbers, usually fewer than 1 cell/10 g (cf. Section IV,A,3). These cells would be destroyed in thorough cooking and are of no special significance. The spores that are indirectly responsible for food poisoning are of uncertain origin. There is no evidence that the vegetative cells sporulate in meat; more likely, the spores represent extrinsic contamination. Besides being present in low numbers (cf. Tables 15.7 and 15.8), spores do not germinate and multiply at temperatures below 20°C (doubling time 6 hr or more, according to pH). Consequently, it seems needless to consider numerical criteria for *C. perfringens* in carcass meat; the species can become important only through temperature abuse, and control is best effected through control of temperature conditions.

As salmonellae are readily destroyed by thorough cooking, problems arise largely through cross-contamination from raw meat to food surfaces where multiplication is possible. Relatively small numbers of cells may be significant because the organism is an invasive pathogen. In principle, therefore, confirmation of their absence from raw meat is desirable, but unfortunately no practical system of examination can ensure this.

In such a situation, the usual recourse is to use an "indicator" for *Salmonella,* i.e., some form of "coliform" or Enterobacteriaceae count. Unfortunately, several investigations on a substantial scale have demonstrated that there is no correlation between coliform or Enterobacteriaceae level and *Salmonella* level on carcass meat. One reason might be the well-authenticated occurrence, as a usual minor part of the *Pseudomonas–Acinetobacter–Moraxella* association, of some Enterobacteriaceae that are cold-tolerant and that apparently do not indicate fecal pollution. The only practical way to diminish the *Salmonella* problem by bacteriological examination of carcass meat seems to be to examine for *Salmonella* directly and to set guidelines. These should include a sampling plan calculated to reject a sufficient proportion of material to make improvements in production, transport, and slaughter hygiene economically worthwhile. The nature of the plan will depend critically on the method of sampling. ICMSF (1974) described different techniques for obtaining field samples from carcasses (e.g., cutting, swabbing, or rinsing).

Because there are few organisms and negligible growth within refrigerated carcass meat, sampling of the surface is usually sufficient. The surface of meat is very uneven and discontinuous, and bacteria gradually penetrate it during storage. Large numbers of bacteria may reside in natural cavities such as hair follicles on pig skin. Hence, the greatest numbers of bacteria are usually recovered by cutting out and homogenizing an area of surface with several millimeters of underlying tissue. However, this procedure is often not permissible or practicable; also it can be applied only to small areas, whereas it is desirable to sample large areas because the contamination of the surface is far from uniform. Consequently, bacteria are removed from the surface by simpler techniques such as swabbing, washing, or scraping or by a combination of these methods, which can be applied over larger areas. Unfortunately, used alone, they recover only a proportion, sometimes as little as 10%, of the bacteria present.

The even simpler adhesion and impression methods recover still less of the total flora. Moreover, besides being small, the percentage recovery varies greatly from one occasion to another. These difficulties are aggravated by the heterogenous nature of the surface—lean, fat, or skin. Only if washing is prolonged or if several swabs are used successively on the same area can a majority of the bacteria be recovered. By contrast, a relatively homogenous product like comminuted meat can be mixed (not very effectively!) and sampled directly to get a more nearly complete and representative sample.

Accordingly, the number of bacteria apparently associated with spoilage will be markedly affected, to the extent of several-fold, by the method of sampling used. For example, in comparing carcass counts with those in mince, sampling of the former (e.g., by swabbing) is likely to recover a far smaller proportion of the cells actually present than is sampling of the latter. Similarly, in setting numerical criteria, the associated sampling procedure will be a most influential factor.

There is now a tendency to swab large areas of the surface of carcasses with larger swabs to obtain more representative samples than is possible by excising and comminuting small areas. Swabbing the surface of a carcass produces a sample that represents a relatively large weight of the meat (for example, approximately 100 kg with a large beef carcass), which might require a radical change of sampling plan.

The incidence of salmonellae in beef and mutton is usually much lower than in pork. Felsenfeld *et al.* (1950) reported only 0.2% positive among beef samples (weight not stated). Application of the plan* $n = 5$, $c = 0$ would exert almost no control on that situation ($P_a = 0.99$); the plan $n = 60$, $c = 0$ would reject 1 in 10 of such lots. However, the above example represents low levels of contamination, and higher levels are sometimes reported (Hobbs and Wilson, 1959; see also Table 15.5). Using a swabbing technique covering the entire carcass (in eight eighths), Weissmann and Carpenter (1969) found that 56% of pork and 74% of beef carcasses sampled were positive—the beef samples presumably represented an area approximately five times greater than the pork. In this situation, even the plan $n = 2$, $c = 1$ would exert a high degree of control. Evidently, as has been emphasized before (ICMSF, 1974, pp. 73 and 75), the desirable choice of sampling plan depends on the circumstances, especially the amount of material represented in the sample examined.

The number of cells of the *Pseudomonas—Acinetobacter—Moraxella* association, conveniently estimated by plate count at 25°C, is clearly a reasonable index of the likelihood of spoilage. Because the number recorded varies so much with procedure (cf. above), a prime requirement for any criterion would be agreement about the procedure to be adopted, and this is still lacking.

It is important to distinguish between viable counts per gram and counts per square centimeter. In cutting out a sample and homogenizing (ICMSF, 1974, p. 140), the result varies according to the thickness of sample and the area of surface included because nearly all microorganisms are on or near the surface. A sample piece 5 cm^2 in area $\times$ 0.2 cm deep, and 1 g in weight, will yield a count about five times as great as a piece

---

* For definitions and a detailed description of sampling plans, see Book 2 of this series (ICMSF, 1974).

1 cm² × 1.0 cm deep if reckoned per gram but will yield the same count if reckoned per square centimeter. Further, counts per square centimeter are usually based on some surface sampling technique (e.g., swabbing, washing) that removes only part of the total flora, often 20% or less. Consequently, there is uncertainty by a factor perhaps as great as 5, in comparing counts per gram with counts per square centimeter. The importance of this is made evident by the following discussion.

For this criterion, it is advantageous to use three-class plans, where the number of samples falling between the maximum tolerable (M) and the level of no concern (m), gives an indication of need for preventive action.

The maximum tolerable number (M) might be set at a level where spoilage is immediately likely (approximately $10^7/cm^2$) or likely after an undesirably short period of storage (approximately $10^6/cm^2$). The choice of the level of no concern (m) is a matter of opinion even more, depending largely on expectations and the stage in the history of the carcass at which examination is to be made. Ayres (1955) suggested 10^4 to $10^5/cm^2$; the latter level appears the more realistic since a proportion of carcass meat may have counts exceeding $10^4/cm^2$ even immediately after preparation (Tables 15.9 and 15.10); ICMSF (1974) suggested $5 \times 10^5/cm^2$. The implication of such levels can be illustrated in relation to the observations in Fig. 15.1 (Ingram and Roberts, 1976).

Of lamb carcasses, there were none with counts as great as $10^7/cm^2$, and only 1% with counts exceeding 5×10^5, i.e., $p_d = 0$ and $p_m = 1\%$. Table

TABLE 15.9

Surveys of Aerobic Plate Counts on Carcasses before Chilling [a, b]

Product	Country	Number of carcasses examined	Percentage of carcasses with counts/cm² ($\log_{10}$)					
			1–2	2–3	3–4	4–5	5–6	6–7
			%	%	%	%	%	%
Beef	A	726	2.5	31.1	50.8	15.3	0.3	—
Lamb	A	1338	0.1	2.2	30.6	57.5	9.4	0.1
Beef	B	3041	72.9		22.1	4.5	0.5	—
Lamb	B	6180	32.9		43.9	20.0	3.0	0.1

[a] Previously unpublished data of T. A. Roberts and P. M. Nottingham.

[b] Samples were obtained by swabbing. Agar plates in country A were incubated at 25°C, and those in country B at 37°C. Since incubation at 25°C recovers both psychrotrophs and mesophiles and 37°C recovers only mesophiles, the actual bacterial levels in meats from country B would be approximately double those shown.

TABLE 15.10

Comparison of Bacteriological Condition of Retail Cuts with Beef Carcasses [a]

Bacteriological condition	Carcass	Retail cuts
Aerobic plate count	5 [b]	6 [b]
Enterobacteriaceae	3 [b]	4 [b]
Salmonella	< 0 [b]	< 1 [b]
Escherichia coli	20 [c]	50 [c]

[a] Mossel *et al.* (1975).
[b] Log_{10} of the level not exceeded by 95% of samples examined.
[c] Percent positive/cm^2.

4 of ICMSF (1974) indicates that the sampling plan $n = 5$, $c = 3$ would reject virtually none of such material ($P_a = 1.00$), and $n = 5$, $c = 1$ would reject only about one consignment in 100 ($P_a = 0.99$). There would be even fewer rejections with beef, for which counts were about 10 times lower than for lamb carcasses.

If, however, the m level was set at 5×10^4 (Patterson, 1971) and 19.5% of the lamb carcasses had counts exceeding this value, i.e., $p_m = 0.20$, the sampling plan $n = 5$, $c = 3$ would reject one in 100 ($P_a = 0.99$) of such lots, and $n = 5$, $c = 1$ would reject one in four ($P_a = 0.74$). For beef (see Fig. 15.1) with an m level at 5×10^4, p_m would be about 3% for $n = 5$, $c = 1$.

The proposals of ICMSF (1974, Table 24) would apparently be very lenient if they were applied to freshly slaughtered carcasses as in Fig. 15.1. But Fig. 15.1 is based on sampling by swabs, whereas the ICMSF proposal refers to cut and weighed samples. Had the latter procedure been used for Fig. 15.1, the counts might all have been five times greater with perhaps 10% exceeding 5×10^5. In that case, the ICMSF proposal would be less lenient.

Note that these examples should not be regarded as more than illustrations of the implications of the choice of sampling plan, because the data of Fig. 15.1 refer to a restricted choice of carcasses and to restricted areas different on the two types of carcasses. Patterson (1971) suggested that the same microbiological criteria should apply to beef and lamb carcasses.

In most countries, carcass meat is held under cool conditions and is cooked before it is eaten; both these conditions of use are likely to diminish hazards from food-poisoning bacteria and to justify the choice of criteria of relatively low stringency with respect to pathogens. Where circumstances of use are different, more stringent criteria might be desirable.

Microbiological criteria reflecting quality of raw meats are difficult to define because of the extreme heterogeneity of contamination (e.g., Tables 15.6 and 15.9). Differences of 100-fold often occur between individual counts at different sites on the same carcass. Similar differences exist between the same sites on different carcasses (cf. Fig. 15.1), even within the same abattoir on the same day. Such differences are likely to be reflected in the condition of cuts or ground meat, prepared from the different carcasses or joints, although there has been no clear experimental demonstration of this.

The heterogeneity in the total flora is repeated in its individual elements. Counts for Enterobacteriaceae or coliforms, or for *Escherichia coli* or *Staphylococcus aureus,* show similar variability though at successively lower numerical levels. Similarly *Clostridium perfringens* or *Salmonella* may be very unevenly distributed on the carcass and absent from large areas.

In summary, the following can be stated;

1. A wide spread of aerobic plate counts in, e.g., small samples of minced meat is to be expected.

2. An individual small sample may grossly misrepresent even the particular piece of meat from which it came. It is clearly ridiculous to base conclusions on the examination of a few samples: differences among these are likely to result mainly from unknown random variation.

3. To get any valid impression of the result of a particular operation or output of a factory requires examination of numerous samples covering at least the identifiable sources of variation, e.g., day, week, and season (Ingram and Roberts, 1976).

4. Variability, and hence the spread of frequency distribution depends directly on sample size. Statements such as, "85% of samples contained 10^2 or fewer coliforms per gram" or "the log 95 percentile value of plate count is approximately 5," are meaningless without knowledge of the sample size (i.e., number of sample units examined).

B. Vacuum-Packed Primal Joints

1. Effects of Processing on Microorganisms

A rapidly increasing practice is to store and distribute chilled raw meat as primal cuts vacuum-packed in gas-impermeable bags. There are advantages to this in ease of handling, cleanliness, and the preservation of color, and a much longer storage life results from a different microbiological situation. This provides opportunity for full maturation of the meat without significant microbial alteration. For a discussion of vacuum-packaging, see Chapter 11.

2. Spoilage

Inside a vacuum package, the residual oxygen is consumed sooner or later by tissue respiration, and is replaced by CO_2 that effectively reaches a high partial pressure in the limited volume available (Ingram, 1962). Consequently, while conditions within the joint are virtually the same as if it were part of a carcass, conditions at the surface are no longer aerobic, and any bacteria growing there must do so semianaerobically while tolerating increasing concentrations of CO_2. The *Pseudomonas–Acinetobacter–Moraxella* association may develop as residual supplies of O_2 dwindle and is often followed by multiplication of *Microbacterium thermosphactum* (incertae sedis) and Enterobacteriaceae. It ultimately is outgrown, however, by a gram-positive flora of lactic acid bacteria—mainly so-called "atypical streptobacteria" whose identity remains to be satisfactorily defined (Reuter, 1975). For reasons not yet clear and perhaps in unusual circumstances, *M. thermosphactum,* Enterobacteriaceae, or even *Pseudomonas* spp. may sometimes predominate, perhaps depending on the initial contamination and their stage of growth.

By the time such meat is unpacked, cut up, and exposed to air, it normally carries large numbers of bacteria. The type and rate of spoilage are mainly determined by the initial numbers of bacteria and the opportunity for development of the *Pseudomonas–Acinetobacter–Moraxella* association before the meat was vacuum-packaged (Sutherland *et al.,* 1975).

Within vacuum packs the gram-positive bacteria do not multiply on meat of normal pH as rapidly as gram-negative spoilage bacteria on aerobically stored meat. At 0°–2°C numbers may attain 10^7–10^8/cm^2 after about 2 months with only a general souring due to acid production from available glucose. In fact, glucose becomes depleted in the surface layers and numbers do not increase beyond this level (Gill, 1976). In these circumstances, it is unusually difficult to decide whether the meat is spoiled. Most of the usual criteria are inapplicable [e.g., NH_3 production, increased pH, or associated extract release volume (ERV)]. Sutherland *et al.* (1975) recommended titratable alkalinity, but this merely indicates the extent of change that has already occurred and has no value as a *predictor* of spoilage (Dainty, 1971).

On further storage, the odors become more cheesy, because of the presence of short-chain fatty acids (C_2–C_6) of which acetate and butyrate are present in the greatest concentrations. At this stage decreases in ribose concentrations can also be demonstrated.

The foregoing applies to normal meat of pH < 6.0. Meat of higher pH contains less glucose; amino acids are attacked at an early stage of bacterial growth. Consequently, spoilage takes a different form and becomes apparent at lower levels of bacteria. Meat of higher pH is liable to "putrefy" even when vacuum-packed, owing to development of *Pseudomonas* spp. and *Aeromonas* spp. in addition to high numbers of lactobacilli, and therefore, the meat spoils in about half the time. Hydrogen sulfide, trimethylamine, and a number of fatty acids, including acetate, butyrate, isobutyrate, and isovalerate can be extracted from such meat, the concentrations of these fatty acids being much greater than in normal pH meat. Pure culture studies with various lactic acid bacteria and *M. thermosphactum* are consistent, with acetate being derived from glucose and/or ribose and isobutyrate and isovalerate from the corresponding branch-chained amino acids.

Meat spoiled by *Pseudomonas* appears different from that spoiled by the aerobic *Pseudomonas–Acinetobacter–Moraxella* association; e.g., some produce H_2S and so cause greening of vacuum-packed beef if its pH is higher than 6.0 (Nicol *et al.*, 1970). For these reasons, DFD meat is regarded as unsuitable for vacuum-packaging (Bem *et al.*, 1976).

3. Pathogens

No special hazards are associated with vacuum-packaging. The lactic acid bacteria that normally develop are harmless, and the circumstances of relatively low pH and low temperature are unfavorable to recognized pathogens. Organisms resembling *Yersinia enterocolitica* have been isolated from vacuum-packed beef and lamb stored at 1°–3°C, and higher incidence occurred on cuts packed under high vacuum condition (Hanna *et al.*, 1976). The significance of these findings in relation to public health hazard has not been established yet nor is there sufficient published information to justify suggestions for numerical criteria for any pathogen.

4. Control

An aerobic plate count on vacuum-packed primal joints is of little value, as counts of approximately 10^7–10^8/cm² do not necessarily reflect the degree of spoilage. Determination of the number of gram-negative microorganisms initially present on the surface of the meat will give an indication of possibilities for growth of psychrotrophic bacteria prior to vacuum-packaging.

C. Retail Cuts

In principle, the microbiology of retail cuts of meats is similar to that of carcasses. The compostion of the material and the intrinsic contamination are virtually the same. In practice, differences arise because of the greater opportunities for contamination by the extra handling, the greater surface/ volume ratio, and the exposure to more diverse conditions at retail.

1. Effects of Processing on Microorganisms

The starting condition of a retail cut depends first on the previous history of the carcass from which it came. If it has been prepared promptly and hygienically from a carcass cooled quickly, the initial load of bacteria may be $<10^3/cm^2$. If it has been prepared from meat cooled slowly, after some delay, or without special hygienic precaution, counts may approach $10^6/cm^2$, and storage life will be short; this is the situation in the conventional butcher's shop. If cuts are prepared from meat previously "matured" for about a week near 0°C, initial counts are likely to be even higher, exceeding $10^6/cm^2$. The likely nature of the flora under these different circumstances has been broadly indicated above.

The starting condition may, however, depend even more on the hygienic conditions under which the retail cuts are prepared. Where cutting is under sanitary conditions, increases in count seldom exceed 10-fold, but much larger increases often occur, through use of contaminated knives, cutting tables, slicing machines, and other utensils and pieces of equipment. Many countries now require that retail cuts be prepared at temperatures not above 10°C or even 7°C.

Retail cuts have a relatively high surface/volume ratio and a large proportion of cut-muscle surface. If cuts are exposed without special precaution, evaporation is rapid, and quality is spoiled by loss of sheen and changes in color before microbiological spoilage occurs, unless the initial numbers of bacteria are large. Where surface drying occurs, micrococci and yeasts become more numerous in the spoilage flora.

For these reasons, it is increasingly common practice to protect cuts of meat with a transparent wrap to preserve a high humidity at the meat surface. To maintain the meat pigments in an oxygenated state and preserve a bright red color, oxygen-permeable film is generally used. In these circumstances too, the *Pseudomonas–Acinetobacter–Moraxella* develops and causes spoilage.

Retail cuts are packed to an increasing extent in controlled atmospheres containing CO_2, with an extension of the shelf-life at refrigeration temperature (Newton *et al.,* 1977a). See also Chapter 10.

2. Spoilage

From the scanty evidence available, the spoilage of retail cuts resembles that of larger pieces of meat. In the open, or in air-permeable packs, the flora is dominated by nonpigmented pseudomonads. In vacuum-packs these are replaced by lactic acid bacteria.

There are reports (Bem *et al.,* 1976) of souring of retail cuts in oxygen-permeable packages, which may well occur if permeability is low, for if retail cuts are vacuum-packed, they undergo souring spoilage by lactic acid bacteria as do larger joints.

A sophisticated system involves (1) the packaging of retail cuts in gas-permeable packages, so that the pigment can oxidize on exposure to air and (2) the storage of these packages in bulk in anaerobically packed ($CO_2 + N_2$) master-containers to provide long storage life. Little is known of the corresponding microbiology; the relationships described so far for vacuum-packaged meat presumably apply. It has been reported further (for chopped beef) that, after as much as 20 days anaerobic storage, the numbers of bacteria may at first decline with admission of air (Berry and Chen, 1976).

Microbes on anaerobically stored meat do not break down protein significantly until late in spoilage when *Pseudomonas* spp. produce changes in both sarcoplasmic and myofibrillar proteins (Dainty *et al.,* 1975). Initial growth is first at the expense of glucose and ribose and second at the expense of lactate and amino acids (Gill, 1976). The chemical nature of the odors produced has not been studied in detail, but H_2S, ethanol, acetoin, and ethyl esters of short-chain fatty acids can be extracted from such meat. Various amines have been detected in spoiling pork, e.g., ammonia 500 mg, methylamine 1.0 mg, dimethylamine 0.2 mg, and trimethylamine 0.5 mg/kg with traces of corresponding ethyl and propyl derivatives (Patterson and Edwards, 1975). Undoubtedly the picture is similar to that found for fish (Shewan, 1974) where a wide range of volatile compounds has been detected. The process is commonly but mistakenly called putrefaction, although it is clearly different from the classical putrefaction caused under anaerobic conditions by various clostridia at warm temperatures.

3. Pathogens

The pathogenic species in retail cuts are the same as those on the meat carcasses. There is little statistical evidence to suggest whether they are more numerous on the smaller pieces, as might be expected. Table 15.10 suggests levels of approximately ten times greater on retail cuts than on carcasses.

For boneless meat the situation is naturally different, the salmonellae being likely to be distributed in the bulk of the meat. Table 15.11 illustrates the higher rates of contamination of horsemeat as compared with beef, and of boneless as compared with carcass meat.

TABLE 15.11

Distribution of Salmonellae in Frozen Meat from Several Countries [a]

Product	Number of samples	Positive (50 cm^2)	
		Number	Percent
Horsemeat			
carcass	333	47	14
boneless	184	99	54
Beef			
boneless	751	97	13

[a] van Schothorst and Kampelmacher (1967).

There appears to be no special hazard of staphylococci on retail cuts despite the extra handling. Normal handling temperatures are too low to favor staphylococci, which grow poorly below 10°C and produce toxin only slowly below 20°C. Furthermore, growth and production of toxin might be hindered by concomitant growth of the *Pseudomonas–Acinetobacter–Moraxella* association (see Collins-Thompson *et al.,* 1973).

4. Control

The considerations for control appear analogous to those with larger pieces, though counts generally may be higher on retail cuts.

D. Comminuted Meat

1. Effects of Processing on Microorganisms

Comminuted meat is regarded widely as more perishable than intact meat because of the greater availability of meat juice and because surface microbes are distributed throughout the mass during mincing. In fact, experiments show that *Pseudomonas* grow less rapidly within the mass of minced meat than on the surface (Kitchell and Ingram, 1956), doubtless because conditions gradually become anaerobic within. Hence, the count and the floral composition in mince will depend on the proportions of surface and deep material included in samples examined. The numbers of microbes vary greatly, presumably reflecting the situation in the original meat. For example, counts are generally lower in minced beef than in minced pork. Large quantities of minced meat are prepared commercially with relatively low counts, e.g., for hamburgers; but minced meat at retail is often of poor microbiological quality.

TABLE 15.12

Percent Distribution of Bacterial Counts among 74 Raw Beef Patties [a, b]

Count	Percentage of set with mean count ($\log_{10}$/g)						
	<1	1–2	2–3	3–4	4–5	5–6	6–7
Aerobic plate count	—	—	—	4	16	55	24
Coliform	43	41	9	5	1	—	—
Escherichia coli	59	32	4	4	—	—	—
Staphylococcus aureus	43	42	12	1	1	—	—

[a] Surkiewicz *et al.* (1975).

[b] Samples frozen for 3 to 4 weeks prior to examination. Weight of sample unit: 50 g.

Table 15.12 gives data for patties consisting only of ground beef, produced under conditions of good manufacturing practice where no change in bacterial numbers was believed to occur during grinding (Surkiewicz *et al.,* 1975). But these levels, while lower than those in minced meat in shops (cf. Table 15.13), are higher than those on beef carcasses (cf. Table 15.6). The common practice of using mince as a vehicle to sell meat of low bacteriological quality is prohibited in several countries. The addition of 1% of spoiling meat with bacterial counts of about 10^8/g immediately and inevitably gives counts of about 10^6/g in the bulk to which it is added.

TABLE 15.13

Percent Distribution of Counts from Approximately 1000 Nationwide Samples of Minced Beef in the United States [a]

Aerobic plate count, $\log_{10}$/g	% of total number of samples [b]	Coliform count, $\log_{10}$/g (MPN in LST [c])	% of total number of samples
<5	13	<1	8
5–5.7	17	1–2	41
5.7–6	10	2–3	30
6–6.7	26	3–4	17
>6.7	34	>4	4
>7.7	5	—	—

[a] Goepfert (1976).

[b] Weight of sample unit: 50 g.

[c] LST is lauryl sulfate tryptose broth. MPN is "most probable number."

Furthermore, as with retail cuts, contamination may readily occur during trimming and preparation, especially if meat of inferior quality is used. Meat for mincing is normally boned by hand. The degree of comminution, in itself, is not known to be important.

Machines for boning meat have recently been introduced. If the meat is derived from a hygienically acceptable process—particularly if strict temperature control is exercised—such meat is microbiologically equivalent to or better than mince from hand-boned meat.

In fresh minced meat, the flora is mainly composed of Micrococcaceae, but lactobacilli, *Pseudomonas,* and Enterobacteriaceae are also found in significant proportions. Immediately after mincing, minced meat produced under good manufacturing practice will have a plate count (at $22°C$) well below $10^6/g$; a count at $35°$ or $37°C$ will normally be slightly lower. Coliforms, *Escherichia coli,* and fecal streptococci will normally be present but in small numbers—below $10^3/g$. Counts of *Staphylococcus aureus* will normally be below $10^2/g$. Salmonellae are occasionally isolated from raw minced meat, especially pork. The evidence suggests that they usually come with the meat and are not introduced during preparation.

Minced meat with a high initial bacterial load deteriorates quickly during unfrozen storage in air; the count often exceeds $10^8/g$ within 24 hr, according to temperature. At this point the meat may have a pronounced off-flavor, starting with a decline of the meat aroma and gradually developing into an off-odor. The product at retail normally has much higher counts than those on freshly prepared meats, as Table 15.13 illustrates.

Multiplication within the mince may contribute to this difference, especially if temperature control is poor. The number of pseudomonads and/or lactic acid bacteria increases, and if the meat is kept at too high temperatures ($15°C$) Enterobacteriaceae become important. The growth rate in minced meat can be influenced by inoculating it with lactic acid cultures resulting in a lower count of gram-negative organisms, a lower pH, and less proteolysis and lipolysis (Reddy *et al.,* 1975), but only when carbohydrate was added with culture.

It used to be common practice to add SO_2 in concentrations of approximately 0.05% to minced meat to preserve color. Sulfite returns the red meat color even after the meat is organoleptically spoiled. Thus sulfite misleads the consumer as to the freshness of the meat. Though this practice is now regarded as fraudulent in some countries, it may persist even where it is forbidden (e.g., van der Meijs, 1970) because storage life may be extended by up to 50%. A retarding effect on *Pseudomonas* has been demonstrated, also on Enterobacteriaceae in general and *Salmonella* in particular (Moerman *et al.,* 1966). The effect is greater at lower pH (cf. Chapter 10).

2. Spoilage

The general microbiological relations resemble those with unminced meat, but counts in general are 10– to 100–fold greater in commercial minced meat than on the equivalent carcass. The development of the microbial flora depends on whether the meat is packaged anaerobically or aerobically. On the surface of aerobically stored mince, conditions are aerobic and the color (when fresh) bright red; in the interior (unless loosely packaged) conditions become anaerobic owing to tissue respiration, and the pigments are reduced again. On the surface, a *Pseudomonas–Acineto-bacter–Moraxella* association develops; in the interior a gram-positive lactic acid flora predominates (Jaye *et al.,* 1962). There is no indication that the nature of these associations is different on minced meat from the corresponding ones on unminced meat, except in the higher frequency of adventitious contaminants, e.g., Enterobacteriaceae and *Aeromonas*. The proportion of psychrotrophs increases after preparation: whereas counts at 20° and 35°C differ only slightly in fresh minced meat, numerous examinations of commercial material reveal counts at 20°C ten or more times greater than those at 35°C, the latter temperature being evidently too high for the bulk of the spoilage flora.

It is reasonable to assume that chemical changes in the surface layers of minced meat resemble those in aerobically stored meat, and those in the center resemble those in anaerobically stored meat, although there is little experimental substantiation.

The nature of the eventual spoilage differs according to the nature of the flora, as determined by the conditions just indicated. Amines are produced and the pH rises on the surface, with lactic acid production and falling pH in the center. When air is excluded, bacterial counts can become very high without obvious spoilage.

3. Pathogens

Among the frequent contaminants of minced meat are various organisms significant in public health. Nearly 50% of commercial samples of minced beef contain *Clostridium perfringens*.

Salmonellae are not common in minced beef, despite the relatively high counts of bacteria, in general, and, still more, of Enterobacteriaceae. Among samples of about 10 g, usually considerably less than 1% have been found positive for salmonellae. Minced pork often has a higher incidence of salmonellae, which reflects the high incidence on pork carcasses; up to 25% of samples of pork sausages and sausage meat have been found to contain salmonellae, though in very small numbers (Roberts *et al.,* 1975; van Schothorst *et al.,* 1970). Viruses have been reported in minced meat, but their significance is still uncertain.

Up to 2000 cases of food poisoning were reported annually in Germany before 1936, most of them alleged to be caused by the consumption of raw minced meat. As soon as the Minced Meat Ordinance came into force, the outbreaks due to minced meat dropped sharply to 3 (including some 60 cases) in 1937 (Lerche *et al.,* 1957). This ordinance prescribes that minced meat be chilled and that it be offered for sale solely on the day of production.

There have been some widespread outbreaks of salmonellosis associated with minced meat, but the indications were that in a number of cases those affected had eaten it raw. Salmonellosis can be caused by inadequate cooking which permits salmonellae to survive, or by cross-contamination from the raw meat to a cooked food (see U.S. Department of Health, Education and Welfare, 1975e).

4. Interrelations

The interrelations on the surface are those already described for chilled retail cuts, while below the surface the interrelations are those described under chilled vacuum-packed meats.

In one nationwide U.S. survey (Goepfert, 1976), there was no positive correlation between plate counts at 35° or 20°C, Enterobacteriaceae or coliforms, and the occurrence of *Escherichia coli* or *Salmonella.* In another there was a good correlation between plate counts at 20°C and Enterobacteriaceae (van der Meijs, 1970). In Goepfert's survey, *E. coli* occurred with *lower* absolute frequency in association with high than with low plate counts at 20°C.

5. Control

Because the high microbial counts at retail lead to a short storage life, one might seek to limit the level of bacterial count, low enough to guarantee a reasonable storage life; e.g., 24 hr at 5°C, which corresponds roughly to an initial psychrotrophic count about $10^6/g$. This could have some significance with open packs, where spoilage is dominated by the surface flora of *Pseudomonas–Acinetobacter–Moraxella* leading to aesthetically objectionable off-odors. With vacuum packs, where spoilage is a souring caused by lactic acid bacteria, there is less evident need for control other than organoleptic, this type of change being acceptable in many foods. In either case, it is clear that a microbiological criterion would have little relevance to health hazard, which is probably negligible.

The microbiology of minced meat could be regarded in two quite separate ways. First, criteria similar to those for carcass meat might apply. If mincing is hygienic, it should not greatly change the microbial content. Second, however, it is highly desirable to provide some commercially useful outlet for scrap meat. In scrap meat, counts are likely to be very high but restricted to harmless psychrotrophs if refrigeration has been satisfactory. To check whether refrigeration has been adequate, a *mesophilic* count could be useful. Apart from this, it is doubtful if any criteria other than organoleptic are necessary, especially if such meat is directed to heat processing.

Much has been written about bacteriological criteria for minced meat in recent years especially in North America, revealing a lamentable confusion of ideas and techniques. For example, proposals for criteria have ranged from 5×10^4 to 1×10^7/g for plate-count, and from 0 to 10^2/g for *Escherichia coli* (Westhoff and Feldstein, 1976). While incubation temperatures for plate counts extend from 17°C (Mossel and Ratto, 1973) to 35°C, early investigators favored 37°C. Much of this confusion arises from ambiguity in objective: e.g., the lower incubation temperatures are clearly desirable to evaluate spoilage under refrigeration, the higher to evaluate pathogenic hazard.

For *Salmonella,* direct bacteriological examination may be necessary. Working with ground meat samples, of weight not clearly stated but probably 25 g, Williams and Spencer (1973) reported that about 2% were positive for *Salmonella* with normal pork meat and 4% with similar meat of lesser quality. Pivnick *et al.* (1976) reported contamination of the same order for ground beef.

With the sampling plan $n = 5$, $c = 0$ (ICMSF, 1974), $p = 0.02$ leads to $P_a = 0.90$ or rejection of one lot in ten; $p = 0.04$ implies rejection of one lot in five. Such a proceeding would exert pressure for improvement; but it could not, for statistical reasons (see ICMSF, 1974, Chapter 5), give much immediate protection to the consumer, for which purpose insistence on cooking should be more effective.

In Canada (Pivnick *et al.,* 1976), microbiological standards have been proposed for ground beef. Two sets of standards have been considered, one for frozen and one for nonfrozen products, the latter being the more lenient. The criteria are based on the three-class sampling plan for plate counts at 35°C, for *Escherichia coli* and for *Staphylococcus aureus* suggested by the ICMSF (1974).

V. FROZEN RAW MEAT

A. Effects of Processing on Microorganisms

Numbers of bacteria on the meat may be increased by contamination during the preparation and freezing operations. However, freezing meat kills a small proportion of all microorganisms present; more die during frozen storage, but this occurs rather slowly (approximately 5% of survivors per month) at temperatures around $-20°C$. Gram-negative rods are more susceptible than gram-positive cocci; spores remain unaffected; and vegetative cells of *Clostridium perfringens* die rapidly. Regardless of such destruction, bacterial levels may be very high in frozen thawed beef. Aerobic plate counts averaged 10^6 and some exceeded $10^9/g$ (Mossel *et al.*, 1972) among 115 representative samples of South American frozen beef, carefully thawed under controlled conditions. The conditions of thawing are most important, especially if temperatures and times are not carefully controlled. Long periods of time are required to thaw large masses of meat. When warm water is used to speed thawing, parts of the meat almost immediately reach temperatures at which growth of mesophilic pathogens becomes possible. In addition, warm water used for thawing itself rapidly becomes a favorable nutrient medium in which contaminants develop. If, to avoid this, thawing temperatures are kept low, there may be development of psychrotrophs on the surface. For these reasons, bacterial numbers commonly increase 10- or 100-fold during commercial thawing, increases which more than offset any decreases directly resulting from freezing.

There is little difference between thawing in air or water provided that conditions are good, but it is undesirable to use temperatures above 10°C (Bailey *et al.*, 1974). See also Table 15.14.

It is often stated that thawed meat is more perishable than fresh meat

TABLE 15.14

Effect of Thawing Times and Temperatures on Aerobic Plate Count of Beef Hind Quarters of Weight Approximately 60 kg [a]

| Thawing conditions | | 35°C [c] | | 25°C [c] | | 1°C [c] | |
| Temperature | Time [b] | | | | | | |
(°C)	(hours)	Frozen	Thawed	Frozen	Thawed	Frozen	Thawed
5	105	3.66 [d]	3.78	4.01	4.64	3.21	4.24
10	65	3.33	4.38	4.30	5.49	3.70	5.23
20	40	3.32	5.26	3.89	5.81	3.27	5.32
30	25	3.75	6.70	4.63	7.37	4.20	6.96

[a] James *et al.* (1977).
[b] Time from $-30°$ to $0°C$ at thermal center.
[c] Plate incubation temperature.
[d] Log_{10} aerobic plate count.

that has never been frozen, especially because of the drip exuded from thawed meat, which constitutes a medium favorable for microbial growth. But the growth rate of bacteria on thawed meat is broadly no different from growth on equivalent meat that has never been frozen. (See also Chapter 1.) Meat that has been frozen and thawed in commercial practice, however, commonly carries many more bacteria than the equivalent fresh meat, leading to a shorter storage life.

B. Spoilage

Properly frozen and stored, meat is not spoiled by microorganisms, but two malpractices make spoilage possible. First, exposure to temperatures in the range $-5°$ to $-10°C$ permits slow development of molds, which form small spots on the surface. *Cladosporium herbarum* is especially noticeable, producing black colonies. Usually there is little off-odor or off-flavor. Second, if a high microbial load develops before freezing, the meat may slowly spoil during frozen storage. Although these microorganisms cannot multiply when frozen and gradually lose their viability, their enzyme systems are relatively resistant and may remain active down to about $-30°C$. This is especially the case with lipases and lipoxidases that require minimal amounts of water. Frozen meats should not carry large numbers of microorganisms that will develop if thawing and refreezing take place under uncontrolled conditions. It is also important to store frozen meats (particularly fatty meats) at or below $-20°C$ if they are to be held for long periods.

Spoilage usually occurs after thawing, and its onset and nature are decisively influenced by the conditions of thawing just indicated. When the flora of the raw meat has not been changed greatly as a result of freezing and storage, and thawing has been carried out under cool conditions, spoilage is likely to follow much the same course as if meat had not been frozen, and will depend on the nature of the meat (as already described in Sections IV,A–IV,D).

Biochemical alterations in spoilage likewise depend on the circumstances and the consequent nature of the spoilage flora. If spoilage is caused by molds, musty odor development and lipolysis are likely to be the principal changes. It is difficult to distinguish whether rancidity has been caused by bacterial or tissue enzymes or by auto-oxidation.

Where freezing, storage, and thawing have been satisfactory, spoilage is likely to follow a normal course: ammoniacal spoilage under aerobic conditions; lactic acid souring under anaerobic conditions.

C. Pathogens

The pathogens of frozen meat are the same as those of the meat before freezing, but their numbers may be very much changed. It is a mistake to assume that, because freezing kills bacteria, frozen meat is likely to be free from pathogens. First, not all microorganisms are killed; second, freezing is only part of the history. For example, vegetative cells of *Clostridium perfringens* rapidly decrease in number in frozen meats, but spores remain unchanged and multiplication can be rapid during thawing under warm conditions.

The occurrence of salmonellae in frozen beef and especially in veal has been a source of much concern in international trade. For example, of 115 representative samples of frozen boneless beef received in the Netherlands from South America in 1968 to 1971, 16 (or 15%) were positive for *Salmonella* in 50-ml portions of drip collected during thawing (Mossel *et al.*, 1972). Where action has been taken to deal with these problems, the situation has improved in recent years.

D. Interrelations

In all the foregoing cases, spoilage developed under circumstances unfavorable to mesophilic pathogens and has no public health significance.

Where thawing involves excessive temperature abuse clostridial putrefaction with its associated risks is conceivable. Such a situation might be indicated by relatively high levels of aerobic plate counts at 37°C in comparison to those at 20°C.

E. Control

Microbiological criteria applied to frozen meats should be similar to those applied to fresh meats, assuming the samples are obtained from the frozen block or from meat thawed under conditions that did not allow bacterial growth.

Because the incidence of salmonellae is often higher on frozen boneless meat than on carcass meat, and because frozen meat may be a vehicle to bring alien serotypes of *Salmonella* into an importing country, there is good reason to examine samples for salmonellae.

The criteria suggested by ICMSF (1974, Table 24), which apply to frozen carcass meat, assume (a) a "normal" situation where the flora is similar to that of carcass meat that has never been frozen and (b) meat is sampled in the frozen state and thawed quickly. Spoilage occurs at about the same level, hence the M values are the same ($10^7/g$ or cm^2). Reflecting the small diminution in numbers due to freezing and storage, m values are 5×10^5 instead of 10^6, and the sampling plan for *Salmonella* is the same ($n = 5$, $c = 0$) as for chilled carcass meat.

For boneless frozen meats, comminuted frozen meats, and frozen edible offal, where the incidence of *Salmonella* may be greater, a less stringent sampling plan is indicated until improvement in microbiological quality of these products is found.

Sampling of frozen meats for bacteriological examination presents special problems. Since swabbing or scraping is ineffective, the preferred method is to bore with an augur, transport the frozen shavings, and thaw rapidly during homogenization in making the first dilution. But, because thawing is carried out quickly, such examination neglects the consequences of normal thawing and is liable to yield an optimistic impression of the micro-biological status of meat that having been frozen must then be thawed.

Most cells surviving freezing are "damaged," and though they may grow on favorable media including thawed meat, they cannot grow in the presence of salts or similar inhibitors included in selective media. Hence, selective counts for staphylococci or Enterobacteriaceae, for example, may be low, unless preliminary resuscitation is used (see Chapter 12).

VI. LOW-TEMPERATURE RENDERED MEAT

A. Effects of Processing on Microorganisms

Meat trimmings with a high proportion of fat are sometimes rendered in a tank held at or slightly below 48.8°C. The mixture of melted fat and protein is pumped to a centrifuge that separates the fat from the protein fraction. The protein is then rapidly chilled to about 10°C, boxed, and frozen for later use in manufactured meat products, such as sausages and patties. The low temperature of the rendering process does not alter the protein. The flora of the scraps entering the melt tank is that of the raw meat after cutting. Because the scraps are primarily trimmings from the outer surface of carcasses and primal cuts, they frequently have high numbers of gram-negative spoilage organisms such as *Pseudomonas*. Since these are primarily psychrotrophs with maximal growth temperatures at or slightly below 30°C, they cannot grow at the temperature of the melt tank. However, their numbers may not be measurably reduced either, so that their level may remain that of the original trimmings throughout the process.

B. Spoilage

This product is highly perishable because of its high bacterial level and high degree of comminution. It should be kept frozen and should be cooked promptly upon being combined with other products for further processing.

Obviously, if *Clostridium perfringens* and other anaerobes grow to very high numbers in the melt tank, they must be changing the nature of the substrate (putrefaction). However, there have been no studies on this question.

C. Pathogens

The melt tank at 48°C offers ideal anaerobic growth conditions to *C. perfringens,* whose maximum growth temperature is about 50°C. If the process permits particles of meat and fat to remain at this temperature for several hours, *C. perfringens* may grow to reach numbers as high as $10^9/g$. However, the organisms are nearly all in the vegetative stage, so they die when the meat is subsequently frozen, leaving only a few spores.

If the temperature of the melt tank is 45°C, *Staphylococcus aureus* could grow, but it is unlikely to produce toxin under the anaerobic conditions of the process, and competitive organisms would probably keep it in check.

Although *Salmonella* is unlikely to grow at the higher temperatures, it will not die under the conditions imposed on it. Its level would remain at about that of the raw scraps entering the system.

D. Interrelations

If the temperature remains at or only slightly below 48°C, the system will select anaerobes or facultative anaerobes that can grow at this relatively high incubation temperature. Other mesophilic bacteria will remain alive, or die very slowly.

E. Control

Microbiological control measures should take into account the temperature characteristics of the surviving flora. The trimmings and scraps should be handled rapidly and held in refrigeration. Any obviously dirty or spoiled material should be discarded or used for inedible purposes. The melt tank should be designed so that particles will not remain at a high temperature longer than necessary to separate fat from protein. Hence, the melt tanks now in use should be replaced by heat exchangers in which the product flows continuously without the delays inherent in a melt tank.

VII. DRIED MEATS

Preservation of meat by drying predates recorded history. It is sometimes combined with salting, which was possibly a secondary development. This section refers to meat dried without the addition of salt, where microbes are inhibited by a low water activity (a_w) achieved simply by the removal of water. Salted products are considered later (Section VIII).

It is useful to distinguish between traditional products and those prepared using modern technology. Although the basic microbiological principles of preservation are the same for both, the associated circumstances are so different as to lead to different problems in practice.

A. Effects of Processing on Microorganisms

1. Traditional Technology

The traditional products are mostly prepared in dry climates, sometimes even in polar regions, where dried meats play a major role in providing animal protein for the inhabitants. Well-known examples are charqui and carne secca (South America). Drying is carried out on wooden frames in the open air with coverings of gauze to protect against insects and birds. The drying process may require 5 to 14 days, according to temperature, humidity, and size of the pieces of meat.

When drying conditions are favorable, the surface of the meat dries rapidly. The microbiological requirement is to dry the *interior* of the meat before bacterial growth occurs. This can be done only if the pieces are thin and well separated. If the pieces are too large, or are packed closely together, the packed areas dry relatively slowly, forming wet spots where there is likely to be development of external mesophilic contaminants that are anaerobic or facultatively anaerobic, e.g., clostridia, Enterobacteriaceae, fecal streptococci, and staphylococci.

Exposure to dampness can permit superficial mold growth on the surface. Microbiological details are scarce. Traditionally, dried meat is spoiled much more often by insects than by microbes.

2. Modern Technology

In industrialized countries, substantial quantities of meat are dried under controlled conditions, the products being used in fabricated foodstuffs, notably soup mixes.

Sophisticated methods of drying aim to minimize or avoid the possibility of bacterial growth during the drying process. Two methods are available. The first is to dry at temperatures above 50°C that are too high for growth of the relevant bacteria (Sharp, 1953). After removal of fat, the lean meat is cooked and minced, spread in thin layers, and dried under controlled ventilation in hot-air tunnels. The air-dried product consists of granules of 5–10 mm diameter, with a water content less than 15%. Dried fat may be added as desired.

The drying tunnel provides warm, moist, aerobic conditions, which remain until drying is well advanced. Therefore, pieces must be small, and efficient separation is essential to ensure drying throughout in an hour or two. Drying is needlessly prolonged otherwise, and bacterial growth may occur in wet spots. The bacteriological quality of the product may be very good or very bad, depending entirely on the care given to these critical aspects of processing.

In the second and more modern process, cooked or uncooked meat is freeze-dried. Pieces as large as chops or steaks can be dried, and hence this process has a wider range of uses than the air-dried cooked mince process. The microbiological problems of the two processes differ correspondingly. During freeze-drying microbiological problems are minimized by subzero temperatures that are maintained until the water content is reduced to a level at which microbial growth is impossible. When the process is properly carried out, the numbers of microbes undergo little change during drying. Microbiological quality, therefore, depends on the state of the meat entering the drier and on avoidance of contamination between drying and packaging.

3. Dried Blood Plasma

Dried blood plasma is used as an additive in meat processing and in nonfood products. After the addition of citrate or phosphate, the blood is centrifuged and dried. There may be a large increase of saprophytic bacteria during the separation process. Ammonia is sometimes added as a preservative but is removed during subsequent drying—usually spray-drying—when counts fall slightly. Inoculated mycobacteria die more rapidly in phosphate-treated than in citrate-treated plasma (Kotter and Terplan, 1960). In material prepared satisfactorily, the plate counts on the final product are of the order of 10^4–10^5/g. As a rule, Micrococcaceae and *Bacillus* spp. are found, and sometimes also corynebacteria and lactobacilli. But gram-negative organisms are usually absent because the temperature used during spray-drying is not below 60°C. Clostridia, too, are usually absent.

4. Gelatin

Gelatin is widely used in foods prepared in households and is an ingredient in commercially prepared foods such as bakery products and ice cream. It is marketed either as leaves or powder. Gelatin is produced from skin, cartilage, and bones. Skin is treated with lime, and fat is removed by extraction with a solvent. Bones are treated with hydrochloric acid. The gelatin is then removed by extraction with boiling water, the solution filtered, and the water removed by drying under vacuum. Sulfur dioxide or H_2O_2 are commonly used to blanch gelatin, and both have a marked bactericidal effect. Upper limits for SO_2 residues have been established by national standards in some countries.

Gelatin produced under good manufacturing practice normally carries a low bacterial load (less than 10^3 bacteria/g). The flora mainly consists of sporeforming bacteria, micrococci, and fecal streptococci. In several countries, there are legal standards concerning bacteria in gelatin. Limits are specified for *Clostridium perfringens, Staphylococcus aureus,* and *Salmonella* (ICMSF, 1974) but may also include aerobic plate counts, Enterobacteriaceae, and fecal streptococci.

B. Spoilage

The key to stability and safety of dried-meat products lies in the water relations. An average piece of lean meat might contain about 75% water, 3% intramuscular fat, and 24% dry matter—not fat (mostly protein). Calculated on a fat-free basis, the normal water content of about 75% must be reduced to about 20% to prevent growth of resistant bacteria or yeasts and molds and to about 15% to prevent growth of xerotolerant molds over periods of several months. At intermediate water contents and a_w levels, molds grow progressively more quickly as shown in Table 15.15.

Dried meat with 40% fat would require a water content of about 9% to be safe from mold growth. With 12% moisture it would have a_w approximately 0.75, and mold growth might become apparent after several

TABLE 15.15

Time (in Days) for Appearance of Molds on Dried Meat, at Different Temperatures and Humidities [a]

°C	Relative humidity (%)			
	90	85	80	75
20	5	7	15	40
28	7	11	15	30
37	6	9	12	36

[a] Macara (1943).

weeks. Exposure to an atmosphere of 85% RH would raise the water content by a further 8% and lead to molding in about 2 weeks. For bacterial growth to become possible, the water content would have to be raised to about 30%, representing a water uptake of about one-fifth of the weight. It is clear that properly dried meat is microbiologically stable unless relatively large uptake of moisture has taken place upon exposure to moist conditions.

The principal spoilage organisms are molds,—a relatively restricted group capable of growing at relatively low water activities (e.g., *Aspergillus glaucus*). Most mold species, and nearly all bacteria, are unlikely to take part in spoilage unless there has been a large uptake of water. Examples of the times taken for molds to appear under different circumstances are given in Table 15.15. Systematic surveys of the nature of the fungal flora are not available.

When molds spoil dried meat, the most obvious alteration is in appearance. Musty odors and off-flavors may be produced, but there is no information regarding the nature of the biochemical changes involved. Lipolysis may be important.

The general relations in the spoilage of reconstituted meat are similar to those with fresh meat. Bacteria are the likely cause.

C. Pathogens

The risk from pathogens coming directly from the live animal in the traditionally dried products appears relatively small. Any virus present would probably survive for long periods in the dry product, though the significance of this is uncertain. The problem is largely eliminated by precooking before drying.

The risk of secondary contamination (e.g., with salmonellae) during unsuitable preparation, storage, and handling is much greater. Furthermore, if the product is exposed, it may be accessible to vermin and domestic animals. Safe packaging is, therefore, the primary problem, and all possible measures to prevent infestation by vermin should be taken.

For dried meats, the following pathogenic microorganisms may be important:

1. Clostridia, *Salmonella,* and other Enterobacteriaceae are likely to be introduced by contamination associated with meat.

2. Staphylococci and *Bacillus cereus* are likely to be introduced by contamination during preparation and drying.

3. Mycotoxin-forming fungi might conceivably be present but have not been reported for this type of product.

Multiplication of the bacteria will be prevented as long as the meat is kept dry. Therefore, these bacteria may be tolerable in appropriately small numbers. The most dangerous circumstances arise when water has been added to reconstitute the dried meat, especially if warm water is used to accelerate reconstitution. When multiplication of *Clostridium perfringens* and other mesophiles might be a hazard, the period of presoaking in water before cooking should not exceed 4 hr. Where risk appears likely, one should apply microbiological criteria corresponding to Cases of high risk, and the situation resembles that of fresh meat kept under warm conditions (ICMSF, 1974).

D. Interrelations

Spoilage of dried meat by molds has no known significance for public health. Involvement of mycotoxigenic species has not been reported. Bacterial pathogens remain inhibited until the water content has been raised considerably above that at which molds can grow and spoil the product.

E. Control

The microbiological problems of dried meat resemble those for the dried forms of other perishable foods. During storage of dried meat, bacteria in it gradually lose their viability, though the change is negligible for spores. At 15% RH with a_w about 0.3, the plate count falls about 10-fold per month, but less than 2-fold per month, if the dried meat is stored under nitrogen (Sharp, 1953). The decline is slower at lower temperatures and at lower water content.

It is difficult to make satisfactory microbiological examination of dried meat because it is heterogenous in structure and quality and cannot be perfectly mixed before withdrawing samples. Further, the samples themselves cannot be homogenized unless they have been thoroughly rehydrated, which takes several hours unless warm water is used, when there is a risk of appreciable multiplication of some of the microbes. Recommended procedures have been to shake in water with broken glass for 1 hr, representing a compromise between imperfect homogenization and increase of bacterial numbers. However, modern methods of homogenizing, e.g., in a stomacher, will probably give satisfactory results.

Many of the cells that survive drying are damaged in a manner resembling that caused by freezing (see previous discussion) and similarly may require resuscitation before exposure to selective media (Kampelmacher *et al.*, 1969, pp. 69–164).

In summary, to obtain dried meat of good microbiological quality, it is evidently necessary that technical control be exercised in several respects, which follow:

1. Use of only meat of suitable microbiological quality;
2. Control of contamination during preparation and transport to the ·drier;
3. Strict control of temperature/time relations during drying and avoidance of wet spots in the drying mass;
4. Drying to a sufficiently low water content relative to lean content;
5. Protection, by suitable packaging, of the dried product from reabsorption of moisture;
6. Careful reconstitution to minimize possibilities of bacterial multiplication.

VIII. RAW CURED MEATS, INCLUDING FERMENTED AND DRIED PRODUCTS

For a discussion of the effect of a_w and curing salts on microorganisms in foods see Chapters 4 and 8, respectively.

At concentrations used in raw cured meats, common salt (sodium chloride) is not bactericidal but for some species is bacteriostatic. The tolerance of microorganisms to common salt differs. Growth of some is inhibited by 2% salt; others are able to grow in saturated salt solution. Salt-tolerant microbes multiply in high and low salt concentrations; obligate halophiles can grow only if the salt concentration is above 10%. Salt-tolerant microbes are micrococci, *Sarcina,* some spirillae and vibrios, flavobacteria, and strains of the lactobacilli, which propagate in salted foods.

Salt-tolerant lactobacilli apparently come from the processing environment (Kitchell and Shaw, 1975). It seems likely that other types also may be enriched from the environment by selection in curing brines or added with the salt itself. In such cured products, enteric pathogens should be controlled by combined effects of salt, nitrite, pH, and low temperature. If the product is unrefrigerated a hazard may arise from staphylococci that are salt tolerant, especially where the competing flora have been removed by cooking.

Nitrite and nitrate are used as curing agents to impart the characteristic color, flavor, and microbial stability to meat products. Nitrate may act as a reservoir for nitrite when nitrate-reducing bacteria are present but exerts little bacteriostatic or bactericidal effect per se in the concentrations used. In most of the products, nitrate may be replaced by lower concentrations of nitrite. The bacteriostatic effect of nitrite depends on the pH of the meat. With a pH drop of one unit the bacteriostatic effect enhances about 10-fold, reaching an optimum around pH 5.

At pH 5.7–6.0 sodium nitrite at 0.02% (200 ppm) inhibits the growth of *Acinetobacter–Moraxella, Flavobacterium, Pseudomonas, Enterobacter, Escherichia,* and some micrococci, some bacteria being more resistant than others.

The prime microbiological characteristic of cured meat is that it does not putrefy. Production of NH_3 and so forth by the psychrotrophic gram-negative association characteristic of cool conditions can be arrested at a_w about 0.96; under warm conditions, multiplication of putrefactive anaerobes and Enterobacteriaceae is prevented at a_w 0.92, which level distinguishes between high- and low-a_w cured products (cf. following discussion). In the low-a_w group, the only known hazard is from *Staphylococcus aureus,* which grows to a_w 0.85, although production of enterotoxin at this low level has never been observed. Danger from mycotoxinogenic fungi is theoretically possible but has not yet been demonstrated. While these limits may be useful as generalizations, however, they are frequently too low in practice, for they refer to limiting a_w values under near optimal conditions. In practice, combination with additional factors such as low pH, nitrite, and low temperature (especially together) may inhibit the relevant organisms at a_w levels well above those indicated (cf. Chapter 13).

Spoilage and food-poisoning potential of unheated products varies with the class of products. Dry-cured hams may spoil primarily from growth of clostridia near the bone (Mundt & Kitchen, 1951), although other organisms may be involved (Giolitti *et al.,* 1971; Kemp *et al.,* 1975). The slow penetration of curing agents and the extremes in concentration (1 to 7% brine in the same ham) present opportunities for growth of a variety of organisms. Fortunately, they do not ordinarily cause spoilage or public health problems despite a 10-day drying period at about 35°C designed to kill salt-weakened *Trichinella spiralis* (Zimmermann, 1971). However, home-cured hams have caused botulism in France (Sebald, 1970) and other countries.

Fermented sausages undergo a cure/fermentation period of hours to days at temperatures of 10° to 30°C depending on the product. The emulsion fermented from the natural flora of the meat and from contaminants from the plant environment, generally reach pH 5.0 to 5.5. Adding a fermentable carbohydrate will drop the pH to 4.5 (Christiansen *et al.,* 1975), and the pH will drop more rapidly if the carbohydrate-supplemented emulsion is inoculated with a suitable starter culture. The acidulant, glucono-δ-lactone (GDL), may also be added.

In some starter cultures mixtures of organisms are used, e.g., *Micro-*

coccus spp. to reduce nitrate to nitrite and *Lactobacillus* spp. or *Pediococcus* spp. to produce lactic acid (Nurmi, 1966). The curing salts do not inhibit the starter cultures (Zaika *et al.*, 1976).

Spoilage organisms, mainly gram-negative rods, are inhibited in the salty, essentially anaerobic emulsion; and nitrite may augment the inhibition. The desired fermentation is carried out by lactic bacteria including fecal streptococci. Some fermented sausages are smoked and may be heated to 50°–70°C. Following fermentation they may require refrigeration. Others are more completely fermented and subsequently dried, with or without smoking, at controlled humidity and at temperatures not conducive to the growth of pathogens in the semi-inhibitory material. Drying prevents growth of molds on the surface.

Among these products, for reasons just given, a distinction is made between high- and low-a_w meats, with the boundary set arbitrarily at a_w 0.92, which corresponds approximately to a salt/water ratio of 10%.

A. High-a_w Products

Typical of this class of cured meats are (a) bacon and raw ham and (b) semidry fermented or unfermented comminuted sausages. The differences between these are not only in their manufacture but also in their microbiology.

1. Effects of Processing on Microorganisms

The older curing procedures of dry-salting or brining introduced many salt-tolerant microorganisms from the salt and the environment, but the evidence suggests that the flora of the original meat was the most important. More than 5% salt in water permitted salt-tolerant micrococci on the surface to outgrow the pseudomonads (cf. Table 15.4). A curing period of 1 or 2 weeks was necessary, during which the micrococci reduced nitrate to nitrite, which in turn produced the red color characteristic of these meats.

The salt-tolerant micrococci were weak in proteolytic and deaminating power so that spoilage was slow. Their numbers remained very high for as long as 10 days before the meat developed cheesy odors and flavors from lypolytic enzymes. The mesophilic spores in properly cured meat were unable to develop even if the temperature was uncontrolled.

Modern processing, improved hygiene, and storage conditions have improved the microbiological quality of such raw, high-a_w products, and some (e.g., sliced bacon) may have a storage life at temperatures below 5°C of several months.

Raw cured, high-a_w meats comprise different sorts of sausages with a short storage life. These are (a) semifinished products like the German "Bratwurst" for which a further processing (heating) is necessary before consumption and (b) fermented sausages to be eaten without further processing. The fermentation may be natural (that is the microbial flora selected by curing produces the desired organoleptic properties) or the fermentation may occur only after the addition of starter cultures. Commercially available starter cultures are strains of bacteria isolated from sausages undergoing a natural fermentation. They are gram-positive microorganisms like *Micrococcus, Pediococcus, Streptococcus, Leuconostoc,* hetero- and homofermentative lactobacilli, streptobacteria, and also yeasts *(Debaryomyces kloeckeri).* Microorganisms that are not present in meat but are claimed to have a physiological effect on the human gut (e.g., *Lactobacillus acidophilus*) are also used. It is not clear on what biochemical or ecological principles these additions are based. The fermentation of carbohydrates, from whatever source, has some effect on acid formation and lowers the pH. Mere formation of acid is, however, not sufficient for development of the desired aroma. Lipolytic activity is more important, especially for the formation of carbonyls that lead to aroma development. On the other hand, high concentrations of salt and low pH inhibit these activities.

Certain additives may be used to diminish the risk of faulty production and the development of pathogenic organisms. Some (e.g., GDL) speed up pH reduction. Sometimes SO_2-generating compounds (up to 450 ppm) are used in unfermented sausages and seem especially effective against Enterobacteriaceae (Dyett and Shelley, 1966). But since the toxicological safety of sulfite is debatable and, as SO_2 makes the meat look fresher than it is, addition of sulfite is prohibited in most countries. SO_2 is used only in products cured without nitrate or nitrite.

2. Spoilage

A distinction should be made between the initial flora of cured meats and the flora that develops during curing and fermentation and that gradually becomes the dominating flora in the products. The former consists mainly of the gram-negative bacteria described above, while the developing flora consists of gram-positive bacteria (mostly micrococci, but also lactobacilli, streptococci and *Microbacterium thermosphactum* [*incertae sedis*]).

Modern processors of bacon and ham employ brines with added nitrite and use mechanical injection of smaller pieces of meat so that diffusion of salt is rapid. Curing times can be short and bacterial numbers low. Such material is often sliced and vacuum-packed (see following discussion). In particular, these techniques permit the use of increasingly lower salt concentration (desired commercially) to the point where the control of internal mesophilic bacteria becomes questionable, without refrigerated storage.

In fresh raw sausages with pH below 5.8 and a_w above 0.95 (fresh pork sausages, e.g., the German "frische Mettwurst") there is normally no multiplication of Enterobacteriaceae, whose numbers at $+7°C$ are in the range between 10^3 and $10^5/g$ for 2 weeks, after which they slowly disappear.

The characteristic surface flora of bacon and ham consists of micrococci and coagulase-negative staphylococci, with a subordinate element of gram-negative rods and lactobacilli. The gram-positive microflora that develops in or on cured, unheated meats with a high a_w will gradually cause changes in the organoleptic qualities of the meats (Giolitti *et al.,* 1971). The surfaces will become slimy, and the flavor and odor will change, although not necessarily to the point of repugnance. The off-odor is normally characterized as "sweet," "flowery," or "acid," indicating that the attack is not on the meat protein but rather on carbohydrates and fats. Salt accelerates rancidity, but this is to some extent counteracted by the presence of nitrite.

For fermented sausages the organoleptic changes, especially the development of an acid taste are desired and characteristic for these products.

3. Pathogens

The pathogens of most concern in cured, high-a_w meats are *Salmonella* spp. and *Staphylococcus aureus.* Growth of both is restricted by a rapid drop in pH. For products heavily infected with *S. aureus,* there may be a risk after storage above 15°C, as staphylococci can multiply to $10^6/g$ even at pH 5.6 to 5.8. Toxin production, however, by *S. aureus* is diminished under the microbiologically competitive and anaerobic conditions in sausages.

In raw bacon and ham, *S. aureus* has been found frequently, but never in large numbers, and no cases of poisoning have been recorded from these products, when eaten just after they are cooked. *Salmonella* may or may not die out (Baran and Stevenson, 1975).

Even though they occur frequently in sausages, *Bacillus cereus, Streptococcus* spp., and *Clostridium perfringens* are not currently identified as causing outbreaks of disease among those who consume sausages (Nurmi, 1966). *Clostridium botulinum* has been found in pigs and bacon, and may develop in sausages if the concentrations of curing salts are reduced too much. The part played by nitrite is of special interest in view of the desire to minimize its use (Krol and Tinbergen, 1974; Tinbergen and Krol, 1977). Botulism was at one time so frequently associated with sausages that the disease was named for the food: *botulus* is Latin for sausage.

4. Interrelations

The salt-tolerant microflora in cured meats suppresses the growth of gram-negative proteolytic microorganisms and pathogens such as salmonellae. *Staphylococcus aureus* may grow and produce toxin before the combined effects of nitrite and low pH arrest its growth.

5. Control

A goal of the processor is to develop a characteristic microflora that contributes normal flavor to the cured meat. To measure the progress toward this goal, tests for pH, a_w, percent salt, and organoleptic quality would be valuable but plate counts would not.

A count of *Staphylococcus aureus* should not exceed 10^3/g during or immediately after the processing of the cured meats. In products with a very low salt concentration (i.e., high a_w), an Enterobacteriaceae count or a presence-absence test for salmonellae would indicate whether there is a potential hazard of food poisoning.

B. Low-a_w Products

This group of meats includes certain types of sausages prepared by salting, ripening with cold smoking, and drying. The final product has a pH of 5.8–6.0, and a brine concentration of 13–16%, which corresponds to 25–28% water and 3.5–5.5% salt (Takács and Zukál, 1962). Hungarian salami is characterized by long slow ripening; the water content should be below 28–30%, the protein content at least 19.5%, the fat content below 47%, and the salt content 4–4.4%. The pH is 5.9–6.2 but sometimes rises to 6.5 (Takács *et al.*, 1963). It also has an outer layer of mold, which adds to the typical flavor of this product.

Other types of salami contain even more salt, e.g., farmer salami with a brine concentration at or above 30%. These salamis have excellent microbial stability even in tropical areas.

The "Bündener Fleisch" (i.e., meat from the district Graubünden) is a delicacy produced in many Swiss valleys. It is generally made by subjecting beef to a slow curing and drying process. The water activity is usually very low (below 0.88). The biltong produced in Africa is a similar salted product of even lower a_w, often approaching 0.60. The North American country style ham is subjected to a very long ripening and drying period at temperatures that reach 30°C after the salt has fully equilibrated.

1. Effects of Processing on Microorganisms

The fermentation used during ripening may be natural or generated by starter cultures (e.g., lactobacilli, *Pediococcus*). A well-ripened raw sausage is characterized by its organoleptic, chemical, and microbiological properties. In perfectly ripened raw sausages, lactobacilli, salt-tolerant nitrate-reducing micrococci, and fecal streptococci can be detected. The plate count decreases during the ripening process from 10^{10} to 10^6/g, according to the degree of drying. Lactobacilli number between 10^4–10^5/g, while the salt-tolerant micrococci are below 10^5/g. Coliform bacteria are absent from sausages of this type, after a length of time that depends on the speed of ripening and drying. Persistence of coliform bacteria indicates that there was a large number in the sausage before ripening, or there was a faulty ripening procedure. Distribution of plate counts in these sausages is shown in Table 15.16.

Yeasts, present before filling sausage meat into casings, can multiply in the stuffing material near the casing. *Debaryomyces kloeckeri* belongs to the normal microflora and plays a role in the ripening, color, and flavor formation.

Molds are sometimes found on the surface of ripened raw sausages. Some contribute characteristic flavor and texture. Most common are *Penicillium* (various spp.) and *Scopulariopsis alboflavescens*.

2. Spoilage

Their ability to tolerate low water activity makes molds of the *Aspergillus glaucus* group the most frequent cause of microbial spoilage of biltong. Trials with biltong inoculated with *A. amstelodami, A. chevalieri, A. repens,* or *A. ruber* gave results shown in Table 15.17, indicating a_w 0.70 as the critical level below which microbial growth is delayed virtually indefinitely. Although species capable of growing at a_w below 0.70 exist, they do not grow in dried meat (compare Table 15.15).

If the lactobacilli in the stuffing material are too numerous and their fermenting activity too great, the pH drops below 5.4, inhibiting the nitrate reductase so that the required red color does not develop. Instead, the grey

TABLE 15.16

Distribution of Aerobic Plate Counts in Hungarian Sausages Processed by Curing, Cold Smoking, and Drying [a]

Number of samples tested	% Samples with (in thousands)			
	<5	5 to 50	50 to 500	>500
775	10	26	34	30

[a] Takács (1964).

TABLE 15.17
Effect of Equilibrium Relative Humidity (ERH) on Mold Growth on Biltong [a]

ERH (%)	Weeks to develop mold at 25°C	Amount of growth
>80	< 2	Abundant
78	< 4	Abundant
72	<10	Slight
70	>10	None

[a] van der Riet (1976).

or brown metmyoglobin forms. Direct addition of nitrite to the mix will result in a rapid formation of stable nitrosomyoglobin. Enzymes from lactobacilli can develop an off-odor (sharp, biting) and off-flavor (acid, biting, bitter, cheesy). The characteristic aroma is not different from that of high-a_w products, i.e., the odor of carbonyls from lipolytic activity.

3. Pathogens

Although mycotoxin-forming molds can grow on certain types of sausages, mycotoxins apparently do not penetrate into the stuffing. Aflatoxin B_1 has been found in country cured hams heavily infected with molds (Frank, 1972). *Aspergillus glaucus* can produce aflatoxin on biltong down to a_w 0.85, but this happens so slowly that the material would be spoiled much earlier by the nonpathogenic species (van der Riet, 1976).

Staphylococcus enterotoxin can be a risk in low-a_w sausage, particularly those made rapidly by fermentation at relatively high temperatures, but toxin does not form during storage of the finished product, where the a_w is sufficiently low to prevent microbial growth and toxin formation. The risk lies instead in the period of fermentation when a_w is still high. Any preformed enterotoxin will carry through into the finished product, where the staphylococcal cells gradually die out as fermentation slows, and drying reduces the a_w. Food poisoning from badly manufactured Genoa salami has been reported in the United States (U.S. Department of Health, Education and Welfare, 1971a; Barber and Deibel, 1972). Salmonellosis from unheated dried sausage has occurred in Canada.

4. Interrelations

Sausages or other cured meats are not fully ripened if gram-negative rods can be detected. Catalase-positive bacteria disappear during the processing, or their number is low and their enzymatic activity inhibited. The metabolites of catalase-negative bacteria, primarily lactobacilli, inhibit the proteolytic bacteria.

5. *Control*

Because the flora includes more useful than harmful microbes, aerobic plate counts have no significance. To evaluate the bacteriological state of these products, examine for pathogens and for the microflora indicating contamination. The microbiological examination should take into account the uneven distribution of types of microorganisms (varying degrees of oxygen tension in the products).

Prevention of the formation of staphylococcal enterotoxin during the fermentation of sausages depends mainly on the following (National Academy of Sciences/National Research Council, 1975):

1. Keeping the numbers of staphylococci in the original ingredients as low as possible through proper care of meat trimmings in order to minimize bacterial growth and through scrupulous sanitation and personal hygiene during manufacture.

2. Holding the ground meat in the refrigerator for a day or two to give lactic bacteria a chance to grow before "green room" fermentation begins.

3. Assuring rapid acidification through the addition of an active starter culture and a fermentable carbohydrate, and/or a chemical acidulant to inhibit growth of staphylococci in the fermenting sausage.

Samples of the meat should be taken at the height of fermentation activity and analyzed for numbers of *Staphylococcus aureus*. If there are several hundred thousand per gram at this point in the process, the sausages eventually completed may contain enterotoxin. Each sausage fermented in the casing is a world unto itself, so that some may ferment normally, whereas others may permit rapid staphylococcal growth. Sampling to evaluate the safety of a batch, therefore, has a risk of probability of acceptance of toxic material.

Analysis for *S. aureus* in the finished sausage is less informative because most would be dead. On the other hand, analysis for enterotoxin *in the centimeter of the meat nearest the surface* may be necessary if the lot is questionable. Enterotoxin forms only near the surface where there is sufficient oxygen. Routine analysis for enterotoxin, however, is not recommended.

C. Packaged Cured Meat Products *

Packaged, unheated cured meats present no greater microbiological problems than unpackaged, but are discussed separately because they are gaining a wider popularity—especially in sliced forms—and there is a gen-

* See also Chapter 11.

eral tendency for the public to suppose that all packaged products have a longer shelf-life than the identical unpackaged product. The difference between a packaged and an unpackaged meat depends greatly on the permeability of the film used. If it is moisture-proof and sufficiently permeable for O_2 and CO_2, meats with an initially high a_w will spoil as quickly as will unpackaged meats since there is no growth-inhibiting effect of surface drying observed with the latter. The flora remains similar, but the aerobic plate count may increase in numbers even more rapidly than in the unpackaged product.

If the film is impermeable for O_2 and CO_2, the flora is quite different, though the numbers of microorganisms may not be seriously affected; bacteria such as the lactic organisms that do not cause drastic organoleptic changes are favored, producing a longer storage life.

The entrapped atmosphere under a gas-impermeable vacuum package containing bacon gradually loses residual O_2 and gains CO_2. The bacon comes to resemble meat in a cured sausage; and the microbiology is similar, a flora of micrococci and fecal streptococci giving way to lactobacilli. The predominant lactobacilli belong primarily to the "atypical" group related to *Lactobacillus casei* and *L. plantarum* (Reuter, 1975), but it is not yet clear whether they are identical with sausage strains. Ultimately, spoilage is by souring, but numbers of lactobacilli may exceed $10^7/g$ for a week or more at 5°C, before this happens. Yeasts may develop in large numbers ($10^4–10^6/g$) if the pH of the bacon is below about 5.8. Molds do not grow.

In adequately cured meat, the clostridia are inhibited, whereas in improperly cured meat their numbers can increase (Roberts and Smart, 1976). If bacon is stored at 30°C, staphylococci can grow to high numbers ($10^6/g$) (or become the dominant organism), but at 20°C and below they are outgrown by micrococci (Cavett, 1962).

The shelf-life of meat products with a_w below 0.92 (e.g., salami) is not necessarily affected by packaging, except that the film will protect them from being contaminated, especially with molds. If salamis are vacuum-packaged, surface growth of molds present before packaging will be inhibited or restricted.

D. Sliced Meats

During slicing, meat may be heavily contaminated by the slicer, other equipment, and by human contact. Frequent and efficient cleaning and sanitizing, as well as good personal hygiene, are essential (Chapter 14). The contaminating flora may simply reduce shelf-life or may be pathogenic with a risk of multiplication where temperature is favorable and a_w sufficiently high. Sliced vacuum-packaged meats with a_w below 0.85 present no public health hazards because bacteria cannot grow at the low a_w, and mycotoxigenic fungi either cannot grow or do so slowly with limited access to oxygen. Gas-packaging with pure nitrogen does not extend the storage life over that of vacuum-packaged sliced meats, but the slices remain separated in the package—a convenience to the consumer.

E. Natural Casings

The nature of the casing (sausage skin) is important to the technological and hygienic quality of the finished sausages. Natural casings (i.e., animal intestines) are used in large quantities. Sometimes they will be eaten (e.g., the small intestines from sheep on raw sausages, frankfurters, and other cooked sausages); and sometimes they serve only as a container in which the product may ferment and/or be stored until consumption.

For edible casings, the intestines are emptied and washed, then the mucous membrane and the serous and muscular membranes are removed. The casings are then salted either in crystalline salt (dry salting) or in saturated salt solution (wet salting), then stored for later use. The natural flora of the intestine dies out, and only salt-tolerant or obligately halophilic species remain. In a satisfactory product, the plate count at 30°C falls between 10^4 and 10^7/g. The surviving organisms are primarily *Bacillus,* but also frequently *Pseudomonas, Clostridium, Micrococcus,* and *Gaffkya.** *Proteus* and a few lactobacilli may also appear in wet-salted casings (Riha and Solberg, 1970).

Putrefactive clostridia in sheep casings may cause swells in canned frankfurters, particularly if they were retorted at marginal temperatures and times. *Bacillus* may also weaken casings in canned sausages, without producing swells.

If the salt concentration is too low or the storage temperature too high, proteolytic organisms *(Proteus, Pseudomonas)* can spoil the casings. Heavy growth of *Serratia* can cause harmless discolorations, which can be washed away. Often localized spoilage appears as black spots, which can be decolorized by hydrogen peroxide. Such spots are mechanically less resistant, so the casings can be torn easily during further processing. Mycobacteria and many viruses are not completely inactivated by salting. Salmonellae are virtually eliminated 21 days after dry salting of naturally contaminated casings at 6°C, or in 24 hr at 25°C in saturated brines at pH 4 or 10 (Gabis and Silliker, 1974).

* Not recognized in Bergey's Manual (Buchanan and Gibbons, 1974).

Before use, the casings are washed and may be soaked in solutions of either hydrogen peroxide, sodium peroxide, tartaric acid, or lactic acid (Lerche *et al.,* 1957; Zakula *et al.,* 1964). These treatments reduce the plate count and effect a complete kill of *Salmonella* (Gabis and Silliker, 1974).

IX. COOKED, UNCURED MEATS

Only meats cooked under industrial and semi-industrial conditions will be discussed here.

Some meat regarded as cooked is in microbiological terms uncooked. The color change from red to grey associated with cooking, occurs at temperatures approaching 60°C, the precise level depending on duration of treatment. The center of a steak 15 mm thick, grilled "rare," barely attains 40°C even momentarily. Temperatures in the range 40°–60°C, especially of short duration, will not eliminate even relatively sensitive vegetative bacteria. Beef is the meat most commonly eaten "rare." The greatest hazards to the consumer of rare meat are salmonellae or parasites (see also ICMSF, 1978).

Heat treatment varies according to the expected storage life of the products. (a) Some meats are pasteurized at 60°–75°C to inactivate bacteria, yeasts and molds, parasites, and most viruses (cf. Chapter 1); (b) certain canned, uncured meats are heated more severely to center temperatures of 115°–120°C to inactivate spores of *Clostridium botulinum* (a "botulinum cook"); (c) finally, some canned products, e.g., soups, meat, and vegetable dishes, are heated at even higher temperatures or for longer times in order to kill spores of thermophilic bacteria. As meat often forms a minor ingredient in such foods, they will not be discussed here (see Chapter 1).

Raw meat used for the production of uncured, cooked meat carries with it a microbial flora that has not been substantially changed by any preprocessing treatment (see Section II). The vegetative organisms are of little concern, except for the spoilage they can cause before the cooking step.

In pasteurized or moderately heated meat, fecal streptococci may be important. With more severely heated products, only bacterial spores remain. Clostridia are the important species in anaerobic packs except for nitrate-reducing *Bacillus* spp. in cured meats where nitrate is used. The gram-negative enteric pathogens, e.g., *Salmonella* are absent from such products, except when postprocessing contamination occurs (Howie, 1968).

A. Lightly Heated (Pasteurized) Meats

Typical industrial preparations are cooked sausages, patés, and similar products, as well as a range of ready-to-eat dishes or parts of such dishes. Some require cooking before serving, others can be eaten without further heat treatment.

Unless they are immediately marketed and consumed (which will rarely be the case), lightly heated meat products should be kept at refrigeration temperatures so that spores of mesophilic *Bacillus* and *Clostridium* spp. cannot germinate and multiply. Some lightly heated meat products are kept frozen (e.g., the prepared dinners), ready to be heated in an oven. Heated preparations are made in food service establishments, often from raw meat, for entrees or snacks (see Chapter 29).

1. Effects of Processing on Microorganisms

Gram-negative rods are most sensitive to heat treatment, followed by gram-positive rods and cocci; the fecal streptococci and certain lactobacilli are among the most resistant vegetative bacteria.

2. Spoilage

In products pasteurized after packaging (e.g., sausages, patés), spoilage depends on the surviving flora and conditions of storage, especially the temperature. Frequently the survivors that may grow and spoil the product are psychrotrophic strains of micrococci, streptococci, and lactobacilli, and occasionally *Microbacterium thermosphactum* (incertae sedis).

Cooked meat products are excellent media for the growth of bacteria, molds, and yeasts; hence recontamination is a problem. During handling subsequent to the heat treatment, the surface of the meat can be contaminated with mesophilic gram-negative rods (e.g., Enterobacteriaceae), gram-positive cocci and rods, yeasts or molds, or any combination of these. Contamination with bacteria takes place from hands and surfaces in contact with the cooked meats, and with molds from the air. Vacuum-packaging will extend the shelf-life because any aerophilic contaminants including molds will not grow. In vacuum-packs the predominant flora is lactic acid bacteria.

In preparations composed of several ingredients—e.g., in ready-to-heat-and-eat dishes—the microflora depends less upon that from the meat; the main contribution of spoilage microorganisms comes from other ingredients, e.g., from vegetables, spices, and condiments.

Where pasteurized meats are sliced and packaged after heat processing (e.g., roast pork, roast beef) the microflora on the slices consists of that in the equipment used for slicing. As this process normally is carried out in a room at approximately 10°C, the flora will be psychrotrophic and even if the sliced, cooked product is stored at low temperatures rapid growth can occur.

The storage life of pasteurized meat products varies according to the degree of the heat treatment, and the possibilities of recontamination. Sliced cooked meats or other meats handled after cooking may have a storage life of less than 1 week even at temperatures below 5°C.

The spoilage pattern of cooked meats where recontamination can be eliminated or restricted is normally nonproteolytic. A sour taste and odor will gradually develop, the rate dependent on the storage temperature. If recontamination with proteolytic bacteria has taken place, proteolysis will occur, and the odor will be repugnant.

If the meat is stored under warm conditions (> 20°C), where spores, especially *Clostridium* spores may germinate and grow, the spoilage will be putrid, and gas may develop. On the other hand, a food containing sufficient *Clostridium perfringens* to cause food poisoning will often be found organoleptically acceptable.

3. Pathogens

Clostridium perfringens is the pathogen of most concern in pasteurized meats. Because food poisoning by *C. perfringens* requires ingestion of many vegetative cells (approximately $10^6/g$), outbreaks can take place only after temperature abuse of cooked meat, such as holding it at approximately 35°–45°C for several hours. Because the meat is cooked there may be no accompanying spoilage species to make the meat inedible, or repress *C. perfringens* by competition. Hence, for poisoning to occur, the period of multiplication usually has to be preceded by sufficient cooking to eliminate everything except spores. Therefore, although the enterotoxin actually arises from sporulating *vegetative* cells in the intestine, this type of food poisoning originates from the presence of *spores* in or on the meat. The spores can germinate readily in warm, cooked meat, the cooking having provided heat activation. Table 15.18 shows the relation between

TABLE 15.18

Temperature and Doubling Rate for *Clostridium perfringens* [a]

°C	39	38	37	36	35	33	30	25	20
Doublings per hour:	4.5	4.1	3.7	3.2	2.8	2.2	1.4	0.5	0.2

[a] Ingram (1972b).

temperature and doubling rate for vegetative cells of *C. perfringens* at pH near 7 (Ingram, 1972b).

Food poisoning from *C. perfringens* frequently is reported in large institutions where dishes are prepared immediately before a weekend, cooled inadequately, and served after light cooking on Saturday and Sunday (Dam-Mikkelsen *et al.*, 1962). Survival and outgrowth of *C. botulinum* under similar conditions seem possible in principle but rarely occur.

Staphylococcus aureus will be killed normally by the heat treatment given to these products, but preformed heat-stable enterotoxin, if present, would persist in the meat after cooking and cause food poisoning. More often, however, staphylococci are introduced after cooking as workers slice or otherwise handle cooked meats. *Salmonella* will be destroyed during cooking, and cases of salmonellosis due to the ingestion of pasteurized meats are caused by cross-contamination after cooking. This is likely to happen in establishments where *Salmonella*-contaminated raw meats (or other foods) directly or indirectly come in contact with the cooked meat, as would be the case where the same personnel handle the raw meat and then handle cooked meat. The cooked meat can also become cross-contaminated with *Salmonella* via improperly disinfected utensils and equipment.

4. Interrelations

Processing eliminates the competing flora, enabling surviving pathogens or pathogens entering as a result of recontamination to develop. When such microorganisms grow in the meats, and where their metabolic activities are so slight that they cannot be detected organoleptically, they may cause disease upon ingestion.

5. Control

Proper frozen or refrigerated storage is essential to prevent microbial growth. In many cases pasteurized meat products have such a short storage life that the usual microbiological examinations cannot detect faulty processing before the products are eaten. For some products, however, like certain sausages, patés, and precooked roast beef, determination of indicator organisms and particularly pathogens like *Salmonella, Staphylococcus aureus,* and *Clostridium perfringens* may be useful. A plate count would demonstrate the degree of cooking of the product and the subsequent storage temperature and time. Indicators like Enterobacteriaceae would reveal postprocessing contamination, as would the presence of *S. aureus* and *Salmonella. Clostridium perfringens* will often be present because its spores can survive the time–temperature effect of pasteurization, and in addition are common in ingredients that are added to the cooked meat product.

B. Fully Retorted ("Botulinum-Cooked") Meats

1. Effects of Processing on Microorganisms

Uncured meats and meat products heated for the purpose of storage for long periods at ambient temperature must be heat processed to inactivate botulinal spores or other bacterial spores of even higher heat resistance (cf. Chapter 1). They must be protected against postprocessing contamination by packaging in tin cans, aluminum cans, glass jars, semirigid metal foils, or strong plastic. The meat or meat product is normally heat processed after packaging, but aseptic packaging of heat-treated meats are gaining in importance.

Most canned meat products fall in the low-acid category. A few less important specialty items, like vinegar-pickled meats fall into the high-acid group, and will not be discussed here.

Meat products that require a 12D process (a "botulinum-cook") are meats in gravy, meat pastes, stews, hashes, and meat soups. Hygienically handled meat normally contains less than 1 *Clostridium* spore/g, and fewer than 10 *Bacillus*/g. Head meat, diaphragm, and pig skin, which may be used in certain low-quality meat products, frequently carry higher counts of spores.

The main source of spores is frequently nonmeat ingredients (e.g., spices, vegetable and animal proteins, starches) used in many canned meat products. Many spices contain up to 10^6 spores/g, and decontaminated spices or spice extracts are recommended for the production of canned meats (see Chapter 24). Other sources of spores are casings for sausages, vegetables, brine, and even the container (can or glass jar), if it is not cleaned before it is filled.

2. Spoilage

Canned, uncured meats may undergo microbial spoilage from three different sources.

a. Growth of bacteria before heat processing. If a meat paste is kept at a relatively high temperature before it is filled into cans or glass jars, vegetative cells may multiply and cause deterioration of the paste, but the heat processing may be so severe that the canned meat is found sterile. Microscopic examination of the contents will reveal numerous dead microorganisms. If the microorganisms that develop between closing of the container and heat processing are gas forming, cans may be found "blown," but again no living organisms will be found.

b. Survival of heat resistant spores. Some mesophilic spores have a heat resistance higher than that of *Clostridium botulinum,* e.g., strains of *C. sporogenes* and putrefactive anaerobes, for which $D_{121.1}$-values of 1 min have been recorded. A heat process designed to kill 10^{12} botulinal spores ($F = 2.5$) would not destroy 10^3 or more spores with such heat resistance. Many uncured, canned meats are heated in practice to inactivate a large number (10^6 or more) of *C. sporogenes* spores, which would require an F_o value of about 4–6 (see Chapter 1).

After inadequate processing, surviving spores might germinate and grow in the cans, particularly after prolonged storage at ambient temperature— or after a storage or incubation between 30° and 40°C. *Clostridium sporogenes,* a proteolytic gas former, can cause cans to swell or even burst.

c. Postprocess leakage. Canned meats are prone to spoilage from microorganisms that may enter the can through pinholes (either from a can defect or from external corrosion), through faulty seams (from improper manufacture or closure), or through the mastic that seals the ends to the body.

When the hot can leaves the retort and enters the cooling tank, or otherwise becomes wet as it cools, the vacuum that forms may suck water, with suspended bacteria, through any orifice, and even through the mastic, that is still warm and therefore semiliquid. Any bacteria from hands, machines, or conveyor belts may enter.

Canned meats that were cooled in river water were responsible for outbreaks of typhoid fever that occurred in the United Kingdom, the largest in Aberdeen in 1964 (Howie, 1968). To avoid such postprocessing contamination, cans should be cooled in water of drinking water quality or in chlorinated water. Handling of cans after processing should be carried out under strictly hygienic conditions, as stipulated in international codes (Food and Agriculture Organization/World Health Organization, 1976b) or national regulations.

3. Pathogens

Clostridium botulinum is the organism of most concern for low-acid canned meats; hence all processes for uncured, canned meats are based on its heat resistance, but many processes are more severe to ensure against the survival of more heat-resistant spoilage organisms.

Commercially canned meats have rarely been involved in outbreaks of botulism. Canned liver paste has caused cases of botulism (of which one was fatal) in Canada and New York. Cans contained both type A and type B spores (Thatcher *et al.*, 1967). The cause was severe underprocessing, although the heat processing was originally designed to give a botulinum cook. Processing conditions had been changed without consultation with a recognized authority and resulted in a process that was inadequate to destroy *C. botulinum*. Beef stew containing type A toxin resulted in two cases of botulism with one death (U.S. Department of Health, Education and Welfare, 1974a).

Spores of other pathogens like *Clostridium perfringens* and *Bacillus cereus* have a considerably lower heat resistance than those of *C. botulinum*. Therefore, given a proper botulinum cook, there is no risk of spore survival and outgrowth. Theoretically they could (like other pathogens such as *Salmonella* or *Staphylococcus aureus*) gain access from cooling water to the meat through a seam after processing.

4. Control

The microbiological control of canned meats has been discussed extensively by ICMSF (1974) and will not be repeated here. No realistic sampling plan will detect the presence of *C. botulinum;* measures to control this organism should be taken during processing.

X. COOKED, CURED MEATS

Heated, cured meats can conveniently be divided into three groups, according to the temperature during processing: (a) pasteurized, cured meats, (b) hermetically sealed, shelf-stable meats (also termed commercially sterile meats or ¾ conserve), and (c) fully sterilized, cured meats. The latter category is not common and will only briefly be discussed here. Table 15.19 gives data on these cooked cured meats.

A. Pasteurized Meats

Mildly heated (pasteurized), cured meats fall into two classes—those heated in the final container, e.g., sausages in casings and cans, hams and shoulders in foils or in cans, and those that are manipulated after heating, e.g., portioning, slicing, or skinning frankfurters. Those in the last category are frequently repacked prior to sale. Both must be stored under refrigeration.

1. Effects of Processing on Microorganisms

The water activity of most of these products is high with the exception of certain lightly heated and dried sausages (Cabanossi-type and many

TABLE 15.19

Various Types of Cooked, Cured Meats, F Values, Nitrite Content, and Shelf-Life [a]

Heat treatment	Nitrite content before heat treatment (ppm)	Destroyed during heat treatment	Category and shelf-life
$65°$–$75°$C reached in center	125	Vegetative cells with the exception of heat resistant micrococci and fecal streptococci. Viruses will be destroyed by a sufficiently long process.	Semiconserves or pasteurized meats. Shelf-life 6 months at or below 5°C.
$F_0 = 0.4$–0.6	125	As above and mesophilic *Bacillus* and some *Clostridium*	¾ conserves or commercially sterile. Shelf-life 6–12 months at 15°C or longer at ambient temperatures (depends on combination of heat, salt, and nitrite).
F_0 above 2.5 often 3.0–5.5	50	As above and most mesophilic clostridia.	Fully sterilized. Shelf-life up to 4 years at 25°C. Abiotic deterioration may occur.
$F_0 = 12.0$–15.0	(50)	As above and thermophilic *Bacillus* and *Clostridium*.	Tropical conserves. Shelf-life maximum 1 year at 40°C. After this time abiotic deterioration may occur.

[a] Takács (1977).

Mortadella-types). The salt level for the different products ranges from 1.2 to 4.5%. The brine concentration (i.e., the percent salt in the aqueous phase) will depend on the water content. As nitrite is used normally as a part of the curing salt mixture (in many countries regulated to constitute 0.4–0.6% of the mixture), the amount of added nitrite may vary from 60 to 200 ppm. Microbial reduction of nitrate will increase the nitrite levels beyond those shown.

Cooking reduces the amount of nitrite in the product, depending on the heat treatment and the presence of reducing substances, so that immediately after heat processing 50% or less of the original nitrite will be found. This will diminish more rapidly and completely at increased storage temperature. The final pH depends largely on the initial pH of the meat, which ranges from 5.6 to 6.4. The inclusion of polyphosphates may increase the pH by 0.3, and the final pH after processing may vary between 6.1 and 6.5.

Heat treatment is carried out in retorts with water or hot air, or in open kettles with water, at temperatures between 70° and 85°C, resulting in final temperatures in the center of the product between 65° and 75°C.

For meat products taken out of the wrapping or container in which they have been cooked, and either sold unwrapped or put into a new package (e.g., vacuum-packaged after slicing), the meat will be recontaminated often with 10^1 to $10^3/cm^2$ of a cold-adapted flora, usually *Lactobacillus,* occasionally fecal streptococci or *Leuconostoc* (Kempton and Bobier, 1970). Postprocessing contamination may also occur with psychrotrophic gram-negative rods and mesophilic Enterobacteriaceae.

2. Spoilage

The microflora of products sold in the package in which they have been heated comprises spores of *Bacillus* and *Clostridium,* which often come from additives like spices (Surkiewicz *et al.,* 1976) rather than from the meat. Some lactobacilli and fecal streptococci may also survive. Canned hams frequently contain the latter, but their significance is debatable. Some authors consider fecal streptococci as food-poisoning organisms; many others regard them as a normal part of the flora of pasteurized, cured meats.

The plate count after a proper heat processing will be approximately 1 to 1000/g of meat, the lower figure being frequently found in whole pieces of meat like hams, shoulders, and shanks, and the higher in comminuted meats like sausages and luncheon meats. In canned whole meats stored under refrigeration, counts may not change for 6 months or more. Spores may germinate and then die out. The shelf-life of comminuted meats varies. Normally, the heat-resistant lactic acid bacteria and fecal streptococci start multiplication, even under refrigeration, primarily causing discolorations, and later, changes of flavor and odor.

Products handled after heat treatment will be contaminated with a flora that grows well despite the salt and nitrite, increasing about tenfold per week at 5°C and one-hundred-fold at 10°C. Air-impermeable plastic pouches for such products produce an ecosystem that prevents growth of obligate aerobes and CO_2-sensitive organisms. The active flora will therefore consist of microorganisms resistant to the 30–50% CO_2 atmosphere developing in the pack and capable of growing at low temperatures. The flora is dominated by lactic acid bacteria, e.g., *Leuconostoc,* (if sugar was added to the product), or by *Microbacterium thermosphactum.** Some products show no evidence of spoilage despite bacterial populations of 10^8/g, while others do (Qvist, 1976).

Postprocessing growth of molds on meat in cold stores can be controlled by treating the surfaces with sorbic acid.

The shelf-life of high-a_w products that are sold unpackaged or in permeable films is only slightly longer than that of fresh meat. Psychrotrophic rods on the surface will cause slime formation. Catalase-negative bacteria accumulating hydrogen peroxide can oxidize nitrosohaemochrome (formed during heat processing from nitrosomyoglobin) to brown metmyoglobin or to oxidized porphyrins, some of which are green in color.

During spoilage, cured pasteurized meats that lack fermentable carbohydrate (e.g., cured hams) usually become putrid, and the pH remains neutral (Gardner and Patterson, 1975). Products that do contain fermentable carbohydrates may become acid or sticky, or if they are vacuum-packed, they may swell the package from carbon dioxide generation by *Leuconostoc* spp.

3. Pathogens

The pathogen of most concern in pasteurized, cured meats is *Staphylococcus aureus*. It rarely survives a proper heat treatment but is frequently a contaminant from the hands of persons who handle and package the meats after processing. Its growth is not controlled by the usual concentrations of salt and nitrite, and enterotoxin forms at temperatures suitable for its aerobic multiplication.

Cured, cooked meats, especially vacuum-packaged, have seldom been vehicles of salmonellosis. Davidson and Webb (1973) showed that salmonellae inoculated into such meats would grow at warm ambient temperatures for several days. The good record of these products with respect to *Salmonella* is probably due to an effect of salt and nitrite combined with

* *Incertae sedis* (Buchanan and Gibbons, 1974).

low-temperature storage, where a competing flora may develop before any salmonellae can multiply to a number sufficiently high to cause disease.

Botulism has not been associated with cooked, cured meats of this type, although inoculation experiments have shown that there is a potential for growth and toxin production from *Clostridium botulinum* type A and B (Pivnick *et al.,* 1967; Greenberg, 1972; Bard, 1973). The rate of growth and toxin production depends on temperature, salt concentration, and the amount of added nitrite. Inhibition was related to added nitrite rather than to residual nitrite present at the time of abuse.

4. Interrelations

The question of competitive growth has been described (see preceding discussion). Pathogenic bacteria are prevented from growing primarily by storage at low temperatures, but the fact that millions of mildly heated cans of sausages have been vended without refrigeration, but have not caused foodborne disease, shows that not only salt and nitrite, but also the normal spoilage flora influences the growth of the pathogens. In vacuum-packs, the concentrations of salt and nitrite select a specific microflora that rapidly develop CO_2, which, in turn, suppresses the growth of gram-negative bacteria (e.g., coliforms, *Salmonella*) and staphylococci.

5. Control

The second book of this series (ICMSF, 1974) presents a sampling plan for canned cured pasteurized meats like hams and shoulders. These products are not sterile and should be stored at below 6°C.

Certain countries recommend incubation tests of 5 days at 35°C, after which the product is examined for clostridia and coliforms, but the validity of such tests for canned perishable meats seems doubtful.

Aerobic plate counts on vacuum-packaged cooked, cured meats sampled immediately after the slicing operation give good indication of the sanitation and hygiene of the operation (Qvist, 1976). However, such counts made during or after storage are meaningless except in a research study of the flora of such meats in storage. At later stages, only determination of certain specific organisms (staphylococci, fecal streptococci), and organoleptic examinations, are relevant.

B. Hermetically Sealed, Shelf-Stable, Cured Meats

Shelf-stable, cured meats play an important role in international trade. These are hams and shoulders of $\frac{1}{2}$ to 4 lbs, corned beef, luncheon meat, sausages, and other minced meat products packed in hermetically sealed containers, which are normally rigid, but sometimes semirigid.

They are usually termed "commercially sterile" or "$\frac{3}{4}$ conserves," indicating that they are not necessarily sterile but are shelf-stable under normal distribution and storage in temperate climates.

1. Effects of Processing on Microorganisms

The stability of these products is based on a combination of heat treatment, salt and nitrite concentration, and storage temperature. Nitrate was used formerly for long-term curing of meats before canning, but modern technology employs a very short period between curing and canning, so that the conversion of nitrate to nitrite is minimal. Polyphosphates may be used to reduce shrink, and ascorbic acid may be added to stabilize color. Many of these products contain added animal and vegetable proteins, starches, and spices. The use of many of these additives is regulated by national legislation in many countries.

The salt concentration is from 3.5 to 5.5% in the aqueous portion of the meat, just as one would expect to find in pasteurized meats.

Shelf-stable cured meats are normally heat processed in retorts with water or steam at temperatures between 100° and 115°C, resulting in center temperatures in the product between 95° and 112°C. Expressed as F-values the heat treatment frequently corresponds to $F_0 = 0.4-0.6$, but F_0-values as low as 0.05 are used. Some products like corned beef may require a higher F_0-value, primarily because the amount of added nitrite is less. Part of the nitrite may disappear during precooking of corned beef, thus leaving a smaller amount of nitrite to be active during the final processing and during storage.

Clearly, the F_0-values indicated above do not safeguard against spoilage or outgrowth of botulinal spores in the product. The excellent record of these products is due to the combined effect of heat and curing ingredients. As the thermal process increases the following occurs: (a) more spores are killed; (b) the survivors suffer increased damage, and (c) the salt, nitrite, and perhaps reaction products of nitrite (cf. Chapter 8) prevent growth from the heat-damaged spores.

The microbiological problems in producing commerically sterile, cured, canned meats are not much different from those of uncured meat subjected to a botulinum cook (see previous discussion). The curing before heat processing may change the microflora to more salt-tolerant species of gram-positive rods and cocci, but these will be completely inactivated during the first stages of the heat processing, leaving spores of *Clostridium* and *Bacillus* in the meat. The origin of these may be the meat itself but is more commonly the nonmeat ingredients. Spore counts are always higher in comminuted meats than in whole pieces of meat, e.g., hams and shoulders.

2. Spoilage

The causes of spoilage are (a) a high level of mesophilic anaerobes, (b) an inadequate heat process, (c) a low concentration of curing ingredients, or (d) a combination of these factors. In addition, spoilage can take place before the canning operation and in the can from leakage.

The first signs of spoilage will be seen after 1 to 2 weeks if spores (normally *Clostridium sporogenes* and other putrefactive anaerobes) have survived the heat process, if the concentration of curing salts is too low to prevent their germination and outgrowth, if factors like pH and Eh are favorable to growth, and if the cans are stored at temperatures favorable to microbial growth. The first signs will be swelling of the cans due to gas formation. The contents normally will be more or less liquified, often with an extremely disagreeable rotten odor. High numbers of sulfite-reducing clostridia will be present.

In some cases, spoilage may be due to *Bacillus* spp. This is normally less conspicuous, and in many cases remains unnoticed, if the cans are inspected only visually. Cases of swelling due to *Bacillus* have been reported, but only certain *Bacillus* spp. can form gas in such a menstruum' (e.g., N_2 and CO_2) (Jensen, 1954; Simonsen, 1968). In certain sausage products, the bacilli may hydrolyze starch, which alters the product to a paste-like texture.

In unspoiled cans, viable but dormant spores may be recovered to an extent depending upon initial contamination and heat processing. These will not germinate and grow, if the concentration of curing salts is inadequate.

3. Pathogens

The pathogen of most concern in commercially sterile meats is *Clostridium botulinum*. It seems, however, that the problem is only theoretical. Billions of cans of hermetically sealed, shelf-stable, cured meats have been produced over the years without causing a single case of botulism. As the $D_{121.1}$-value of *C. botulinum* is 0.2, this might suggest that a heat process to an F_0-value of 0.1 gives less than a decimal reduction of such spores. The fact that no cases of botulism or any recovery of botulinal spores from shelf-stable cans of cured meats, have been reported may be due to two facts: (a) the low incidence of botulinal spores in ingredients for these products, and (b) a very strong synergistic effect of salt and nitrite. Nitrite is claimed to have an especially antagonistic effect on outgrowth of spores.

4. Control

Microbiological control of shelf-stable, cured meats is discussed in detail in the second book of this series (ICMSF, 1974). Takács (1969) also has proposed microbiological criteria for canned meat products.

C. Fully Sterilized Cured Meats

Fully sterilized cured meats fall into three types: (a) those that have salt and nitrite levels so low that they have minimal effect on microorganisms, (b) those that are cured with salt but no nitrite, and (c) those intended for tropical areas.

The first two categories are termed full conserves and are subjected to a heat treatment that would at least kill all botulinal spores and give a good assurance that the more heat-stable putrefactive anaerobes are inactivated. For that purpose F_0-values of 3 to 5.5 are employed (Wirth et al., 1971). These products have good stability in storage at 25° to 35°C.

The tropical conserves are heated to $F_0 = 12 - 15$ to exclude survival and outgrowth of thermophilic spores, and therefore can be stored at temperatures above 40°C. It is uncommon and probably unnecessary to give even lightly cured meat products a cook of that order because increasing heat damage to thermophilic spores renders them increasingly sensitive to sodium chloride (Briggs and Yazdany, 1970).

XI. CHOICE OF CASE

For an understanding of "Case," see Table 1 and the brief description in Appendix IV. ICMSF has already recommended some sampling plans and microbiological criteria for chilled and frozen raw meats and for processed meats (cf. Tables 24 and 25 in ICMSF, 1974).

Varying technological procedures, however, may change the health hazard of a particular product, and a more detailed description is desirable, on the basis of the classification given in Section I,D.

Since the manner of handling to be expected may greatly vary, there is much overlap among the Cases, as indicated in Fig. 15.4. Consequently, the majority of the products could have been mentioned in each of the categories listed. As this is impossible, a very few examples were selected for illustration. See also Table 15.20.

TABLE 15.20

Cases of Health Hazards for Various Meats

Type of hazard	Reduce	No change	Increase
No direct health hazard	(1) Raw sausage (a_w 0.88+) Dried meats [a] Brawns (pH 4.5) Cured meats in larger pieces [b] Country-cured ham [a] Carcass meat [b] Meat salads (pH 4.5) Minced meat [a]	(2) Raw sausage Country-cured ham Minced meat Carcass meat Frozen meat	(3) Country-cured ham Raw sausage Cooked sausage Corned beef, when sliced
Low indirect health hazard	(4) Meat salads (pH 4.5) Carcass meat [a] Primal joints [a]	(5) Minced meat [a, c] Fresh mettwurst Frankfurt type sausage Cooked sausage	(6) Minced meat [c] Scraped raw beef
Moderate direct limited spread	(7) Fresh blood sausage [b] Raw sausage for frying [b] Ready-to-eat meats [b]	(8) Fresh mettwurst (pH > 5.8) [b] Liver paté Cooked sausage Cooked ham	(9) Hamburger patties [c] Lightly cooked, uncured meats [c] Meat salads (pH > 5.0) Semifinished products
Moderate direct potential extensive spread	(10) Minced meat [b] Retail cuts [b] Viscera [b] and as in (4)	(11) Minced meat [a, c]	(12) As in (9) Canned meats (leakages)
Severe direct	(13) —	(14) Frying sausage [b] Patties [b]	(15) Canned meats (leakage, under-cooking)

[a] Packaged in impermeable films.
[b] Heated prior to consumption.
[c] When eaten raw or insufficiently cooked.

CASES IN RELATION TO VARIOUS MEATS

fresh, unprocessed — processed

carcass meat primal joints | deboned meat, retail cuts minced meat, viscera | not heated | lightly heated | fully pre-served (vollkon-serve)

not cured | cured | cured | not cured

unfrozen chilled | frozen | unfrozen | dried | $a_w < 0.92$ | $a_w > 0.92$

packaged | unpackaged | packaged | unpackaged | pH > 5.8 | pH < 5.8 | un-packaged | packaged

EXAMPLE:

| carcass meat primal joints | deboned meat, retail cuts, minced meat viscera | "biltong" | ham (country-cured) salami | semi-finished products "brat-wurst" | raw sausage brawns | frank-furter, cooked sausage | canned ham | ready to eat meat dishes | canned meat and gravy, corned beef |

CASE:

| 1 10 11[a] | 1 10 12[a] | 1 10 11[a] | 1 10 12[a] | 5 8 11 | 1[b] 10 14[a] | 5 8 11 | 2 8 11[a] | 2 8 | 2 5 8 | 2 15[a,c] |

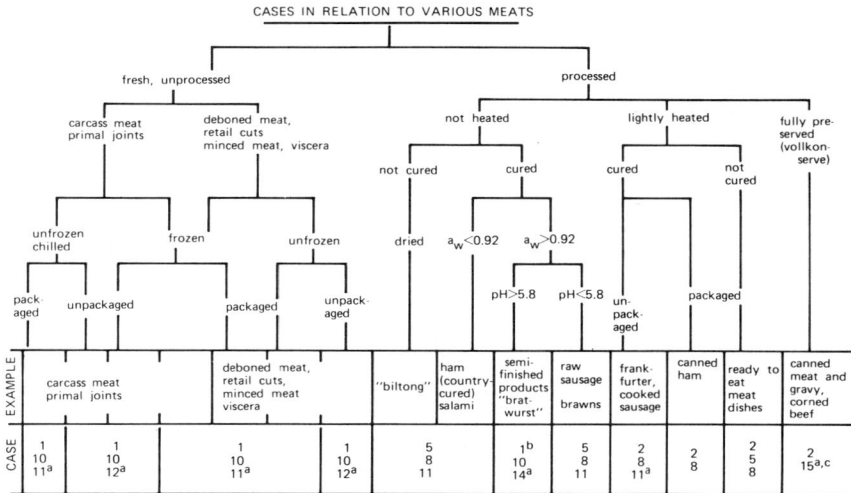

Fig. 15.4. Cases in relation to various meats; a: when eaten raw or insufficiently cooked; b: when heated prior to consumption; c: suspect for leakage or under-cooking.

Several sampling plans are applicable among raw and processed meats since the handling and manner of preparation for consumption may be quite different for the same product. Raw meat may contain *Salmonella* which leads to Case 10. Most of the processed meats are known to have caused staphylococcal or *Clostridium perfringens* food poisoning. There-fore, Case 8 was applied. Plate count (Case 1) for raw fermented products would be meaningless. Plans for Case 2 in canned products must include first of all incubation tests rather than cultural examination.

Generally conditions that *reduce* hazards are the following: (1) packag-ing in impermeable films;* (2) heating prior to consumption; (3) low pH; (4) low a_w; (5) curing; (6) cold storage.

Conditions that *increase* the health hazards are the following: (1) no packaging; (2) insufficient heating, or none; (3) near neutral pH; (4) high a_w; (5) no curing; (6) storage at elevated temperatures.

---

* Packaging may reduce or increase hazards, dependent on the film, pH, a_w, and the contamination associated with the packaging process.

16

Poultry and Poultry Meat Products

I. INTRODUCTION

Large amounts of poultry and poultry meat products are consumed in all countries and transported in international trade. Processors and governmental officials in importing countries are interested in the microflora of these products as indices of previous sanitation and storage, remaining shelf-life, and danger to public health. This chapter describes (a) sources of microorganisms that contaminate carcasses and products, (b) operations that increase or decrease the spread of microorganisms during processing, (c) factors that affect the chances of survival and the rate of growth of microorganisms on carcasses and in poultry products, and (d) the significance of laboratory findings.

A. Definitions

Poultry meat is the muscle tissue, attached skin, connective tissue, and edible organs of avian species that are commonly used for food. In international trade, carcasses or parts of chickens, turkeys, ducks, and geese constitute the bulk of shipments. Other poultry (cornish game hens, guinea fowl, peacocks, pigeons) are commonly processed in some countries, and game birds (quail, partridge, pheasant, and grouse) are sometimes processed for food service establishments.

B. Important Properties

The water content of the edible portions of carcasses is about 71% for broiler chickens, 66% for roaster chickens, 56% for hen chickens, and 58% for medium fat turkeys (Mountney, 1976). The flesh of young birds has a higher proportion of water than that of old birds.

The approximate protein and fat contents of chicken fryers are 20.5 and 2.7%, respectively; of chicken roasters, 20.2 and 12.6%; of turkeys, 20.1 and 20.2%; and of duck, 16.1 and 28.6% (Burton, 1976). Unlike red meat, in which fat is distributed throughout the tissues, most of the fat in poultry is found just under the skin and in the abdominal cavity. The amount of fat varies with age, sex, anatomy, and species.

In addition to the nutrients in poultry meat, other properties influence the growth of microorganisms. The water activity (a_w) is about 0.98 to 0.99 depending on whether and how long the meat has been stored in dry air. The pH of chicken breast muscle is 5.7 to 5.9, while that of leg muscle is 6.4 to 6.7. As chickens mature, the pH of the skin increases to an average of 6.6 for 9-week-old chickens, and 7.2 for 25-week-old chickens (Adamcic and Clark, 1970). The redox potential of poultry meat is similar to that of the meat of mammals (Chapter 15). The skin, which harbors many microorganisms, nevertheless serves as a physical barrier to microorganisms that might otherwise contaminate the underlying muscle. Because of their composition and other properties, both poultry muscle and skin are excellent substrates for a wide variety of microorganisms.

C. Methods of Processing and Preservation, and Final Products

Freshly laid fertile eggs are collected, fumigated, and incubated at a constant temperature for 21 days for chickens and for 28 days for turkeys. After they hatch, chicks or poults are delivered to farms, reared there until ready for slaughter (6 to 15 weeks), and then hauled to a processing plant. At the plant, they are killed and bled by severing the carotid arteries. Feathers, heads, feet, and viscera are removed, and the carcasses are then washed, chilled, and transported to markets packed in ice, refrigerated, or frozen. These processes are illustrated by a flowsheet in Fig. 16.1. A certain proportion of dressed birds are further processed; the main types of these processes are illustrated in Fig. 16.2. Some typical processed products are listed in Table 16.1.

II. INITIAL MICROFLORA

Microbial contamination of the egg can occur either in the ovaries or oviducts during development of the egg, or later as a result of penetration of the eggshell (see Chapter 19). After eggs hatch, the young acquire additional microorganisms from various other sources that are described in the following discussion.

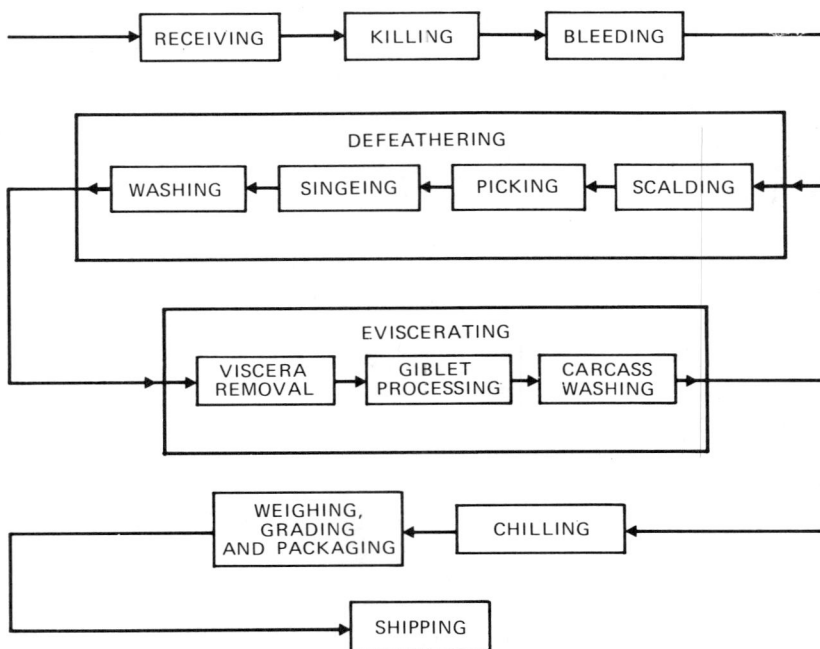

Fig. 16.1. General process flowsheet for poultry processing.

A. Infection and Contamination of Eggs

Some organisms—*Salmonella, Escherichia coli, Mycoplasma,* vibrios, and enterococci, for instance—can infect the ovaries and the oviducts of hens. The organisms pass from the ovary to the ovum or from the oviduct to the yolk, albumen, or membranes during egg formation. In some countries, breeder flocks are tested, diseased birds (reactors) are removed, and new breeder stock is quarantined. Such a program can eliminate breeder flocks as a source of disease organisms.

Eggshells are contaminated with organisms shed by the intestines when the eggs pass through the cloaca and when they contact nest materials, litter, or incubator surfaces as described in Chapter 19, and by Board (1969) and Williams *et al.* (1968). Gram-negative bacteria that get into the egg by means of either ovarian transmission or penetration can reach the yolk and subsequently infect the developing embryo. Shell contamination can be reduced by providing an ample quantity of clean nest materials, by gathering eggs frequently, and by fumigating eggs promptly after gathering them (Williams, 1978).

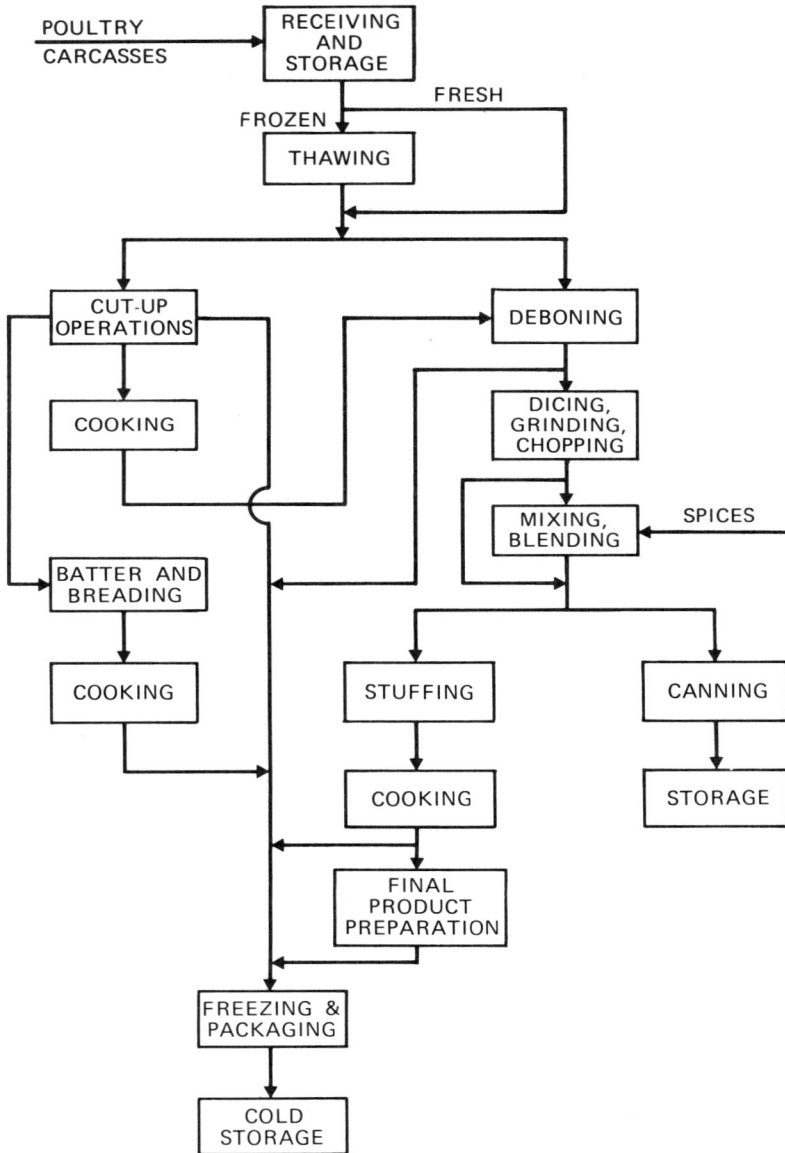

Fig. 16.2. General process flowsheet for further processing of dressed poultry.

TABLE 16.1

Commercial Poultry Meat Products

Raw fresh	Raw frozen	Heat processed	Other processed
Eviscerated whole poultry	Eviscerated whole poultry	Soups	Freeze-dried, diced poultry meat
Poultry parts—legs, thighs, breasts, wings	Poultry parts—legs, thighs, breasts, wings, backs	Pot pies	Dehydrated soups
"New York" dressed (uneviscerated) poultry	Edible viscera—livers, gizzards, hearts, turkey tom-toms	Hash	Smoked whole or sliced poultry [a]
Effilee (birds without intestines only)	Stuffed (filled with dressing) poultry	Complete, prepared frozen dinners	Smoked chicken rolls [a]
Edible viscera—livers, gizzards, hearts, turkey tom-toms	Fillets, roasts, steaks, patties, sausage, loaves, chicken sticks	Luncheon meat	Smoked poultry sausage [a]
		Franks (chicken sausage)	Smoked chicken spread [a]
		Jellied chicken loaf	Fermented poultry sausage
		Rolls	
		Fried chicken	
		Chicken liver-fat paté	
		Breaded edible viscera	
		Barbecued poultry	
		Poultry salads	
		Diced poultry meat	
		Poultry dressing	
		Poultry casseroles	
		Canned poultry products	
		Chicken chow mein	
		Creamed poultry	

[a] Also heat processed but not always to time–temperature exposures that are lethal to many microorganisms.

B. Incubator and Hatchery Contamination

Contaminated eggshells, fecal matter, and fluff from infected, newly hatched birds contaminate incubators, and air currents can distribute microorganisms throughout a hatchery. Newly hatched birds can become colonized from ingestion of feces, from breathing contaminated dust or aerosols, from organisms reaching conjuctiva, or from contact with floors and walls of the incubator (Williams, 1978). Incoming eggs are frequently fumigated with formaldehyde to kill many of the contaminants. Negative air pressure is provided in dirty areas, and incoming air is filtered to minimize transfer of organisms.

C. Healthy Poultry

A healthy bird, like any other animal or man, carries millions of microorganisms in its intestines and on its skin. This flora is not just a chance contamination of the animal but a constant entity that, once established, helps to inhibit pathogenic microorganisms. The gut, for instance, is an important source of enteric bacteria. Psychrotrophic bacteria occur on the feathers and feet of birds (Barnes, 1960).

D. Feed and Rendered By-Products

Feed becomes contaminated with microorganisms when bacterial spores survive rendering and when rendered animal byproducts are recontaminated from airborne and soilborne microorganisms or from microorganisms that remain in box cars from previous shipments of feed or animals or from previous lots of food in grain elevators or feed bins. Many vegetative cells as well as spores persist for long periods in dry feedstuffs. If feed becomes wet, bacteria and molds can grow (Chapter 17). Some molds (*e.g., Aspergillus flavus*) caused illness in poultry following production of mycotoxins in damp fields (Blount, 1961; see also Chapter 23).

Poultry feed, rendered animal byproducts, dried egg powder unfit for human consumption, and fish meal are well-known sources of salmonellae (Chapter 17). Recycled poultry waste that has not been heat treated can spread pathogens. Contaminated feed infects live birds whose fecal droppings then seed farms with salmonellae. Thus, eliminating salmonellae from feed is essential if poultry are to be reared free of these organisms.

E. Drinking Water

Water is a source of many spoilage and some pathogenic microorganisms. *Pseudomonas, Chromobacterium, Acinetobacter, Moraxella, Aeromonas, Alcaligenes, Yersinia, Micrococcus,* and *Bacillus* are natural flora of surface water. Enterococci, *Proteus, Enterobacter, Escherichia,* and *Salmonella* are common contaminants (Holden, 1970). These organisms can survive in water troughs or ponds from a few days to several weeks. Survival times vary with species and strains and with temperature, intensity and duration of sunlight, pH, amount of dissolved oxygen, and concentration of organic matter. Although the water source itself might not be contaminated, farm water usually becomes contaminated by dust, litter, feed, feathers, feet, beaks, and feces while in troughs (Patterson and Gibbs, 1977). Water spilled on litter encourages multiplication of coccidia, fungi, and bacteria, and aids the development of parasitic worms; drains under water troughs help to prevent this problem. Periodic cleaning of water troughs eliminates bacterial build-up in the water and reduces the chances of infection from this source.

F. Soil, Litter, Dust, and Air

Soil is a reservoir of many genera of bacteria (e.g., *Acinetobacter, Actinomyces, Alcaligenes, Bacillus, Chromobacterium, Clostridium, Cytophaga, Enterobacter, Escherichia, Flavobacterium, Micrococcus,* and *Moraxella*), yeasts, and molds. Bacterial spores can survive in soil for years; some vegetative cells can survive for weeks or months. Salmonellae, for instance, have survived in soil for over 200 days (Bryan, 1968). The nature of the soil, intensity and duration of sunlight, pH, amount of moisture, and the level and variety of the bacterial population affect survival; the amounts of rainfall and wind affect transmission. Thus, the types to which birds are exposed by contact with soil and dust will vary somewhat among different parts of the world.

Litter becomes contaminated from droppings, feathers, and soil. As the time of using the same litter is extended, the water activity, ammonia level, and pH increase, resulting in unfavorable conditions for some organisms—salmonellae, for instance (Turnbull and Snoeyenbos, 1973). However, old litter and litter containing water, droppings, or wet feed are good media for growth of yeasts and molds. The kind and number of microorganisms found on feathers and skin of live birds, and subsequently on dressed carcasses, are greatly influenced by the type and condition of litter on which birds are grown and will be very different if they are grown without litter.

Some brooding operations produce considerable dust from soil, litter, and feathers and thus enhance the probability of transmitting microorganisms. Mold spores, yeasts, and cocci are the types most often found drifting in the air of houses used for brooding and rearing poultry, but other kinds of microorganisms can be found as well, especially if the environment is dirty and the humidity is high.

G. Other Reservoirs and Vectors

Some insects are intermediate hosts for blood parasites and intestinal parasites; all are reservoirs or vectors of microorganisms. Rodents and other small mammals can spread microorganisms within poultry pens from their feet, hair, and feces (Lofton *et al.*, 1962). Certain wild birds, (e.g., sparrows, blackbirds, pigeons, sea gulls) frequently gain entrance to houses or the general environment inhabited by domestic fowl, and can transmit salmonellae and other microorganisms from house to house and farm to farm (Snoeyenbos *et al.*, 1967). Reptiles and amphibians are also reservoirs of certain infectious microorganisms (*Aeromonas, Salmonella,* including *S. arizona*) as are common farm animals (*Erysipelothrix rhusiopathiae, Escherichia coli, Pasteurella multocida, Salmonella, Yersinia*). Farm workers can easily spread infectious agents with their shoes or equipment; and they can mismanage flocks in ways that can spread microorganisms. But these sources are of minor importance compared to those previously described (see Sections II,A–F).

H. Transmission

Birds infected with pathogens spread the causative microorganisms to their pen mates and to other carcasses during processing. Many microorganisms, including pathogenic species, are shed in feces. Examples are *Alcaligenes, Campylobacter fetus, Citrobacter, Chlamydia psittaci, Clostridium perfringens, Cryptococcus neoformans,* * *Enterobacter, Escherichia coli, Lactobacillus, Proteus, Proteus (Providencia), Salmonella, Streptococcus faecalis, Streptococcus faecium,* and *Yersinia.* Lactobacilli, clostridia and fecal streptococci, however, comprise the bulk of the fecal flora (Barnes, 1960). Fecalborne organisms can adhere to feathers or feet when birds walk, roost, or set on contaminated surfaces and can also be inhaled with air that contains feces-contaminated dust or aerosols.

Some pathogenic organisms shed in nasal secretions are *Chlamydia psittaci, Haemophilus, Pasteurella multocida,* and *Staphylococcus aureus;* the last named is commonly present in skin and arthritic lesions and in bruised tissues (Hamdy *et al.*, 1964).

* Not recognized in the 8th edition of Bergey's Manual (Buchanan and Gibbons, 1974).

Poultry can become infected by drinking water, eating feed, or pecking in soil or litter previously contaminated by infected birds. Young infected chicks reared on litter spread salmonellae rapidly to pen mates. The infectious dose of salmonellae for day-old chicks may be small; older chickens are much more resistant (Snoeyenbos *et al.,* 1969). Cannibalism in an infected flock spreads many pathogenic and saprophytic organisms that are usually present in the intestinal tract or on the skin. Excessive heat, cold, or other stress factors lower resistance to infections, and wind and rain spread contamination.

Broilers and, to a large extent, turkeys are reared in confinement; thousands are often housed in a single house. In modern industrial practice, feed and water are usually provided by automatic devices that can readily transmit infectious agents from bird to bird.

Disease transmission can be reduced by adequate spacing of houses; exhausting sufficient amounts of air; filtering incoming air; removing dead birds promptly; providing devices that keep birds out of feeders and water troughs; cleaning watering devices daily; using pelleted feed; controlling access of birds, small rodents, reptiles, and people to flocks; testing fecal matter as well as dead or live birds for specific pathogens (e.g., salmonellae); cleaning, disinfecting, and fumigating houses between rearing of successive flocks; using clean clothing; and disinfecting workers' boots. Countries in which these practices prevail report low incidence of salmonellae in live birds and on processed raw poultry (Lundbeck, 1974; Zecha *et al.,* 1977). These measures, no doubt, have an effect on the presence and number of other microorganisms as well.

In experimental work, *Salmonella* infections of young chickens have been diminished significantly by inoculating crops of day-old chicks with material from the crop and intestinal tract of adult cocks (Nurmi and Rantala, 1973). Intestinal isolates grown anaerobically were more effective than those grown aerobically in preventing colonization of *Salmonella* (Rantala, 1974).

I. Hauling and Unloading

No direct evidence shows that the number of infected birds in a poultry flock increases during transit to processing plants (as it does in swine and cattle), but contamination with organisms of fecal origin does increase (Seligmann and Lapinsky, 1970; Patterson and Gibbs, 1977). Microorganisms in feces and on feathers are undoubtedly spread from bird to bird. During transit and prior to unloading, birds defecate and subsequently step or lie in the fecal matter. During unloading, this fecal material can be spread to feathers. During hanging and bleeding, wing flapping creates aerosols and dust that are deposited on the feathers and skin of nearby birds or carcasses. Salmonellae, for instance, have been isolated from the air in poultry-plant unloading zones (Zottola *et al.*, 1970).

The same trucks and cages are frequently used to transport birds from different farms to one or more plants. Microorganisms are thus transferred from farm to farm. The European Economic Community requires that cages and truck beds be cleaned and disinfected after each use, to minimize this problem.

III. CHILLED RAW POULTRY AND POULTRY MEAT PRODUCTS

Processing operations differ somewhat with the type of poultry being processed, the volume of birds involved, the type of processing equipment used, and the intended disposition of the product. In general, poultry is processed as follows (Fig. 16.1): birds are removed from cages, hung by the feet on shackles of a conveyor, stunned by electric shock, killed by cutting the carotid arteries, and allowed to bleed. Next, they are scalded, defeathered, and washed. Heads, hocks, shanks, and oil glands are cut off, and viscera are drawn, inspected, and removed; lungs are vacuumed away, and necks are cut off. The carcasses are then usually spray-washed and chilled. After chilling they are graded and either packaged, packed in crates with ice, or further processed. Cleaned, edible viscera and the neck are sometimes stuffed into the body cavity, and the carcass is packaged and stored in a refrigerator.

A. Effects of Processing on Microorganisms

Some processing operations promote a significant increase of contamination or even permit multiplication of contaminating microorganisms, and others promote a significant decrease of contamination. Tables 16.2 and 16.3 show the levels of aerobic organisms, and Tables 16.4 and 16.5 show the serotypes and incidence of salmonellae at various stages of processing. The significance of these data is discussed below.

1. Further Contamination

Certain steps of the processing operation transfer microorganisms from heavily contaminated sites to lightly contaminated sites or introduce additional contaminants.

TABLE 16.2

Geometric Means (Log$_{10}$) of Bacterial Counts Per Gram of Neck Skin at Different Stages of Processing Chickens by Immersion Chilling Procedures [a]

Plant [b]	After picking	After evisceration	After spray washing	After immersion chilling	After packaging
Aerobic plate counts [c]					
1	5.24	5.42	5.15	4.51	4.72
2	4.32	5.14	5.16	5.06	5.02
3	4.43	5.44	5.00	5.14	5.15
4	5.36	5.37	5.25	5.05	5.04
5	5.15	5.34	5.25	4.99	4.98
Coliform counts [c]					
1	4.74	4.56	4.40	3.44	3.57
2	3.41	4.02	3.70	3.29	3.10
3	3.44	4.82	4.33	4.12	4.08
4	4.50	4.62	4.24	4.03	4.23
5	4.21	4.59	4.59	4.08	4.14

[a] Commission of the European Communities (1976).

[b] Plants 1, 2, and 3 used counter-current immersion chillers, and plants 4 and 5 used through-flow immersion chillers.

[c] During 8 sampling periods, 10 samples of neck skin were collected with the passage of 100 carcasses between samples from each sample station at each plant. These 10 samples were pooled for examination.

a. Incoming Poultry. Incoming infected flocks are the principal source of most microorganisms found on poultry carcasses. Salmonellae, for instance, are spread from carcass to carcass during processing (Bryan *et al.,* 1968a; Morris *et al.,* 1969; Notermans *et al.,* 1975a; Finlayson, 1978). The introduction and spread of salmonellae in poultry plants are illustrated in Table 16.4. The feathers, feet, and bodies of the birds, and outsides of cages are contaminated with bacteria (such as *Acinetobacter, Moraxella, Pseudomonas, Corynebacterium, Micrococcus, Staphylococcus, Flavobacterium*) and yeasts (Barnes, 1960, 1975). Fecal matter is sometimes extruded from the cloacal opening when birds are stunned by electric shock and when the carcasses are struck by the rubber fingers in picking machines.

b. Utensils and Equipment. Microorganisms are transferred from contaminated surfaces of feathers, feet, and skin to other surfaces of the same

TABLE 16.3

Geometric Means (Log$_{10}$) of Bacterial Counts Per Gram of Neck Skin at Different Stages of Processing Chickens by the Air-Chilling Procedures [a]

Plant [b]	After picking	After evisceration	After spray washing	After air chilling	After packaging
Aerobic plate counts [c]					
1a	4.60	4.92	4.64	4.54	4.52
2a	4.84	5.21	4.66	4.73	4.74
3a	6.01	5.98	5.84	5.65	—
4a	6.22	6.24	6.39	6.12	—
5a	5.13	5.76	5.26	5.15	5.42
6a	5.36	5.37	5.25	5.37	5.15
Coliform counts [c]					
1a	3.31	3.70	3.80	3.71	3.53
2a	3.29	3.56	3.39	3.63	3.28
3a	4.96	5.07	4.70	4.68	—
4a	5.51	5.57	5.66	5.23	—
5a	4.26	5.06	4.49	4.41	4.88
6a	4.50	4.62	4.24	4.44	4.19

[a] Commission of the European Communities (1976).

[b] Plants 1a, 3a, 4a, 5a, and 6a used tunnel chilling, and plant 2a used a chilling room.

[c] During 8 sampling periods, 10 samples of neck skin were collected with the passage of 100 carcasses between samples from each station at each plant. These 10 samples were pooled for examination.

and different carcasses by utensils and equipment. Such transfers start when birds are hung on metal shackles and killed with knives; it proceeds as feathers are removed by rubber fingers in picking machines, as feet and heads are cut off by machines or knives, as evisceration is done by knives or machine, as carcasses slide across tables and down chutes and contact the sides or bottoms of chill tanks and coolers, and as they are rehung onto metal shackles. The transfer continues as chilled carcasses are dropped onto and slide across tables, are weighed before packaging, and are dropped into chutes that direct them into bins or shipping boxes. Edible viscera are contaminated during processing by surfaces of chutes, tables, gizzard skinning machines, knives, and scissors.

Equipment that is not cleaned and disinfected at the end of a day's operation will have residual fat, blood, and meat that will be transferred to products processed the following day. Such contaminants, when mixed with

TABLE 16.4

Salmonella Serotypes Isolated from Turkey Carcasses and from Equipment During Progressive Stages of Processing 15 Consecutive Flocks[a,b]

SITE	Day I (A)	(B)	(C)	Day II (D)	(E)	Day III (F)	(EF)	Day IV[c]	Day V (G)	(H)	Day VI (I)	(HI)	(J)	Day VII (K)	(L)	(KL)	Day VIII (M)	(N)	Day IX[c]	Day X (O)
Fecal droppings, farm	–	e	g	g	i,j	b	–	–	k	–	b	–	–	b	–	–	–	–	–	b
Water troughs, farm	–	–	g	g	i	–	–	–	k	–	–	–	–	–	–	–	–	–	–	b
Fecal droppings, truck	–	–	–	g	–	–	–	–	b	l,n	–	–	–	–	–	–	–	–	–	b
Picker 1	a	–	a	a,g	–	–	c,j	g	m	–	i	i	m	–	–	m,p	b,p	p	–	–
Picker 2	–	e	a,e	–	–	–	b,c	–	m	–	–	m	m	–	–	b,g	b	–	–	–
Picker 3	–	–	e	e	–	–	–	–	m	–	–	–	–	–	–	–	–	–	–	–
Spiral Picker	b,c	a,c	b,d	d,f	–	–	a,b,d	b	–	–	–	l	–	–	–	g	g	a	–	b
Chute	–	a,e	b,d	g	–	–	c	–	–	–	–	–	–	–	–	g	c	q	–	b
Table (picking)	–	b,e	f	g	–	–	b	–	–	–	–	l	b	–	–	g	g	c	–	b
Carcasses after picking	–	–	b,e,f	g	a,b,c,h	b,g	–	–	l,m	n,o	g	–	d,g	b	g	g	c	a	–	b
Carcasses after washing	–	–	–	–	–	–	b	–	l,m	g,l	–	–	–	e	g	–	–	–	–	b
Gutter	–	–	f	–	–	–	b	–	l	–	–	b	b	–	–	–	g	c	–	b
Knives	–	–	f	–	–	–	–	–	l	–	–	–	–	–	–	–	–	–	–	–
Head remover	–	e	f	b,g	–	–	e,i	–	–	n	–	n	c	–	–	–	–	–	–	–
Trussing table	b,d	–	f	g	–	–	–	–	–	n	–	b	b	–	–	–	–	–	–	–
Spin Chill 1	–	–	b	g	–	–	–	–	–	–	–	–	–	–	–	–	g	–	–	–
Slide	a	–	f	f	–	–	–	–	–	–	–	–	–	–	–	–	g	–	–	–
Spin chill 2	–	–	–	–	–	–	–	–	–	–	–	–	–	–	–	–	g	–	–	–
Spin chill 3	–	–	–	–	–	–	–	–	–	–	–	–	–	–	–	–	p	–	–	–
Chute and grade table	–	–	f	g	a	–	e	–	–	–	–	–	–	–	–	–	–	–	–	–
Carcasses before icing	–	–	f	g	–	–	–	–	–	–	–	–	–	–	–	–	–	–	–	–
Chill tank[d]	–	e	f	g	–	–	–	–	–	–	–	–	–	–	–	–	–	–	–	–
Carcasses before packaging[d]	c	e	f	g	–	–	i	–	–	–	–	–	–	–	–	–	–	–	–	–
Grade and packaging table[d]	–	–	–	–	–	f,j	–	–	–	–	b	b	–	–	–	–	–	–	–	–
Scales[d]	–	–	–	–	–	–	–	–	–	–	b	–	–	–	–	–	–	–	–	–
Baggers[d]	–	–	–	–	–	–	–	–	–	–	b	b	–	–	–	–	–	–	–	–

[a] Bryan et al. (1968a); A–O: indicate flock. [b] Key: a, S. infantis; b, S. anatum; c, S. chester; d, S. bredeney; e, S. typhimurium; f, S. cerro; g, S. san diego; h, S. derby; i, S. newington; j, S. senftenberg; k, S. halmstad; l, S. muenchen; m, S. stanley; n, S. saint paul; o, S. montevideo; p, S. blockley; q, S. schwarzengrund; I, negative for salmonellae; all blank spaces indicate that no samples were taken. [c] After cleanup. [d] Isolations made on the following day.

TABLE 16.5

Percent of Positive Isolations for Salmonellae on Surface of Poultry during Various Stages of Processing

Process step and sampling point	Investigations at turkey processing plants			Investigations at chicken processing plants			
	A [a]	B	C	C	D	E	F
During picking	—	—	—	—	18	—	—
After picking	—	63	12	—	13	11	11
After pinning	—	—	2	—	—	—	—
After washing	—	18	—	—	5	—	—
During rehanging	—	—	—	—	8	—	—
During evisceration	—	—	—	—	17	—	—
After evisceration	—	—	—	33	4	11	15
Before inside washer	—	—	4	—	—	—	—
After inside washer	14	—	—	20	—	—	—
After outside washer	4	—	—	—	—	10	12
In chill tank	12	—	—	—	—	13	5
After chilling	8	10	—	—	11	—	—
Final product	—	17	2	13	13	15	4

[a] Reference and number of samples for each investigation: A, Wilder and Mac-Cready (1966) (132–348 samples per sampling station, 2 plants); B, Bryan et al. (1968a) (33–58 samples per sampling station); C, Morris and Ayres (1960) (50 samples of turkeys per sampling station; 24–30 samples of chickens per sampling station); D, Morris and Wells (1970) (120–155 samples per sampling station); E, Commission of the European Communities (1976) (400 samples each per station, 5 plants using immersion-chilling methods); F, Commission of the European Communities (1976) (320–480 samples per station, 5 plants using air-chilling methods).

water, can serve as nutrients for multiplication of microorganisms, particularly pseudomonads (Barnes, 1960; Holden, 1970).

c. Process Water. Potable (including chlorinated) water can introduce additional microorganisms into the plant environment (Barnes, 1960; Poynter and Mead, 1964; Clark and Lentz, 1969). *Pseudomonas* and *Aeromonas* were the predominant flora isolated from portions of water mains and well water before the supply entered a poultry processing plant (Barnes, 1960). If suitable nutrients are present, pseudomonads multiply, producing a slime that can build up on equipment and water taps.

d. Ice. The bacterial flora of ice consist mostly of *Acinetobacter, Moraxella, Pseudomonas, Corynebacterium* and cocci. Many psychrotrophic bacteria that contaminate processed carcasses come from ice (Barnes, 1960; Clark and Lentz, 1969).

e. Workers. Microorganisms are transferred from contaminated poultry carcasses to hands of workers and then to surfaces of other carcasses. Workers can shed microorganisms from their skin, hair, nose, and throat as they work, particularly if they have skin or respiratory tract infections or if they practice poor personal hygiene. Workers risk getting pathogens on their hands, in cuts or other lesions, and inhaling them from contaminated carcasses. They may become infected and bring these organisms back into the processing environment. Carcass contamination from workers, however, is minor compared to that from live birds. For instance, during a 15-year survey of over 41,000 fecal specimens from poultry plant personnel, only 0.085% yielded salmonellae (Lundbeck, 1974). The plants surveyed also reported a low incidence of salmonellae on incoming birds and processed products.

f. Aerosols and Dusts. Aerosols and dusts are generated primarily during killing, picking, and spray-washing of carcasses. Air in the unloading and picking areas has a higher bacterial count than that in other areas of processing plants (Kotula and Kinner, 1964; Patterson, 1973). Mean values for aerobic plate counts per cubic meter of air were 17,700 in shackling areas, 77,700 in picking areas, 8100 in eviscerating areas, and 2200 in storage areas; corresponding values for molds were 2200, 1800, 1300, and 1100; for yeasts 570, 3900, 350, and 140; and for psychrotrophic bacteria 1600, 600, 140, and 180 (Kotula and Kinner, 1964). Aerosols produced in spray-washers and coolers can contain salmonellae (Zottola *et al.,* 1970; Notermans *et al.,* 1974). Splashing of organs or discarded parts in water-filled troughs, splashing of carcasses at the base of chutes or on table tops, spattering of blood and other body fluids, and release of pressure in compressed air lines used during offal removal can also create aerosols. Aerosols can be carried about a plant by air current, but this source of carcass contamination is probably of minor importance.

g. Vectors. Insects, rodents, or birds that gain access to the plant can also contaminate the processing environment, particularly if such vectors have previously been in contact with cages, truck beds, waste offal, feather or scrap meat piles, or the picking area. Contamination from vectors is undoubtedly minor compared to that from incoming birds.

2. Scalding

Carcasses are scalded to facilitate removal of feathers. Methods include immersion, hot water spray, steam, and simultaneous hot water spray and picking. Immersion scalding, the method most generally practiced, usually consists of a 2- to 3-min immersion in water at 60° to 63°C. Water in scald tanks is continually replaced at a rate of 0.2 to 1 liter per carcass per minute.

Soil, dust, and fecal matter from the feet, feathers, skin, intestinal tract, and respiratory tract are continually released into scald water. Aerobic plate counts of scald water, however, are usually less than 50,000/ml (Walker and Ayres, 1956, 1959; Schmidhofer, 1969; Mead and Impey, 1970; Mulder and Veerkamp, 1974; Lundbeck, 1974). Continuous overflow of scald water and introduction of replacement water remove many contaminants that reach scald water, and the heat of the water kills others. After initial increases in bacterial numbers during start-up, the count per milliliter of scald water remains relatively constant throughout the day (Schmidhofer, 1969; Veerkamp, 1974; Mulder, 1976). During scalding at 60°C for 115 sec, *Escherichia coli,* Enterobacteriaceae, psychrotrophic, and aerobic plate counts decreased significantly (Table 16.6; Mulder and Dorresteijn, 1977). During scalding at 53°C for 128 sec, Enterobacteriaceae and psychrotrophic counts decrease significantly, but there is little change in *Escherichia coli* or aerobic plate counts (Table 16.6). Aerobic plate counts of the skin of broilers immediately after scalding are usually less than $16,000/cm^2$ (Walker and Ayres, 1956; Clark and Lentz, 1969; Surkiewicz *et al.,* 1969; Notermans *et al.,* 1975a). *Clostridium, Micrococcus, Proteus, Pseudomonas, Salmonella, Staphylococcus,* and *Streptococcus* are sometimes isolated from scald tank water or from carcasses or air sacs immediately after scalding (Fahey, 1955; Walker and Ayres, 1956; Surkiewicz *et al.,* 1969; Mead and Impey, 1970; Lillard, 1971; Lillard *et al.,* 1973; Mulder and Dorresteijn, 1977).

Bacteria attached to the skin are more heat resistant than the same types that are not attached (Notermans and Kampelmacher, 1974, 1975). Salmonellae have survived scalding at 55°C for 105 sec, but they were not isolated from chickens after scalding at 60°C for 200 sec (Notermans *et al.,* 1975a; Mulder and Dorresteijn, 1977). In other studies, however, salmonellae have been isolated from turkey carcasses receiving a 60°C scald (Bryan *et al.,* 1968a).

TABLE 16.6

Comparison of Microbial Count (Log$_{10}$) on Chickens Scalded in 60°C Water for 115 Sec and Those Scalded in 53°C Water for 128 Sec [a, b]

Organism or group	Before scalding at 60°C		After scalding at 60°C		Before scalding at 53°C		After scalding at 53°C	
	Mean	S.D. [c]	Mean	S.D.	Mean	S.D.	Mean	S.D.
Escherichia coli	3.5	1.0	0.8	0.7	>4.0	—	3.4	0.8
Psychrotropic plate count	4.3	0.3	<2.0	—	4.2	0.4	<2.0	—
Aerobic plate count	7.9	0.4	4.9	0.3	7.9	0.4	6.5	0.6
Entero-bacteriaceae	4.9	0.8	<2.0	—	6.1	0.5	4.2	0.8

[a] Notermans *et al.* (1975a); Mulder and Dorresteijn (1977).
[b] Skin around cloaca sampled; 8 samples tested for each organism before and after each temperature scald.
[c] S.D. = Standard deviation.

Scalding carcasses at 58°C or higher, then abrading by mechanical picking, removes the yellow epidermal layer (cuticle) of a bird's skin. Scalding at 54°C or lower, then picking by machine, does not have this effect. The cuticle-free skin serves as a more suitable substrate for spoilage organisms (Ziegler and Stadelman, 1955; Essary *et al.,* 1958; Clark, 1968; Berner *et al.,* 1969).

The combination spray scalder and picker (Veerkamp and Hofmans, 1973), the steam hot-water scalder (Patrick *et al.,* 1972), and the batch-type scalder using steam at subatmospheric pressure (Klose *et al.,* 1971; Kaufman *et al.,* 1972; Lillard *et al.,* 1973; Patrick *et al.,* 1973) apparently reduce bacterial counts more than immersion scalders do, because of their higher temperatures and higher ratio of water per carcass. These scalders are still in the experimental stages of development.

3. Picking (Plucking, Defeathering)

The aerobic plate count, but not the psychrotrophic count, significantly increases after picking (Walker and Ayres, 1956, 1959; Clark and Lentz, 1969). Higher aerobic plate and staphylococcal counts are related to spread of microorganisms within the picking machines and to inadequate cleaning of rubber fingers in these machines (Simonsen, 1975).

Salmonellae are isolated from carcasses more frequently after picking than after any other operation (Bryan *et al.*, 1968a; Morris and Wells, 1970; Notermans *et al.*, 1975a). For instance, during the processing of 14 turkey flocks that were naturally infected, the rubber fingers of the first picker in a series of four were contaminated with salmonellae during the processing of 12 (85.7%) of these flocks; the second picker was contaminated with these organisms during processing of 11 (78.6%) of these flocks; the third picker was contaminated with them during the processing of 7 (50%) of these flocks; and a final spiral picker was contaminated with salmonellae during the processing of 12 (85.7%) of these flocks. A chute and a table, on which carcasses slid after picking, were contaminated with salmonellae during the processing of 10 (71.4%) and 12 (85.7%) of these flocks, respectively. Sixty-three percent of the carcasses sampled were contaminated with salmonellae after they were processed by this series of equipment (Bryan *et al.*, 1968a). Isolates were usually the same serotypes that were associated with incoming flocks (Table 16.4). Studies with an indicator strain showed that cross-contamination by *E. coli* also occurs during feather removal (van Schothorst *et al.*, 1972; Mulder *et al.*, 1977a).

Thus, picking operations spread microorganisms from a few contaminated carcasses to many carcasses. It is during this stage of processing that many fecal and other types of bacteria become attached to surfaces or enter feather follicles, and therefore are difficult to remove during subsequent processing (Notermans and Kampelmacher, 1974; Notermans *et al.*, 1975d). It is true that the levels of organisms that remain on the skin of processed birds reflect the bacterial population of the equipment surfaces and of the washing and chilling suspensions to which carcasses are exposed. However, these levels reflect even more closely the bacteriological quality of the carcasses immediately after picking (Commission of the European Communities, 1976; McMeekin and Thomas, 1978). (See Tables 16.2 and 16.3.)

4. Evisceration

During evisceration, microorganisms can be transferred from carcass to carcass by workers and equipment (Galton *et al.*, 1955b; Wilder and MacCready, 1966; Bryan *et al.*, 1968a).

Manual opening of the abdominal cavities and manual evisceration give rise to considerable cross-contamination, especially when intestines are cut. Vacuuming of the vent and gut cavity removes some fecal and other contamination. No differences in aerobic plate counts of samples taken before and after machine evisceration with properly maintained and continuously cleaned equipment has been observed (Simonsen, 1975).

5. Spray Washing

Spray washing of carcasses after picking and evisceration removes organic material and the superficially lodged microorganisms that are acquired during evisceration (Barnes, 1975), causing a 50 to 90% decrease in aerobic plate, Enterobacteriaceae, and coliform counts (Stewart and Patterson, 1962; Sanders and Blackshear, 1971; May, 1974; Mulder and Veerkamp, 1974; Mulder, 1976). Salmonellae contamination is also significantly decreased after spray washing (Byran et al., 1968a; Morris and Wells, 1970). Chlorination of spray water at a level of 40 to 60 ppm can cause additional reduction of microbial counts (Sanders and Blackshear, 1971). Certain *Pseudomonas* spp., however, can contaminate carcasses during washing (Lahellec et al., 1973).

6. Cooling

Chilling delays growth of psychrotrophic bacteria (thus prolonging self-life) and prevents the growth of most foodborne pathogens. Carcasses are chilled by immersion in tanks containing slush ice or cold water without ice, by sprays of cold water, or by circulation of cold air or other gases (Barnes, 1974). In Europe, wet chilling is usually used for poultry to be frozen, and air chilling for poultry not to be frozen. In North America, nearly all poultry are chilled by wet methods. Typical temperatures of chicken carcasses at various stages of processing during either immersion or air cooling are shown in Table 16.7

a. Slush-Ice (Static-Tank) Chilling. Slush-ice chilling, with approximately equal amounts of ice, water, and carcasses, takes advantage of the cooling capacity of melting ice. Carcasses are held in tanks for 4 to 24 hr, during which time psychrotrophic bacteria can multiply on the carcasses and in the water (Barnes and Shrimpton, 1968). Aerobic plate counts range from 10^3 to 10^6/ml in aerated chill-tank water (Walker and Ayres, 1956, 1959). Chlorination at 5 to 20 ppm is used in some plants to prevent multiplication of psychrotrophic bacteria in static-tank chill water (Barnes, 1965). Static slush-ice chilling has been replaced by other methods for cooling chickens but is still used as a final chill for turkeys.

b. Continuous Immersion Chilling. Carcasses are immersed in water and tumbled, with the aid of mechanical agitators, through one or more

TABLE 16.7

Temperatures of Poultry Carcasses at Different Stages of Processing [a, b]

Plant [c]	After evisceration (°C)	After spray washing (°C)	After chilling (°C)	After packaging (°C)
Air chilling				
1a	27.50	26.25	12.20	13.38
2a	30.50	32.50	2.00	3.44
3a	39.83	38.81	8.21	—
4a	28	27	5	—
5a	34.36	32.12	0.25	2.83
6a	34.00	32.00	7.64	8.00
Immersion chilling				
1	30.50	29.50	7.25	9.25
2	31.63	26.63	10.05	—
3	28.64	27.50	7.71	9.28
4	34.00	32.14	8.93	8.71
5	30.00	28.38	7.81	9.13

[a] Commission of the European Communities (1976).

[b] Average of 8 experimental periods. Temperature of 10 carcasses placed into an insulated box registered by thermocouples inserted among the carcasses in the center of the box after an equilibration period of 30 minutes. In plant 4a, temperatures of carcasses were measured by thermocouples inserted into the deep breast musculature.

[c] Plants 1a, 3a, 4a, 5a, and 6a used tunnel chilling and plant 2a used a chilling room; plants 1, 2, and 3 used counter-current immersion chillers, and plants 4 and 5 used through-flow immersion chillers.

tanks. The water can flow with or against the direction taken by the carcasses. Prechilled water or ice is put into the last portion of the tank or into the last tank, if a series of tanks is used. Continuous immersion chilling has a distinct advantage over static-tank chilling because it cools chicken carcasses to an equilibrium temperature of about 4°C in less than 1 hr. Mechanical agitation removes some of the organisms from the carcasses; however, some may be transferred to other carcasses.

According to Thomson et al. (1974), 12 investigations showed that the aerobic plate count or the quantity of indicator or tracer organisms, particularly those of fecal origin, was reduced during continuous immersion chilling, but 3 investigations showed that one or more of these counts increased. The following factors account for these differences: (a) the amount of bacterial contamination on carcasses before cooling; (b) the amount of water overflow per carcass; and (c) the ratio of the number of carcasses to the volume of water in the chiller. If the amount of water

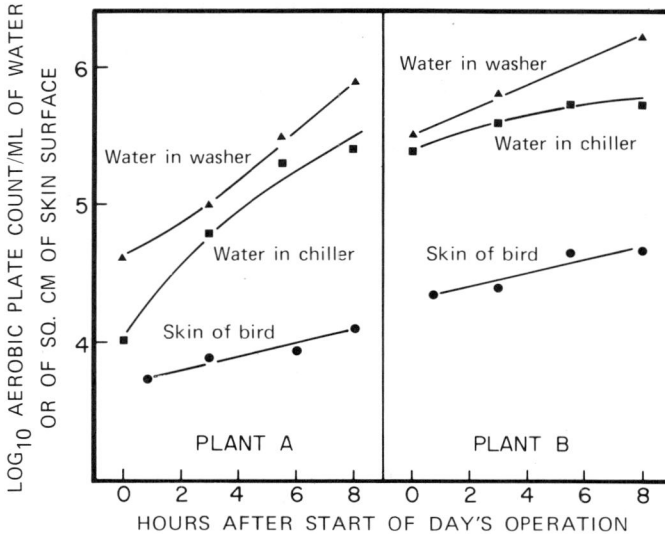

Fig. 16.3. Changes in aerobic plate count in water of washer and chiller and on skin of finished birds during the day's operation. (First samples were taken from the washer and chiller before the birds enterd the tanks.) (From Clark and Lentz, 1969.)

used per carcass is insufficient, microorganisms accumulate in the chill water (Fig. 16.3), and the numbers found on carcasses increase rather than decrease. Carcasses cooled later in a day's operation are, therefore, exposed to a higher concentration of microorganisms (Clark and Lentz, 1969). Continuous and liberal replenishment of cooling water aids in the cleaning of the skin (Mulder and Veerkamp, 1974).

Counter-flow immersion chilling (in which carcasses at the end of the chilling process come into contact with the cleanest water) effectively decreases counts on carcasses or minimizes cross-contamination (Mulder and Veerkamp, 1974). An adequate overflow of water also removes organic matter that would otherwise reduce the effectiveness of the chlorine some firms add to the water (see Section III,D,2).

c. Spray Chilling. Spray chilling has been tested experimentally (Grossklaus and Levetzow, 1967; Leistner et al., 1972; Leistner, 1973; Mulder and Veerkamp, 1974). In the various investigations, between 0.3 and 15 liters of 0° to 11°C water per carcass were sprayed for about 15 to 30 min to reduce carcass temperatures, usually to 7°C or below. These procedures effectively wash carcasses, but the large amount of water necessary for effective cooling has prohibited their commercial use.

d. Air Chilling. Cold air chilling dries the skin and thus retards microbial growth. When air chilling is used, birds are scalded at only 50°C to avoid skin damage and color change. This scald temperature, however, does not reduce enteric organisms appreciably (Buchli *et al.*, 1966; see also Table 16.3). Bacteria can circulate in currents of cold air and contaminate carcasses (Ayres, 1955). Bacterial counts on air-chilled carcasses are sometimes higher than those on water-chilled carcasses (Buchli *et al.*, 1966; Mulder, 1971; Mead, 1975; Thomson *et al.*, 1975; see also Table 16.4), but not always (Berner *et al.*, 1969; Commission of the European Communities, 1976).

e. Carbon Dioxide Snow Cooling. Carbon dioxide snow cooling costs considerably more than conventional chilling. It retards bacterial growth both by its initial cooling effect and by the inhibitory effect of the CO_2 atmosphere during shipment and storage. (See Chapter 10.)

7. Weighing and Packing

Microbial counts can increase after chilling because of transfer of microorganisms during weighing and packaging (Thomson *et al.*, 1975). When chickens were weighed before evisceration, bacteria built up on the surface of the scale contributing a 10-fold increase in bacteria on the skin of birds at point of contact and becoming a focal point of cross-contamination (Simonsen, 1975). Significant differences in aerobic plate counts have been demonstrated among different weight groups of poultry carcasses, depending on the distance between the scales and the packaging table; the longer the distance, the higher the counts (Simonsen, 1975). Weighing and packing can also increase slightly the percentage of carcasses contaminated with salmonellae (Bryan *et al.*, 1968a,b; Morris and Wells, 1970).

8. Uneviscerated (Effilee or New York-Dressed) Poultry

Carcasses are sometimes not eviscerated, or only intestines are removed, and the feathers are either dry-picked or left unpicked (*e.g.*, some game birds). Cooling is in air blast refrigerators. Most of the bacterial growth is in the intestines; skin surfaces are too dry to permit rapid bacterial growth.

9. Edible Viscera (Offal, Giblets)

During evisceration, edible viscera (hearts, livers, and gizzards) are removed from carcasses and conveyed to a viscera-preparation and wrapping area, where they are cleaned and sorted, then packaged in bulk, or put into bags with the necks. Then the bags are inserted into the cavity of dressed carcasses. Equipment used to convey, process, or wrap edible viscera is frequently contaminated with salmonellae and other microorganisms (Galton *et al.,* 1955b; Wilder and MacCready, 1966). Thus, the potential for cross-contamination is apparent for these products.

10. Cutting and Other Handling

Microorganisms from eviscerated poultry carcasses and parts are transferred to the surfaces of cut-up or boned parts, or to other further-processed products, by workers' hands and gloves. In further-processing operations, the hands of the workers, their rubber gloves, and wire-safety gloves were contaminated with salmonellae more than 30% of the time when these workers were handling raw turkey (Bryan *et al.,* 1968b). The serotypes found on the hands and gloves were the same as those isolated from the incoming carcasses. Workers also carry other microorganisms that can be transferred to carcasses and parts being handled.

Equipment and utensils (particularly knives, tables, conveyors, and scales) that touch every piece of poultry also cross-contaminate clean pieces of meat during further processing. In one plant, for instance, 10% of the turkey carcasses were contaminated before and 21% after further processing; in another, 15% were contaminated before and 37% after further processing. These differences are statistically significant (Bryan *et al.,* 1968b). Aerobic plate counts of chicken carcasses increased sixfold in processing-plant cutting areas and about eightfold during cutting and preparation in retail stores (May, 1962).

11. Mechanical Deboning and Mixing

Mechanically deboned poultry meat is produced from necks, backs, turkey skeletal frames, and other parts that are essentially by-products of cut-up operations. The resulting meat fraction is very finely comminuted, providing a large surface area on which microorganisms can cling and a rich medium for the growth of microorganisms. The types present are determined primarily by the microflora that was present on the whole carcass. Aerobic plate counts (mainly psychrotrophic bacteria) are usually in the range of 10^5 to 10^6/g, and coliform counts in the range of 10^2 to 10^3/g (Mulder and Dorresteijn, 1975). Salmonellae, *Staphylococcus aureus, Clostridium perfringens,* enterococci, yeasts, and molds are frequently isolated from these products. Deboning and comminuting should preferably be at or below 10°C to reduce microbial growth on surfaces of equipment.

B. Spoilage

Spoilage of chilled raw poultry is manifested by off-odors fo' slime formation, and various discolorations. Before off-odors a~~ppear, the~~ water-holding capacity of the meat increases (Adamcic and Clark, 1970). The pH rises, and ammonia and related compounds increase as bacteria multiply; but such changes are not marked until the level of psychrotrophic bacteria exceeds 10^8/g.

1. Raw Eviscerated Carcasses

Factors that influence bacterial growth on eviscerated carcasses are (a) the numbers and types of psychrotrophic spoilage organisms present immediately after processing, (b) the storage time and temperature, (c) the type of tissue (skin or muscle), (d) the pH, (e) the redox potential, (f) the type of packaging, and (g) the presence or absence of carbon dioxide. Psychrotrophic spoilage bacteria enter processing plants on birds' feathers and feet and in small numbers in water and ice supplies. They multiply on soiled surfaces of equipment, in chill-water tanks, and on the surfaces of the birds.

Acinetobacter, Flavobacterium, and *Cytophaga* are the genera most commonly present on the surfaces of recently killed poultry. But the main causes of spoilage are pseudomonads [*Pseudomonas fluorescens, P. putida, P. fragi (incertae sedis;* Buchanan and Gibbons, 1974) and related species], and to a lesser extent *Alteromonas putrefaciens* [proposed name; excluded from genus *Pseudomonas* because characteristics not in accord with the generic definition (Buchanan and Gibbons, 1974); previously called *Pseudomonas putrefaciens*], *Acinetobacter,* and *Moraxella* (Barnes and Impey, 1968; McMeekin, 1977). A high initial load of these organisms will allow a shorter lag time and therefore a shorter shelf-life (Fig. 16.4).

Temperature affects the species that are able to thrive. For example, in turkeys, *Pseudomonas* species predominate at 1°C, whereas *Enterobacter liquefaciens* and atypical lactobacilli predominate at 10° and 20°C (Barnes and Shrimpton, 1968). Psychrotrophic pseudomonads thus become the predominant flora on the aerobic surfaces of poultry stored at low temperatures. In general, psychrotrophic microorganisms can multiply between −3° and 34°C (Farrell and Barnes, 1964), although many cannot grow at or above 32°C. Therefore, they can multiply the entire time carcasses are held at commonly used refrigerator temperatures. When their numbers reach 10^7 to 10^8/cm^2 of skin surface, off-odors develop; and when they reach more than 10^8/cm^2, slime forms (Fig. 16.5), and the carcass is un-

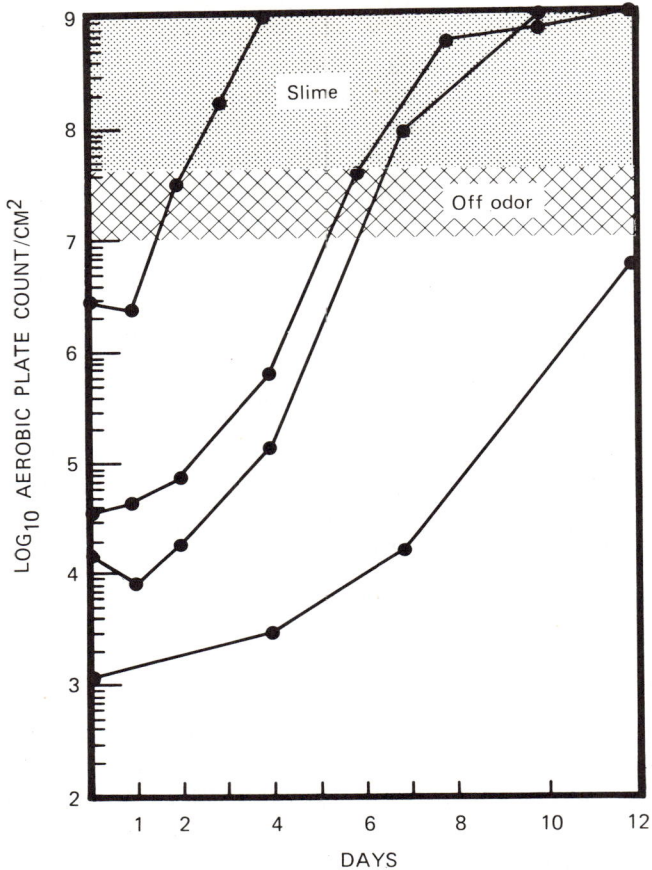

Fig. 16.4. Effect of initial load of microorganisms on chicken meat on time required for development of off-odor and slime at 4.4°C. [From Ayres *et al.* (1950) and Ayers (1959).]

acceptable for food (Ayres *et al.,* 1950). Air-cooled, eviscerated turkeys developed off-odors in an average of 7.2 days at 5°C, 13.9 days at 2°C, 22.6 days at 0°C, and 38 days at −2°C (Barnes, 1976); the longest storage life for unfrozen poultry is obtained at −2°C (Table 16.8).

The few bacteria in the interior of poultry muscle are of types that can multiply only slowly or not at all during low-temperature storage. The important microbial changes take place on the surfaces.

Acinetobacter and *Alteromonas* (*Pseudomonas putrefaciens*) grow better in leg muscle where pH is 6.4 to 6.7 than in breast muscle where pH is 5.7 to 5.9. *Pseudomonas* spp. can grow well in either (Fig. 16.6).

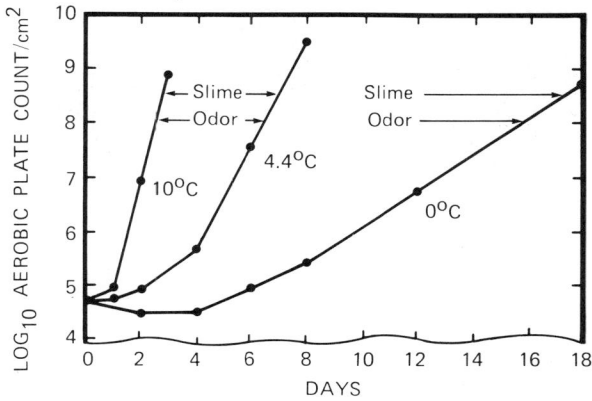

Fig. 16.5. Effect of storage temperature on growth of bacteria on chicken meat. (From Ayres *et al.,* 1950.)

TABLE 16.8

The Effect of Temperature on the Growth Rate of a Typical Mesophile *(Escherichia coli)* **Compared with That of a Typical Spoilage Organism** *(Pseudomonas* **sp.)** **and Consequent Shelf-Life of Chicken Carcasses** [a]

Temperature (°C)	Generation time (hr)		Days of shelf-life [d]
	E. coli [b]	*Pseudomonas* sp. [c]	
− 2	No growth	36.4	30.3 + 4 days lag [e]
0	No growth	13.8	11.5 + 2 days lag
5	No growth	7.4	6.2
10	20	4.7	3.9
15	6	2.2	—
20	2.8	1.4	—
25	1.4	0.9	—
30	0.6	1.0	—
36	0.4	—	—

[a] Barnes (1974, 1976).

[b] Data of Ingraham (1958).

[c] Data of Farrell and Barnes (1964).

[d] Time for *Pseudomonas* sp. to grow from $100/cm^2$ to $10^8/cm^2$.

[e] Time for growth to commence.

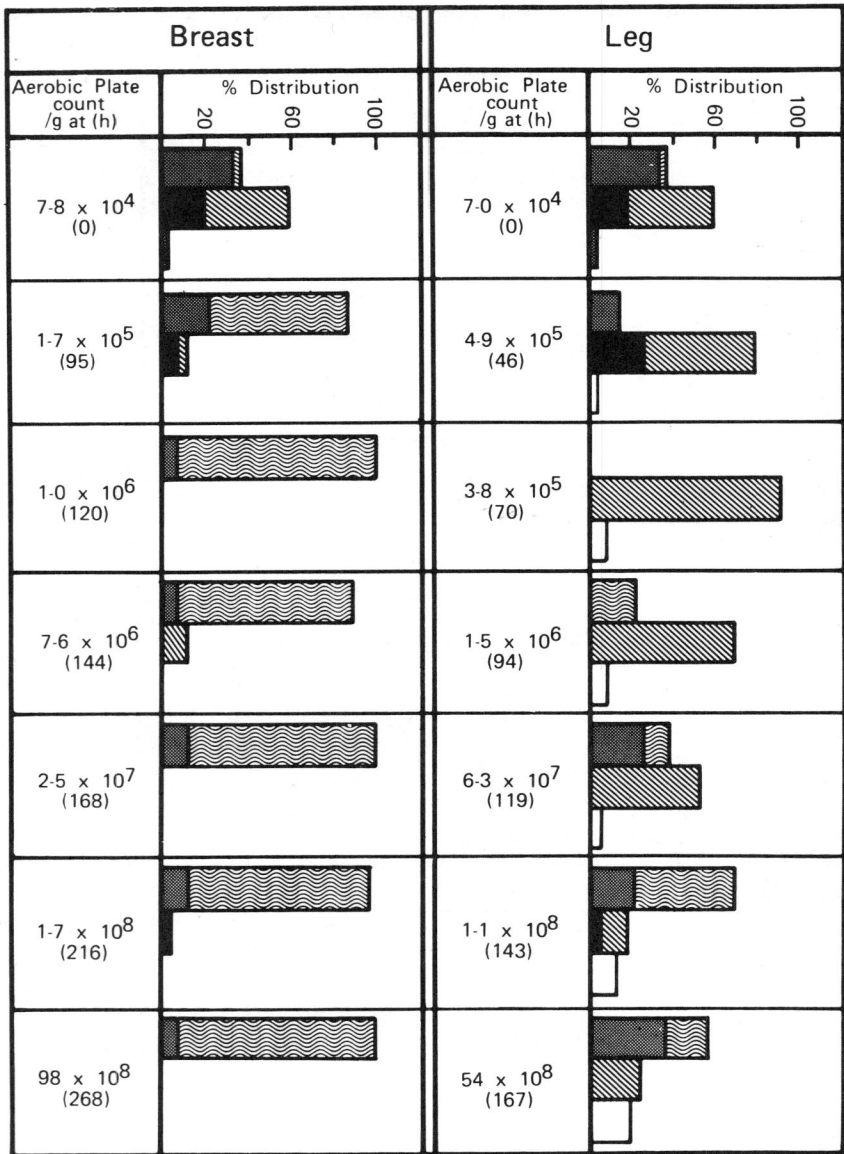

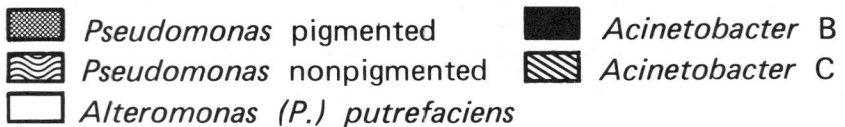

Fig. 16.6. Changes in the distribution of microorganisms during storage at 1°C of minced breast and of leg muscle. (From Barnes and Impey, 1968.)

Pseudomonas spp. are the principal causes of spoilage on carcasses in oxygen-permeable films; *Alteromonas, Microbacterium thermosphactum (incertae sedis;* Buchanan and Gibbons, 1974), and atypical lactobacilli are principal causes of spoilage on carcasses wrapped in oxygen-impermeable films. Reduced redox potential and accumulations of carbon dioxide around carcasses wrapped in the impermeable films inhibit the growth of pseudomonads (Chapters 10 and 11; see also Shrimpton and Barnes, 1960). In vacuum-packaged chickens, *Enterobacter* spp. were the principal causes of spoilage, whereas pseudomonads were responsible for spoilage in control packs (Arafa and Chen, 1975). Other facultative bacteria, including other Enterobacteriaceae, can survive and grow in vacuum-packaged poultry products.

2. Uneviscerated Poultry (Effilee)

The kinds of microorganisms that multiply on uneviscerated poultry depend on temperature. Mesophilic bacteria in the intestines produce hydrogen sulfide, which diffuses into the muscle tissue, and which, in the presence of air just under the skin, combines with heme pigments of blood to produce a green discoloration. Psychrotrophic spoilage microorganisms on the skin are unable to multiply as long as the surface is kept dry and intact. But, as soon as cuts are made to remove the head, neck, and viscera, spoilage organisms are transferred to muscle surfaces and can subsequently multiply and produce off-odors (Barnes, 1976).

3. Edible Viscera

Muscle tissues have a longer shelf-life than giblets, but a difference in shelf-life between poultry packed with or without giblets was not observed (Taylor *et al.,* 1968). At the time spoilage of chicken kidneys, livers, and gizzards stored at 4°C was evident, aerobic plate counts had reached 10^9/g (May *et al.,* 1962).

C. Pathogens

Poultry is a common vehicle of foodborne illness (Todd, 1978). The human pathogens most frequently associated with poultry are *Salmonella, Clostridium perfringens,* and *Staphylococcus aureus* (Hobbs, 1971). Others, e.g., *Yersinia* and *Campylobacter,* have been isolated from poultry products, but the importance of poultry in their spread has not been determined (Leistner *et al.,* 1975; Bruce et al., 1977).

1. Salmonella

Salmonellae are frequently isolated from raw poultry carcasses, parts, and edible viscera during processing and from chilled poultry products at various stages of processing and marketing (Tables 16.4, 16.5, and 16.9). They reach poultry from the sources and by the modes described in Sections II, III, A, and III, B. When present on broiler carcasses, they are usually in relatively small numbers, i.e., about 10–20 but occasionally more than 1400/100 g of skin (Notermans et al., 1975b; Mulder et al., 1977b). Because of the low numbers of salmonellae, it is essential that large surface areas be sampled.

Outbreaks of salmonellosis can follow inadequate cooking. Or raw poultry can serve as a source of cross-contamination of cooked poultry or other foods processed in the same kitchen. (See Chapter 29.) Tests for the incidence of salmonellae on poultry carcasses give an indication of the hazards that can follow subsequent mishandling (e.g., inadequate cooking and cross-contamination).

2. Clostridium perfringens

Clostridium perfringens is frequently found on surfaces of raw poultry but usually in small numbers (Strong et al., 1963; Hall and Angelotti, 1965; Mead and Impey, 1970; Bryan and Kilpatrick, 1971). It reaches poultry surfaces from fecal matter, soil, and dust (see Sections II, III, A and III, B). Raw poultry is ordinarily stored at temperatures too low to permit this species to grow; also, competitive psychrotrophic organisms interfere. Additionally, the redox potential of raw poultry skin is far from the optimum for C. perfringens. Because of these factors, qualitative or quantitative specifications for C. perfringens in raw poultry are of little value for routine investigations. Large numbers of vegetative forms of C. perfringens ($> 10^6$), which can develop only on or in cooked poultry products, are required to produce gastroenteritis. (See Section V and Chapter 29.)

3. Staphylococcus aureus

Live poultry carry staphylococci in bruised tissues, infected lesions, nasal sites, skin surfaces, and arthritic joints, and therefore the meat is frequently and unavoidably contaminated during slaughter and processing. (See Sections II, III,A, and III,B.) Many of the strains isolated from poultry are nontypable with the international set of phages, selected for their activity against human strains; nonetheless, many of the poultry forms are enterotoxigenic (Gibbs et al., 1978a). Low storage temperatures and a competitive flora do not favor staphylococcal development in raw poultry

TABLE 16.9

Prevalence of Salmonellae in Poultry [a]

Type of product	Number of samples analyzed	Positive for Salmonella		Country	Reference
		Number	Percent		
Raw products					
Broiler chickens	297	30	10	United Kingdom	Dixon and Pooley (1961)
Broiler chickens treated with chlorotetracycline	80	14	18	Canada	Thatcher and Loit (1961)
Chickens	525	88	17	United States	Wilson et al. (1961)
Chicken parts	404	61	15	United States	Wilson et al. (1961)
Chicken giblets	93	22	24	United States	Wilson et al. (1961)
Turkeys	146	18	12	United Kingdom	Dixon and Pooley (1962)
Fryer chickens	264	72	27	United States	Woodburn (1964)
Poultry	217	8	4	United Kingdom	Galbraith et al. (1964)
Chickens	1530	245	17	Netherlands	van Schothorst et al. (1965)
Chicken hens	770	41	5	Netherlands	van Schothorst et al. (1965)
Chicken carcass with neck or edible viscera packed inside	237	119	50	United States	Wilder and MacCready (1966)
Whole carcass after chilling	348	27	8	United States	Wilder and MacCready (1966)
Chickens, packing area, plant A	169	37	22	United States	Glezen et al. (1966)
Chickens, packing area, plant B	817	4	< 1	United States	Glezen et al. (1966)
Chickens	531	0	0	Northern Ireland	Patterson (1967)
Chickens, New York dressed	272	0	0	Northern Ireland	Patterson (1967)
Chilled turkey carcasses	58	10	17	United States	Bryan et al. (1968a)
Chilled turkey carcasses, plant A	149	15	10	United States	Bryan et al. (1968b)
Chilled turkey carcasses, plant B	59	9	15	United States	Bryan et al. (1968b)
Further-processed turkey products, plant A	217	46	21	United States	Bryan et al. (1968b)

(Continued)

TABLE 16.9 *(Continued)*

Type of product	Number of samples analyzed	Positive for Salmonella		Country	Reference
		Number	Percent		
Further-processed turkey products, plant B	119	44	37	United States	Bryan et al. (1968b)
Chicken carcass (after giblets added)	4450	2	< 1	United Kingdom	Tucker and Gordon (1968)
Ducks (after chilling)	140	16	11	Northern Ireland	Patterson (1969)
Chickens	171	35	21	United States	Surkiewicz et al. (1969)
Chickens (in crates)	155	20	13	United States	Morris and Wells (1970)
Frozen, whole poultry	90	22	24	United Kingdom	Hobbs (1971)
	149	22	15	United Kingdom	Hobbs (1971)
Frozen poultry pieces	100	13	13	Denmark	Hobbs (1971)
	100	0	0	United Kingdom	Hobbs (1971)
Further-processed turkey products	35	3	10	United States	Zottola and Busta (1971)
Chickens	495	28	6	Northern Ireland	Patterson (1972a)
Giblets	265	28	11	Northern Ireland	Patterson (1972a)
Turkeys	100	0	0	Northern Ireland	Patterson (1972a)
Ducks	597	39	7	Northern Ireland	Patterson (1972a)
Frozen poultry	101	36	36	United Kingdom	Hobbs (1972)
Frozen chickens	532	20	4	Denmark	A. Suzuki et al. (1973)
	332	19	6	Hungary	A. Suzuki et al. (1973)
	2219	207	9	China	A. Suzuki et al. (1973)
	2728	293	11	United States	A. Suzuki et al. (1973)
	340	46	14	Bulgaria	A. Suzuki et al. (1973)
	153	27	18	Canada	A. Suzuki et al. (1973)
	137	39	29	Netherlands	A. Suzuki et al. (1973)
	88	22	27	Other countries	A. Suzuki et al. (1973)
Poultry carcasses	2989	65	2	United States	Goo et al. (1973)
Poultry viscera	724	10	1	United States	Goo et al. (1973)

Product				Country	Reference
Chickens (mostly livers)	8415	—	< 1	Sweden	Lundbeck (1974)
Chickens	25	0	0	Denmark	Siems et al. (1975)
	50	24	48	Poland	Siems et al. (1975)
	25	2	13	Turkey	Siems et al. (1975)
	30	5	17	Belgium	Siems et al. (1975)
	39	5	17	France	Siems et al. (1975)
	298	187	63	Germany	Siems et al. (1975)
	95	61	64	Netherlands	Siems et al. (1975)
	90	28	31	United States	Comptroller General of the United States (1974)
Chicken skin	7218	326	5	Netherlands	Voeten et al. (1974)
Chickens	146	42	29	West Germany	Oberhauser (1975)
Turkeys	55	4	7	West Germany	Oberhauser (1975)
Chickens	4420	456	11	5 European Countries	Commission of the European Communities (1976)
		16	15	Canada	Consumers' Association of Canada (1976)
Turkeys	119	27	32	Canada	Consumers' Association of Canada (1976)
Ducks	53	30	57	West Germany	Reusse et al. (1976a)
Frozen chickens	99	52	52	Netherlands	van Schothorst et al. (1976)
Chickens	496	49	9	Greece	Vassiliadis et al. (1976)
Chickens, cut-up	69	24	35	Canada	Duitschaever (1977)
Chickens	100	9	9	India	Praminik and Khanna (1977)
Chickens	240	107	45	United States	Cox et al. (1978)
Turkeys	1240	145	12	Canada	McBride et al. (1978)
Chickens	157	36	23	Canada	Pivnick et al. (1978)
Cooked products					
Heat-processed turkey rolls	37	0	0	United States	Bryan et al. (1968c)
Heat-processed turkey rolls	38	0	0	United States	Zottola and Busta (1971)

[a] Bryan et al. (1979).

441

products. Thus, a microbiological specification for staphylococci in raw poultry has little value.

D. Control

The composition and number of microorganisms on eviscerated raw poultry reflect the various sources of contamination, the processing operations that spread or decrease contamination, the temperature and length of storage, and the effectiveness of plant sanitation. Certain processes or storage procedures are designed to minimize the number of microorganisms present on the carcasses at the end of processing, and certain microbial tests are useful for quality control or the detection of pathogens on raw poultry.

1. Sanitation

Spray-washing of carcasses at various stages of processing removes a portion of the bacterial population from poultry surfaces. Thorough cleaning and rinsing of the equipment are essential to remove microorganisms from equipment surfaces and to prevent buildup. Hypochlorite or other approved disinfectant kills many of the remaining bacteria. (Principles of cleaning and disinfection are described in Chapter 14.) Operations and sanitary procedures should comply with the Codex Alimentarius Code of Hygienic Practice for Poultry Processing (Food and Agriculture Organization/World Health Organization, 1976c).

2. Chlorination

In some plant operations, water for washing and chilling poultry is chlorinated. Thomson et al. (1974) have reviewed numerous investigations on the effectiveness of chlorine in reducing counts and preventing the spread of specific bacteria or classes of bacteria. Results have varied. In some studies, chlorination reduced aerobic plate counts 50 to 90% and increased shelf-life; but, in other studies, little or no reduction in counts was observed. Sometimes, chlorination has reduced the percentage of carcasses contaminated with salmonellae or other specific organisms; other times it has not. Most contamination of carcasses occurs during defeathering, before much chlorinated water is used (Mead, 1974; see also Tables 16.4 and 16.5). The greatest limitation of chlorine is its rapid inactivation by organic material. It appears to have little effect on the microflora already on carcasses (Barnes, 1965; Patterson, 1968; Mead, 1974). Chlorine, however, prevents a build-up of bacterial slime on equipment surfaces and eliminates microorganisms, particularly psychrotrophic bacteria, present in the incoming water supply or in the water of chill tanks (Barnes, 1965;

Mead and Thomas, 1973; Mead *et al.*, 1975). Concentrations of 45 to 50 ppm of total residual chlorine were recommended for coolers using 5 liters of water per carcass; concentrations of 25 to 30 ppm were effective for coolers using 8 liters of water per carcass (Mead and Thomas, 1973). There has been considerable debate on the value and hazard of chlorination in poultry operations (Cunningham and Lawrence, 1977). Some countries require it or promote its use; others prohibit it (Simonsen, 1975; see also Chapter 14).

3. Packaging

Packaging prevents contamination and loss of water and, if the package is an oxygen-impermeable film (vinylidene chloride–vinyl chloride polymer), it retards microbial growth and extends shelf-life (Shrimpton and Barnes, 1960; see also Chapter 11).

4. Carbon Dioxide

Atmospheric concentrations of 10 to 25% carbon dioxide (CO_2) delay the growth of pseudomonads, molds, and some other types of spoilage organisms when the product is held at 4°C or lower (Ogilvy and Ayres, 1951; Thomson, 1970). The increase in shelf-life is proportional to the concentration of CO_2 up to 25%, at which point discoloration occurs. (See Chapter 10.)

5. Temperature

The lower the storage temperature, the longer the shelf-life (see also Chapter 1). Most pathogens cannot grow below 6°C, but psychrotrophic bacteria grow at and below this temperature, some even at or below O°C. At −2°C, psychrotrophic bacteria have a lag period of about 4 days and spoil the poultry only after about 30 days (Barnes, 1974; see also Table 16.8). Therefore, to increase shelf-life and to minimize the opportunities for pathogens to multiply, poultry should be stored below 3°C, or if feasible at −2°C. A lower temperature will freeze the product; however, some processors do permit the product to freeze, then ship it as it begins to thaw. The shelf-life is extended by the amount of time it remained frozen.

6. Quality Control and Safety Assurance

The microbial condition of poultry can be assessed at any stage of processing or distribution. The type of sample and analysis depends on the information desired, e.g., (a) the expected presence and number of foodborne pathogens and the risk the product offers, (b) the effect of a given process on the spread of microbial contamination or on survival of pathogens or spoilage organisms, or (c) the potential storage life at a given temperature. The time of sampling—during processing, immediately after raw processing, after further processing, after storage, or during retail sales or preparation—will significantly influence the results obtained. Other factors that need to be considered in the selection and performance of microbiological tests for poultry are described in the following subsections.

a. Microorganisms of Concern. An aerobic plate count incubated at 35°C where mesophilic organisms can grow, but where most psychrotrophs cannot (Table 16.8), can measure the overall cleanliness of poultry meat. Tests for fecal indicators such as *Escherichia coli,* coliform group, or Enterobacteriaceae can evaluate the amount of fecal matter present. But since raw poultry surfaces invariably have such organisms in abundance, and since there is poor correlation between their occurrence and the occurrence of specific pathogens such as *Salmonella,* their measurement is rather pointless. Direct tests for specific pathogens, such as *Salmonella,* are more important and useful. The sampling plan must be able to detect a relatively low incidence of contamination (ICMSF, 1974).

An aerobic plate count incubated at 17°–30°C, where both psychrotrophs and mesophiles grow (or even more to the point, incubated at 0°–5°C where psychrotrophs can grow and mesophiles cannot) can illustrate the degree of spoilage and the expected shelf-life. However, by the time results are available (4–5 days at 17°–30°C, 7–14 days at 0°–5°C) the meat is no longer in the same condition as it was. For this reason plate counts of this kind are not recommended to evaluate the acceptability of a lot of chilled poultry meat. They are invaluable, however, to illustrate the microbiological acceptability of a processing operation.

b. Distribution of Microorganisms. Fecalborne microorganisms are spread over skin during scalding and picking and on inner and outer surfaces during evisceration and further processing. Chill water, which contains bacteria, particularly pseudomonads, penetrates the skin mainly by way of the feather follicles and cut surfaces. The primary route into areas under the skin is through the body cavity openings (Sanders, 1969). Spoilage bacteria grow mainly on the skin surface, on cut muscle surfaces under neck flaps, and in the feather follicles. They are found in largest numbers on the neck skin, and to a lesser extent on the back and sites near the vent; fewer are found on the breast (Patterson, 1972b; Barnes, 1975). Therefore, results obtained will depend on the area sampled. Few microorganisms are found in the muscle tissues or internal organs. Deep muscle tissues are generally free of bacteria. Samples that contain meat only will give much lower counts than those that contain skin and meat, which in turn will be much lower than those that contain skin only.

c. Attachment of Microorganisms. During processing, particularly during picking, bacteria may become attached to carcasses (Notermans and Kampelmacher, 1974, 1975). Attachment depends on the presence of flagella and fimbriae, on microbial activity, on the temperature and pH of the water, and on the nature of the surface. Detachment occurs only when skin particles containing bacteria are removed by mechanical force. So, a sampling method that removes only the easily detached organisms will not reveal as many positive tests or show as high numbers as a method that removes both attached and unattached organisms.

d. Methods of Sampling Poultry Surfaces. Samples should be taken from surfaces of skin or of the skin itself to determine the number or presence of microorganisms in raw poultry. Differences of a 100-fold often occur between individual counts at different sites of the same carcass and of the same site on different carcasses (Patterson, 1972b). The variation is greater as the area sampled is diminished. Counts of pathogens show greater variability because they are absent from most areas of a carcass. So, as the area sampled is decreased, the frequency of positive tests for pathogens is lessened.

Poultry skin is difficult to sample because it is rough, somewhat greasy, and contains many feather follicles. Results must be interpreted according to the following methods of sampling:

1. *Swab.* There are several variations of the swab method. A sterile cotton or alginate swab is rubbed over the skin; then, the tip of the swab is broken into a tube containing either a sterile diluent (quantitative determination) or into an enrichment medium (qualitative determination) (Fromm, 1959). This method has been used frequently to evaluate the number of organisms on poultry surfaces and to recover salmonellae, clostridia, and staphylococci (Galton *et al.,* 1955b; Ayres *et al.,* 1956; Bryan *et al.,* 1968a,b,c; Patterson, 1972b; Gibbs, 1973). Variation of results is due to the nature of the surface; type of swab; the pressure on the swab; the duration, rotation, and speed of swabbing; and the number of times the same area is swabbed. People use swabs in different ways, so results are not necessarily reproducible among samples, laboratories, or different workers in the same laboratory. All bacteria are not removed by this method; only 38% of the aerobic cells were removed from a 16-cm$_2$ surface area of chicken skin with one swabbing, 43% with the second, 9% with the third, 8% with the fourth, and 2% with the fifth (Patterson, 1971). Alginate swabs dissolve and release all microorganisms, but have some antibacterial effect; and cotton swabs retain some cells, causing reduced counts. Thus, no differences in aerobic plate counts or Enterobac-

teriaceae counts or in isolation of *Salmonella* were found in comparative tests with the two types (Notermans *et al.,* 1976).

Modified swabbing techniques have been introduced to standardize results, namely, the use of sterile metal or paper templates to outline specific areas to be swabbed (Thomson *et al.,* 1976); standardized time of swabbing (i.e., 15 sec); and repeated swabbing of the same area.

2. *Whole carcass rinse.* Whole carcasses are shaken for 30 sec with approximately 1000-ml dilution water or enrichment broth in a sterile polyethylene bag (Surkiewicz *et al.,* 1969; Blankenship and Cox, 1976). Although chickens can be easily shaken with broth in these bags, it is somewhat difficult to sample turkeys by this procedure, owing to their size and weight. Unattached microorganisms are removed from internal and external surfaces by this method. The whole carcass rinse gave significantly higher numbers of salmonellae than either rinsing or blending neck skin from the same carcass (Cox *et al.,* 1978). An internal cavity spray rinse method has also been devized (Blankenship *et al.,* 1975).

3. *Drip.* Counts on the drip exuding from chilled or thawed whole poultry and parts shows a high correlation with swab counts (Mercuri and Kotula, 1964). Microorganisms originally present in the cavity and cut surfaces, as well as unattached organisms on the skin, are those most likely recovered by the drip method.

4. *Spray.* This method involves washing a circumscribed area of skin surface with a spray diluent. The front rim of a sample collector attached to a spray gun is pressed firmly against the test surface to make a liquid-tight seal, and a stream of washing solution impinges under pressure against the test surface. The washings flow into a collection tube. In one investigation, this method has given higher recoveries than those with blended skin or muscle, or than those with swab samples (Clark, 1965a,b) but not in another investigation (Patterson, 1971). Availability and calibration of spray guns are limiting factors. Microorganisms in feather follicles, and those firmly attached, are probably not removed.

5. *Macerated skin.* In this method, a weighed sample of aseptically sampled skin is blended with a buffered diluent for 1 to 2 min (Fromm, 1959; Avens and Miller, 1970a,b). Skin can also be mixed in a Colworth Stomacher for the same period. These methods break up clumps of bacteria and remove microorganisms present in feather follicles or attached to skin, as well as those on the surface. Maceration gives counts that are at least equal to, and generally higher than, those of rinse methods (Notermans *et al.,* 1975c; van Schothorst *et al.,* 1976). However, maceration is often inefficient because skin tends to retain bacteria. Shaking the skin with an abrasive material such as sterile, rough sand or glass beads in the diluent removes additional organisms (Cox *et al.,* 1976).

6. *Agar contact.* This method calls for pressing a sterile, nutrient agar surface against the surface to be sampled (Patterson, 1971). Rough surfaces, angular areas, heavily contaminated zones, and/or surfaces contaminated by spreading bacteria or molds are not suited to this technique. Microcolonies, not individual cells, are sampled so that counts will be lower than those obtained by swab, rinsing, or blending techniques. Agar contact is not suitable for sampling poultry carcasses but is useful for surfaces of cleaned equipment, walls, or floors.

7. *Skin scraping.* The area of skin is encircled with the open edge of a sterile cylinder or can. Then, 25 ml of 0.1% peptone water is poured into the container, and the meat surface is scraped with a sterile spatula to release the maximum number of organisms (Williams, 1967).

8. *Sampling comminuted poultry products.* The comminuted product is mixed, and sample units are blended with diluent and tested. The sample unit size varies (usually 10 to 50 g).

e. Culturing Conditions. The media used, the temperature of incubation, the type of atmosphere, and the duration of incubation determine which microorganisms are recovered from samples. For analytical methods, see ICMSF (1978).

IV. FROZEN RAW POULTRY AND POULTRY MEAT PRODUCTS

Whereas chilled poultry has a shelf-life of about 12 days, frozen poultry at or near −18°C will not spoil at all from microbial activity. Freezing can be by exposure to a continuous blast of cold air in tunnels or rooms, by immersion in cold brine, by exposure to still cold air, by exposure to liquified or solidified gasses (such as nitrogen or carbon dioxide), by plate freezing (usually prepared dinners), or by various combinations of these methods. Typical products are listed in Table 16.1.

A. Effects of Processing on Microorganisms

Freezing and frozen storage reduce the number of viable microorganisms on poultry; some are killed while others are only damaged sublethally (Chapters 1 and 12). At temperatures below −10°C, these sublethally damaged microorganisms die in time, but above this temperature, recovery can take place (Mulder, 1973).

Aerobic plate counts on poultry skin may decrease by 10 to 95% as a result of freezing; further death then occurs during frozen storage but at a

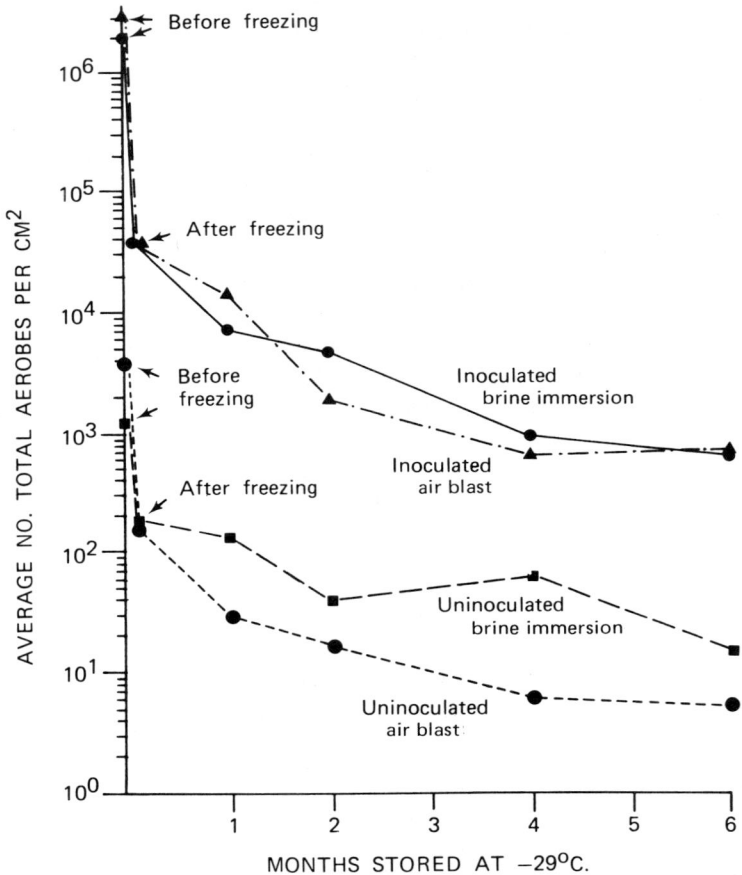

Fig. 16.7. Effect of method of freezing and frozen storage on total numbers of aerobic bacteria on turkeys. (From Kraft *et al.*, 1963.)

slower rate. Figure 16.7 demonstrates a decrease in aerobic plate count on the skin of turkeys by 84 to 98% during brine immersion and by 95 to 99% during air blast freezing. In the same investigation (Kraft *et al.*, 1963), fluorescing bacteria were reduced 99.9% by both methods of freezing. Coliforms, enterococci, and staphylococci were reduced 99, 97, and 96%, respectively, by the air-blast method. Another study, however, demonstrated no decrease in the aerobic plate count and only a slight decrease in the Enterobacteriaceae count when poultry was stored at −20°C (Notermans *et al.*, 1975d). The bacteria that were reduced in number by freezing were largely those in the water film around the carcass and not those attached to or in the skin (Notermans *et al.,* 1975d). Enough viable bacteria remain on the surfaces of frozen poultry to multiply and to produce spoilage when thawed and subsequently stored. Micrococci and enterococci survive well in frozen poultry for considerable periods of time (Straka and Combes, 1951; Wilkerson *et al.,* 1961). Freezing does not kill all salmonellae, and, therefore, they are frequently found on frozen poultry (Gunderson and Rose, 1948; Kraft *et al.,* 1963; Bryan *et al.,* 1968c; see also Table 16.7).

B. Spoilage

Frozen poultry does not usually present microbial spoilage problems while in the frozen state. Some yeasts and molds, however, can grow on frozen carcasses at temperatures as low as $-7°C$. Organisms commonly implicated are *Cladosporium herbarum,* causing black spots; *Thamnidium elegans* and *Thamnidium chaetocladioides,* causing whiskerlike growth; and *Sporotrichum carnis,* causing white spots. Properly stored products ($-18°C$ or lower) can spoil from microbial activity only before freezing and after thawing. If products are allowed to thaw and remain at refrigerator temperatures, the product becomes organoleptically unacceptable after a period of time because of growth of psychrotrophic microorganisms. The rate of spoilage of thawed poultry is the same as that of chilled poultry (Elliott and Straka, 1964) and is affected by the same factors (see Section III,B). From the bacteriological standpoint, the industry practice of freezing birds then thawing them for retail sale has a distinct advantage over using chilled birds because it puts all the chilled shelf-life at the retail level (Elliott and Straka, 1964). Enzymatic damage leading to off-flavor occurs in products that are stored frozen for prolonged periods (Khan, 1964).

C. Pathogens

Raw frozen poultry products harbor the same pathogens as chilled products and present the same problems of cross-contamination after they thaw. Thaw water is particularly hazardous from a public health standpoint because it contains those pathogens that were not attached to the skin. Many surfaces can become contaminated, particularly by salmonellae, from this source (van Schothorst *et al.,* 1976). Those organisms attached to the skin will be killed during cooking.

D. Control

The microbiological quality and safety of frozen poultry depends first on the factors previously described for chilled poultry (Section III,D,6). Chilled carcasses must be packaged, then frozen promptly, and held at or near −18°C. Carcasses must be thawed in such a way that decomposition is limited and growth of pathogens is prevented (Chapter 29, Section III,B). Both thawed poultry and thaw water can be sources of salmonellae, and should be isolated from contact with other foods.

Frozen products either can be thawed and then sampled like fresh products or sampled by aseptically chipping bits from the surface with a sterile meat cleaver or hammer and chisel. Larger blocks can be sampled by a sterile plug cutter with the aid of an electric drill (Barnes *et al.,* 1973). Measured amounts of these frozen pieces or drillings are blended with diluent or enrichment broth for microbiological analyses. Tests to identify salmonellae in thawed poultry products should be done by the rinse or drip-water methods.

V. HEAT-PROCESSED POULTRY MEAT PRODUCTS

Various heat-processed poultry meat products are produced by several heating methods (Table 16.1). These include retorting, water-bath pasteurizing, baking, boiling, barbecuing, and frying. Raw poultry products are either packaged and then heat-processed or heat-processed and then packaged. The type of storage required depends upon the intensity of the heat process.

A. Effects of Processing on Microorganisms

Most of the microorganisms in poultry products destined for cooking come from the raw chilled poultry meat itself. However, spices, unless they have been sterilized, may contain as many as 10^6 bacteria/g, primarily spores, including *Clostridium perfringens* and *Bacillus cereus* (Kinner *et al.,* 1968; Lillard, 1971; see also Chapter 24). The vegetative cells from the poultry will usually die in the cooking, whereas the spores from the spices may not. Therefore, the spices may contribute more than the meat does to the microbial content of the final cooked product.

Canning operations are similar to those of other nonacid products. In some plants, the poultry meat is packed raw into cans with heated broth; in others it is cooked first, then packed into cans. The cans are hermetically sealed under vacuum and retorted at temperatures near 115°C for sufficient time to achieve "commercial sterility" and then cooled. The time–temperature combinations used for retorting should destroy all vegetative cells and all spores that can germinate and multiply during subsequent storage.

Poultry meat is sometimes put into water-impermeable plastic bags and heated in water baths for several hours. For instance, immersing 9-pound turkey rolls in water baths at temperatures of 62.8°C and above for 4 hr reduced *Salmonella senftenberg* 775W from an original level of 10^6–10^7/cm^2 to a nondetectable level (Bryan *et al.,* 1968c). During these cooking procedures, the temperature of the geometric center of a roll was at or above 60°C for at least 40 min. The same result was obtained when the roll was heated in 76.7° and 82.2°C water baths until the geometric center reached 65.6°C. Salmonellae, however, were isolated from the same product cooked in water baths at 54.4° and 60°C for 4 hr.

Salmonellae were not isolated from 37 turkey rolls that were commercially processed in water baths (Bryan *et al.,* 1968c), nor from 38 frozen (cooked) turkey rolls, turkey in gravy, and turkey specialty products (Zottola and Busta, 1971). Staphylococci were not recovered from turkey rolls cooked to 71.1°C (Woodward, 1968). However, some turkey rolls are opened for addition of spices, then closed and shipped chilled. During this handling, contamination can occur (Kinner *et al.,* 1968). Vegetative cells of *Clostridium perfringens* were killed in chicken thighs and breasts cooked in water at 82°C for 20 min and at 93°C for 15 min. Heat-sensitive *C. perfringens* spores were reduced to low levels after 43 min at 82°C and killed after 38 min at 93°C. Heat-resistant spores, however, were not reduced in number after 50 min at 82°C. They were reduced, but not eliminated, after 45 min at 93°C (S. E. Craven *et al.,* 1975).

Vegetative bacteria and some but not all spores on poultry surfaces should be killed during baking, but those in internal portions may survive, depending on time–temperature exposure. For instance, no viable salmonellae or *Staphylococcus aureus* were isolated from inoculated turkey rolls that were cooked in an oven to a temperature of 65.6°C or higher at the geometric center (Wilkinson *et al.,* 1965). *Streptococcus faecalis* did not survive 71.1°C. Baked products are subject to postprocess contamination during slicing and packaging.

Barbecued poultry is heated on a grill or in rotisseries often at lower temperatures but for longer periods of time than those applied during baking. The lethality to microorganisms is similar to that of baking. High aerobic plate counts are frequently found after the barbecued products have been contaminated then stored for several hours at abusive temperatures so that growth occurs (Pivnick *et al.,* 1968; Seligmann and Frank-Blum, 1974).

During the frying of chicken parts in cooking oil, temperatures at the geometric center usually reach 93 °C or higher, which would be lethal to vegetative stages of bacteria but not to all spores. These products are also subject to postprocess contamination during meal preparation and packaging.

Poultry products that are not already packaged when they are subjected to heat treatment can become contaminated during subsequent handling. Contamination can come from workers who have salmonellae or other microorganisms on their hands from handling raw poultry carcasses or parts; or who are intestinal, respiratory, or skin carriers; and who have poor personal hygiene or food-handling habits. Contamination can also come from equipment.

B. Spoilage

Thermally processed poultry products should have low aerobic plate counts. Shelf-life is related to the time–temperature values achieved during processing, the extent of postcooking contamination, the type of packaging, and temperature and time of storage. For instance, after 20 days of storage at 4° to 6°C, psychrotrophic microorganisms in barbecued chicken increased from $< 10^2$ to $10^5/g$; at 0 to 2°C, they increased only slightly. These counts increased to $10^6/g$ after storage for 6 days longer in either temperature range (Mulder and Gerrits, 1974).

C. Pathogens

Thermally processed poultry products should be free of salmonellae, whose presence indicates process failure or postprocess contamination. Such contamination can take place if cooked poultry is processed further with the same equipment or in the same area as raw products, or if workers who previously handled raw poultry handle cooked product without washing and sanitizing their hands.

Cooked poultry products can become contaminated with *Staphylococcus aureus* during slicing, salad making, or packaging. Workers are an important source of staphylococci in cooked products, although cross-contamination from raw poultry products also occurs (Gibbs *et al., 1978b*). Such contamination, followed by inadequate cooling at processing plants or during subsequent storage, can lead to outbreaks of staphylococcal food poisoning (Chapter 29).

Cooking kills competitive organisms but allows heat-resistant *Clostridium perfringens* spores to survive; it drives off oxygen, thus lowering the redox potential of the meat and skin; and it heat-shocks spores, causing them to germinate when temperatures become favorable. *Clostridium per-fringens* contamination can occur either during handling after heat process-ing or from spores that survive heat processing. Tests for *C. perfringens* in cooked poultry products indicate the extent to which they have survived heat processing and the amount of subsequent contamination and multi-plication.

D. Control

To destroy vegetative cells of pathogens such as *Salmonella,* the prod-uct should be heated until the center of the thickest part of the food reaches 68.3° to 73.9°C, or until the center reaches 60°C and remains at that temperature for several minutes. In turkeys and large chickens, the temperature at the center often continues to climb for a few minutes after heating stops.

To sample cooked products, 25 g or larger portions of the food are usually removed aseptically, as described in ICMSF (1974). A high level of coliforms or *Enterobacteriaceae* or a high aerobic plate count indicates contamination occurred, possibly coupled with storage abuse that per-mitted growth.

VI. DEHYDRATED POULTRY MEAT PRODUCTS

Several methods have been used to prepare dehydrated poultry meat products (Mountney, 1976). Finely comminuted chicken meat has been sprayed-dried for use in soups. Chunk-size poultry meat has been dried in conventional air driers, and thin layers of ground cooked meat, in ovens or in vacuum chambers. Poultry meat products have also been dried on heated rollers and by heating in edible oil.

In preparation of freeze-dried poultry meat, carcasses are skinned, boned, and then cooked. After the meat has drained and cooled, it is diced, then frozen in a wind tunnel or sharp freezer, and put into a chamber under vacuum where the temperature is raised so that sublima-tion occurs. The pressure is then equalized with nitrogen, and the product is immediately packaged in an oxygen- and moisture-impermeable film or can.

A. Effects of Processing on Microorganisms

Some microorganisms are killed during heating in ovens and oil, but the more sophisticated drying methods are designed to preserve cellular structure, and most microbes will survive. Few are killed during spray-drying, for instance; many are destroyed during the cooking step preceding drying; some die during the freezing stage of freeze-drying. Postprocessing contamination during packaging is usually not a serious problem.

B. Spoilage

Dried poultry meat containing < 10% moisture and packaged in water-impermeable material will not support microbial growth and can be held for prolonged periods at ambient temperature. If the moisture content is slightly above 10%, molds and possibly yeasts may grow and spoil the product.

C. Pathogens

Spores of *Clostridium perfringens* survive cooking, freezing, and drying, and they germinate and multiply when reconstituted products are improperly stored. Cooked and dehydrated products can be contaminated by salmonellae, staphylococci, and other microorganisms if they are processed in an environment where cross-contamination from raw poultry can occur.

D. Control

Microorganisms are controlled by cooking, by the low water activity of the product, and by packaging that keeps out moisture and microbial contaminants.

VII. CURED AND SMOKED POULTRY MEAT PRODUCTS

Turkeys, geese, cornish game hens, pheasants, ducks, and chickens are sometimes smoked and sold as gourmet foods. Products include whole birds, breast meat, rolls, slices, sausage, and spreads. Among the usual curing agents are salt, sugar, ascorbic acid, nitrite, nitrate, polyphosphate, spices, and water (Chapter 8). The carcass or part is either immersed in brine or injected with brine pumped, through hollow needles into the muscle. After they remain in the curing brine 18 to 24 hrs (or after needling), carcasses are soaked in cold water for a few hours and then drained. Meats intended to be "ready-to-eat" are held in a smokehouse at 76.7° to 85°C until the internal temperature reaches 71°C. Products intended to be cooked before consumption may be smoked at lower temperatures. The finished meat is then packaged and refrigerated.

A. Effects of Processing on Microorganisms

Curing selectively alters the microbial flora of poultry products, favoring the survival of micrococci and fungi. The heat generated during smoking usually kills most non-spore-forming microorganisms that are present on the surfaces of poultry, but heating failures, mild heat treatments, or cold smoking procedures permit survival. The surface also dries, thus reducing the water activity. Wood smoking deposits phenolic compounds and acetic acid on surfaces and lowers the pH of the surface layers.

B. Spoilage

Unpasteurized, cured, and smoked poultry products have a relatively short shelf-life and will be spoiled by molds if stored in gas-permeable films. Vacuum-packed, smoked chickens stored at 1°C, had a shelf-life of about 84 days when pasteurized at 70° or 80°C and of about 104 days when pasteurized at 99°C (Mulder, 1974).

C. Pathogens

Curing in brine containing 9 to 10% sodium chloride inhibits the growth of all foodborne pathogens except *Staphylococcus aureus* and toxigenic molds. Milder cures, however, are often used for poultry meat products, so other pathogens can sometimes grow if present. *Clostridium perfringens* can survive the temperatures reached during smoking. Therefore, a low storage temperature is essential (Mulder, 1974).

D. Control

Microorganisms are controlled in cured and smoked poultry products to some extent by a reduced water activity and by the effects of the elevated temperatures achieved during smoking. The main function of curing and smoking poultry, however, is to provide flavor, not to preserve. Pasteurization of vacuum-packed, smoked chickens for 20 min at 70°, 80°, or 99°C substantially increases shelf-life (Mulder, 1974). Products should be stored at or below 6°C.

VIII. IRRADIATED POULTRY MEAT PRODUCTS

At the time of publication of this book, irradiation of poultry is not practiced commercially, but irradiation of chicken at a maximum dose of 700 krad has been given unconditional acceptance by a FAO/IAEA/WHO

Expert Committee (1977). Packaged poultry can be irradiated either when chilled or frozen. Products irradiated at this dose must be kept refrigerated or frozen because it gives only a pasteurization (radurization) treatment. Toxicological data do not indicate any health hazards resulting from the ingestion of irradiated chicken (FAO/IAEA/WHO Expert Committee, 1977) (see Chapter 3).

A. Effects of Processing on Microorganisms

Gamma rays from Co [60] or Cs [137] or fast electrons of up to 10 MeV energy can be used to irradiate packaged poultry products. A dose of 200–700 krad prolongs shelf-life, and a dose of 500–700 krad significantly reduces the number of pathogenic microorganisms. A dose of about 100 krad is required to effect a 10-fold reduction in numbers of Entero-bacteriaceae on broiler carcasses (Mossel et al., 1968; Mulder, 1975). The same dose results in a 10-fold reduction in numbers of salmonellae, and a dose of 250 krad results in a 25-fold reduction of these organisms on frozen carcasses (Mulder, 1975).

Irradiation is less effective in killing salmonellae on frozen carcasses than on chilled carcasses (Matsuyama et al., 1964). Freezing, however, reduces salmonellae by a half-$\log_{10}$ cycle. If frozen products are to be irradiated, 500 krad or larger doses should be given (FAO/IAEA/WHO Expert Committee, 1977).

B. Spoilage

Following irradiation and storage of poultry in refrigerators, Moraxella becomes the primary flora at the point of spoilage. Enterococci possess considerable resistance to irradiation and are also found in large numbers in spoiled products. In packages in which the atmosphere is anaerobic, lactic acid bacteria are an important cause of spoilage. Changes in color, odor, and flavor have been reported as undesirable effects in irradiated poultry (above 700 krad) treatments. These deleterious changes are less for irradiated frozen carcasses.

C. Pathogens

Radiation doses of 500–700 krad are likely to reduce salmonellae on poultry by a factor of approximately 10,000 (International Atomic Energy Agency, 1968). This should be sufficient for control because there are usually less than 100 salmonellae per gm of skin or per 500 ml of thaw water (Mulder et al., 1977b). Thus, irradiation (if costs can be sufficiently reduced) offers a method of eliminating salmonellae from raw poultry carcasses. *Clostridium perfringens* gastroenteritis becomes of concern if the subsequently cooked product is not properly stored, since *C. perfringens* spores survive pasteurizing doses of irradiation.

D. Control

Packaging before irradiation will prevent subsequent contamination. Storage of poultry at 6°C or lower will prevent multiplication of most surviving microorganisms, including most pathogens, and will inhibit the outgrowth of spores.

Tests for the presence of salmonellae or Enterobacteriaceae should be performed because these organisms are more resistant to irradiation than other non-spore-forming organisms and indicate a direct or indirect health hazard.

IX. CHOICE OF CASE

For an understanding of "Case" see Table A.1 and the brief description in Appendix IV. For more details, see the second book of this series (ICMSF, 1974).

Fresh or frozen poultry is invariably cooked before it is eaten, therefore, all microorganisms except spores are destroyed. Thus, for indicating spoilage and predicting shelf-life of raw poultry, an aerobic plate count incubated at 0° to 5°C, or even at 17° to 25°C, would be useful and Case 1 would apply. To determine the degree of fecal contamination of a freshly processed carcass, the coliform group of Enterobacteriaceae would be of value, and Case 4 would apply. To test for the presence of *Salmonella,* Case 10 would apply.

Ready-to-eat cooked poultry products such as turkey rolls, if held chilled will support growth of spoilage organisms; therefore, aerobic plate counts at 17° to 25°C would be useful (Case 3). Fecal indicators can be useful to evaluate the sanitation of preparation, but will not grow if product is properly refrigerated (Case 5). Staphylococci or salmonellae introduced into poultry rolls during addition of spices after the cooking step would not grow under refrigeration, and Cases 8 and 11 would apply, respectively. Broiled chicken, frequently held warm at the retail level, could support growth of spoilage organisms (Case 3), indicator organisms (Case 6), staphylococci (Case 9), or salmonellae (Case 12).

Fully cooked dehydrated poultry products in the normal course of events would be neither further cooked nor reconstituted and then abused.

Therefore, tests for spoilage organisms (Case 2), indicator organisms (Case 5), *Staphylococcus aureus* or *Clostridium perfringens* (Case 8), or *Salmonella* (Case 11) would apply.

Irradiated poultry products require Cases that are appropriate for raw poultry products. Case 1 is used for evaluating spoilage and Case 10 for determining whether salmonellae are present (ICMSF, 1974). Case 4 becomes appropriate when Enterobacteriaceae are used as an indicator test; some genera in this group, e.g., *Enterobacter* spp., are more irradiation resistant than *Salmonella*.

17

Feeds of Animal Origin
and Pet Foods

I. INTRODUCTION

Animal byproducts, widely used in animal feeds, are direct sources of disease organisms for animals, and indirect sources for human beings (Beasley *et al.,* 1967; Ladiges and Foster, 1974; Thornton, 1972). *Salmonella* is the principal genus of concern.

II. MEALS DERIVED FROM WARM-BLOODED ANIMALS

In past centuries, diseased animal carcasses were stripped of their hides then fed directly to dogs or other animals, or they were buried without treatment in shallow graves that were easily located by scavenging animals. These practices caused outbreaks of several types of animal disease, and farmers turned to burying carcasses in quicklime. Still later, when it was discovered that bacteria caused disease, the cooking or high-temperature rendering of carcasses and offal came into general use.

Meals are prepared from meat, offal, bones, blood, or feathers, or combinations of these. They constitute the principal protein source in mixed feeds. Their use in fertilizer is now a thing of the past, in view of their recognized value in animal feeds. Some meals, prepared from healthy carcasses, are ingredients of various sausages for human consumption. Bones or bone meals from healthy animals are used for making gelatin.

A. Important Properties

Meat bone meal contains about 60% protein, 12% fat, 22% ash, and 5% water. For feather meal, the respective percentages are about 85%,

5%, 1%, and 10%. These moisture levels are too low to support growth of even the most halophilic microorganisms. If the meal becomes wet, rapid multiplication of various species will take place, depending on a_w, temperature and other factors described in Volume I.

B. Initial Microflora

The initial microflora in the manufacture of animal and poultry meals is that of the meat (Chapter 15) or the poultry (Chapter 16) used for their manufacture, except that the levels of microorganisms are usually much higher on these "inedible" meat products. They frequently contain large numbers of pathogenic bacteria, fungi, viruses, or parasites.

C. Effects of Processing on Microorganisms

In most countries, carcasses of animals that die other than by slaughter, or that are diagnosed as diseased at slaughter, must be incinerated or cooked under steam pressure above 120°C. This kills all microorganisms, including spores. In other parts of the world diseased carcasses are not so treated, and *Bacillus anthracis* has been isolated from samples of bone meal (Morehouse and Wedman, 1961; Davies and Harvey, 1972).

Hot rendering, generally between 115° and 150°C, produces fats and the high protein meals that are the subject of this discussion. The temperatures of hot rendering usually destroy all microorganisms (Hess *et al.,* 1970). However, after the heating process there are many opportunities for recontamination, *i.e.,* during cooling, draining, drying, milling, screening, sorting, mixing, storing, and bagging (Tittiger, 1971). Because the product is destined for animal consumption, health authorities and the industry traditionally have been careless of sanitation principles that are a matter of course in processing of human food. As a result, microbial levels in the finished product are frequently high; in one study the aerobic plate count was 3×10^4 to 3×10^6; sulfite-reducing clostridia, 10^4; Enterobacteriaceae, 10^3; enterococci, 10^2 to 3×10^3; and fungi, 10^3 to 10^4/g (Milanović and Beganović, 1974). But the most serious problem is the post-heat contamination with *Salmonella* that can occur at the various steps previously listed.

D. Spoilage

Spoilage is not a problem with the various meals, because the a_w is too low for the growth of microorganisms. If the product becomes wet, it will spoil, most commonly by molds capable of growth at relatively low a_w. Mycotoxins could form from such growth (Gedek, 1973).

E. Pathogens

Salmonellae can grow wherever there is enough moisture and a trace of nutritive material. They then may drop into the dry meal where they survive but do not grow further. Such contamination can occur at any of the various steps after the hot rendering—cooling, draining, drying, milling, screening, mixing, storing, and bagging (Gray et al., 1960; Orthoefer et al., 1968). Percolators are an especially important source of salmonellae in certain types of rendering plants (Timoney, 1968). The organisms can grow and establish a focus of infection in the processing plant wherever there is enough moisture and a trace of nutrients. They cannot grow in the dried meal because of low a_w but can remain viable indefinitely.

Table 17.1 lists a number of surveys of the incidence of salmonellae in various animal by-products and mixed feeds. There is wide variation in the reported incidence, due in part to real variation, but also in part to the size of sample and method of analysis (see ICMSF, 1974; Tompkin and Kueper, 1973; Reusse et al., 1976b). Salmonellae in dried products of this kind are injured, and therefore require a period of resuscitation such as a preenrichment step affords (see also Chapter 12).

Reports have described the spread of specific serotypes of salmonellae from one rendering plant to the animal population throughout a whole region, country, or even the whole world (Rowe, 1973).

Intensive efforts toward improved hygiene will reduce Salmonella contamination. One important step is to maintain complete physical separation between the area where carcasses and other raw materials are unloaded and prepared for rendering (the "unclean" area) and the area where only heated material is handled (the "clean" area). But it is expensive to alter existing plants not designed originally with this in view. Also, the "unclean" area can be so highly contaminated that normal hygienic practices are inadequate to prevent the entry of Salmonella into the "clean" area. Even when there is no physical contact, aerosols, flies, and dust serve to inoculate pockets of condensate or other moist spots in the "clean" area in or near the processing line (Clise and Swecker, 1965). Or dust and flies can directly contaminate the flow of dried product (Quevedo and Carranza, 1966; Loken et al., 1968).

Heating the final product is an alternative that effectively destroys the organisms (Kampelmacher et al., 1965; Carroll and Ward, 1967; Nape and Murphy, 1971). Just exerting physical pressure on the meal, as in pelleting, creates enough heat to reduce the level of Salmonella 100-fold to 10,000-fold (Edel et al., 1966; Galton et al., 1955a; Kielstein et al., 1970; Stott et al., 1975). Expansion and extrusion are even more efficient (Crane et al., 1972). Ionizing irradiation (Dammers et al., 1966; Mossel et al.,

TABLE 17.1

Incidence of *Salmonella* in Feeds of Animal Origin

Material	Number of samples analyzed	Positive samples (%)	Reference
Dog food	98	26.5	Galton *et al.* (1955a)
Dog food	143	19	
Feather meal	524	6.8	
Fish meal	164	11	Morehouse and Wedman (1961)
Poultry by-products	141	33.3	
Tankage	316	14.6	
Meat meals	55	47.3	Leistner *et al.* (1961)
Animal by-products	61	77.0	
Pet food	111	0	Galbraith *et al.* (1962)
Fish meal	8271	0.08	
Whale meal	217	1.4	Karlsson *et al.* (1963)
Meat meal	95	11.6	
Animal by-products	71	60.6	Clise and Swecker (1965)
Blood meal	29	10.3	Moyle (1966)
Animal by-products	193	10.9	
Animal by-products	869	31	Allred *et al.* (1967)
Fish meal	805	4.7	
Meat and bonemeal	224	26	Timoney (1968)
Animal by-products	1395	17	Loken *et al.* (1968)
Animal by-products	1379	30.7	Pomeroy *et al.* (1969)
Fish meal	111	36	Morris *et al.* (1970)
Animal by-products	301	9	Tittiger and Alexander (1971)
Meat and bonemeal	982	21.3	Skovgaard and Nielsen (1972)
Feather meal	99	27	
Fish meal	30	3.3	
Meat and bonemeal	242	7	
Blood meal	36	5.6	Patterson (1972c)
Poultry offal meal	101	9.0	
Feather meal	414	7.2	
Animal by-products	183	81	Tompkin and Kueper (1973)
Meat and bonemeal	82	21	Stott *et al.* (1975)
Fish meal	1470	5	Reusse *et al.* (1976b)

1967; Münzner, 1974) and fumigation with ethylene oxide (Tucker *et al.*, 1974) will also destroy *Salmonella* in these products.

F. Control

To eliminate *Salmonella* from dried animal by-products the "unclean" raw-material area should be physically separated from the "clean" finished-product area by a leak-proof wall or floor. The two areas should have completely separate equipment and personnel for processing and maintenance. The processed-material area should be kept scrupulously clean (Volume I, Chapter 14) and free of flies, rodents, and other vermin. Sweepings should be reprocessed. Persons who enter the "clean" area should don clean clothing, wash hands, and clean the soles of the shoes. Steam, water, and dust should be kept away from the processed material.

Samples of the finished product and the environment should be taken at frequent intervals for analysis for *Salmonella*. Positive findings should be followed by investigation to eliminate the source of the organisms. In the Netherlands, regulations require analysis for *Clostridium perfringens* and a check of the thermocharts, as measures of the effectiveness of the cook. In certain countries, a test for *Enterobacteriaceae* is used to locate critical control points along the processing line (Quevedo, 1965). The latter test can also be applied to the final meal or feed as a measure of cross-contamination and growth (van Schothorst *et al.*, 1966). Some control authorities prefer to examine for *Salmonella* itself instead of relying on indicator tests (Reusse *et al.*, 1976b). The risk of accepting lots that contain *Salmonella* depends largely on the method of sampling, as demonstrated in fish meal (Jacobs *et al.*, 1963).

To prevent growth of molds, and the possibility of mycotoxin production, the finished dried product should be protected from moisture.

III. MEALS DERIVED FROM FISH

Between 30 and 40% of the total world catch of fish is used to manufacture animal feeds. The greater tonnage comes from processing whole fish that are not suitable for human consumption because they are too bony, too oily, or otherwise unsatisfactory. In the United States, for example, the entire menhaden catch goes to the rendering plants. A secondary source is the waste from fish and shellfish operations.

Fish solubles or fish concentrates are either a by-product of fish meal or a primary product of enzymic digestion of entire fish. Fish oil, a valuable by-product of fish-meal manufacture, is sometimes diverted to human use. Lesser amounts of fish and waste are canned as pet foods.

A. Important Properties

Fish meals are used widely in animal rations for their high protein content (60–70%). The fat level varies but usually lies at about 10%. Fish meal contains all necessary nutrients for microbial growth except moisture, which generally lies between 7 and 10%.

B. Initial Microflora

Fish may harbor many different kinds of microorganisms, mostly reflecting the microbiology of the aquatic environment (see Chapter 20).

C. Effects of Processing on Microorganisms

The most widely used method for preparing fish meal is known as wet rendering. The whole or chopped fish is carried slowly by a screw conveyor through a heated vessel; at the same time, steam is usually injected into the mass. The cooked material then passes into a screw press from which about 50% of the water and most of the oil drains off. The press cake then passes into a very large cylindrical hot-air ("flame") rotary drier. The temperature of this fish mass is raised high enough (110°C) to dry the product to a final moisture content below 10%. The material cools during further processing. The meal is sometimes cured before grinding and bagging by stacking it in a shed to allow oxidation to proceed. In most cases, antioxidants such as butylated hydroxytoluene (BHT) or ethoxyquin are mixed with the meal as it leaves the dryer. Stabilized meal directly from the dryer passes through a hammer mill, which reduces the particle size, and then into bags or bulk storage. [In some parts of the world (e.g., Angola), the cooked, pressed fish are simply allowed to dry in the sun.] The press liquor that contains water, oil, and suspended solids is screened, and then, oil and stickwater are separated by centrifuging. Screened solids are added to the press cake, oil is cleaned and refined for industrial and food uses, and stickwater is acidified after hot holding to encourage protein breakdown, then reduced by evaporators (triple effect usually) to 50% solids. The resultant liquid, known as fish solubles, is sometimes added back to the fish press cake but at other times is sold as a liquid-feedstuff component. The heat treatment reduces the number of microorganisms to a rather low level (the actual number depending on the initial flora and the time–temperature combination used), but in any case, all Enterobacteriaceae are eliminated. During subsequent handling recontamination by Enterobacteriaceae and *Salmonella* may take place in varying degrees. (For a more detailed discussion of the effect of processes on microorganisms, see the analogous paragraph, Section II,C).

Milanović and Beganović (1974) found that fishmeal from five factories had a mean aerobic plate count of 10^6, a count of sulfite reducing clostridia of approximately 3×10^3, counts of Enterobacteriaceae and enterococci ranging from 10^2 to 10^4, and counts of molds from 10^2 to 10^5/g. Contamination with Enterobacteriaceae may differ widely among countries of origin as demonstrated by van Schothorst *et al.* (1966) and Reusse *et al.* (1976b).

An increasing amount of ensiled fish is being produced as an animal feed, and the extent of current interest in the product suggests that production will increase even more. Ensiling involves liquefying the fish under acid conditions so that the final pH is below 4.5, producing a bacteriologically stable product. In one system, the chopped or comminuted fish is mixed directly with mineral (e.g., sulfuric) or organic (e.g., formic) acid and allowed to liquefy at temperatures above $20°C$. Another process involves mixing a fermentable carbohydrate such as molasses or cereal meal with minced or chopped fish, inoculating with lactic acid bacteria (e.g., *Lactobacillus plantarum* or *Streptococcus lactis*) and allowing fermentation to proceed optimally at temperatures near $30°C$. In both procedures, oil is removed from the final product by skimming or centrifuging.

D. Spoilage

Fishmeal is a microbiologically stable product because its a_w is below that which will support growth. Thus, in most cases, microbial spoilage is not an important factor. Only if the product becomes wet (e.g., during transport or storage) will rapid multiplication of bacteria and deterioration of the product take place. There is no evidence of significant microbiological problems with fish oils, first, because they are unsuitable as growth media, and second, because they undergo extensive physical and chemical refining.

E. Pathogens

Fishmeal has been recognized as a source of *Salmonella* for animal feeds since the early 1950s, when several serotypes, such as *Salmonella agona,* were introduced into many countries by the importation of Peruvian fishmeal. Investigation showed that Enterobacteriaceae, including *Salmonella,* could be isolated from the product at all stages of processing

(Quevedo, 1965). The extremely poor sanitation of fishmeal-rendering plants has contributed to the spread of contamination. Table 17.1 lists the percentages of fishmeal samples found positive for *Salmonella* in several investigations.

If a fishmeal were to become wet, it would probably mold. Such a product could contain a mycotoxin (Gedek, 1973).

F. Control

Until recently, it was common to find the floors, walls, and equipment of fishmeal plants covered with a layer of fine meal, and fish caught in equipment or lying in corners. It was believed, in the past, that the heat of cooking and rendering would produce a "sterile" product, and that, therefore, good sanitation was unnecessary. Now, however, under pressure from regulatory agencies, and faced by buyer specifications that *Salmonella* be absent, fishmeal manufacturers in advanced countries, at least, have come to recognize that they are engaged in a type of food processing. Many have taken steps to protect the product from *Salmonella* contamination with varying degrees of success. Control procedures for fishmeal manufacture are essentially the same as those for animal by-products (Section II) and keeping the meal dry during transport and storage will prevent microbial growth.

IV. PET FOODS

Commercially processed pet foods include high moisture canned products, intermediate moisture "semimoist" (a_w 0.80 to 0.90) products, and dried products ($a_w < 0.60$). In some countries, raw animal offals are used directly as pet foods. These products will not be discussed, and it is also unnecessary to discuss the microbiology of canned pet foods, which should comply with the specifications that appertain to other low-acid canned foods. Indeed this is essential because human beings sometimes eat canned pet foods.

A. Important Properties

Dry and semimoist pet foods are mixtures of cereals and products of animal origin. On occasion these pet foods contain "feed grade" nonfat milk, dried yeast, and soybean meal. The major source of protein among the dry pet foods is rendered animal by-products; that among semimoist foods is raw animal offal.

B. Initial Microflora

Many raw materials are used in pet foods; for discussion of their microflora, see Chapters 15, 16, 20, and 23. Materials of low quality are used, and salmonellae are introduced frequently with products of animal origin.

C. Effects of Processing on Microorganisms

In dry meals the two major ingredients, cooked cereals and meatbone meal (or other products of animal origin) is sometimes mixed without heat treatment, resulting in microflora that are more or less a reflection of the initial flora of the various components. In most cases, however, the material is heated during an expansion–extrusion process in which the temperature rises above 100°C, killing all vegetative microorganisms including salmonellae. The intermediate moisture foods may or may not be expanded, and the temperatures achieved during processing vary. One of three processes is used: (a) a single-stage process in which the material is mixed and then passed through a cooker extruder, (b) a two-stage process in which preliminary pasteurization of meats and liquid ingredients is followed by cooking and extruding, and (c) a cold extrusion process that is preceded by pasteurization. In most processes, the highest product temperature reached is around 95°C. The required stability is achieved by a combination of low a_w, low pH, and preservative chemicals. The reduction of water activity is achieved by humectants some of which have additional antimicrobial activity (e.g., propane 1,2-diol, butane 1,3-diol, sorbitol, or diol esters). The microbial flora of the end product depends on the initial flora and the processing techniques that have been used. Because these can vary so greatly, it is difficult to describe the "normal" microbiological condition of these intermediate moisture pet foods.

D. Spoilage

As long as the water activity of dry or semimoist pet foods remains low, spoilage is no problem, but moistened and rehydrated foods will spoil rapidly due to fungal or bacterial growth.

E. Pathogens

Many of the ingredients of pet foods (e.g., animal by-products, nonfat dry milk, dried yeast, soybean meal) have been reported contaminated with salmonellae. Although the heat applied in the expansion and extrusion process is sufficient to destroy salmonellae, contamination of the finished product does occur, presumably from cross-contamination of the extruded or pelletized product from the raw ingredients. Table 17.1 lists the percentages of pet foods found contaminated by *Salmonella* in a few surveys.

Human beings can be infected when *Salmonella*-contaminated pet foods are brought into the kitchen. Although some people eat such foods, the greater risk is from cross-contamination to human foods in which the organisms can grow, and therefore, become more highly infective. Pets with clinical or subclinical disease, which they acquire as a result of eating contaminated pet food, can spread pathogens to man. Pet foods are commonly rehydrated at the time of feeding, then held for an extended period of time at temperatures that permit growth of microorganisms. Thus, the risk of infection in pets increases as does that to human beings (Morse *et al.,* 1976).

A case of *Salmonella havana* infection in a 2½ month-old child led to an investigation (Pace *et al.,* 1977). Twenty-five samples of dried dog food, representing four different manufacturers and two retail store brands, were examined for *Salmonella*. Serotypes of *Salmonella* were isolated from seven of eight bags of commercially processed dried dog food. Each of eleven samples, produced by one manufacturer, contained one or more serotypes of *Salmonella*. Eight of them contained *S. havana* strains that had antibiotic sensitivity patterns similar to or identical to those of nine of ten *S. havana* isolates recovered from the dog food, from the child, and from her mother.

Semimoist pet foods are not a serious health problem, because during the manufacturing process the product is cooked sufficiently to destroy enteric pathogens such as *Salmonella,* the water activity and pH are reduced, then humectants and preservatives are added so that growth of microorganisms is completely prevented. These foods are fed to the animal without rehydration, so that they do not support the growth of microorganisms, and remain safe.

Improperly processed canned pet foods could theoretically be a danger to elderly and impoverished people who eat them. However, when they are processed and prepared under good manufacturing practice such canned foods are commercially sterile and as safe to eat as canned low-acid foods meant for human use.

F. Control

In the production of dry unheated pet foods, monitoring and control of raw ingredients are essential to avoid health hazards. In the production of other pet foods, a safe product can be obtained by proper technology and good manufacturing practice, with particular attention to preventing recontamination, as described in foregoing sections (II, E; II, F; III, F).

V. CHOICE OF CASE

For an understanding of "Case" see Table A.1 and the brief description in Appendix IV. For more details, see the second book of this series (ICMSF, 1974).

Animal feeds and pet foods are not considered in ICMSF (1974). However, sampling and analysis are desirable; *Salmonella* testing is the most important, and indicators are next.

A. Meals Derived from Warm-Blooded Animals

In the normal course of events, meals from animal by-products will remain dry until fed to animals. Therefore, salmonellae will not grow but will remain relatively stable during storage, and Case 11 would apply. *Clostridium perfringens* or Enterobacteriaceae when used as indicators would warrant Case 5.

B. Meals Derived from Fish

In the normal course of events, meals from fish will remain dry until fed to animals. Therefore, salmonellae will not grow, but will remain relatively stable during storage, and Case 11 would apply.

C. Pet Foods

Dry pet foods frequently are rehydrated then permitted to stand at ambient temperature for several hours. Therefore, salmonellae, if present, would grow and become more infective, and Case 12 would apply.

Intermediate (semimoist) pet foods need not be rehydrated, yet they will not support the growth of microorganisms. Therefore tests for *Salmonella* would be Case 11.

18

Milk and Milk Products

I. INTRODUCTION

This chapter deals with the microbiology of dairy products only to an extent sufficient to give the reader some appreciation of the relationships of microorganisms to the products. Texts by Foster *et al.* (1957) and Hammer and Babel (1957) and other selected references cited throughout this chapter provide further information. Dairy products as vehicles for the transmission of diseases will be discussed primarily from the standpoint of current problems. A review by Snyder *et al.* (1978) of relatively recent outbreaks and the reviews by Marth (1969) and Minor and Marth (1972b) on salmonellae and staphylococci, respectively, in dairy products, will provide further detail. Similarly, processing technology will be treated briefly but in a manner that will assist the reader to understand the microbiology involved. More details may be found in the texts by Campbell and Marshall (1975) and Harper and Hall (1976). The microbiology of butter is included in Chapter 25. Compositional standards for various dairy products have been developed under the joint FAO/WHO Food Standards Program of the Codex Alimentarius Commission. Lists of these and of those still in various stages of development are available (Food and Agriculture Organization/World Health Organization, 1978).

A. Importance

Microorganisms are important in dairy products for three principal reasons:

1. Microorganisms produce desirable flavor and physical characteristics in many dairy products during their manufacture, e.g., in various cheeses, fermented milks, and cultured cream butter.

2. Dairy products may become contaminated with pathogens or microbial toxins and thereby serve as vehicles for the transmission of disease to humans and other animals. The nature of the organisms involved must be determined and measures for their control must be applied.

3. Many microorganisms are capable of causing off-flavor and physical defects in dairy products. The likely sources of these organisms and the factors .influencing their growth and destruction must be known. Furthermore, newly emerging technologies of food processing and handling must be checked to be certain that they will not lead to contamination and subsequent survival or growth of undesirable microorganisms.

B. Products Considered

1. *Raw Milk* is that obtained from the producing animal but not yet processed.

2. *Market milks* are pasteurized fresh milk products consumed primarily in fluid form, e.g., pasteurized, "sterilized," or ultra-high-temperature (UHT) treated milk, low-fat milks, skimmilk, flavored milks, and creams.

3. *Edible ices and ice mixes* (ice cream and ice cream mixes) are milk products intended for consumption in the frozen or partially frozen state. These include dried products to be reconstituted and frozen and liquid products to be frozen to yield edible ices.

4. *Concentrated milks* are those from which part of the water has been removed, e.g., evaporated milk, sweetened condensed milk, or concentrated milk for recombination and direct consumption.

5. *Dried milks* are dried whole milk, skimmilk, buttermilk, and whey. Low-heat and instantized milks are special forms of dried milks.

6. *Cultured or fermented milks* are fresh milk products intended for consumption after lactic fermentation, e.g., cultured buttermilk, yoghurt, and acidophilus milk.

7. *Cheeses* considered are unripened, soft, and semisoft ripened cheese, hard and hard-grating cheese, processed cheese, and cheese foods.

II. RAW MILK

A. Definition

The following discussion relates primarily to cow's milk (although much of it is applicable to the milk of sheep, buffalo, and goats) and is restricted to the handling of milk on the producing farm and while it is in transit to the processing plant. In many countries, milk is defined by law or regulation, and the composition of milk varies accordingly. Generally, it is stated to be the lacteal secretion, practically colostrum free, obtained from the milking of one or more healthy cows and containing at least 3.25% milkfat and not less than 8.25% nonfat milk solids. The composition of milk varies widely among breeds and even within breeds of cows. Milk for consumption in the fluid form (market milk or bottled milk) usually is standardized to at least 3.25% milkfat and not less than 8.25% nonfat milk solids by the addition of fresh or dried skimmilk, cream, or milk of higher or lower fat content. The gross composition of such standardized milk is approximately 3.25% milk fat, 3.2% protein (casein and lactalbumen), 5% lactose, and 1% ash. A more detailed outline of the constituents of milk is given by Campbell and Marshall (1975).

B. Important Properties

The composition of milk makes it an optimum medium for the growth of many microorganisms. However, freshly drawn milk does, to a limited extent, possess a "germicidal" or "bacteriostatic" property (Auclair and Berridge, 1953; Auclair and Hirsch, 1953). The level of inhibitory activity varies in milk from different animals and from different quarters of the same udder. However, this temporary retardation of growth is of little importance in modern milk-handling practices. Thus, growth of microorganisms in milk is inevitable unless it is frozen. Even though growth is slow at temperatures between 0° and 5°C, undesirable changes may be readily detectable in a few days. The extent of such changes will depend on the types of organisms present and their numbers. Moreover, heat treatments may kill some types of bacteria (e.g., certain psychrotrophs) but permit their enzymes to remain active and cause undesirable changes in products during storage.

C. Initial Microflora

Microorganisms in raw milk come from several sources that may be conveniently grouped as follows:

1. Interior of the Udder

Milk is retained in the udder largely by the combined capillary forces of the lacteriferous duct network and the sphincter muscle at the lower end of the teat canal. During the milking process, its removal from the udder is accomplished by hormonal influences assisted by intermittent pressure applied on the teat that forces milk through the teat opening. Since the interior of each quarter of the udder is not sealed from without, organisms having access to the teat may invade the teat opening and migrate to the interior. Inherent antimicrobial influences undoubtedly restrict both the number and types of organisms that make up the so-called normal microflora of the udder. In normal udders, organisms other than bacteria are rarely found. The number of bacteria in aseptically drawn milk varies widely from cow to cow and even among quarters of the same udder. Generally, they range from a few hundred to a few thousand per milliliter. Occasionally, some quarters yield milk free of detectable organisms. The species of bacteria likely to be present are limited to a few genera. Micrococci generally predominate followed by streptococci and the diphtheroid *Corynebacterium bovis.* Abnormal conditions in the udder resulting from infection and disease or poor milking practices may greatly affect the number and types that may be present.

Mastitis, an inflammatory disease of the mammary tissue, may lead to the development of millions of the infectious organisms per milliliter of milk or exudate from diseased quarters. Detailed discussion of this disease may be found elsewhere (Schalm *et al.,* 1971; International Dairy Federation, 1975; National Mastitis Council, 1978). Briefly, the more common causative types are *Streptococcus agalactiae, Staphylococcus aureus,* coliforms, *Pseudomonas aeruginosa,* and *Corynebacterium pyogenes (incertae sedis;* Buchanan and Gibbons, 1974). Types less common include *Clostridium perfringens, Mycobacterium* spp. (Tucker, 1953), *Nocardia asteroides* (Pier *et al.,* 1958) and *Mycoplasma* spp. (Schalm *et al.,* 1971). The numbers of organisms generally are highest in milk from animals in the early acute stage of the disease. Some animals suffer from chronic mastitis characterized by periodic acute and subacute clinical forms of the disease with accompanying fluctuations in bacterial numbers secreted in the milk. In advanced stages of untreated mastitis, the milk becomes progressively abnormal, eventually consisting of a serous, often bloody, exudate. At that point, few if any organisms can be detected. In milk from mastitic cows, the organisms may be easily observed microscopically in stained preparations where the organisms are often engulfed by polymorphonuclear leucocytes. The latter, while few in number in normal milk, increase markedly soon after infection, often exceeding several million per milliliter. Accordingly, the somatic cell count (including leucocytes) of milk (Ullman *et al.,* 1978) is commonly used as a screening test to detect abnormal milk in deliveries to processing plants.

In addition, sick cows may shed other bovine or human disease agents in the milk, e.g., *Mycobacterium bovis, Brucella abortus, Brucella melitensis, Brucella suis,* and *Coxiella burnetii.*

The streptococci found in normal-appearing milk drawn from cows showing no clinical symptoms of mastitis are restricted primarily to the mastitis streptococci: *Streptococcus agalactiae, S. dysgalactiae,* and *S. uberis.* Various factors such as injury to udder tissue by faulty milking practices account for sporadic occurrence of acute symptoms. The lactic streptococci—e.g., *Streptococcus lactis, S. cremoris,* and *S. lactis* subspecies *diacetylactis,* the organisms most commonly found in raw milk supplies—are not found in the udder.

Many species of micrococci have been reported in aseptically drawn milk (Breed, 1928; Abd-El-Malek and Gibson, 1948). The taxonomy and nomenclature of the aerobic gram-positive micrococci have changed over the years; aseptically drawn milk may yield strains belonging to all three species currently recognized in each of the genera *Micrococcus* and *Staphylococcus* (Buchanan and Gibbons, 1974). Heat-resistant (thermoduric) strains (see Section III) of certain species of *Micrococcus* have been isolated from the udder, occasionally in large numbers. Bryan *et al.* (1946) reported invasion of the udder by such organisms due to use of unclean milking machines. As many as 200,000 per milliliter of milk persisted over a long period, but after continued use of properly cleaned and sanitized machines, the animals gradually rid themselves of the thermoduric types in 1 to 4 months.

2. Exterior Surfaces of the Animal

Materials such as soil, bedding, feed residues, manure, and so forth, commonly found in the environment of the animal, are present on the surface of the udder, teats, and coat of the cow to a greater or lesser extent. Numerous microorganisms of various types are associated with these materials. Members of the genus *Bacillus* from the soil, clostridia from silage fed to animals, coliforms common in manure and soiled bedding, and other types easily find their way into milk. The contribution of these sources to the total numbers of microorganisms found in freshly drawn milk may range from fewer than 100 to several thousand per milliliter depending on the care in cleaning the animal surfaces prior to milking. Of greater importance are the *types* contributed from these sources. For example, certain clostridia present in large numbers in silage are the cause of serious gassy defects of hard tight-bodied cheeses such as Swiss or Emmentaler, Edam, and Gouda. Clostridial spores survive pasteurization and unless effective measures for their control are taken, serious economic losses may occur.

3. Milk-Handling Equipment

It is well recognized that equipment generally contributes the largest proportion of microorganisms found in raw milk supplies (Smith, 1919, 1920; Jensen and Bortree, 1946; Hoy and Rowlands, 1948; Churchill and

Mallmann, 1950; Dahlberg *et al.,* 1953; Richard and Auclair, 1967). Equipment used to handle milk on farms and in transport include various components of milking machines such as teatcup inflations, milk tubes and airline hoses, pails, strainers, milk cans, churns, surface coolers and other types of coolers, bulk-milk refrigerated cooling tanks, milk transport pipelines, tank trucks used to transport milk from farm to plant, and other equipment and accessories.

Milk residues left on surfaces after faulty cleaning provide ample nutrients for growth of many types of microorganisms. Ambient temperatures, at which such equipment remains between periods of usage, usually are favorable for growth. Furthermore, surfaces often remain wet or moist for long periods. If these conditions prevail they permit buildup of enormous numbers of microorganisms. During subsequent use of the equipment, the organisms pass into the milk. The types and numbers of organisms introduced from this source largely depend on the efficiency of cleaning and sanitizing (Thomas *et al.,* 1950). Occasional neglect of adequate cleanup results in rapid yet short-lived increases of the more rapidly reproducing types such as the lactic streptococci, coliforms, and certain other gram-negative rods of the genera *Pseudomonas, Alcaligenes Flavobacterium,* and *Chromobacterium.* These organisms are heat sensitive and are readily destroyed by sanitizers containing chlorine. Hence, continued and adequate cleaning effectively removes them from the surfaces and prevents their buildup. On the other hand, under conditions of frequent and persistent neglect, milk stone gradually forms on equipment surfaces. The more resistant and slower growing organisms such as micrococci, enterococci, and certain lactobacilli become embedded and persist in the milk-stone matrix, eventually reaching very large populations. It is not uncommon for bacterial counts of milk from inadequately cleaned and sanitized equipment to reach several hundred thousands per milliliter. Many of these organisms are thermoduric and therefore may become troublesome in pasteurized products.

4. Miscellaneous Sources

The air of the milking environment is relatively insignificant in terms of its contribution to the total microbial content of milk. Even hand milking into open topped pails rarely contributes more than 25 organisms per ml of milk (Ruehle and Kulp, 1915). Under extremely dusty conditions counts may increase somewhat. Far more important are the *types* likely to be added to milk from this source. Organisms commonly found in air are resistant forms. Micrococci and spores of the genera *Bacillus* and *Clostridium* are commonly found. Many of these survive heat processes and may cause flavor and physical defects in processed products.

Milk-handling personnel may contribute various organisms, including pathogens, directly to milk. Micrococci and staphylococci from skin and upper respiratory tissues may gain entrance especially during hand milking. Numerous milkborne outbreaks of typhoid, diphtheria, sore throat, scarlet fever, salmonellosis, and other enteric and diarrheal infections have been caused by raw milk contaminated by milk handlers and other personnel. The extent to which other sources become significant in contamination depends largely upon the diligence of individuals responsible for the care of animals, the milking operation, cleaning and sanitizing of equipment, and various other associated tasks.

D. Spoilage

Various undesirable and detectable organoleptic and physical changes in raw milk are caused by microorganisms (Hammer and Hix, 1916; Punch *et al.,* 1965). Souring by *Streptococcus lactis* was detected by taste at populations of 30–90 million per ml, and ropiness by *Alcaligenes viscosus** at 15–44 million per ml of milk. Slight coagulation, "sweet curdling," caused by an organism that produces a rennetlike enzyme occurred at 1.25–4.9 million per ml. Numerous other defects may occur, including malty, rancid, yeasty, bitter, fruity, and putrid flavors and purple and reddish colors. Generally, when actively growing types of organisms capable of causing changes in flavor and physical appearance (such as coliforms, *Alcaligenes, Flavobacterium* and *Pseudomonas*), reach population levels of 5–20 million per ml, change is evident or imminent.

E. Control

Prevention of microbial contamination of milk requires several control measures. Animals must be maintained in a healthy condition, and illnesses should be treated immediately, if necessary by a veterinarian. Regularly, at each milking, a few streams of milk from each quarter should be drawn onto a fine screen or the surface of a dark plate (component parts of a device known as a strip-cup) to detect small clots, which if present indicate abnormal secretion probably due to mastitis. Prompt antibiotic therapy usually will bring the infection under control. The udder and rear quarters of the cow should be cleaned just prior to milking to remove adhering extraneous materials. Teat surfaces should be carefully cleaned. Following milking the teats should be dipped in a "teat-dip" solution containing

* Reclassified as *Alcaligences viscolactis* in the Seventh Edition, but not recognized in the Eighth edition of Bergey's Manual (Buchanan and Gibbons, 1974).

about 1% iodine (iodophor). Since injury to udder tissue, particularly that near the base of the teat cistern, can easily occur due to faulty vacuum control during machine milking, frequent maintenance checks of the machines should be made.

Milk-contact surfaces of all equipment used in handling milk should be carefully cleaned with an appropriate cleaning solution as soon as possible after each use. Occasional use of an acid-type cleaner will remove milk-stone from equipment surfaces. Cleaned equipment should be arranged and stored to permit water to drain away. Immediately before use, equipment should be rinsed with a sanitizing solution. Chlorine and iodine sanitizing solutions are commonly used for this purpose (see Volume I, Chapter 14).

To avoid excessive airborne contamination, movement of materials such as bedding and forage should be delayed until after milking. Walls, ceilings, and floors of the milking area should be kept free of loose materials likely to be swept into the air by air currents. The hands and clothing of milk-handling personnel should be clean, and such personnel should be well versed in sanitary milk production practices and proper use and care of milking machines and other equipment.

Generally, the freshly drawn milk should be cooled immediately to 5°C or lower to slow the growth of microorganisms. Freezing should be avoided. If the initial number of psychrotrophs is kept to a minimum by adherence to sanitary milk-handling practices, milk can be safely held on the farm 2 days before their numbers reach troublesome levels. The growth of many mesophiles and thermophiles likely to be present will be completely inhibited in properly cooled milk.

Microbiological hazards of raw milk relate primarily to human disease and spoilage. The literature is replete with accounts of milkborne disease outbreaks due to consumption of raw milk. Experience has made it clear that measures short of pasteurization cannot be relied on to provide safe milk for human consumption. Consequently, except in local, remote, or some undeveloped areas of the world, fluid milk supplies available to the consuming public are pasteurized. Certified raw milk (see Section III) as sold in the United States also is an exception. Microbiological examination of raw milk measures the effectiveness of sanitation and good milk-handling practices. Some tests may be applied to detect early symptoms of mastitis. Routine microbiological examinations provide a continuing historical record of the bacterial counts of milk received from producing farms. Counts usually are obtained at weekly or biweekly intervals and in most legal jurisdictions must meet prescribed standards for the mesophilic aerobic plate count (standard plate count). Raw milk standards ranging from less than 100,000 to 200,000 per ml are not uncommon for raw milk destined for pasteurization and subsequent consumption in fluid form (market milk). The most useful tests are the aerobic mesophilic plate count and the thermoduric count; the latter is merely a plate count obtained after laboratory pasteurization of the sample (Vedamuthu *et al.*, 1978). In some countries, the standards that must be met for milk supplies used in manufacturing other dairy foods are less severe, generally ranging from 500,000 to 1 million per ml. In many countries, such distinction in bacteriological quality of milk for different uses does not exist. Often indirect measurements of bacterial numbers are used, such as the methylene blue and resazurin reduction tests (Flake *et al.*, 1978). There is always some margin between the standards specified and the numbers associated with defects. However, low populations can rapidly reach levels at which defects may be detected. Accordingly, it is only prudent to keep counts as low as practicable.

III. MARKET MILKS

A. Definitions

With certain exceptions, market milks are heated milks and creams intended for consumption in the fluid state. Their composition, usually expressed in terms of percent milk fat and percent non-fat-milk solids (NFMS), is commonly specified in laws or regulations of national and local governments and, therefore, varies accordingly. The figures in Table 18.1 are typical.

Some raw milk is sold commercially for fluid consumption. This practice

TABLE 18.1

Typical Composition of Market Milks

Product	Milk fat (%)	NFMS (%)
Whole milk	3.25	8.25
Skimmilk	0.5	8.25
Lowfat milk [a]	0:5–2.0	8.25
Flavored milks	0.5–2.0	8.25
Cream	18–20	—
Heavy cream	30–35	—
Whipping cream	35	—
Cereal cream	10–12	—

[a] It is common practice in some countries to add dried skimmilk to lowfat milks bringing their NFMS content to about 10%.

is not necessarily confined to areas where pasteurized milk is unavailable. Unfortunately, there still remains an unenlightened segment of the population (perhaps imbued with the belief that "natural is best,") that actually prefers raw milk, despite the preponderance of evidence showing that raw milk is a common vehicle for transmission of disease to humans as well as to other animals. "Certified" milk is the outstanding example of attempts to produce clean, safe and nutritious raw milk for direct consumption. It is produced in the United States under very strict conditions of sanitation in accordance with standards adopted by the American Association of Medical Milk Commissions (1976). In spite of the extreme and costly efforts to prevent all possible contamination of the product, outbreaks of illness transmitted by certified milk continue to occur. The most recent outbreak caused numerous cases of salmonellosis in California (U.S. Department of Health, Education and Welfare, 1974b). Where pasteurized milk is unavailable, or where refrigeration facilities are inadequate, it is common to boil milk to control both health and spoilage hazards.

B. Important Properties

As in the case of raw milk, market milk products are highly susceptible to microbial growth. To whatever extent the "germicidal property" (see Section II,B) may have been present, its influence for all practical purposes is eliminated by heat treatments given to these products.

C. Processing

When raw milk reaches the processing plant, the usual practice is to pass it through a centrifugal clarifier or cloth filter and then through a plate-type cooler into a storage tank where it is maintained at a temperature not to exceed 7°C until further processing. The clarification process removes suspended particulate matter. Some of the milk is separated into two fractions, cream and skimmilk, in a centrifugal separator similar in construction to that of a clarifier. Milk, cream, and skimmilk are used to prepare and adjust the composition of the various market milk products. This process is termed standardization.

Heat treatments (pasteurization or ultra-high temperature) eventually are applied. Homogenization of milk often is coupled with pasteurization. Basically the homogenizer is a high pressure pump that forces the milk through small orifices under high pressure. The fat globules, now reduced in diameter to about 2 μ or less remain in stable suspension throughout the milk. On the other hand, unhomogenized milkfat globules range in size from about 0.1 to 16 μ, and when milk is left to stand undisturbed they rise to form a compact cream layer several centimeters in depth. The sweeping action of the rising fat globule clusters carry many microorganisms upward into the cream layer.

Specifications for the time and temperature of heating and for construction and operation of the heating equipment differ somewhat, depending upon the requirements of the regulatory agency concerned and on the type of milk product. For example, specifications of the Food and Drug Administration (U.S. Department of Health, Education and Welfare, 1978f) for products labeled pasteurized and ultra-pasteurized [synonymous with the term ultra-high-temperature heat treatment (UHT) commonly used in the United Kingdom and Europe] follow:

Pasteurized
  145°F (62.7°C)–30 min
  161°F (71.7°C)–15 sec
  191°F (88.4°C)–1 sec
  204°F (95.6°C)–0.05 sec
  212°F (100°C)–0.01 sec

Ultra-pasteurized (i.e., UHT)
  280°F (138°C)–2 sec

The temperature requirement of the first two temperature–time specifications listed above must be increased by 5°F (2.8°C) if the product being pasteurized contains 10% or more of milkfat or if it contains added sweeteners.

In addition, an in-bottle or in-can heat process of about 105°–110°C for 30–40 min is common in the United Kingdom and Europe. The product so treated is labeled "sterilized," even though it is not sterile. Since the term has been used for many years, to change its designation would not be worth the confusion that would be likely to occur.

Following heat treatment, the products are cooled to 5°C or below, packaged in final containers by automatic filling or bottling machines, and then stored and distributed under refrigeration.

UHT milk (138°C–2 sec), aseptically packaged and distributed without refrigeration, has achieved considerable success in European markets. However, its shelf-life, often stated to be 3–6 months, would not be the same as that expected of a "commercially sterile" product, i.e., a product that is free of both pathogens and of nonpathogens capable of growing under normal ambient temperatures (without refrigeration) during storage and distribution (see Chapter 1, Section V,C,2). A UHT-heat treatment somewhat more severe than 138°C for 2 sec, coupled with aseptic packaging, would be required for milk to be considered commercially sterile. In any event, interest in such a product has stimulated considerable current research. Enzymes are particularly troublesome. Whereas most psychrotrophs are easily destroyed by the UHT treatment, their lipase and protease enzyme systems may not be. These may cause product deterioration during a long storage period (Kishonti and Sjöström, 1970).

D. Initial Microflora

The initial microflora of market milk products is that of the raw milk from which these products are prepared. The types of microorganisms likely to be present in raw milk as it arrives at processing plants, and the control measures used to minimize numbers present have been discussed previously (see Section II,C and E).

E. Effects of Processing on Microorganisms

1. Milk-Contact Surfaces of Equipment

The number of microorganisms likely to be added to milk will range from practically none from properly cleaned and sanitized equipment to many thousands from unclean surfaces. Such additions may increase bacterial counts (agar plate method) several-fold above those obtained on milk newly received at the plant. Furthermore, agitation such as pumping may increase the count, simply because it breaks up bacterial clumps into smaller but more numerous units capable of forming colonies on agar plates (see Section III,E,2).

2. Separation and Clarification

The separation process produces three fractions: skimmilk, cream, and sediment (separator or clarifier "slime"). Many organisms are physically removed from the milk and concentrated with other particulate matter in the slime, which contains millions of bacteria per gram. The bacterial counts on the cream fraction may often be higher than those on the skimmilk or whole milk (Leete, 1925). The total bacterial count of the skimmilk and cream fractions combined often exceed somewhat that of the whole milk which was separated. Since the slime fraction will contain large numbers of organisms, it becomes obvious that the combined population in the cream and skimmilk should be less than that in the milk, assuming that the separator was properly cleaned and sanitized. The explanation lies in the distinction between bacterial "content" and bacterial "count" as determined by the usual agar plate method. Agitation during separation (and clarification) breaks up bacterial clumps, thus increasing the number of bacterial colonies that will develop on agar surfaces. Separation and clarification, therefore, increase the count but not the bacterial content.

3. Pasteurization

Pasteurization heat treatments (see Section III,C) are the minimal required temperature–time combinations. However, in actual practice the temperature of exposure often is increased by several degrees, to be certain that the minimal heating is attained (see Chapter 1, Section V,C,4). The minimal pasteurization specifications of 62.7°C for 30 min and 71.7°C for 15 sec, commonly referred to as low-temperature-holding (LTH) and high-temperature–short-time (HTST), respectively, are sufficient to destroy pathogens likely to be present, as well as most spoilage organisms (see Sections III,F and G). The more severe "sterilizing" and UHT processes will destroy all but the most heat-resistant bacterial spores.

Low-temperature-holding and HTST treatments at or near the minimum will permit the survival of a rather large group of heat-resistant mesophilic types collectively known as thermodurics. These organisms will not grow at the pasteurization temperature but will grow when the temperature returns to within their growth temperature range. The most commonly encountered thermodurics in milk are species of *Micrococcus,* although certain enterococci (heat-resistant strains of *Streptococcus fecalis* and *S. faecium*), aerobic sporeformers (particularly *Bacillus subtilis* and *B. cereus*), and *Lactobacillus casei* may be present in large number (Thomas *et al.,* 1950, 1967; Deibel and Hartman, 1976). Thermodurics may represent the dominant microflora of raw milk supplies from farms where cleaning of milk-handling equipment has been neglected persistently. The relative ineffectiveness of moderate-pasteurization heat treatments in killing these organisms may make it difficult for market milk processors to meet bacterial-count standards for pasteurized products.

Generally, thermodurics are relatively inert and grow slowly in milk at 5°C, in contrast to psychrotrophic types; but if they are present initially in excessive numbers, sufficient growth may occur to cause spoilage well within 10 to 14 days of refrigerated storage.

4. Homogenization

Homogenization of raw milk markedly accelerates enzymatic (milk lipase) hydrolysis of milk fat thus causing rapid development of rancid flavor. Rancidity thus induced may be confused with rancidity induced by bacterial lipases. Homogenization has little effect on the microbiology of market milk products, except that it breaks up clumps of bacteria (see Section III,E,2).

5. Cooling and Bottling or Filling

The cooling, bottling, and filling operations can introduce microorganisms from equipment such as pumps, pipelines, valves, balance or surge tanks, and bottles or other final containers. The air also may contribute to a minor degree. The types added will vary depending largely upon the cleaning and sanitizing practices prevailing. For example, sporadic lapses in care of equipment favor development of rapidly growing gram-negative non-spore-forming bacteria such as *Pseudomonas, Alcaligenes, Chromobacterium, Flavobacterium,* and coliforms and other members of Enterobacteriaceae. Persistent neglect will encourage growth of slower-growing thermoduric types such as micrococci in the dried milk–water residues ("milk-stone") which build up on poorly cleaned surfaces. Aseptic packaging eliminates post pasteurization contamination.

F. Spoilage

Microbial spoilage of refrigerated market milk products is manifested primarily by development of off-flavors, often described as "unclean," "putrid," and "fruity". Physical changes such as ropiness and partial coagulation are less common defects. The intensity of flavor defects depends on the extent of microbial enzymatic decomposition of milk protein, fat, and to some extent lactose. Usually, the organisms involved are postpasteurization, gram-negative, psychrotrophic contaminants belonging to the genera *Pseudomonas, Flavobacterium, Chromobacterium, Alcaligenes,* and coliforms (Thomas and Sekhar, 1946; Erdman and Thornton, 1951; Thomas and Druce, 1969; Schultze and Olson, 1960a,b). This is in sharp contrast to the brief report of Credit *et al.* (1972), who found psychrotrophic bacilli to be the major types in pasteurized milks found spoiled after storage at 4.5°C for 30 days. The very active non-spore-forming psychrotrophic spoilage types might have been overlooked. These organisms, if present, would have reached peak populations early in the storage period and likely would have been overgrown by the bacilli after 30 days of storage.

The time required for flavor change to occur depends on the number and types of organisms present and the storage temperature. For example, a recent study by Hankin *et al.* (1977) showed the average time required for spoilage of HTST pasteurized milk when stored at 1.7°, 5.6°, and

10°C was 17, 12, and 6.9 days, respectively. Pasteurized milks from modern well-operated plants can be expected to have well over 10 days of shelf-life under refrigeration.

Recently, considerable attention has been focused on psychrotrophic types that survive pasteurization (Washam *et al.,* 1977; Weckbach and Langlois, 1977; Grosskopf and Harper, 1969, 1974; Shehata and Collins, 1971, 1972). Species of *Bacillus* are the most frequently encountered; *Bacillus cereus* is particularly troublesome in milk and cream because of its lecithinase enzyme that acts on the phospholipids of the fat globule, forming small proteinaceous fat particles (Stone, 1952). These adhere to surfaces of glasses, cups, etc., and present an unsightly appearance. The defect is known as "bitty." Flavor defects occur as well due to this organism. The thermoduric psychrotrophs generally grow under refrigeration much more slowly than do those that are heat sensitive. Often, defects become evident only after several weeks. However, if higher temperatures of storage prevail, the "bitty" and various flavor defects may occur in 10 days or less. "Sterilized" products often are involved (Grosskopf and Harper, 1969; Mourgues and Auclair, 1973).

G. Pathogens

Pasteurized market milk products present little or no hazard of milk-borne disease. Minimum-pasteurization heat treatments specified by laws or in regulations of government authorities generally allow sufficient margins of safety to assure destruction of pathogens likely to be present in raw milk. Numerous studies that support this conclusion are listed in Table 18.2.

Although pasteurization effectively destroys pathogens of concern in milk, the enterotoxins of *Staphylococcus aureus* easily withstand such treatment. Also, while pasteurization effectively destroys most pathogenic viruses in milk, there are a few exceptions. For example, some foot-and-mouth-disease virus in naturally infected milk survived 72°C for 15–17 sec (Blackwell and Hyde, 1976; Dhennin and Labie, 1976). In the study by Blackwell and Hyde (1976), at least 10^4 reduction of active virus particles occurred; residual activity remaining depended on the initial level of virus.

While foot-and-mouth-disease virus causes occasional skin infections in humans (Larkin, 1973), primary concern centers in preventing the infection of cattle. Milk from infected cattle may contain large numbers of the virus and thus can be a significant vehicle in its transmission; hence, the efficacy of pasteurization against the virus becomes important. Recently,

TABLE 18.2

Research Studies Showing That Pasteurization Destroys Pathogens in Milk.

Pathogen	Reference
Mycobacterium tuberculosis	Harrington and Karlson (1965)
M. bovis	
M. avium	
M. fortuitum	
Salmonella typhi	Tanner and Dubois (1925)
Shigella paradysenteriae	Daoust *et al.* (1959); Tanner and Dubois (1925)
Escherichia coli	Read *et al.* (1961)
Streptococcus pyogenes	Weber (1947)
Corynebacterium diphtheriae	Daoust *et al.* (1959)
Brucella abortus	Foster *et al* (1953); Park *et al.* (1932)
B. suis	
B. melitensis	
Staphylococcus aureus	Zottola and Jezeski (1969)
Coxiella burnetii	Enright *et al.* (1956); Enright (1961)
Poliomyelitis virus	Kaplan and Melnick (1952, 1954)
Coxsackie viruses	Kaplan and Melnick (1952, 1954)
Oncogenic viruses	Sullivan *et al.* (1971)

the effect of UHT pasteurization exposures on foot-and-mouth-disease virus in artificially infected milk was determined (Cunliffe *et al.,* 1979). Treatments at 130°, 138°, and 148°C for 2–3 sec yielded milk that did not infect cell cultures. However, cattle infection occurred from milk treated at 130°C and in one of five trials from milk treated at 138°C. No infections occurred from milk treated at 148°C. The authors suggest that the versatility of UHT technology makes the employment of mobile units at sites of foot-and-mouth-disease outbreaks feasible to provide safe disposal of infectious milk as well as to salvage an otherwise valuable product.

There are no simple and rapid methods for detecting pathogenic viruses in milk. Accordingly, most information available on their significance in foods (including milk and milk products) is epidemiological. Much additional research is needed in this area.

Producing animals may be a significant source of other pathogens found in milk. Tubercle bacilli are often shed in milk from infected cows. *Coxiella burnetii,* the cause of Q fever in humans, is typical of pathogens associated with feces, dust, and other extraneous matter in the environment of infected animals. Other sources of pathogens are infected milk handlers such as milkers and others who may have direct contact with equipment either on farms or in processing plants. For further discussion of milk-borne disease see the review by Snyder *et al.* (1978).

H. Control

Adequate pasteurization is the first positive measure to assure the virtual absence of pathogens and to reduce the numbers of spoilage organisms to an insignificant level. After pasteurization, market milks should be protected against recontamination. Rapidly growing mesophilic non-spore-formers, as well as psychrotrophs, are ubiquitous in the milk plant environment; consequently, they may contaminate surfaces of milk-handling equipment. Essential protective measures include (1) appropriate controls of the pasteurizing process (see Chapter 1, Section V,C,4); (2) proper cleaning and sanitizing of all milk-contact surfaces immediately prior to their use for pumping, transporting, and storing the pasteurized product (see Chapter 14); (3) prevention of airborne contamination; and (4) maintenance of adequate refrigeration during storage and distribution.

Testing of market milk products for the presence of pathogens is not warranted, first, because pasteurized products rarely cause disease, and second, because simple, rapid tests are unavailable. Nevertheless, such products are regularly tested for aerobic plate count and coliform group, often several times daily during processing. Such tests maintain a constant focus on overall microbial quality and sources of contamination and leave an historical record. Variations from the normal or expected test results serve to alert the processor that some facet of the process is out of control and needs correction. For such a purpose, the coliform test is particularly valuable, because coliform organisms are destroyed by proper pasteurization, and almost invariably occur on improperly cleaned milk-contact surfaces of equipment. Selective enumeration of the psychrotrophs is too time-consuming to be generally applicable because of the prolonged incubation period required (7–10 days at 5° to 7°C). Aerobic plate counts at 30° or 32°C, where essentially all of the significant organisms present can grow, provide for a continual check on numerous factors that affect the bacterial levels of processed products. Regulations frequently require that aerobic plate counts remain below 10,000 or 20,000 per ml and coliform group below 1 or 10 per ml. Numerous other analytical methods have been proposed and used especially for the examination of raw milk supplies. Randolph et al. (1973) have compared many of these procedures.

IV. CONDENSED AND DRIED MILK PRODUCTS

A. Definitions and Uses

Several types of condensed-liquid milk products are prepared commercially by evaporation of water from milk under reduced pressure. Dried products are prepared by a spray or drum process. With the spray process, products are first condensed to about 40–45% total milk solids before drying. With the drum process, products usually are not concentrated before drying.

The composition of condensed and dried milks may differ in different countries depending upon regulations that pertain; however, adoption of standards developed through the Joint FAO/WHO Foods Standards Program of the Codex Alimentarius Commission (Food and Agriculture Organization/World Health Organization, 1978) has fostered considerable uniformity among nations relative to the composition of dairy products and other foods as well.

Concentrated milk is a distinctive name for a product marketed in the usual milk bottles or cartons and intended for reconstitution (dilution with water) to the concentration of whole milk by the consumer. Generally, it is produced as a 3:1 concentrate containing 10–12% milkfat and about 36% total milk solids. In the late 1940s and early 1950s, this product received considerable attention and promotion, but for various reasons it never received the marketing success expected of it.

Bulk condensed milk or skimmilk is a 3 or 4:1 concentrate (depending upon usage) of whole milk or skimmilk. It is shipped in bulk, often in large tank transport trucks and used primarily for the preparation of fortified low fat milks, ice cream, candy, bakery products, infant formulas, etc.

Sweetened condensed milk contains 8% milk fat, 28% total milk solids and sufficient added sweetener (usually sucrose) to prevent or retard spoilage. Two types, differing in sugar content, are produced commercially. One contains added sugar sufficient to reduce the water activity (a_w) to 0.86 or slightly below. This corresponds to about 62.5% sugar in the water phase. The product, intended primarily for home cookery, is packaged in small sealed cans, but requires no terminal heat treatment. The other type product is marketed in bulk for use in various foods such as candies and bakery goods. The sugar content varies depending upon the desires of the user, but, generally, it is less than that in the canned product.

Evaporated milk also is a distinctive name to designate condensed whole milk containing 7.5% milkfat and 25% total milk solids. It is packaged in small cans, hermetically sealed, and heat processed to make it "commercially sterile" (see Chapter 1, Section V,C,2), therefore, it need not be refrigerated during storage and distribution. Evaporated milk is used primarily in the home for purposes such as cookery, reconstitution and drinking, infant formulas, and in coffee and tea. Some evaporated milk is used in institutional and restaurant cookery.

Dried milks are usually spray- or drum-dried products that contain not more than 5% moisture, usually less. Dried skimmilk, known as nonfat dry milk (NDM), dried buttermilk, and dried whey are the principal products. These products also are used by the dairy industry and other food processors as a source of milk solids in various food formulations. Depending upon the level of heat treatment given skimmilk before drying, NDM is designated as either a "high-heat" or "low-heat" product.

Inclusion of NDM in bread dough improves the flavor, texture, and crust color of the loaf; high-heat NDM is preferred because low-heat NDM impairs loaf volume. Low-heat NDM is favored to supplement the milk-solids content of other dairy products and as an ingredient of other foods. Also, considerable low-heat NDM is further processed to produce a granular, readily wettable powder that may be easily reconstituted and consumed in liquid form. This product is known commercially as *instant nonfat dry milk.*

Dried whey and *dried buttermilk,* which are by-products of the cheese and butter industry, also are used in various dairy products and other foods as sources of milk solids. Some are used in animal feeds.

B. Initial Microflora

With the exception of sweetened condensed milk, the initial flora of condensed and dried milks is derived from the raw milk (see Sections II,C and D). Sugar used in sweetened condensed milk may be a source of molds, osmophilic yeasts, and sporeforming bacteria, some of which are highly heat resistant. Milk-contact surfaces of preprocessing equipment may contribute additionally to the flora of these products.

C. Effects of Processing on Microorganisms

1. Condensed Products

The following publications provide details of the technology involved in the preparation of condensed milk products: Webb (1970), Pallanch (1971), Harper and Hall (1976), and Campbell and Marshall (1975). The microbiology of these products is treated extensively in the texts by Foster et al. (1957) and Hammer and Babel (1957). Briefly, milk, skimmilk, and other products are heated (forewarmed) before they are condensed. The extent of heat treatment applied will vary depending upon the intended use of the product. In any case, the minimum must be that which will satisfy the requirement for pasteurization. For example, condensed skimmilk used in baking or for preparation of high-heat dried skimmilk requires a forewarming treatment of at least 80°–85°C for 10 to 20 min or the equivalent. In the preparation of other products such as ice cream, cottage cheese, various fluid milks, or instant NDM, the heat treatment may be much less, e.g., 72°–75°C for 16 sec to 10 min.

After it is heated, the hot milk is drawn into evaporators that operate at reduced pressures. In preparation of sweetened condensed milk, a boiled 50% sugar syrup follows the milk into the evaporator. Evaporators consist of up to four vacuum chambers or "pans" arranged in a series termed "effects." Operating temperatures decrease successively from as high as 80°C in the first effect to as low as 38°C in the fourth effect. Moisture is removed from the boiling milk until the desired concentration of milk solids is reached.

The milder forewarming or pasteurization heat treatments permit survival of thermoduric bacteria, including non-sporeformers. The more severe heat treatments (80°C or more for 10 min or longer) will permit only resistant spores to survive. Since vacuum-pan operating temperatures may be as low as 38°C, even in single or double effect systems, bacterial growth can occur. If the units are improperly cleaned and sanitized before use or if forewarming or pasteurization of the incoming milk is faulty, a variety of bacterial types may be introduced or will survive. Low operating temperatures, particularly when associated with inadvertent shutdowns, may easily permit buildup of mesophilic or thermophilic types. The extent of their growth will depend largely on the operating time of the condensing units and the length of time that the product may be held in storage, in balance tanks, or in other equipment within the growth temperature range.

In the preparation of evaporated milk, whole milk is forewarmed, condensed, and standardized to the desired concentration of milkfat and total milk solids. Small amounts of emulsifiers and stabilizers (up to 0.1% of either or both) may be added. The product is homogenized and dispensed into cans that are then sealed and heat treated by steam under pressure in either a batch-type retort or a continuous sterilizer. The heat treatment must be sufficient to attain commercial sterility (see Chapter 1, Section V,C,2). Various factors determine the extent of heat treatment necessary, usually 116°–118°C for 15–20 minutes. As in the case of other foods processed to commercial sterility, all pathogens are destroyed. Extremely resistant spores such as those of *Bacillus stearothermophilus* might survive but would be of significance only during storage of the product under abnormally high ambient temperatures.

Concentrated sour milk, sour skimmilk, or sour buttermilk are preserved by the acid present. Prior to the condensing step, a culture of *Lactobacillus bulgaricus* and a mycoderm yeast are added, the latter to enhance growth of the lactobacilli. Incubation is continued until about 2% acidity, calculated as lactic acid, is reached (18–24 hr). After the condensing step, the acidity is about 6%. These products may be acidified instead by direct addition of acid.

2. Dried-Milk Products

Several milk products, including whole milk, skimmilk, whey, buttermilk, cheese, and cream, may be prepared in dry form. All of them may be spray dried, although considerable quantities of whey and buttermilk are drum dried. The following brief description of the preparation of nonfat dry milk illustrates the application of the two processes. Pallanch (1971), Webb (1970), Harper and Hall (1976), Campbell and Marshall (1975), and Hall and Hedrick (1966) provide details of the various drying processes.

In spray drying, concentrated product (at about 40–45% total solids) is repasteurized and then atomized by one of several devices (depending on dryer design and product) into a drying chamber of hot air in which the air may flow in the same, opposite, or a combination of directions relative to the flow of the atomized milk particles. The drying air is heated to 150°–260°C by passing it over a natural gas flame, electrically heated elements, or steam coils. Moisture is removed as particles move through the hot air and collect at the base of the drying chamber. The dried powder is cooled to 38°–40°C as it moves from the drier to the sifter (to remove large clumps) and then to packaging equipment where it may be placed in bags, cans, large bulk containers, or bins.

The extent of bacterial destruction during drying depends upon the types of microorganisms present, and upon the drying temperature as indicated by the temperature of the exit air. For example, in a recent study, Thompson *et al.* (1978) determined the effect of exit-air temperatures of 93.3°, 82.2°, and 71.1°C on the survival of *Bacillus subtilis, Micrococcus flavus* (reclassified as *Micrococcus luteus,* Buchanan and Gibbons, 1974), and *Escherichia coli.* The percentage of survivors increased as the exit-air temperature was decreased. At 71.1°C survival of *B. subtilis* and *M. flavus* was about the same, 22 and 27%, respectively. At 93.3° and 82.2°C survival of *B. subtilis* was about 12 and 14%, respectively, and of *M. flavus* about 1 and 2.5%, respectively. Survival of *E. coli* was much lower, ranging from 0.02% at 93.3°C to 0.46% at 71.1°C. *Micrococcus flavus* and *B. subtilis* are examples of common thermoduric types in raw milk. *Escherichia coli* is typical of heat-sensitive types that are easily destroyed by minimal-pasteurization treatments. Therefore, its survival during drying illustrates that the drying process cannot be a substitute for pasteurization, or good sanitation. It demonstrates particularly that processors should carefully protect the product against contamination between the pasteurizer and the drier. These studies further substantiate the reports by Foster *et al.* (1957) and Koegh (1966) that the microbial content of dried milk is influenced more by the types of organisms than by the number in the original milk.

During storage of dry milks, the surviving organisms slowly die. The rate of death varies primarily with the type of organisms (Thompson et al., 1978; Crossley, 1962; Higginbottom, 1953; Haines and Elliott, 1944). For example, Thompson et al. (1978) found the proportion that died after 32–36 weeks ranged from about 32–40% for *Bacillus subtilis,* 54–97% for *Micrococcus flavus,* and 98+% for *Escherichia coli.*

Large volumes of whey and buttermilk and some skimmilk are drum dried. Basically, the liquid—usually not concentrated—is forewarmed at about 85°C for 10 min and then fed into a trough formed at the top of two closely aligned rotating drums. The drums are heated internally to about 150°C by saturated steam. The position and rotational speed of the drums are adjusted to permit a thin film of dried milk to form on the drum surfaces as they rotate. The dried film is scraped away by knives and is conveyed to a mill where it is reduced to a fine powder. The milk boils during the time that it remains in the well or trough at the top of the rotating drums. Forewarming and boiling destroy all except resistant spores, provided population levels of heat-sensitive types are not so great as to permit some survival.

Instant dry milk evolved from the desire for a market for dried milk for beverage purposes. Normally dried milks do not wet or disperse easily when added to water; rather, the particles partially solubilize and aggregate into large sticky lumps containing unwetted powder. The instantizing process rewets surfaces of dried particles in steam or atomized water droplets, causing them to agglomerate in clusters, and then dries the clusters. The wetting process raises the moisture content to about 12–14%, which facilitates change in the lactose from a glassy to crystalline form. The product is then redried to 5% moisture or less. Instant NDM is granular, free-flowing, and readily disperses in water. The principal microbiological problem associated with instant NDM occurs after it is reconstituted, since the a_w of the dried product is too low to permit growth.

D. Spoilage

Condensed milks, with the exception of sweetened condensed milk and sour condensed products (5–6% lactic acid), are favorable media for growth of many microorganisms found in the environment of milk-processing plants. Therefore, spoilage can be prevented only by sterilization (as in evaporated milk) or it can be delayed by refrigerated storage. Keeping quality or shelf-life varies from a few days to a week or more depending upon the severity of the heat treatment used in processing and upon the extent to which postpasteurizer contamination is avoided. Eventually, condensed milks spoil, developing "fruity" or "putrid" flavors (similar to that of spoiled pasteurized milks), "bitty" coagulation (similar to that of "sterilized" milk), or an acid-type coagulation due to lactose-fermenting sporeformers or non-sporeformers.

Canned sweetened condensed milk with its high sugar content and low a_w keeps almost indefinitely without refrigeration. Cans may swell occasionally due to growth of osmophilic yeasts. Mold growth also may occur on the product surface if the can is underfilled, for the large head-space affords enough oxygen for growth. Molds will have entered the product between the pasteurizer and the can closing machine (or through a defective seam or pinhole), since the molds most likely to be evident (*Aspergillus* and *Penicillium*) are heat sensitive. Such contamination can arise from filling equipment and nonsterile cans.

On the other hand, bulk products, especially those of lower sugar content, are more susceptible to microbial growth. Mold growth often occurs as "buttons," i.e., small areas of mycelial growth on the surface of the stored product. Species of *Aspergillus* and *Penicillium* usually are involved.

Spoilage of evaporated milk may occur in the unlikely event of under-processing, which would permit survival and outgrowth of organisms normally destroyed during the retorting of the cans after closure. Faulty can seals may readily permit entrance of a variety of organisms that subsequently may grow and cause defects (e.g., gassy, yeasty, and various disagreeable flavors). Unusually high storage temperatures may permit growth of *Bacillus stearothermophilus* and *B. coagulans,* if present.

The high acidity of condensed sour milk, sour skim milk, and sour buttermilk (5–6%) provides for long shelf-life; however, in time, growth of oxidative molds may become evident in localized areas. The molds utilize the lactic acid and thus raise the pH in the area of growth to levels that will permit growth of other organisms that may be present. From those limited areas, growth will spread and spoilage will develop rapidly.

Instant NDM when reconstituted is as fully perishable as other pasteurized liquid milks and is subject to the same types of defects on being held at various temperatures. In fact, spoilage may be even more rapid due to contamination during reconstitution from water and utensils, particularly those used in homes (Olson and Neilsen, 1955). In view of the fact that instant NDM is commonly used in institutional feeding, only carefully cleaned and sanitized utensils should be used to reconstitute and store this product.

E. Pathogens

Few outbreaks of microbial foodborne illness due to the consumption of condensed or dried milks have been reported in recent years. However, these are sensitive products and if abused may present serious health hazards. For example, NDM was the vehicle in the transmission of two series of outbreaks of staphylococcal food poisoning, one of which occurred in England in 1953 (Anderson and Stone, 1955) and the other in Puerto Rico in 1956 (Armijo *et al.,* 1957). About 2000 persons were involved in the English outbreaks and about 1000 school children became ill in the Puerto Rican outbreaks. Also numerous cases of salmonellosis occurred in the United States in 1964–65 due to NDM contaminated with *Salmonella newbrunswick* (Collins *et al.,* 1968); the NDM was produced in a single plant but was instantized in other installations. Illnesses were reported nationwide.

During the investigation of these outbreaks and the surveys of plant processes, several faulty practices were discovered that led to significant changes in subsequent processing procedures. Some milks were inadequately pasteurized before they entered condensing units; some showed heavy growth of coagulase-positive staphylococci in balance tanks or surge tanks prior to the atomization of the milk in the driers. Undoubtedly sufficient growth occurred to produce hazardous levels of enterotoxin that carried over into the dried milk. The faulty practices causing the Puerto Rican staphylococcal poisonings and the U.S. salmonelloses were not clearly delineated. In the investigation of the latter outbreak, *Salmonella newsbrunswick* was found in NDM and in various environmental samples from the plant, although the means of entry into the plant were not determined. In the Puerto Rican investigation, there was some indication that extensive staphylococcal growth had occurred in the milk before it was condensed and dried.

Control measures now commonly practiced in the industry include (a) assurance of heat treatment equal to or greater than that necessary for pasteurization, e.g., installation of flow-diversion valves and proper pressure differential between heating plates (see Chap. 1, Section V,C,4); (b) cooling condensed milk promptly unless drying is commenced immediately after condensing; and (c) holding condensed milk in balance tanks or surge tanks prior to drying at 65°C or more, or if such temperature is not maintained, emptying and cleaning the tanks after each 3 or 4 hr of operation to prevent microbial growth. In addition, the *Salmonella* outbreak in the United States led to an extensive and continuing routine lot-by-lot sampling and testing program of finished product as well as of "environmental samples" taken at strategic locations in plants. Also, cleaning and sanitizing practices have been improved with emphasis to avoid any accumulation or persistence of moisture on or above product-contact surfaces. Product lots from which samples are found positive for *Salmonella* must be reprocessed or otherwise disposed of. The infrequency of outbreaks in recent years attributable to condensed- or dried-milk products indicates the effectiveness of these control measures.

The *Salmonella* problem in dried milk has not disappeared, however; an outbreak was reported in 1979 (U.S. Department of Health, Education and Welfare, 1979). Furthermore, organized routine sampling and testing programs of the Dairy Grading Branch of the U.S. Department of Agriculture continue to reveal a low but persistent level (generally less than 1%) of *Salmonella*-positive samples.

F. Control

The aerobic plate count of line samples taken periodically during condensing and drying operations and of the finished product at intervals during the day should be a sufficient check on the adequacy of operations in controlling microbial buildup.

In view of the rather prodigious efforts that have been expended to eliminate salmonellae from dried milks, it is likely that the current incidence of contamination represents an almost irreducible minimum unlikely to change in the foreseeable future. Therefore, appropriate sampling and testing programs, based primarily on usage of the product (ICMSF, 1974; National Academy of Sciences/National Research Council, 1969; U.S. Department of Health, Education and Welfare, 1978e), undoubtedly will continue as an important mechanism to prevent contaminated product from reaching the market. Consumers of NDM include an extremely sensitive population, namely, the very young, the infirm, and the aged. For this reason, sampling plans should be appropriately stringent.

V. ICE CREAM AND EDIBLE ICES (FROZEN DAIRY DESSERTS)

A. Definitions

Frozen dairy desserts include ice cream, ice milk, sherbet, ices, and various modifications such as parfait, mousse, custard, bisque, and several novelty products. Their composition is fairly well standardized either by laws or regulations of various countries or by custom. Compositional standards under the Food Standards Program of the Codex Alimentarius Commission (Food and Agriculture Organization/World Health Organization, 1976d) are currently being developed.

Ice cream contains 8–12% milkfat, 10–11% nonfat milk solids, 15–

17% sugar, and 0.3–0.5% stabilizer or emulsifier or both. *Ice milk* contains 2–6% milkfat and slightly more nonfat milk solids and less sugar than ice cream does. *Ices* are merely frozen mixtures of water, sugar (about 30%), citric acid, and flavoring. *Sherbet* contains 1–2% milkfat, 1–2% nonfat milk solids, and 26–35% sugar. Frozen dessert *novelties* are variously shaped and flavored forms of these products, often "on a stick," such as chocolate-coated ice cream bars and "popsicles." *Parfaits, custards,* and *"French" ice cream* are basically ice cream containing about 1.5% egg yolk solids. *Mousse* is whipped cream that has been flavored, colored, sweetened, and frozen. *Mellorine* is similar to ice cream but milkfat is replaced by vegetable fat.

B. Initial Microflora

The microflora of frozen dessert mixes before pasteurization is that of the several ingredients. In this chapter, see milk and cream (Section III,E), condensed milk and dried milk (Section IV,C), buttermilk (Section VI,C), and in other chapters, butter (Chapter 25), sugar and chocolate (Chapter 26), fruits and nuts (Chapter 21), and eggs (Chapter 19).

C. Effects of Processing on Microorganisms

Modern technology for preparing frozen desserts is presented in detail by Arbuckle (1972). Abbreviated versions may be found in the texts by Campbell and Marshall (1975) and Harper and Hall (1976), and in the following paragraphs. The various steps are generally the same for all frozen desserts.

All ingredients can be made available in liquid form that facilitates automation of the processing operations. Milkfat is provided primarily by fresh cream, butter oil, sweet butter, and to a lesser extent, whole milk and other dairy products. Nonfat milk solids are added in the form of condensed milk, dried whey, buttermilk, and nonfat dry milk. Sugar is either sucrose or a blend of sucrose and corn syrup solids (enzyme or acid hydrolysate of corn starch). A small amount of emulsifier improves whipping properties; addition of stabilizer improves body and texture and prevents rapid melting of the frozen product. The ingredients are mixed together, and after 15 to 20 min to permit time for the stabilizer to hydrate, the mix is pasteurized. Pasteurization is either the Low-Temperature-Holding procedure (LTH) at 68.3°C for 30 min, the High-Temperature–Short-Time method (HTST) at 79.5°C for 25 sec, or the Ultra-High-Temperature method (UHT) at 138°C or higher. Ice cream mix is always homogenized, often as a step in the pasteurization process in which the homogenizer (basically a high-pressure pump) functions as the metering pump for the HTST pasteurizer. Homogenization breaks up the fat globules, body and texture is improved, and clumping of fat globules is prevented. Flavors or flavoring substances, fruits, candies, chocolate, nuts, etc., may be added before or after pasteurization and homogenization, depending upon whether these processes are detrimental to them. Color is generally added after pasteurization. After pasteurization, the mix is cooled and "aged" for about ½ hr to allow for complete hydration of the stabilizer and for other physical changes to occur, such as the absorption of protein on the fat globules, that improve whipability and smoothness.

The minimum LTH and HTST heat treatments permit survival of thermoduric non-spore-formers and spores of many aerobic and anaerobic spore-formers. Only the more resistant spores survive UHT processes. Ices because of their low pH (3.5 or less) are not pasteurized.

Fruits and nuts, flavors, colors, and so forth, may contribute significant contamination, particularly if added after pasteurization; however, their procurement from reputable suppliers of high-quality products will minimize this problem, providing they are carefully stored and handled in the plant before use. Poor-quality products or postpasteurization contamination from equipment surfaces (vats, pumps, pipelines, and packaging materials) will introduce coliforms and other types that may make it difficult to meet minimum bacterial count standards. Furthermore, gram-negative sucrose fermenters from various fruits may give false coliform counts in ice cream. The sucrose present in 1.0 or 0.1 ml of mix added to petri dishes used in the coliform test may be converted to sufficient acid by these organisms to yield red to purple colonies, typical of coliforms, on selective coliform media (Barber and Fram, 1955).

After the "aging" period, ice cream mix usually is promptly frozen, unless it is to be used for soft-serve ice cream. For the usual hardened ice cream this is a two-step process. In the first step, partial freezing occurs while air is beaten into the freezing mix. The increase in volume of the ice cream from the incorporation of air ("overrun") usually ranges from 80 to 90% of the volume of the mix. In the second step, the soft, partially frozen mix, now at −5° to −8°C, is packaged and immediately placed in a "hardening" room or freezing tunnel where its temperature drops to −25° to −30°C.

Whereas liquid ice cream mix readily supports microbial growth, the low temperature and low water activity of ice cream prevent it. However, microbial death in ice cream is minimal and, therefore, is not a dependable control measure.

Soft-serve ice cream usually is drawn from the freezer at about $-6°$ to $-7°C$ with an overrun of about 40%.

D. Spoilage

Pasteurization treatments of ice cream mixes are almost always more severe than minimal requirements. Consequently (provided the ingredients were of good bacteriological quality), microbial destruction is extensive, and survivors are primarily spores. Accordingly, aerobic plate counts after pasteurization are frequently no higher than a few hundreds per milliliter of mix.

If the mix is frozen promptly, spoilage is impossible, for microorganisms cannot grow in the frozen product. But if there is a delay between pasteurizing and freezing, spoilage can occur. Such delays are unusual in the manufacture of hardened ice cream but mix for soft-serve ice cream must be transported, often long distances, by trucks to retail soft-serve stores or "stands" where it is kept until soft frozen and dispensed to consumers. Both contamination and temperature abuse of the mix easily may occur during this sequence of events. Furthermore, refrigerator space usually is limited and adequate facilities for cleaning and sanitizing the freezer and associated equipment often are lacking or are at best marginal. Under these conditions and especially if faulty practices had occurred previously, the mix may become heavily contaminated. Sufficient growth may occur to raise bacterial counts far above legal limits and cause spoilage.

A study by Martin and Blackwood (1971) was particularly pertinent. Aerobic plate counts on mix immediately after UHT-pasteurization at 140.6°C were below 80 per ml. Shelf-life of the mix at 4.4°, 10°, and 15°C was 3–4 weeks, 2–3 weeks, and 1–2 weeks, respectively. The aerobic plate counts on the mix stored at 4.4°C even for only 2 weeks exceeded the usual bacteriological criteria for ice cream. In a further trial, UHT pasteurized mix, made from high quality ingredients, was aseptically sampled at the pasteurizer, thus precluding any post pasteurization contamination. No organisms were evident when counts were made immediately after pasteurization and fewer than 20 per ml were detected after storage at 4.4°, 10°, or 15°C for 8 weeks. Mix from this same batch was inoculated with about 1 million spores of *Bacillus cereus* and again UHT pasteurized at 137.7°C: *B. cereus* is a well-known food-poisoning organism and its spores are fairly heat resistant. About 10,000 spores per ml survived and persisted but with no change in numbers for 6 weeks at 4.4°C (see also Fig. 13.1), and at this temperature the number gradually declined during the final 2 weeks of the 8-week storage period. At 10° and 15°C, counts after 1 week had increased well above 20 million and 100 million per ml, respectively, and curdling and proteolysis had become visibly evident at 6 and 3 weeks, respectively.

E. Pathogens

The same dangers of illness caused by drinking raw milk are inherent in ice cream either made from raw milk and cream, or handled under insanitary conditions (see Section III,G). With few exceptions, outbreaks that have occurred in recent years have been caused by ice cream made not in commercial establishments but rather in homes where a combination of faulty practices occurred such as use of raw milk, cream, and eggs, inadequate heat treatment, and contamination from infected persons.

As in the case of milk (Section III,G), outbreaks of disease from ice cream are rare in those areas of the world where pasteurization, sanitation, and handling are under regulatory control. The occasional outbreak has been salmonellosis or staphylococcal poisoning (Snyder *et al.,* 1978; U.S. Department of Health, Education and Welfare, 1967–1979). In several instances, raw or improperly pasteurized eggs containing *Salmonella* were added to the mix. (However, in recent years, pasteurization of liquid eggs or liquid eggs for drying, has been required in many countries; see Chapter 19.) Staphylococcal food poisoning has occurred due to contamination of the mix with *Staphylococcus aureus* and subsequent temperature abuse that permitted growth sufficient to yield toxic levels of enterotoxin. Also use of dried whey as a source of milk solids may present a special hazard, for failure of starter cultures to grow normally during cheese manufacture may permit uninhibited staphylococcal growth and enterotoxin formation. Enterotoxin will be in both the whey and the cheese and being heat stable will carry over into dried whey, which then may easily be responsible for the presence of enterotoxin in ice cream. Finally, pathogens, if present, may survive for many months in ice cream (Wallace, 1938).

F. Control

The results of Martin and Blackwood (1971) (see D, above) serve as excellent examples of the benefits of (a) selection of low-count ingredients, (b) use of pasteurization heat treatments severe enough to reduce bacterial populations to very low levels (but not so severe that they are detrimental to the product), (c) avoidance of postpasteurization contamination and, (d) constant maintenance of low temperature during storage of ice cream mix. Failure to adhere to these practices may lead to high bacterial counts, spoilage, and perhaps public health problems.

For the most part, bacteriological sampling and testing of the usual hardened ice cream or of ice cream mix for aerobic plate counts and coliform group at various stages of preparation and handling is routine to maintain a check on plant sanitation. In addition a test for *Staphylococcus aureus* might be useful for control of soft-serve ice cream. Variations from counts normally experienced point to the need for remedial measures. Obviously, an investigation of a foodborne disease outbreak must include sampling and analysis for the appropriate pathogen, such as *Salmonella* or *S. aureus*.

VI. CHEESE

Beneficial microbial activity plays a major role in the sequence of events leading to the production of cheese. Hundreds of named varieties are produced throughout the world (Sanders, 1953). The characteristics that distinguish one from another, as well as differences within a single variety, are the results of modifications made in one or more of several basic steps in cheesemaking. These basic steps are (a) adding to milk of bacterial cultures that produce lactic acid, (b) coagulating of milk, (c) expelling whey from the coagulated mass (shrinking or pressing), (d) salting, and (e) ripening. The microbiology involved in each of these steps is complicated. In this chapter, only a general description of the normal and abnormal microbial activities can be given. For further detail and additional references, see the following texts, reviews, and monographs: Foster *et al.* (1957); Hammer and Babel (1957); Reinbold (1963, 1972); Reinbold and Wilson (1965); Emmons and Tuckey (1967); Olson (1969); Eekhoff-Stork (1976); Kosikowski (1977).

A. Definitions

Basically cheese is a concentrated form of milk obtained by coagulating the casein. This entraps most of the milkfat and some of the milk sugar (lactose), water, and serum proteins (albumin and globulins); most of the water and water soluble constituents are expelled as whey during further manipulation of the coagulated mass or curd. Following is a general classification of cheese (adapted from Foster *et al.,* 1957):

Natural cheese
    Unripened, soft: Cottage, Cream, Neufchatel
    Ripened
        Hard grating cheese: Romano, Provolone, Parmesan
        Hard: Cheddar, Swiss, Emmenthaler, Gruyère

Semisoft: Roquefort, Gorgonzola, Blue, Muenster, Gouda, Edam
Soft: Camembert, Brie
Pasteurized processed cheese, cheese foods, and cheese spreads
Whey cheese: Mysost

As in the case of other dairy products, compositional standards for many cheeses are specified by regulatory agencies of many countries. For example, standards of identity for over 50 types of cheese have been specified by the U.S. Food and Drug Administration (U.S. Department of Health, Education and Welfare, 1978g). Usually the composition of cheeses, expressed in terms of moisture content and the percent of milkfat in the total milk solids, varies greatly among the different cheese varieties. In an unripened cheese, such as cottage cheese, the moisture is about 80%; the milkfat may range from 0.5 to 4%. For ripened cheese, the moisture ranges from about 35% in hard grating cheeses to about 55% in soft cheeses, while the percent of fat in the milk solids may range from 32 to 50%, respectively.

B. Initial Microflora

The microflora in the raw milk when it reaches the cheese plant constitute the initial flora. Contamination from surfaces of equipment used in subsequent handling of the milk will add additional numbers and types of organisms. Furthermore, storage of the milk for excessive periods, particularly at temperatures above 4.4°C will permit relatively rapid bacterial growth (see Section II).

C. Effects of Processing on Microorganisms

1. Clarification

Milk for some cheese, notably Emmenthaler and the U.S. variety Swiss is clarified. Such cheese characteristically contains eyes or holes, produced by gas formed primarily by the propionic acid bacteria during ripening. The eyes should typically be of proper size and number. An excessive number of eyes is a serious defect in such cheese varieties. Particles trapped in the cheese weaken the curd structure in the immediate area, and serve as ready foci for accumulation of gas and the consequent development of numerous small eyes throughout the curd mass. Clarification removes excessive particulate matter and thus helps to prevent excessive eye formation.

2. Heat Treatment

Raw milk, pasteurized milk, and milk that has been given subpasteurization heat treatment are all used in cheesemaking. In pasteurization, the minimum HTST treatment (71.7°C–16 sec) is applied, since more severe heating yields soft and fragile curd. In the United States, milk for cheese intended for consumption as natural cheese must be either pasteurized or held for at least 60 days at or above 1.7°C. The purpose of this require-ment is to reduce the possibility that cheese might become a vehicle for transmitting foodborne pathogens (see Section VI,E).

Pasteurization destroys not only pathogens but also many spoilage or-ganisms and some inherent milk enzymes; the latter in raw-milk cheese play a useful role in ripening. Thus, ripening cheeses made from pas-teurized milk (particularly, Cheddar cheese and related varieties) is slower than ripening cheeses made from raw milk. Cheeses made from pasteurized milk also fail to develop the "full bodied" well-aged flavor of raw-milk cheeses. On the other hand, raw milk may contain pathogens as well as spoilage organisms that produce defects of flavor and body. Subpasteuriza-tion, a reasonable compromise, ranges from 64° to 70°C for 15 to 20 sec, depending upon the type of cheese made. This amount of heat destroys most pathogens; and the public health risk is reduced further by the re-quired (in the United States) 60-day holding period at or above 1.7°C. For example, subpasteurization reduces *Staphylococcus aureus* to insignifi-cant levels (Zottola *et al.,* 1965; Zottola and Jezeski, 1969) and destroys certain spoilage organisms such as coliforms that cause gassy defects in many cheeses. And finally, such practice makes possible the consistent day-to-day production of cheese of uniformly good quality and flavor ap-proaching that of raw-milk cheese.

3. Addition of Starter Cultures

Following heat treatment, if applied, milk is pumped to cheese vats where starter cultures are added. Starters are one or more strains of one or more species of bacteria whose main function is to produce acid, primarily lactic acid, which, in turn, (a) promotes curd formation by the coagulating enzymes such as rennet, (b) promotes drainage of whey from curd par-ticles (thus shrinking them and later, in the case of certain cheese types, aiding the particles to fuse together to form desired cheese body and tex-ture), (c) destroys, retards, or prevents growth of spoilage organisms and pathogens, and (d) influences the nature of enzyme-induced changes dur-ing ripening. The amount and type of starter, and the form in which it is added to the milk (i.e., liquid, frozen, concentrated and frozen, lyophilized, dried, etc.), vary with the type of cheese. Modern starter-culture tech-nology has been described in detail by Sandine (1979) and Kosikowski (1977).

Lactic starter cultures are of two types: The first is made up of single, multiple, or mixed strains of *Streptococcus cremoris, S. lactis,* and *S. lactis* subsp. *diacetylactis*. Single strains or mixtures of strains may be used as starters for the preparation of different ripened and unripened cheese varieties and fermented milks (see Section VII). They produce $L(+)$ lactic acid from lactose. In addition, *S. lactis* subsp. *diacetylactis* uses citric acid to produce flavor compounds, principally diacetyl and CO_2. These cultures are restricted to those cheeses that require moderate curd-cooking temperatures, i.e., up to about 40°C. The second type, often referred to as thermophilic and used where curd-cooking temperatures are high (45°–54°C), consists of single or mixed strains of *Streptococcus thermophilus* and *Lactobacillus bulgaricus,* depending upon the type of cheese made. Mixtures of *S. lactis* and *S. thermophilus* may be used for cheese cooked at intermediate temperatures. For those cheeses in which eye formation is important, *Propionibacterium freudenreichii* subsp. *shermanii* is used, although not exclusively for all varieties of such cheese. The functions and uses of various bacterial species in starter cultures are given in Table 18.3.

In addition, the mold-ripened cheeses, Roquefort, Gorgonzola, Stilton, and Blue—often referred to as blue-veined cheeses—require the mold

TABLE 18.3

Functions and Some Uses of Bacterial Species as Starters in Cheese-Making [a]

Culture	Function	Use
Propionibacterium freudenreichii subsp. *shermanii*	Flavor and eye formation	Emmenthaler, Swiss, and some related varieties
Lactobacillus bulgaricus *Lactobacillus lactis* *Lactobacillus helveticus*	Acid, flavor	Bulgarian buttermilk, Yoghurt, Kefir, Kumiss, Swiss, Emmenthaler, and related varieties, some Italian cheeses
Lactobacillus acidophilus	Acid	Acidophilus milk, Emmenthaler,
Streptococcus thermophilus	Acid	Swiss, Yoghurt, some Italian cheeses
Streptococcus lactis, subsp. *diacetylactis*	Acid, flavor	Sour cream, butter, cultured buttermilk, cottage cheese dressing
Streptococcus lactis *Streptococcus cremoris*	Acid, flavor	Cultured buttermilk, sour cream, cottage cheese, Cheddar, many cheese varieties
Leuconostoc cremoris	Flavor	Cultured buttermilk, cottage cheese, butter

[a] Adapted from Kosikowski (1977).

Penicillium roqueforti in their manufacture. Similarly, the soft Camembert and Brie cheeses require *Penicillium camemberti*. Usually the inoculum, as an aqueous or powdered spore-mycelium preparation, is added to the milk in the vat or is sprayed or dusted over the surfaces of the cut-curd particles (as for the blue-veined cheeses). On the other hand, suspensions are sprayed or rubbed over the surfaces of the wheels of Camembert, Brie, and other similar varieties. Lactic cultures also are added to milk for these cheeses.

The enterococci—*Streptococcus faecalis* and *Streptococcus durans* (the latter now included as a strain of *Streptococcus faecium;* Buchanan and Gibbons, 1974)—have been proposed as supplementary starter cultures for Cheddar cheese (Dahlberg and Kosikowski, 1948). Their contribution to organoleptic properties of Cheddar has been studied recently by Jensen *et al.* (1975).

4. Curd Formation

After addition of starter cultures, the milk is held at a favorable temperature to permit a small amount of acid production by the culture before addition of a coagulant (usually rennet, although use of other coagulants is increasing). The rennet, aided by the acid, precipitates the casein in the form of an aqueous gel structure. The milk is left undisturbed then for a period of time depending upon the type of cheese. This may be for as long as 16 hr in the case of long-set cottage cheese or less than an hour for blue cheese.

5. Shrinking the Curd

Whey removal and shrinking of the curd usually begins with cutting the curd mass into small pieces (1–2 cm^2) and may continue for several hours (as for Cheddar) while the curd is still in the vat; alternatively the cut curd may be "dipped" into forms after about an hour (as for Camembert).

In any event, acid production by the lactic culture proceeds, and whey continues to be expelled from the curd. During this period, the curd and whey mixture is held and stirred at a cooking temperature appropriate for the starter culture and the cheese being made. Heating and stirring aids whey expulsion and helps to produce the curd characteristics desired. For example, in making Cheddar a temperature at setting (addition of rennet) of about 32°C is gradually raised to 38°–39°C, where it is maintained until the whey is drained away. For Swiss cheese, the temperature is raised gradually from the setting temperature of 35°C to 50–53°C where it is maintained until the whey is removed.

Progress in curd shrinkage and acid production is monitored by titration of the acid in a sample of the expelled whey with a standard solution of sodium hydroxide. Acidity is calculated and expressed as percent lactic acid. The acidity attained at the various stages of the curd shrinkage process determines in large measure the acceptability of the finished cheese.

One of the major problems encountered during this stage of cheesemaking is inadequate development of acidity by the starter culture. This may be due to several factors, including the presence of residual antibiotics present in the milk from cows having mastitis and, therefore, undergoing antibiotic therapy (Jezeski et al., 1961; Reinbold and Reddy, 1974). Antibiotics such as penicillin in relatively low concentration (e.g., 0.05–0.2 IU penicillin) may slow or completely stop acid production. Poorly selected cultures, faulty culture management, and small residual levels of sanitizing agents in the milk are other common causes of "slow starters."

However, the most devastating and probably most frequent cause of slow acid development is the infection of cultures with bacteriophages. *Streptococcus lactis* and *S. cremoris* are particularly susceptible to lysis by these viruses. Fortunately, strains differ in their susceptibility to infection. Accordingly, knowledge of the sensitivities of each strain used is essential in developing starter handling systems. Control measures have been outlined in detail by Sandine (1979): (a) Strains used for multiple-strain cultures should have different phage-sensitivity patterns. Thus, if one strain is attacked, nonsensitive or immune strains in the starter will continue to grow and produce acid. Also, such cultures should not be lysogenic.* (b) Starters containing unrelated phage-sensitive strains should be rotated on a day-to-day or vat-to-vat basis. (c) Chlorine-based sanitizers should be used in equipment cleaning and sanitizing programs. (d) Isolated and specially constructed culture-preparation rooms and equipment having air-filtering mechanisms should be used to prevent airborne phage contamination: (e) Phage-inhibitory media should be used for preparation of bulk starters.

Lactic phages survive minimum-pasteurization heat treatments (Zottola and Marth, 1966). Accordingly, bulk starter milk must be heated to 82°–88°C for at least 1 hr to assure their destruction not only in the milk but in the head space air of the bulk-starter tank as well.

Starter cultures are often tested to determine various characteristics. These tests have been outlined and further referenced by Sandine (1979).

When curd particles or the curd mass have become sufficiently firm, they are placed in forms to shape the cheese. This may be accompanied by application of varying degrees of pressure depending upon the cheese type.

---

* In simple terms, a lysogenic strain is one that has phage particles attached to the cells, but is apparently unaffected. The cells may lyse later, or may transfer the phage to a susceptible strain.

At some point salt is added to enhance flavor; additionally it helps control the growth of many spoilage organisms.

6. Ripening

Most cheeses undergo a period of ripening or curing during which flavor and body characteristics continue to develop and eventually reach those typical of the variety. The agents responsible for transforming fresh curd into finished cheese are primarily enzymes from several sources: (a) microorganisms originally present in the milk; (b) starter cultures and other types that grow in the interior of the cheese and on the surface; and (c) those enzymes that are inherent to milk.

During the ripening period, changes occur in the fat, protein, and other constituents that are reflected in flavor development and physical or body characteristics (i.e., elasticity, firmness, etc.). Ripening periods may range from a week or two to a year or more. Generally, the softer higher moisture cheeses undergo relatively short ripening periods. In contrast, well-aged sharp-flavored cheeses may require a year or more. The conditions of temperature and humidity during ripening also differ depending on cheese type. High humidity is necessary for surface-ripened cheeses to encourage growth of organisms on the surfaces. Low humidity is necessary for most hard cheeses where internal enzymatic activity is encouraged and surface growth is discouraged or prevented.

Blue-veined cheeses are purposely made to have an open internal texture that permits entrance of oxygen to support mold growth and permits release of CO_2 produced by the mold. "Spiking" helps such gaseous exchange: metal rods about 3 mm diameter and about 20 mm apart are passed through the cheese wheels at or near the beginning of the ripening period.

D. Spoilage and Other Defects

All cheeses are subject to microbially induced abnormal fermentations, which cause spoilage and abnormal physical appearance. Spoilage is usually manifested by abnormalities such as putrid, unclean, yeasty, fermented, and rancid flavors. The types of organisms involved in spoilage of ripened cheese are restricted largely to those that can oxidize lactate and various protein and fat degradation products, since little or no lactose remains. Molds yeasts and anaerobic spore-formers are involved most often.

Mold growth occurs most frequently on cheese surfaces and often extends into the interior of the cheese along slight cracks or fissures causing discoloration and an unsightly appearance. Considerable physical loss may be incurred because of the need to remove affected portions. Species of *Penicillium* are common offenders. Waxing or plastic-coating the cheese, clean storage shelves, and careful control of humidity will help to minimize this problem.

Abnormal gas formation or "gassiness" is common to many cheeses. This may be evident by the presence of numerous small gas holes throughout the cheese structure; sometimes massive internal voids develop, causing the cheese to be grossly misshapen or bulged and even to split open ("blown" defect). Paradoxically, the gassiness defect, when present in Swiss or Emmenthaler, is caused most frequently by the propionic acid bacteria that produce the typical eye structure of these cheeses. Gassiness of the "blown" defect type is troublesome particularly in tightly structured cheese such as Emmenthaler, Swiss, Provolone, Gouda, and Gruyère. Clostridia, primarily *Clostridium tyrobutyricum* and *Clostridium butyricum,* are most frequently involved. Since silage is a common source of clostridia in milk, feeding silage to cows whose milk is used for these types of cheese is discouraged. The defects of Swiss and related cheeses have been extensively treated by Reinbold (1972), Hettinga *et al.* (1974), Hettinga and Reinbold (1975), and Langsrud and Reinbold (1974).

A complex or family of closely related antibiotics termed nisin produced by *Streptococcus lactis* is antagonistic particularly to spore-formers (Hurst, 1972). In several western countries, nisin has been accepted as a legal food additive and is used to some extent to control the late gas or "blown" defect caused by clostridia in Swiss, Edam, and related cheeses.

Gas formation early in the cheesemaking process may cause defects. For example, growth of coliforms in cottage-cheese milk may produce sufficient gas during the cooking process to cause the curd to float. The resulting cheese will have off-flavor and soft and shattered curd particles. Also, starters with a high proportion of the flavor organisms—*Streptococcus lactis* subsp. *diacetylactis* and *Leuconostoc cremoris*—may cause floating curd and gassiness in cottage cheese as well as in cheese such as Cheddar and related types that undergo a relatively long cooking process. Use of raw milk, inadequate heat treatment, postpasteurization contamination of the milk, and slow acid production by starters are common factors that lead to gas production during this phase of cheesemaking. The references cited at the beginning of this section provide further details on defects of various cheese varieties.

Several factors contribute to the preservation of cheese. The low pH (generally below 5.3) prevents or slows growth of most bacteria. Salting to a level of 1.5 to 5% reduces the a_w because the salt dissolves in the aqueous phase. The low temperature of ripening and the low oxidation–reduction potential also restrain microbial growth. Favorable combinations of these factors in different cheese types prevent spoilage (see also Chapters 1, 4, 5, and 6).

E. Pathogens

In recent years, cheese has been the vehicle of several foodborne disease outbreaks (Snyder et al., 1978). The list includes brucellosis, listeriosis, shigellosis, botulism, salmonellosis, staphylococcal food poisoning, and gastroenteritis due to Escherichia coli. Raw milk, improperly pasteurized milk, or underheated milk have been the usual vehicles for the causative organisms, particularly for those of the first three diseases listed above.

Since brucellosis and listeriosis are zoonoses (and the causative organisms are shed in the milk), they are readily transmitted from cow to man through milk and occasionally through cheese. Diarrheal disease of man caused by species of Shigella transmitted through cheese is not common; however, careless milk-handling practices by infected persons can result easily in contamination of milk and cheese directly (Rubinsten-Szturn et al., 1964). While inadequate heat treatment or recontamination may account, as well, for the presence of Clostridium botulinum, salmonellae, Staphylococcus aureus and enteropathogenic Escherichia coli (EEC) in milk and cheese, additional factors tend to increase the hazard presented by these pathogens.

The ubiquity of C. botulinum in the environment and the heat resistance of its spores makes its occasional presence in milk inevitable. Fortunately, certain of the preservative factors previously mentioned (see Section VI,D) singly or in combination serve to reduce the likelihood of its growth and toxin formation in cheese. On the other hand, in certain fresh cheeses, lightly salted, with relatively closed structure (conducive to anaerobiosis), and high pH (the latter normally so or because of faulty starter activity), the probability of growth of C. botulinum, if present, is materially increased.

Salmonellosis is a common disease of cattle, goats, sheep, and many other animals. Salmonellae are often found in milk from infected animals and in the dairy environment, particularly lactose-fermenting serotypes (Blackburn and Ellis, 1973). Furthermore, while with few exceptions, salmonellae are easily killed by pasteurization, they are probably among the most resistant of the enteric pathogens to drying, freezing, low pH, and storage under dry and cold conditions. Thus, faulty postpasteurization sanitation practices may easily permit these organisms to gain access to milk, to grow during cheesemaking, and to persist during ripening and well beyond the 60-day aging period at or above 1.7°C currently required in the United States for cheese made from unpasteurized milk (Geopfert et al., 1968). Recently, survival of salmonellae in cheese was reviewed and further studied (White and Custer, 1976). *Salmonella newbrunswick* and *S. infantis* survived at least 9 months in 9 of 16 samples of Cheddar cheese when held at 4.5°C. Survival times at 10°C generally were shorter, but in at least one sample of 16, both serotypes survived 9 months. Admittedly, the inoculum for each serotype was high (approximately 3.7×10^5 per ml of cheese milk); however, in the event of contamination and starter failure, it is conceivable that a number approaching such a level could be reached. On the other hand, *S. typhimurium* could not be isolated from cottage cheese whey (pH 4.5–4.6) stored longer than 3 days at 5°C when the population in the whey was approximately 1500/ml or less (Westhoff and Engler, 1973). Of approximately 100,000/ml originally present, none were detected after 3 days at 25° or 35°C. Undoubtedly, the low pH and certain compounds resulting from starter activity were important factors in destroying the organisms.

Staphylococcus aureus is frequently found in milk and dairy products. Moreover staphylococcal infections of the bovine mammary gland (mastitis) represent a significant reservoir of enterotoxigenic strains of *S. aureus* (Olson *et al.,* 1970).

The outbreaks of staphylococcal food poisoning due to cheese, which occurred in the United States in 1958 and 1965 (Hendricks *et al.,* 1959; Allen and Stovall, 1960; Zehren and Zehren, 1968a,b) stimulated numerous studies on the incidence and behavior of staphylococci in milk and cheese. Many of these studies and relatively recent outbreaks of staphylococcal food poisoning from dairy products have been reviewed by Minor and Marth (1972b). Except for a Canadian outbreak from Swiss cheese (Health and Welfare, Canada, 1979), few outbreaks have been reported since 1965. Undoubtedly, a better understanding of the behavior of *S. aureus* in milk before and during cheesemaking has contributed significantly toward control of this problem in cheese. For example, subpasteurization heat treatments (see Section VI,C,2) kill *S. aureus* in milk. Especially important were the findings of Zehren and Zehren (1968b). Their analysis of 2048 individual lots representing over 1 million pounds of Cheddar cheese suspected of containing staphylococcal enterotoxin, and under seizure by the U.S. Food and Drug Administration, revealed 56 lots containing detectable amounts of enterotoxin A. Toxic lots were destroyed and the others were released to commerce. Examination of "make" records showed that each toxic lot had a milling acidity (just prior to placing curd in forms for pressing) of 0.4% or less. Such low milling acidity is evidence of slow acid production by the starter culture. This makes possible relatively unrestricted growth of staphylococci during the $4\frac{1}{2}$ to 5 hr and sometimes longer that the curd remains in the vat at a favorable temperature. If the pH finally attained is higher than normal (5.2–5.3), growth can continue during pressing. Generally, little or no growth occurs beyond the pressing stage, and numbers decline during ripening. In any event, if sufficient growth occurs (1 million or more per g of cheese or ml of whey) detectable toxin should be suspected (Tatini *et al.*, 1971a,b, 1973).

On the other hand, high populations of known active enterotoxigenic strains of *S. aureus* in cheese are not always associated with enterotoxin production. Tatini *et al.* (1973) were unable to detect enterotoxins in blue cheese after an active enterotoxin-producing strain had reached 25–50 million/g of cheese at the end of 24 hr. Similarly, no toxin was detected in Mozzarella cheese in which the same strain had reached 20 million/g. Although the reason for failure to produce detectable toxin in these cheeses is not certain, it is known that low oxygen tension inhibits enterotoxin formation by *S. aureus*. Barber and Deibel (1972) found that enterotoxin in fermented sausage was restricted to an outer ring area of about 1/4 in. where oxygen tension was higher than in the interior of the sausage. Similarly in blue cheese, the CO_2 concentration is high and the O_2 correspondingly low because of active growth of the normally heavy inoculum of *P. roqueforti*. Such a condition early in the cheesemaking process may account partially for the failure of enterotoxin formation.

The observation of Zehren and Zehren (1968b) that low milling acidity was associated with the presence of enterotoxin in the cheese provides an important alerting mechanism to the possibility that a vat of cheese may be toxic. Cheese vats with low acidity should be analyzed for the level of *S. aureus* and perhaps for enterotoxins (Baer *et al.*, 1976; Bergdoll and Bennett, 1976; ICMSF, 1978). However, in this context these two tests have serious disadvantages. The enterotoxin test is complicated and time consuming (3–5 days minimum). Generally, *S. aureus* dies off relatively rapidly during ripening. Therefore, unless analysis is made within the first 24–72 hr, peak populations may be missed. These deficiencies led Cords and Tatini (1973) to apply the thermonuclease assay (TNA) for assessment of staphylococcal growth and likely presence of enterotoxin in cheese. The enzyme, deoxyribonuclease (DNase) is heat stable and persists in cheese during ripening and even after several years of storage. Furthermore, growth to at least 10^6/g of cheese is necessary to produce detectable DNase. Thus, a positive test indicates unusual staphylococcal growth and the possible presence of enterotoxin.

The persistence of the *S. aureus* problem in cheese has emphasized the need for monitoring cheesemaking and finished cheese and the value of TNA for this purpose (Tatini *et al.*, 1976; van Schouwenberg-van Foeken *et al.*, 1978; Stadhouders *et al.*, 1978; Park *et al.*, 1978). For example, Park *et al.*, examined 25 lots of Cheddar cheese from plants showing evidence of faulty sanitary and cheesemaking practices. Twenty-two of these were TNA positive; of these, enterotoxin A was found in six, enterotoxin D in two, and B in one. One would be hard pressed to find a better example of the public health value of a simple and rapid analytical procedure!

Fortunately, growth of *S. aureus* to high populations is required to produce detectable enterotoxin. Therefore, the frequency of the organism in many types of cheeses at levels of a few hundred or a few thousand should not be cause for panic. The organism is ubiquitous in the dairy environment, and the carrier rate in humans is high (often cited at 50% or higher). There seems little to fear from consumption of *S. aureus* even in relatively high numbers. The important practical objective is to keep initial numbers at a minimum and to limit growth subsequently to levels far below that necessary for production of detectable enterotoxin. Modern technology of cheesemaking and current knowledge of the behavior of *S. aureus* are adequate to attain these ends. Adequate heat treatment of the milk (including quality assurance practices that assure that such heat treatment was applied, in fact—see Chapter 1, Section V,C,4), unimpaired starter activity, and strict plant sanitation are essential elements of control.

Enteropathogenic *Escherichia coli* (EEC) is a common cause of gastroenteritis or diarrheal disease in humans and other animals, particularly the young. It has caused many institutional outbreaks in infants, and it is often the cause of "tourista" or "traveler's disease" among tourists and others when visiting countries where enteric infections are endemic or commonplace. It seems incongruous that there is little published evidence to indicate that foods are frequent vehicles for the transmission of EEC infections. Two factors may be responsible for lack of such information: (a) difficulty in diagnosing the disease (Dupont *et al.,* 1971); (b) difficulty in detection, enumeration, and proof of pathogenicity of isolated cultures (Mehlman *et al.,* 1974, 1976).

In any event, a series of outbreaks of EEC gastroenteritis from Camembert cheese imported from one manufacturer occurred in the United States in 1971. These were the first documented outbreaks of foodborne illness caused by *E. coli* in the United States (Marier *et al.,* 1973; U.S. Department of Health, Education and Welfare, 1971b,c; Tullock *et al.,* 1973). A few earlier outbreaks in other countries had been reported (Costin *et al.,* 1964; Gaines *et al.,* 1964). The U.S. outbreaks caused considerable concern about cheese as a vehicle for transmission of EEC diarrheal disease to humans. Also contributing to this concern is the fact that coliforms including *E. coli* are often present in various cheeses, occasionally in high number.

Following the U.S. outbreaks, a series of studies was initiated dealing with (a) further assessment of the inhibition of *E. coli* by starter cultures (Frank and Marth, 1977a,b); (b) the incidence of *E. coli* and EEC in cheese (Collins-Thompson *et al.*, 1977; Frank and Marth, 1978); and (c) the fate of *E. coli* and EEC in Camembert cheese (Park *et al.*, 1973; Frank *et al.*, 1977) and Brick cheese (Frank *et al.*, 1978). Briefly, lactic fermentation of milk is inhibitory to the growth and survival of *E. coli* and EEC. However, growth of the latter organisms generally increases during the first 6 hr, the extent depending upon the amount of starter inoculum and the temperature of incubation. Similarly, in the making of Camembert, marked growth of EEC occurred during the second to fifth hours followed by a rather sharp reduction in population over the next few hours and a decline to fewer than 10–100/g after 7–14 days. In the case of impaired starter activity, populations of *E. coli* approached 10^9/g in 24 hr and then declined shortly thereafter.

In general, semisoft surface-ripened cheeses are particularly susceptible to abnormal fermentations because of slow or lack of acidity development during manufacture and low salt levels in the interior of the cheese during early stages of ripening (Olson, 1969). Brick cheese, a variety having all of these characteristics, was chosen for study (Frank *et al.*, 1978): Two EEC strains increased 1000-fold during the first 7 hr but began to decline slowly after 24 hr. The increase observed was 10-fold greater than that observed with these same strains in Camembert cheese. They also declined at a less rapid rate than in Camembert.

Two recent surveys of soft and semisoft cheeses (Collins-Thompson *et al.*, 1977; Frank and Marth, 1978) failed to find EEC; however, in the latter survey 17% of the cheeses contained fecal coliforms in excess of 10^4/g with one as high as 2×10^6/g. In the former, only 0.8% of the semisoft and 2% of the soft cheeses contained more than 1600 fecal coliforms/g. The survey by Frank and Marth was made exclusively during the summer months, which probably accounted for the difference between the two surveys.

In view of the information to date, there seems little doubt that the occurrence of excessive numbers of fecal coliforms in soft and semisoft cheeses represents a potential EEC hazard. To minimize such hazard, adequate heat treatment of milk (pasteurization preferable), use of active starter cultures, and strict sanitary practices are necessary. Since *E. coli* can grow well on the surface of Camembert (Frank *et al.*, 1977), special care should be taken to prevent surface contamination, and the temperature during the development of the surface mold matt should be as low as practicable, i.e., not higher than 10°–11°C.

The occurrence, mechanism of formation, and catabolism of biologically active amines in foods have been reviewed by Rice *et al.* (1976) and Voigt and Eitenmiller (1978). Two of these, tyramine and histamine, are often found in cheese (Voigt *et al.*, 1974). Both amines are vasoactive in humans. Tyramine in sufficient quantity can cause critical increases in blood pressure especially in persons receiving monoamine oxidase inhibiting (MAOI) drugs. Normally monoamine oxidases (MAO) deaminate tyramine and other amines. Histamine is a strong capillary dilator and can produce hypotensive effects. Symptoms of histamine poisoning include flushing, rapid pulse, fall in blood pressure, and headache. These are also symptoms of scombroid fish poisoning. Normally, tyramine and histamine do not present a hazard to humans unless large quantities are ingested, unless natural mechanisms for their degradation such as MAO are inhibited (e.g., persons on MAOI therapy), or unless the victim is genetically deficient. Nevertheless, tyramine-induced hypertensive crises and histamine poisoning from cheese have been reported. An outbreak of histamine poisoning due to Swiss cheese occurred recently in the United States.

The formation of tyramine and histamine in cheese occurs on decarboxylation of the amino acids tyrosine and histidine, respectively. Coliforms and the enterococci, *Streptococcus faecalis* and *S. faecium,* commonly found in cheese, produce tyrosine decarboxylase. Histidine decarboxylase producers include coliforms, *Lactobacillus casei,* and *Proteus* spp. It does not follow necessarily that high levels of tyrosine- and histidine-decarboxylase activity in cheese are always associated with corresponding quantities of the respective amines (Voigt and Eitenmiller, 1978). Limiting factors are the absence of the amino acids, the presence of decarboxylase inhibitors, and/or the presence of catabolism pathways for the amines. Such pathways may be provided by certain starter cultures, i.e., *Leuconostoc cremoris, S. lactis, S. lactis* subsp. *diactylactis,* and *S. cremoris,* which have mono- or diamine oxidase activity (Voigt and Eitenmiller, 1978), and the suggestion was made that amine contents could be controlled by use of dairy starters having oxidase capabilities.

F. Control

Broad recommendations for routine microbial sampling and testing of cheese are not appropriate. The technologies involved in making the many varieties of cheese differ, and the health hazards presented also differ. Furthermore, our knowledge of microbial hazards inherent in cheese is incomplete. However, judicious analytical programs can be helpful.

With the exception of *S. aureus* and possibly EEC, the rarity of outbreaks due to the bacterial species discussed above would indicate little, if any, value in routine examination of cheese for the presence of those organisms. On the other hand, the potential for staphylococcal growth, particularly in those cheese varieties requiring relatively long cooking temperatures, would indicate the value of frequent analytical checks. At the plant level, should there be any indication of faulty heat treatment or slow acid development, a thermonuclease assay (TNA) (Tatini *et al.,* 1976) on the cheese lot in question would be indicated. A positive test would be a certain indication of need for an enterotoxin analysis. At the wholesaler or buyer level for retail distribution, a TNA or a *S. aureus* count purchase specification should be seriously considered. At the regulatory level, either a TNA or a *S. aureus* count would be indicated where plant inspection or other information signals the existence of a potential problem.

Enteropathogenic *Escherichia coli* presents a potential hazard, particularly in soft and semisoft cheeses. The lack of evidence of recurring outbreaks tempers the extent to which routine monitoring might be applied. The fecal coliform count (Fishbein *et al.,* 1976) would be useful if some break in the operation likely to foster the presence and growth of *E. coli* has occurred. The need for analysis to substantiate the presence of EEC would be indicated by the results obtained (i.e., populations in excess of a few thousand *E. coli* or fecal coliforms per gram).

The ICMSF sampling plan system (ICMSF, 1974) may not necessarily apply in the instances discussed previously, since the distribution of the organism in any lot of cheese involved is almost certain to be uniform and numbers of any significance would be sufficiently great to permit certain detection even if only a single sample from a particular lot was analyzed.

VII. FERMENTED MILKS

A. Introduction

For centuries a significant proportion of milk has been consumed in the form of freshly prepared fermented milk products (in contrast to the ripened and unripened cheeses). Basically, these are milks that have been soured by a lactic acid fermentation. In some fermented milks, a citric acid fermentation produces neutral flavor compounds, primarily diacetyl, and certain volatile flavorful organic acids. In some instances, small amounts of ethyl alcohol are present. Collins (1972) provided details of the metabolic pathways by which the principal flavor compounds are formed.

Several different species of organisms are used in the preparation of fermented milks. Most of them also are components of various cheese starters (see Section VI,C,3). The consistency of fermented milks varies, ranging from the easily pourable to those having a thick, jellylike structure; the latter are eaten rather than drunk. Their composition also varies; some have essentially the same composition as that of whole milk or skimmilk, while others may be fortified with additional milk solids (usually nonfat dry milk). Various flavors and fruits are often added. In many countries, the composition of fermented milks is specified by regulation.

The low pH and other factors provide for a relatively long shelf-life—up to 2 weeks or more. Generally, flavor deterioration during storage is due to activity of the typical culture organisms in excess of that necessary to develop the distinctive flavors and physical characteristics of the product.

Over the years, many have ascribed health benefits to the consumption of fermented milks, particularly those containing *Lactobacillus acidophilus,* i.e., claims far beyond those due to the nutritional attributes of the milk constituents. These began with studies of Metchnikoff (1907, 1908)* and later were furthered by Rettger and Cheplin (1921), Kopeloff (1926), Rettger *et al.* (1935), and others. Reviews have recently been presented by Sandine *et al.* (1972), Speck (1975, 1976b), and Gilliland and Speck (1977). Briefly, information to date shows that many individuals with chronic intestinal disorders benefit from implantation of *L. acidophilus* in the intestine. For this to occur, large numbers of viable cells must be fed and a fermentable carbohydrate must be available to the cells in the intestine. Also, *L. acidophilus* has been shown to be antagonistic to several foodborne pathogens.

In the subsequent discussion, several popular fermented milk products will be treated. Greater detail and sources of additional information relative to health aspects of these products can be found in the references and reviews mentioned. For further discussion of the technology and microbiology of fermented milks, see Kosikowski (1977), Emmons and Tuckey (1967), and Foster *et al.* (1957).

B. Cultured Buttermilk

Prior to the advent of the factory system of buttermaking, butter was churned on the farms, usually from raw naturally soured cream. Much of it was consumed by the farm family, and some was sold locally and elsewhere.

If fermentation of the cream was "normal," associative growth of naturally occurring lactic streptococci and citrate fermenters (*Leuconostoc cremoris*) dominated the fermentation. The result was a pleasant, sour, and flavorful cream with slight effervescence due to a small amount of CO_2 present. When such cream was churned, the liquid *buttermilk* fraction retained most of the acid and flavor components. Such a product was

---

* Metchnikoff erroneously ascribed the benefits derived as due to *Lactobacillus bulgaricus*. This was later clarified when it was shown that *L. acidophilus* rather than *L. bulgaricus* could be implanted in the intestine.

relished as a beverage. If the fermentation was not as described (as often was the case), the buttermilk would often contain off-flavors of varying intensity due to growth of proteolytic and lipolytic spoilage organisms. In a normal fermentation, these would have been inhibited by the low pH and various inhibitory compounds.

As the technology of buttermaking advanced and different types of butter evolved (see Chapter 25), commercial utilization of buttermilk as a beverage was not feasible. However, the lactic streptococci and *Leuconostoc* were soon identified as the organisms responsible for the good flavor of ripened cream. These organisms were prepared in pure culture, and were then introduced as starters to make ripened cream butter and to ferment milk or skimmilk to produce "cultured buttermilk," simulating the buttermilk of yesteryear.

Cultured buttermilk is the most popular fermented milk made in North America and is commonly produced in many other countries. It may be made from whole milk, skimmilk, partially skimmed milk, or reconstituted nonfat dry milk. Skimmilk often is fortified with additional nonfat dry milk to provide desirable viscosity in the final product. The milk is pasteurized at about 85°–88°C for at least 2 min and often for as long as 30 min. After cooling to 21°–22°C, it is inoculated and mixed with about 1% of starter culture. Incubation is continued at the inoculating temperature until the titratable acidity reaches 0.85–0.9% (about 12–14 hr). The product is then cooled with mild agitation to about 5°C, packaged, and refrigerated at 5°C or below until distribution.

The starter cultures used consist of mixtures of selected strains of *Streptococcus lactis, S. cremoris,* or *S. lactis* subsp. *diacetylactis.* The function of the first two listed is to produce lactic acid rapidly and that of the latter is to produce flavor compounds. The flavor compounds are principally diacetyl and volatile acids, including acetic and other organic acids.

A proper balance between the acid producers and the flavor types must be maintained in the starter cultures and during the making of the product. Generally, the lactic streptococci outnumber the flavor producers. Careful control of the incubation temperature at 21–22°C is essential to maintain this balance. Lower temperatures cause insufficient acid development and higher temperatures cause excessive acidity.

Cultured buttermilk containing insufficient acid may taste "flat" because the highly flavored diacetyl and volatile acids do not form in appreciable amounts until late in the fermentation. Starters may become contaminated with yeasts and other spoilage organisms that eventually will cause off-flavors and physical defects in the final product. Normally, there is little danger from pathogens in cultured buttermilk or other fermented milks. The low pH, the level of lactic acid, and the presence of small quantities of other organic acids and other inhibitory compounds provide an unfavorable environment for pathogens and most other undesirable organisms.

Sour creams of various fat percentages are made similarly but with some procedural modifications, including homogenization and sometimes the addition of a small amount of rennet or stabilizer to provide for greater viscosity.

C. Yoghurt

Yoghurt is one of the oldest fermented milks known. It probably originated in the Balkan area; there and throughout Europe it is the most popular fermented milk. In recent years, its popularity has dramatically increased in North America. The name is spelled differently, depending on country or region (e.g., yogurt, yoghurt, yahourt, yaourt, jugurt), but basically it is the same product.

Careful control over the production of yoghurt and the introduction of a variety of flavors have made it highly popular. Its consistency is thick or jell-like; hence, it is eaten rather than drunk.

Yoghurt may be made from milk, skimmilk, or partially skimmed milk. Normally additional nonfat dry milk is added bringing the solids-not-fat content to 11–15%. Up to about 4% sugar is sometimes added. In some countries the composition of yoghurt is specified by regulations.

In the preparation of yoghurt, milk of appropriate composition is pasteurized at 85° to 90°C for 15 to 30 min. It is then cooled to 42°–45°C, at which point a 2% inoculum of a starter culture containing *Lactobacillus bulgaricus* and *Streptococcus thermophilus* is added. At this point, the inoculated milk may be distributed in final containers and incubated, or may be left to incubate in bulk. In either case, growth and acid production occurs rapidly and in about 3 hr a firm gel develops, and the pH drops to about 5.

The yoghurt is cooled and, if in bulk, it is dispensed into final containers. In a few hours, the product again thickens. During incubation, the two species grow in a symbiotic relationship (Pette and Lolkema, 1950a,b,c; Galesloot *et al.,* 1968; Viringa *et al.,* 1968) and in some instances this growth relationship is commensalistic and competitive (Moon and Reinbold, 1976). Briefly, *Streptococcus thermophilus* stimulates *Lactobacillus bulgaricus* by producing formic acid, and the proteolytic activity of *L. bulgaricus* releases certain amino acids, particularly valine, which are stimulating to *S. thermophilus*. Thus acid production is considerably more rapid when the two species are grown together than if grown independently of each other.

Generally, the shelf-life of yoghurt is about 3 weeks during refrigerated storage at 5°C. During this period, acid slowly continues to develop, reducing the pH to 4.0 or lower. This results in a sharp acid taste that is objectionable to many.

Yoghurt may be heated after incubation to destroy the lactobacilli and streptococci, thus preventing further acid development. Through use of proper equipment and aseptic packaging, its shelf-life may be extended beyond that common for the unheated product.

Occasionally, contaminants in the starter culture impair fermentation, producing off-flavors and physical defects. Also, both *Lactobacillus bulgaricus* and *Streptococcus thermophilus* are subject to bacteriophage infection—especially the latter organism (Ciblis, 1970; Galin *et al.,* 1970; Reinbold and Reddy, 1973). In addition, residual antibiotics present in milk may interfere with the fermentation since both organisms are sensitive to penicillin and tetracyclines.

Normally, growth of pathogens in yoghurt is prevented by the low pH. Similarly, coliforms, if present, rapidly disappear. For example, two strains each of *Enterobacter aerogenes* and *Escherichia coli* were rapidly inactivated in 4 days at 7.2°C when added individually to samples of yoghurt (Goel *et al.,* 1971). In several samples, the organisms were undetectable after 24 hr of storage. However, factors that interfere with acid formation can lead to a health hazard. For example, an outbreak of food poisoning occurred in a Strasbourg military hospital when an excess of sugar was added to the milk prior to incubation. This inhibited the normal fermentation and staphylococci grew to high numbers (Mocquot and Hurel, 1970).

Yeasts may also be troublesome, for many are unaffected by the lactic acid; in fact they utilize it and grow well in association with the normal culture organisms, producing yeasty and fruity off-flavors, often with gassiness from CO_2.

D. Acidophilus Milk

As indicated previously, the health promoting attributes of *Lactobacillus acidophilus* have been investigated by many. As a consequence, acidophilus milk has been produced and marketed for many years in many parts of the world. Its popularity, however, has been sporadic.

Lactobacillus acidophilus may be easily overgrown in milk by contaminating bacteria. Therefore, the milk before inoculation is given a high heat treatment (90°–95°C for 1 hr). Some producers sterilize the milk, and others employ two successive exposures of 90°–95°C for 1 hr with an intervening interval of several hours at 40°C between heat treatments to permit germination of spores that survived the first heat treatment. The vegetative cells thus formed are destroyed by the second heating.

A starter inoculum of 1 to 2% is used, and the milk is incubated at 38°–40°C until the titratable acidity reaches about 0.65%. This may require an incubation period of 24 hr or longer, depending upon the characteristics of the culture. The product is cooled, packaged, and stored under refrigeration until used. The value of acidophilus milk depends on the presence of large numbers of viable cells in the product. Unfortunately, its useful shelf-life is limited to a few days since the number of organisms originally present declines rather rapidly, especially at acidities higher than 0.65% and at temperatures below 5°C. At best, the highly cooked and sharp acid taste is not pleasing to most consumers. As a consequence, some producers add a concentrate of *L. acidophilus* cells to cold pasteurized milk which then becomes "sweet acidophilus milk" (Morey, 1954; Duggan *et al.*, 1959). The product is kept refrigerated until used. The cell population of the final product is about 2–4 million per ml, and its taste remains sweet and unaltered from that of the milk before addition of the cells. Recently, such a product has been widely promoted in the United States, Canada, Japan, and to some extent in Europe.

Survival of *L. acidophilus* in sweet acidophilus milk during storage at 4°C recently was determined by Young and Nelson (1978). Counts decreased from 2.6×10^6 to 6.4×10^6 per ml initially to 5.1×10^4 to 3.1×10^6 per ml after 23–24 days of storage. Retail pull-dates (suggested date to be removed from market) for the product varied from 8–13 days. Generally, counts remained at about 1 million or above per milliliter until the pull-date was reached. Contaminating organisms did not appear to be affected by the lactobacilli. Deterioration of the milk occurred in a manner typical of plain pasteurized milk. Also there was no drastic decrease in numbers of *L. acidophilus* during the marked increase in numbers of other organisms after 4–7 days of storage.

In a recent study (Gilliland and Speck, 1977), *L. acidophilus* exerted appreciable antagonistic action during early stages of its growth against *Staphylococcus aureus, Salmonella typhimurium,* enteropathogenic *Escherichia coli,* and *Clostridium perfringens.*

In view of the apparent benefits derived by some from a large intake of *L. acidophilus,* the use of this product seems destined to increase.

E. Other Fermented Milks

Numerous other fermented milks have been described; many are more or less indigenous to certain countries or regions. For example, *kefir,* native to the Caucasus Mountain area in southwestern Russia between the Black and Caspian Seas, is made from the milk of several animals. It is largely a home-prepared product resulting from a mixed fermentation by

S. lactis, L. bulgaricus, and one or more lactose-fermenting yeasts. The organisms grow together in small masses called kefir grains that may be strained from the culture and used again to prepare additional batches. Kefir contains approximately 1% acid, approximately 1% alcohol, plus CO_2; the latter provides an effervescence.

Fermented mares' milk known as *kumiss* has been popular for centuries in various regions of Russia. Its preparation and presumed health-giving attributes were described recently by Berlin (1962). He stated that currently it is made only on collective and state farms mainly for local need, although some is sold in the market. For this purpose, 225,000 mares are kept. Furthermore, kumiss is prepared and used in special sanatoria offering "kumiss treatment" to as many as 11,000 people at one time. This requires milk from about 3500 mares.

Like kefir, kumiss is a product of a mixed lactic and alcohol fermentation. According to Berlin, a bulk starter containing *L. bulgaricus* and yeast(s) is built up over a period of 3–4 days by successive additions of fresh milk to the fermenting culture. The final product is prepared by adding bulk starter in an amount equal to about 30% of the freshly pasteurized milk to be fermented. Incubation is at 26°–28°C for about 1 hr, followed by agitation for 1 hr additional to enhance the growth of the yeasts. The product is bottled and incubated in the bottle for 1½ to 2 hr at 18°–20°C and then stored at 4°–6°C for 12 to 24 hr before use.

Taette is a Scandinavian product made primarily in the home by fermenting milk with a ropey strain of *Streptococcus lactis. Ymer* and *Lactofil* are trade names for a fermented milk made in Denmark and Sweden, respectively (Campbell and Marshall, 1975). The product, resembling a soft pudding, is the result of a mixed fermentation by *S. lactis, S. lactis* subsp. *diacetylactis,* and *Lactobacillus cremoris.* After fermentation, its volume is reduced by about 50% by removal of whey. The remaining curd is mixed with cream and homogenized. The product is eaten with a spoon.

Skyr is an Icelandic product apparently resulting from the fermentation of milk or skimmilk with *Lactobacillus thermophilus* and *L. bulgaricus.* After the fermentation some whey is removed and milk and sugar added. It, too, is eaten with a spoon.

Bulgarian buttermilk, the original product of Metchinkoff's investigation, is a pure culture of *L. bulgaricus* in milk.

VIII. CHOICE OF CASE

For an understanding of "Case", see Table A.1 and the brief description in Appendix IV. For more details, see the second book of this series (ICMSF, 1974).

ICMSF has proposed sampling plans neither for raw nor pasteurized fluid milks, which are rarely shipped in international commerce. However, Cases can be discussed in general terms.

Consumption of raw milk should be discouraged, and, therefore, one should consider here that it is always destined for pasteurization or other processing. There are unavoidable delays (as long as 3 to 4 days) between production and processing during which microorganisms may increase. The aerobic plate count (standard plate count—SPC) is the choice of test for raw milk, and Case 3 would apply.

In the normal course of events, pasteurized milk would be refrigerated at about 5°C until consumption, during which many bacteria enumerated by the SPC at 32°C could grow; thus Case 3 would apply. A test for coliform group as a measure of postpasteurization contamination would warrant Case 5, since these bacteria would survive storage at 5°C, but with few exceptions would grow hardly at all.

The SPC is frequently applied to dried milk; Case 2 applies because during subsequent storage, the levels will not change appreciably. *Staphylococcus aureus* is a hazard in dried milk but expected conditions of use would not increase the hazard. and Case 8 would apply. If, however, the dried milk were to be reconstituted then abused by holding it at a temperature permitting staphylococci to grow, the hazard would increase and Case 9 would apply. If dried milk were to be tested for *Salmonella,* Cases 10 through 15 would apply, depending upon expected use of the product. At one extreme the dried milk may be added as an ingredient to a food destined for cooking before it leaves the processing plant. At the other extreme it may be reconstituted for direct consumption by highly susceptible persons.

Levels of microorganisms in frozen dairy desserts (ice cream, etc.) remain fairly stable during storage. Therefore, for SPC and coliform group, Cases 2 and 5 would apply, respectively. For *Staphylococcus aureus,* Case 8 would apply. On the other hand, if liquid ice cream mix is shipped in commerce, the Cases would reflect the fact that it supports growth of a wide variety of microorganisms; Cases 3, 6, 9, etc. would apply.

Although most microbiological problems in cheese would not lend themselves well to ICMSF sampling plans (see Section VI,F), tests for *S. aureus* and enteropathogenic *Escherichia coli* would in most cases require Case 8 because their levels would likely decrease or remain unchanged during long periods of storage.

19

Eggs and Egg Products

I. INTRODUCTION

This chapter encompasses the microbiology of eggs and egg products, primarily from the domestic chicken. Eggs from other birds such as ducks, turkeys, geese, and guinea fowl appear in international commerce in much lesser tonnage.

A. Important Properties

The principal function of the egg is reproduction. Thus, it must nourish the chick until hatching (21 days for the domestic chicken) and must protect it against microbial entry from a generally filthy environment. The protective devices of the egg are highly complicated.

The egg consists of (1) the cuticle, a largely protein coating on the outside of the shell, (2) the shell itself, mostly calcium carbonate, (3) an outer coarse membrane, (4) an inner fine membrane, (5) the outer thin white,* (6) the thick white, (7) the inner thin white, (8) the chalaziferous layer culminating in the chalaza cords which anchor the yolk in the center of the egg, (9) the vitelline membrane that encloses the yolk, and (10) the yolk itself. Shortly after the egg is laid, water evaporation reduces the volume of the contents, and an air chamber (sac) forms between the inner and outer membranes at the blunt end of the egg.

Table 19.1 gives the proximate analysis of the hen's egg. Fat-soluble vitamins (A, D, E, and K) are present in the yolk; water-soluble vitamins (B complex) are present in either the white or yolk or both. Ap-

* The terms *white* and *albumen* are synonymous.

TABLE 19.1

Composition of the Hen's Egg [a]

Fractions in the entire egg contents	Amount	
	Grams	Percent
Water	38.0	73.6
Solids	13.6	26.4
Organic matter	13.2	25.6
Proteins	6.6	12.8
Lipids	6.1	11.8
Carbohydrates	0.5	1.0
Inorganic matter	0.4	0.8
Total	51.6	100.0

Fractions in the albumen (white)	Percent
Water	87.9
Solids	12.2
Organic matter	11.6
Proteins	10.6
Lipids	Trace
Carbohydrates	0.9
Inorganic matter	0.6
Total	100.1

Proteins in the albumen (white)	Percent of egg white solids
Ovalbumin	54
Conalbumin	13
Ovomucoid	11
Lysozyme	3.5
Ovomycin	1.5
Flavoprotein-apoprotein	0.8
Ovoinhibitor	0.1
Avidin	0.05
Globulins and others	8

Fractions in the yolk	Percent
Water	48
Solids	51.3
Organic matter	50.2
Proteins	16.6
Lipids	32.6
Carbohydrates	1.0
Inorganic matter	1.1
Total	99.3

[a] Adapted from Board (1969).

proximately equal amounts of trace minerals are present in the yolk and white, sometimes in combination with proteins and lipids.

For further information on the chemical composition of eggs, see Shenstone and Vickery (1958), Forsythe (1968), and the reviews by Board (1969), Lineweaver *et al.*, (1969), and Parkinson (1966).

B. Methods of Preservation

1. Shell Eggs

About 90% of eggs eaten are transported to the purchaser in the shell. The interior of most shell eggs contains no bacteria, and thus eggs require no special preservation procedure. Refrigeration (slightly above 0°C) slows the chemical and enzymatic activities that reduce functional quality and flavor, and retards the growth of microorganisms that may have entered the egg. Gross temperature and humidity changes can enhance entry of water and bacteria and thus have an adverse effect on keeping quality.

Dipping or spraying eggs with oil is a widely used commercial procedure to protect them against water and gas exchange.

2. Liquid, Frozen, and Dried Eggs

For commercial use in various foods, the eggs are broken from their shells and may then be mixed whole or separated into whites and yolks. They are sometimes sugared or salted for manufacture of other foods. The eggs are mixed, desugared by enzymic or microbial action (if destined for drying), pasteurized, and held in liquid form, or more commonly frozen or dried.

3. Further Processed Egg Products

Eggs are used in many ways in the preparation of processed foods— primarily in baked goods, confectionery, drinks, special diets, infant foods, sauces, mayonnaise, and noodles. Table 19.2 lists the principal egg products and their relative importance in the United States.

II. INITIAL MICROFLORA

Freshly laid eggs rarely contain microorganisms, although some may enter through the oviduct. Such entry is more common with ducks than with chickens. The usual organisms are diphtheroids or micrococci, and less frequently salmonellae, such as *Salmonella pullorum, S. gallinarum,*

TABLE 19.2

Uses of Egg Products in the United States [a]

Egg product	Use	Percent of total 1966 production [b]
Whole egg		
Frozen	Baked goods, noodles, institutional foods, mayonnaise, pet foods	31.5
Dried	Baked goods, soup, institutional foods, dry mixes, egg nog, ice cream, confections, pet foods	19.0
Total		50.5
Albumen		
Frozen	Baked goods, glue, confections, leather, chocolate	12.2
Dried	Baked goods, glue, chocolate, confections, dry mixes, leather	16.4
Total		28.6
Yolk		
Frozen, salted	Salad dressings, mayonnaise, soups, shampoo, leather	6.9
Frozen, sugared	Baked goods, egg nog, dairy and baby foods, ice cream	5.3
Frozen, plain	Baked goods, noodles, ice cream	4.6
Dried	Dry mixes, mayonnaise, macaroni, noodles, baked goods	4.4
Total		21.2

[a] Adapted from Forsythe (1970).

[b] Egg products make up about 10% of the eggs consumed in the United States; the remaining 90% are shipped in the shell to consumers. Calculated as liquid equivalent.

S. typhimurium, S. enteritidis, and *S. thompson.* Common spoilage organisms never enter by this route (Miller and Crawford, 1953; Jordan, 1956; Philbrook *et al.,* 1960; Board *et al.,* 1964).

The principal sources of microorganisms on the outside of eggs are the intestinal tract of the fowl, the nest box, dust, feedstuffs, shipping and storage containers, human beings, and other creatures. Some eggs come from small family farms, others from large semiautomated batteries of individually caged layers. The nature and amount of shell contamination will vary accordingly (Harry, 1963). Table 19.3 lists the types of organisms found on shells and shell-contact surfaces in one investigation. Note that the gram-negative organisms that are important in shell-egg spoilage are in the minority. In addition, salmonellae are found on the shells, particularly when the hen is infected. The numbers of bacteria vary widely, from 10^2 to 10^7/shell, depending on the amount of feces, dust, or soil present (Board, 1969).

TABLE 19.3

Types of Bacteria Found on Egg Shells, Fillers, and Flats [a]

| | Proportion of isolates | |
Types of bacteria	Number	Percent
Gram-positive cocci		
Micrococcus	282	43.7
Staphylococcus (mannitol —)	26	4.0
Gram-positive rods		
Arthrobacter	82	12.7
Bacillus	17	2.6
Gram-negative rods		
Pseudomonas		
fluorescent	8	1.2
nonpigmented, proteolytic	45	7.0
nonpigmented, nonproteolytic	31	4.8
Acinetobacter–Moraxella (Achromobacter)	11	1.9
Alcaligenes	12	1.9
Escherichia	39	6.0
Enterobacter (Aerobacter)	22	3.4
Aeromonas	6	0.9
Other gram-negative rods	64	9.9
Totals	645	100.0

[a] Adapted from Board *et al.* (1964).

III. SHELL EGGS

A. Factors Affecting Microbial Penetration and Growth

The egg is a remarkable natural package. The contents are as rapidly perishable as milk, yet the fragile shell, if undamaged and dry, will usually keep the egg edible for many months, even when stored at room temperature. Before spoilage or contamination with pathogens can take place, the microorganisms encounter several highly efficient barriers. As a result, nearly all shell eggs in commerce contain no bacteria within the edible portion. In one study (Haines, 1938), 98% of egg whites and 93% of egg yolks were sterile, and in another (Brooks and Taylor, 1955), 90% were sterile. Two factors are responsible: (1) the resistance to penetration of the shell and its membranes, and (2) the multiple factors in egg white that make it a poor medium for growth.

1. Resistance to Penetration

The shell structures can be listed in decreasing order of ability to retard penetration: cuticle > inner membrane > shell > outer membrane (Lifshitz et al., 1964). Obviously, cracks that penetrate the inner membrane allow microorganisms to bypass the barriers and permit immediate entry by spoilage and pathogenic bacteria. In the United States, cracked eggs whose whites are leaking to the outside surface must be rejected as inedible (USDA, 1975b). Similarly, if the shell is very dirty, the challenge is greater, so that bacteria will penetrate sooner and in greater numbers (Rosser, 1942; Hartung and Stadelman, 1963). See also Section III, A.

The outer surface of the shell is covered by the cuticle, also called bloom, a thin stratum of minute glycoprotein spheres. The cuticle extends far down into the pores of the guinea fowl egg but only a short distance into the pores of the chicken egg. With duck eggs, the cuticle tends only to cap the pore canals. The cuticle makes the shell resistant to the entry of water. If it is damaged, there is a greater susceptibility to microbial entry into the contents (Board and Halls, 1973; Seviour and Board, 1972). The amount of infiltration due to damage is probably related directly to the extent to which the pores are plugged with cuticle. It may sometimes be absent, and it is often damaged when the egg rolls down the floor of battery cages; it is removed by certain methods of cleaning dirty eggs. The cuticle is fairly resistant to water or detergents and to gentle rubbing with a cloth, but abrasion will damage it. Even localized damage may permit entry through a few holes. The protection offered by the undamaged cuticle generally lasts at least 4 days, after which it begins to fail, presumably be-

TABLE 19.4

The Effect of Egg-Shell Specific Gravity and Time of Bacterial Challenge on the Time of First Fluorescent Spoilage of Eggs [a,b]

Challenge time (min)	Specific gravity of the shell		
	1.070	1.077	1.085
1	8 [b]	10 [b]	12 [b]
3	4	10	12
5	3	11	12

[a] Sauter and Peterson (1969).
[b] Time is in days.

cause cracks develop when the cuticle dries (Baker, 1974). Eggs without cuticle, or eggs experimentally treated with chemicals for its removal, spoil much faster than normal eggs (Vadehra et al., 1970a).

The calciferous shell has multiple holes through which microorganisms can pass readily. The number and size of the holes is greatest at the blunt end, where permeability is also greatest (Vadehra et al., 1970b). Shells with high specific gravity offer more resistance to bacterial penetration. Table 19.4 shows that spoilage began in 3 days when shell specific gravity was low, but not until 10 to 12 days when the specific gravity was high. The percentages of eggs spoiled after 8 weeks at 10°C showed a similar difference (Table 19.5). The penetration by *Salmonella* is also more rapid through shells of low specific gravity (Table 19.6).

TABLE 19.5

The Effect of Egg-Shell Specific Gravity and Time of Bacterial Challenge on the Total Incidence of *Pseudomonas* Contamination of Eggs After 8 Weeks of Storage [a]

Challenge time (min)	Specific gravity (% infected eggs)		
	1.070	1.077	1.085
1	69.2	43.3	21.5
3	77.5	54.2	26.7
5	84.2	75.8	36.7

[a] Sauter and Peterson (1969).

TABLE 19.6

Proportion of Eggs of Different Shell Qualities Penetrated by Various Salmonellae in 24 Hr [a]

Selected species	Specific gravity of the shell (% infected eggs)		
	1.070	1.080	1.090
Salmonella anatum	19.4	7.5	3.8
S. brandenburg	68.1	17.1	7.2
S. typhimurium	82.1	48.7	21.2
Average of 12 *Salmonella* spp.	47.5	21.4	10.0

[a] Adapted from Sauter and Peterson (1974).

The outermost of the two shell membranes is very porous and does not provide a barrier to bacterial entry, but the inner membrane usually delays entry for a few days because of its fine structure (Gillespie and Scott, 1960; Elliott, 1954; Garibaldi and Stokes, 1958; Board, 1965a). The superior protection offered by the inner membrane is not because of its thickness and weight, for it is one-sixth as heavy and one-third as thick as the outer membrane (Lifshitz and Baker, 1964). The electron microscope shows that the inner membrane has no pores. Although some motile bacteria can penetrate by wriggling through the finely overlapped fibers (Baker, 1974), there is good evidence from electron microscope studies that the bacteria penetrate the membrane through the albuminous cementing matrix. The keratin core and its polysaccharide mantle are unaffected. Zones of hydrolysis surround the bacteria in the membranes, tending to support the theory (Stokes *et al.,* 1956) that enzymes are actively involved in the penetration (Brown *et al.,* 1965).

Bacteria enter only when the outside of the shell is wet, and particularly when it is dirty, and when a drop in temperature occurs. The reduction in temperature causes the air sac to contract and to pull water and bacteria through the shell and against the inner membrane. As the egg ages and the air sac grows in size, this effect becomes more pronounced. This may be one of the several factors that make stored eggs more susceptible to penetration.

Organisms such as *Pseudomonas* may colonize the external surface of the inner membrane, sometimes permitting fluorescent pigments to diffuse into the white before the bacterial cells actually penetrate (Elliott, 1954). Although *Pseudomonas* can grow on the membrane when it is isolated from other parts of the egg and immersed in saline (Board, 1965a; Elliott and

Brant, 1957), antibacterial activity of the membranes has been reported several times in the literature. The presence of lysozyme in both membranes (Vadehra *et al.,* 1972) could in part explain this phenomenon.

2. Antimicrobial Factors in the Albumen (White)

The albumen kills or prevents growth of a wide variety of microorganisms, whereas the yolk, or a mixture of the yolk and albumen does not (Haines, 1939; Brooks, 1960). Table 19.7 summarizes the principal factors in egg white that provide undesirable conditions for most bacteria. Lysozyme, conalbumin, and the alkaline pH are the most important. All the adverse conditions listed apply to the thick white; only the high pH applies to the two layers of thin white (Baker, 1974).

Lysozyme, first named in 1909 because it "lysed" (dissolved) bacterial

TABLE 19.7

Antimicrobial Factors in the Albumen of the Hen's Egg [a]

Component	Activity
Lysozyme (muramidase)	Lysis of cell walls of gram-positive bacteria
	Flocculation of bacterial cells
	Hydrolysis of β-1, 4-glycosidic bonds
Conalbumin	Chelation of iron, copper, and zinc, especially at high pH
Riboflavin	Chelation of cations
Glucose	Repression of respiratory capacity of facultative anaerobes(?)
pH 9.1–9.6 [b]	Enhances chelating potential of conalbumin
	Provides unsuitable environment for many organisms
Low nonprotein nitrogen	Fastidious organisms cannot grow
Avidin	Binds biotin, making it unavailable to bacteria that require it
Apoprotein	Combines with riboflavin
Ovoinhibitor	Inhibits fungal proteases
Ovomucoid	Inhibits trypsin, but does not affect growth of gram-negative bacteria
Uncharacterized proteins	Inhibit trypsin and chemotrypsin
	Combine with vitamin B6
	Chelate calcium
	Inhibit ficin and papain

[a] Adapted from Garibaldi (1960) and Board (1969).

[b] The pH of egg white in newly laid eggs is 7.6–7.8. After a few days at room temperature, during which CO_2 evolves, the pH rises to 9.1–9.6.

cells, has now been identified as a muramidase—so named because it attacks the murein layer or the murein sacculus of the bacterial cells. Gram-positive bacteria are especially sensitive to lysozyme, but gram-negative organisms are less so, possibly because the murein layer or sacculus is protected, or because it occurs in lesser quantity (Board, 1969). The alkaline conditions in the egg tend to make sensitive cells even more susceptible to lysis. The activity of lysozyme ceases as soon as the yolk mixes with the white (Galyean et al., 1972).

Conalbumin is important because it sequesters (chelates) metal ions, particularly iron, copper, and zinc, so that these metals are unavailable to bacteria. Many bacteria cannot grow at all in the presence of conalbumin. Those that can grow usually have a long lag phase and a decreased rate of multiplication in the log phase.

The gram-positive bacteria are more easily inhibited by conalbumin than are the gram-negative bacteria. One group of the gram-negatives, the pseudomonads, often produce in egg white a mixture of green fluorescent chelators, collectively called "pyoverdine." This material competes successfully with the conalbumin for metals necessary for microbial growth, but unlike conalbumin, the pyoverdine releases the metals to the bacteria (Elliott, 1954; Garibaldi, 1960, 1970; Elliott et al., 1964). Pyoverdine is associated with, or may be identical to, fluorescent hydroxymate transport compounds shown to neutralize the effect of conalbumin (Garibaldi, 1970).

Pyoverdine-producing members of the genus Pseudomonas penetrate and grow in shell eggs more quickly than any other group of bacteria, and they are frequently the only organisms present in stored eggs immediately on removal from storage (Lorenz et al., 1952). Other bacteria such as the Acinetobacter–Moraxella group (Achromobacter), Alcaligenes, Flavobacterium, and Proteus usually follow the primary invader (Florian and Trussell, 1957; Elliott, 1958; Ayres, 1960b), presumably because they are able to use the metals held by the pyoverdine in competition with the conalbumin. Salmonellae produce phenolate compounds that act in the same manner, permitting these organisms to enter and multiply (Garibaldi, 1970).

Experimentally added metal salts of iron (Garibaldi and Bayne, 1960), aluminum, copper, manganese, and zinc will saturate the conalbumin and release a variety of metals necessary for growth (Sauter and Peterson, 1969).

With the loss of carbon dioxide to the atmosphere (Fromm, 1963), the pH of egg white rises to 9.1–9.6 after 1–3 days of storage, a level at which most bacteria do not grow well. This pH, furthermore, enhances the chelating power of conalbumin (Board, 1969).

B. Effects of Processing on Microorganisms

1. Collection, Transport, and Storage

Eggs are collected by hand in small farms, or they roll by gravity from caged hens into troughs in large commercial operations. They are then placed by hand into pressed paper or polystyrene trays, and transported in cases to be candled (either singly by hand, or in special candling racks). Then they are washed, and perhaps oiled. Finally they are packed in cases, or in smaller packages, shipped by truck or rail, and stored for a few days or up to 90 days before consumption. For packaging, polystyrene was found to be more hygienic than pressed paper when both were wet; that is, it discouraged spoilage (Pfeiffer, 1972).

Nests must be clean and dry, and surfaces in contact with the shells free from visible feces and other soil, as well as water. Because eggs come from the cloaca and pass by the anus of the chicken, complete freedom from fecal smears is impossible. Wet feces-smeared nests, wet hands of collectors, the laying of eggs on dirty floors, and wet equipment all enhance penetration. A most critical period is during the time that the newly laid egg is cooling in the nest from its original temperature of 40°–42°C to ambient temperature.

Cold-stored eggs brought into a warm, moist atmosphere can become wet from condensation (sweating), and when returned to the cold room may be penetrated by surface bacteria as the air cell contracts from the chilling (Forsythe et al., 1953). Some investigators disagree, because they have not seen an increase in penetration except after several periods of alternating storage temperatures and the sweating that accompanies it (Vadehra and Baker, 1973). Sweating enhances spoilage to a greater degree when the outside of the shell is dirty than when it is clean (Forsythe et al., 1953). This could account for the discrepancies in results.

Penetration increases with duration of contact with contaminated material, and especially with high relative humidity during storage. This is true both of spoilage bacteria (Table 19.8) and salmonellae (Simmons et al., 1970). The relative humidity should be between 70 and 85% (Henderson and Lorenz, 1951). Below 70%, there is an unacceptable loss of weight by evaporation. Above 85%, bacterial penetration is enhanced, and in time, molds may grow, particularly in the air sac.

The storage temperature should be below 15°C, or preferably below 10°C, to inhibit bacterial growth and to slow loss of internal quality. Storage slowly breaks down the egg's natural barriers so that it becomes increasingly susceptible to bacterial entry and growth, whether or not it has been oiled (Elliott, 1954; Brown et al., 1966a).

Eggs are stored with the blunt end up, for this tends to keep the yolk,

TABLE 19.8

Effect of Relative Humidity (RH) and Wet Packing on the Incidence of Infection by Spoilage Organisms of Eggs Stored 3 months [a]

Prior treatment of eggs	Storage conditions	
	20°C, RH 60%	5°C, RH 90%
Not washed, packed dry	13 [b]	26 [b]
Washed, packed dry	22	23
Not washed, packed in wet packing material	16	52
Washed, packed wet	27	49

[a] Adapted from McNally (1952).
[b] % infected

which has a lower specific gravity than the white, from drifting toward the inner membrane. If it does drift upward into contact with the membrane, it offers any microorganisms that may be penetrating the membrane at that spot an opportunity to bypass the protective factors in the white, and thus spoil the egg more rapidly (Board, 1964; Brown *et al.,* 1970).

2. Cleaning Eggs

In most countries, purchasers demand shell eggs that are visibly clean; therefore dirty eggs are unsalable. Eggs can be either dry cleaned or washed. Dry cleaning is usually carried out by means of a stiff brush, sandpaper, or steel wool. Machine dry cleaners are often difficult to clean, and the frequent changing of brushes is expensive. Both dry cleaning and washing remove the cuticle, so that the eggs are more susceptible to penetration and spoilage if they subsequently become wet (Brown *et al.,* 1955). Dry cleaning usually results in a lower incidence of spoilage than does washing (Table 19.9), although it is a less efficient method of cleaning.

Most processors wash all eggs as received to avoid the labor of sorting, (Forsythe, 1970), although in some countries only the dirty eggs are washed. The temperature of the water should be 32°C or higher, and at least 11°C higher than that of the egg; otherwise, bacteria can be drawn through the shell to the underlying membranes (Haines, 1938; Haines and Moran, 1940; Brant and Starr, 1962). Even wiping the egg with a wet cloth can enhance penetration.

Washing in dirty water or in dirty machines increases spoilage. Recirculated water should be strained, and clean water should flow into the system to allow a continuous overflow of dirty water to the sewer. The tanks should be drained, cleaned, and refilled at least daily, and preferably each

TABLE 19.9

Proportion of Slightly Dirty Eggs Penetrated by Spoilage Bacteria during Storage for 9 Months at 1.7°–4.4° and 65–80% Relative Humidity, as Affected by Cleaning Method [a]

Cleaning method	Number of eggs tested	Percentage of eggs penetrated
Dry cleaned with mechanical sander	577	3.47
Washed in detergent, rinsed in water	276	7.25
Washed in detergent, no rinse	286	7.0
Washed in detergent-sanitizer, rinsed in water	278	13.3
Washed in detergent-sanitizer, no rinse	284	4.23

[a] Miller (1959).

4 hr during use. Washing should take place as soon as feasible after collection, because the microorganisms that have had time to penetrate to the inner membrane are not readily removed or destroyed.

Dirty eggs spoil more frequently and more rapidly than clean eggs, and washing usually increases spoilage during subsequent storage, regardless of whether the eggs were dirty before washing (Table 19.10).

A detergent added to the wash water improves the dirt-removing efficiency of the water. Only alkaline detergents are practical, because acid detergents attack the shell. Washing eggs experimentally with 1 to 3% acetic acid destroyed many microorganisms and cleaned the surface of the shell well, but reduced the thickness of the shell, and reduced the egg quality (Heath and Wallace, 1978). Simple alkaline compounds such as trisodium phosphate or sodium metasilicate are as satisfactory for this purpose as the more complicated proprietary mixes (Swanson, 1959). A good detergent will physically remove up to 92% of the bacteria on the shell surface (Forsythe et al., 1953; Bierer et al., 1961a,b).

TABLE 19.10

Effect of Washing Eggs on Spoilage during and after Storage [a]

Original condition	Washed(?)	Percent spoiled
Clean	No	0.55
Dirty	No	12.72
Clean	Yes	5.76
Dirty	Yes	19.93

[a] Adapted from Lorenz and Starr (1952).

A spray of freshly prepared disinfectant (sanitizer) over the egg surfaces after they have been cleaned destroys many of the remaining bacteria. The disinfectant of choice is chlorine at 100 to 200 ppm available chlorine (USDA, 1975a), although some investigators have shown that iodophors, quaternary ammonium compounds (Sauter *et al.*, 1962), or chlorine–bromine compounds (Forsythe, 1970) are more effective, particularly if permitted to remain on the egg shells without a final rinse. Some prefer detergent-sanitizers, because they clean and sanitize in one step (see Table 19.9). Organic matter in the wash water, however, will destroy much of the sanitizer's effectiveness, so a two-step (wash–sanitize) procedure is preferred (see Volume I, Chapter 14).

The effect of adding metals to overcome the chelating capacity of conalbumin has been described (see Section III,A,2). The practical result is that washing eggs in water with a high content of metal salts increases subsequent spoilage dramatically. Eggs washed in natural water containing 4.8 ppm iron showed 6.2% spoilage by pseudomonads on storage, whereas those washed with water containing 0.2 ppm iron showed only 0.8% spoilage (Garibaldi and Bayne, 1962). Whether hard water, high in calcium or magnesium, encourages spoilage has not been reported.

The final step in a washing operation is to dry the surface of the eggs before they are packed into containers. Otherwise, the moisture on the surface will draw in bacteria as the eggs cool again to ambient temperature (Table 19.8). Complete drying is unnecessary if the eggs are immediately broken to make liquid egg, so long as free water from the shell does not enter the liquid egg (USDA, 1975b). The International Association of Milk, Food, and Environmental Sanitarians (IAMFES, 1976d) gives further suggestions on good practice in egg washing. See also the review by Moats (1978).

3. Shell Coatings

Clean shell eggs sprayed with mineral oil (paraffin oil) are protected from loss of water and the associated increase in air cell volume during cold storage. Whether the oil protects against loss in functional quality of the egg (Sauter *et al.*, 1954; Shenstone and Vickery, 1958; Rutherford and Murray, 1963; Iotov *et al.*, 1974; Siddiqui and Reddy, 1974) or whether it increases or decreases the incidence of spoilage during subsequent cold storage (Lorenz and Starr, 1952; Bernstein, 1952) are matters for disagreement among investigators. Oil does not protect aged eggs from their high susceptibility to penetration and growth by spoilage bacteria (Elliott, 1954).

Oil is the only coating material now in general use. Waterglass (sodium silicate) was used for the retention of functional qualities in storage, but its popularity has waned. Its success probably depended on the interaction of the silicate with the shell, to produce an impervious calcium silicate. Experiments have shown that coatings of alginates, polymethacrylic acid, and certain butyl rubbers help to maintain egg quality as well as oil (Rutherford and Murray, 1963). There are reports that corn prolamine, polyvinylidene choloride, epolene wax emulsion, and hydrolyzed sugar derivative plus shellac all greatly retard penetration by *Pseudomonas fluorescens* and *Salmonella typhimurium* (Tryhnew *et al.*, 1973).

4. Pasteurization (Thermostabilization) of Shell Eggs

Surface cleaning, even with detergents and germicides cannot protect eggs against entry of spoilage organisms (Funk *et al.*, 1954) or salmonellae. However, from 1867 until the present, there have been many reports in the scientific and patent literature on the efficacy of heat to kill bacteria on and near the surfaces of shell and membranes and thus eliminate microbial entry and growth. Frequently, heat has been applied with coating materials—especially mineral oil—but more commonly, it has been applied with hot water, sometimes in conjunction with washing (Table 19.11).

All reports have been favorable, concluding that heat will destroy both spoilage and pathogenic microorganisms (i.e., *Salmonella*) on or near the shell and thus reduce hazards from pathogens and spoilage during cold storage. The heat establishes a nearly impervious layer of coagulated protein immediately beneath the shell membranes. This reduces evaporation in subsequent storage. There have been a few reports of loss to functional properties—for example, increased beating time (Goresline *et al.*, 1950; Knowles, 1956)—but most workers have not reported quality changes (Feeney *et al.*, 1953). The heat should be applied within 24 hr of collection because microorganisms that have penetrated to the white already will not be destroyed by the minimal heat applied (Feeney *et al.*, 1954).

C. Spoilage

In Sections III,A,1 and III,A,2, above, the barriers to bacterial entry and growth within the egg have been described, as have factors that permit bacteria to overcome these barriers. The fluorescent pseudomonads, originating in soil or water, are frequently the first to enter and grow because they are motile, produce a fluorescent pigment that competes for metals with the conalbumin of the white, and are resistant to other protective mechanisms of the white. Nearly all spoilage during and immediately after removal from storage is caused by these fluorescent pseudomonads (Lorenz and Starr, 1952). An egg showing bright fluorescence in most or all of the white when examined with a candler emitting long-wave ultraviolet

TABLE 19.11

Recommended Times and Temperatures to Thermostabilize Shell Eggs

Authors	Date	Temperature °C	Time	Heating menstruum, coatings
Atwell and Crawford	(1867)	100+	5–10 sec	Water with glue, lime, corn starch, isinglass
Gray	(1887)	100	5 sec	Water, then grease and soda mix (boiling)
Rylander	(1902)	80	?	Air with salicylic acid, then potassium silicate in water
Bache-Wiig	(1903)	60–66	?	Waste sulfite lye
Jacobsen	(1910)	100	5 sec	Water with 15 g alum per 20 eggs
Clairemont	(1914)	120	Momentary	Coat with cactus juice
Thornburg	(1915)	43–55	Several hours	
Almy et al.	(1922)	320 and 610	5 sec	Air
Henderson	(1916)	99–107	5 sec	Oil
Funk	(1943)	60	10 min	Oil
	(1950)	54	15 min	Water
Murphy and Sutton	(1947)	54	15 min	Water
Salton et al.	(1951)	60	320 sec	Water
		62	128 sec	Water
Goresline et al.	(1950)	54–58	16 min	Oil
Funk et al.	(1954)	54	15 min	Water
		60	5 min	Water
Feeney et al.	(1953, 1954)	100	2–3 sec	Water
Winter et al.	(1954)	67	2 min	Oil
		62	2 min	Water
Bierer and Barnett	(1961)	66	1–3 min	Water
Scott and Vickery	(1954)	57.5	13½ min	Water
		62.5	2.1 min	Water
Vadehra and Baker	(1967)	60	4 hr	Air

light (black light) always contains a high number of bacterial cells. Such eggs are not readily detected using a white-light candler, and the odors of decomposition in the early stages are mild, often detectable only after soft-cooking (Elliott, 1954). There are, of course, other organisms that also are capable of entering as primary invaders. One investigational team listed three strains of *Alcaligenes,* two of *Proteus,* and one each of *Flavobacterium* and a probable *Citrobacter* (*Paracolobactrum*). They also found other strains that would grow in eggs but that were incapable of primary invasion, namely, two strains each of the *Acinetobacter–Moraxella* group (*Achromobacter*), *Alcaligenes, Escherichia, Flavobacterium,* and a probable *Citrobacter* (*Paracolobactrum*) and one each of *Enterobacter* (*Aerobacter*), and a coli-aerogenes intermediate (Florian and Trussell, 1957).

The nature of a rot depends on the bacterial strain or mixture of strains present. For example, the nonproteolytic *Pseudomonas putida* produces fluorescence in the white, whereas the lecithinase-producing *Pseudomonas fluorescens* breaks down the diffusion barrier at the surface of the yolk and turns the white pink. This is probably due to Fe^{3+} ovotransferrin chromogen. *Pseudomonas maltophilia* produces a characteristic "nutty" odor and causes a slight crusting of the yolk with streaks of ferric sulfide on the surface.

The strongly proteolytic organisms, *Proteus vulgaris* and *Aeromonas,* digest the white and blacken the yolk. Some other organisms do not cause macroscopic changes but can form populations as large as those of the "rot producers" (Board, 1965b). They include *Alcaligenes faecalis, Enterobacter* (*Cloaca*) spp., and *Pseudomonas fluorescens.* These organisms might not be detected at candling or when the eggs are broken out, and thus they would contaminate egg products (Johns and Berard, 1945, 1946). Table 19.12 shows the frequency of occurrence of various types of rotten eggs and the genera of bacteria that have been isolated from them.

Molds grow occasionally in the air cells of oiled eggs, but rarely in unoiled eggs. Growth is usually self-limiting because of depletion of oxygen (Brown and Gibbons, 1954).

Refrigeration is of value primarily to maintain egg quality, but it also delays the spoilage of shell eggs that have been exposed to conditions encouraging bacterial entry (Vadehra *et al.,* 1969a). The pseudomonads are favored in cold-stored eggs (Lorenz and Starr, 1952; Ayres and Taylor, 1956).

D. Pathogens

Salmonellae are the only important human pathogens carried by eggs. Some strains of salmonellae can enter the egg through the hen's ovaries

TABLE 19.12

Bacterial Genera Isolated from Various Types of Rotten Shell Eggs [a]

Type of rot, in decreasing order of frequency	Bacterial genera isolated
Green	*Pseudomonas*
Colorless	*Acinetobacter–Moraxella* *(Achromobacter)*
Black	*Pseudomonas* *Proteus* *Escherichia* *Alcaligenes* *Enterobacter (Aerobacter)*
Pink	*Pseudomonas*
Red	*Pseudomonas* *Serratia*

[a] Alford *et al.* (1950); Florian and Trussell (1957).

(Section II), but more commonly they enter from the outside of the shell, in the same way as other bacteria—namely, when the outside of the shell is dirty with feces, when it is wet, and when the egg is cooling.

Table 19.13 shows that washing fails to remove salmonellae from shells and membranes. Sanitizers are only partially effective (Solowey *et al.,* 1946; Mellor and Banwart, 1965), but heating (thermostabilizing) the shell destroys surface salmonellae along with other bacteria (Section III,B,4).

On rich media, and under otherwise optimal conditions, the maximum temperature for growth of salmonellae ranges from 43.2° to 46.2°C (Elliott and Heiniger, 1965), and the minimum temperature for growth, from

TABLE 19.13

Incidence of Salmonella on Shells and Membranes of Clean, Dirty, and Washed Dirty Shell Eggs Received at Six U.S. Processing Plants [a]

Sample	Percentage containing *Salmonella*		
	Clean eggs	Dirty eggs	Washed dirty eggs
Shells, outer surface	1.3	4.7	4.8
Ground shells	2.3	15.7	12.9
Shell membranes	0	9.0	15.4

[a] Adapted from Solowey *et al.* (1946).

5.3° to 6.8° C (Matches and Liston, 1968b). However, under the less than optimal conditions on and in the egg, salmonellae cannot penetrate and grow below 10°C (Stokes *et al.,* 1956; Ayres and Taylor, 1956; Simmons *et al.,* 1970). At temperatures between that of a cool room (15°C) and that of the newly laid egg (40°C), salmonellae can penetrate and grow rapidly (Stokes *et al.,* 1956; Licciardello *et al.,* 1965), secreting their own chelators to compete with conalbumin (Section III,A,2). Their numbers within the egg can reach more than $10^8/g$ (Stokes *et al.,* 1956). Such eggs, eaten raw, would be hazardous, and indeed, sporadic cases of salmonellosis after eating single eggs in shell have been reported occasionally. Eggs boiled long enough to solidify the yolk (for example, 15 to 20 min) are sufficiently hot to kill all salmonellae, but soft-boiled eggs (for example, 4 min) are not (Ministry of Health, 1954; Licciardello *et al.,* 1965). Large-scale samplings have shown that not more than about 0.2% of the shell eggs of commerce have been invaded by salmonellae (Ministry of Health, 1954; Peel, 1976). However, salmonellae do not usually alter the odor or appearance of eggs (Vadehra and Baker, 1973), so that an egg containing millions of salmonellae could pass inspection. One such egg could contaminate an entire batch, or even more than one batch, of liquid egg (Stokes *et al.,* 1956; see also Section IV,A,1).

E. Control

Eggs should be laid in clean nests or in clean cages and kept dry at all stages of handling and transport except for the washing operation. They should be collected at least daily, preferably oftener, and chilled promptly to below 10°C. There should be a candling step to remove as inedible all spoiled, leaking, or otherwise unacceptable eggs, using white-light and/or black-light candlers.

If the eggs are washed, the water should be at 42°C or higher, and in any case the water should be at least 11°C warmer than the eggs. The water should be potable and low in iron content. It should contain an alkaline detergent, such as sodium metasilicate or trisodium phosphate, and should be continuously replenished to allow an overflow. The cleaned eggs should be rinsed in a spray of fresh water containing a suitable disinfectant such as chlorine at 100 to 200 ppm. The washing machine should be emptied, cleaned, and refilled with clean detergent solution at least daily.

The shells should be dried, then recooled to below 15°C, or preferably below 10°C, but they must not freeze because freezing breaks the shell. Movements out of and into storage in a manner to permit condensation on the shell surface (sweating) should be avoided. All surfaces in contact with the shells should be dry.

Equipment and practices should conform to regulations established by suitable authorities (e.g., U.S. Department of Agriculture, 1975a; International Association of Milk, Food, and Environmental Sanitarians, 1976d).

Shell eggs can be thermostabilized (pasteurized) by immersing them into hot water (for instance, 100°C for 5 sec), followed by rapid cooling. Alternatively, the washing operation can be operated at a high enough temperature to thermostabilize the eggs.

These steps will help control both spoilage organisms and salmonellae. In addition, poultry and egg producers should try to break the *Salmonella* cycle by the development of *Salmonella*-free breeding flocks, the fumigation or sanitization of hatchery eggs, and the use of *Salmonella*-free feeds (Marthedal, 1973).

IV. LIQUID AND FROZEN EGGS

Eggs for further processing are broken out either by hand or machine, examined for evidence of spoilage by sight or smell, then homogenized as whole egg or separated into white and yolk. Salt or sugar may be added to yolks destined for further processing. (See Table 19.2 for the relative importance of such products.) All liquid egg products then should be pasteurized, chilled, filled into cans or tanks, and shipped chilled or frozen.

The flora of liquid egg consists of many kinds of gram-positive and gram-negative bacteria that come mainly from the shell which is contaminated with fecal and other matter, second from contaminated an occasional egg, and third from equipment such as breaking utensils, pipes, pumps, filters, pails, churns, and holding tanks. Equipment used continuously for long periods of time is difficult to clean. There can be pockets of liquid and semistagnant films where bacteria will accumulate and multiply.

A. Effects of Processing on Microorganisms

1. Breaking, Separating, and Homogenizing

The primary source of bacterial contamination of liquid egg is the shell. Unless eggs are already clean, they should be washed immediately before the breaking operation. They need not be dried after washing so long as they drain enough that water from the shells does not run directly into the liquid egg product (U.S. Department of Agriculture, 1975b). In one investigation, washing dirty eggs before breaking reduced the aerobic plate count of the liquid egg from 1,500,000/g to 3000/g (Penniston and Hedrick, 1947).

A single spoiled egg can contaminate breaking equipment and add millions of bacteria to the liquid egg. Candling before the breaking operation will detect most spoiled eggs, but some types of spoilage are hard to detect. For example, fluorescent rots by *Pseudomonas* are hard to see using only a white-light candler. Their mild odors are also deceiving (Johns and Berard, 1945; Elliott, 1954; Mercuri *et al.,* 1957). Similarly, the *Acinetobacter–Moraxella* group (*Achromobacter*) can produce colorless rots that can enter the liquid egg undetected. With automatic breaking machines, examination for rots is even more difficult. Efficient use of such machines depends on a uniform supply of sound, clean, unspoiled eggs (Forsythe, 1970). Table 19.14 shows how spoiled eggs can contaminate breaking equipment. First, eggs in four sets of 102 each were inoculated heavily with *Serratia marcescens,* as a "tracer" organism. These four sets of eggs (Samples 1 to 4) were broken on the machine, followed immediately by six additional sets, each consisting of 102 clean uninoculated eggs. *Serratia marcescens* remained in moderate numbers on the equipment and contaminated all of the six clean sets of eggs. Such data support the requirement that equipment contaminated by spoiled eggs be removed, washed, and sanitized before reuse (U.S. Department of Agriculture, 1975b).

Crushing the entire egg, so that the melange becomes intimately mixed

TABLE 19.14

Contamination of Liquid Egg by the Machine-Breaking of Eggs Infected with *Serratia marcescens* [a]

Sample order of eggs	*Serratia marcescens* per ml of liquid egg
1 [b]	350,000
2	24,000
3	62,000
4	55,000
5 [c]	7,900
6	1,600
7	800
8	400
9	1,200
10	800

[a] Kraft *et al.* (1967a).

[b] Each sample represents the contents of 102 eggs. Samples 1 to 4 were from eggs infected with *S. marcescens.*

[c] Samples 5 to 10 were from fresh uninfected eggs broken on the same machine after samples 1 to 4, without cleanup.

with shell, can lead to heavy contamination. However, if the eggs are surface-sanitized, rinsed, and dried immediately before crushing, contamination is minimal and the product acceptable. The use of cracked eggs will increase the rate of contamination by salmonellae and spoilage organisms in the various egg mixes (Baker, 1974).

Broken-out incubator-reject eggs, previously stored at 40°C for up to 18 days, will have a different flora from that usually found in the mix from fresh eggs. The liquid may contain organisms predominantly of one species, such as *Micrococcus* or even *Salmonella*. On the other hand, Dutch investigators (Janssen and Mulder, 1971; Mulder and van der Hulst, 1973) found that liquid egg from incubator rejects contained a wide variety of organisms from families Bacillaceae, Enterobacteriaceae, Lactobacillaceae, Micrococcaceae, and Pseudomonadaceae. Pasteurization destroyed most of the organisms (see Table 19.15). The use of incubator-reject eggs is not permitted in some countries (USA and Canada). In some other countries, incubated eggs that have not remained longer than 6 days in the incubator may be used in the manufacture of pasteurized egg products.

Homogenization of eggs is usually accomplished in a large mixing vat called a churn. Here the individual eggs, white, or yolks are thoroughly mixed, and during the process, any microbial contamination is uniformly distributed throughout the batch.

2. Pasteurization

Pasteurization is carried out primarily to kill salmonellae. All liquid mixes except those intended solely for production of high-acid dressings and mayonnaise ($< pH 4.1$) should be pasteurized.

Salmonellae are fortunately not particularly resistant to heat. However, in egg products, surrounded by proteins and fats, their resistance increases. Unfortunately, the times and temperatures that kill salmonellae in egg products are at or near those that adversely affect the physical and functional properties of the egg products. The albumen (white) is the most sensitive, for it is denatured in a very few minutes at or above 60°C, whereas whole egg is reasonably stable at this temperature.

With a few rare exceptions, the salmonellae are uniformly destroyed by the recommended pasteurization treatments. The principal strain that is not destroyed is *Salmonella senftenburg* 775W, isolated from eggs in 1946. Among hundreds of isolates tested over a period of over 30 years, no one has ever again found this or another strain with comparable heat resistance. Therefore, pasteurization times and temperatures have been designed to destroy the typical, less-resistant strains. The decision was made in the late 1950s that the quality of the pasteurized product should not be sacri-

TABLE 19.15

Number of Various Bacteria in Whole Egg from Incubator-Reject Eggs, before and after Pasteurization for 3 Min at Various Temperatures [a,b]

Bacterial group	Whole egg Before pasteurizing	Whole egg Pasteurizing temperature			Whole egg + 30% sucrose Before pasteurizing	Whole egg + 30% sucrose Pasteurizing temperature		
		63°C	65°C	67°C		69°C	73°C	77°C
Aerobic plate count	7.8	4.4	4.2	4.1	7.4	6.1	4.2	4.2
Enterobacteriaceae	7.3	<1	<1	<1	7.1	<1	<1	<1
Pseudomonadaceae	6.2	<1	<1	<1	—	<1	<1	<1
Micrococci	7.4	4.4	4.3	4.2	—	6.1	4.0	4.1
Fecal Streptococci	7.0	4.0	3.3	<2	—	<2	<2	<2
Lactobacillaceae	2.2	<1	<1	<1	—	<1	<1	<1
Thermophilic bacilli	3.9	3.7	3.8	3.8	—	4.6	4.4	4.2
Escherichia coli	7.1	<1	<1	<1	—	<1	<1	<1
Yeasts and molds	5.5	<1	<1	<1	—	<1	<1	<1
Staphylococcus aureus	<1	<1	<1	<1	—	<1	<1	<1

[a] Janssen and Mulder (1971).

[b] Numbers are $\log_{10}$.

ficed to protect consumers against a strain that occurs so infrequently (Ng, 1966; Lineweaver *et al.,* 1967).

The degree of protection offered by pasteurization is related to the numbers of salmonellae originally present. Most recommendations for egg pasteurization reduce the level of salmonellae in inoculated eggs by 1000- to 10,000-fold (3 to 4 logs: see Tables 19.16 and 19.17). Because the level in the unpasteurized product is usually very low, the usual recommendation of time and temperature gives a broad safety margin, reducing the level to fewer than 1 in 10^3 to 10^5 g. Such a process would eliminate virtually all salmonellae, even if the initial level were $> 10^2/g$. The fact that initial contamination is usually much lower is shown in Table 19.18. In this summary of three investigations, 93% of the positive samples contained fewer than 1 *Salmonella* cell/g, and none contained more than 110/g. That is not to say that a batch of eggs with large numbers of salmonella cannot be found, but it certainly would be exceptional.

One troublesome problem is "tailing." That is, when liquid egg is inoculated with overwhelming numbers of salmonellae (say $10^9/g$) and pasteurized, a few highly resistant salmonellae survive (Dabbah *et al.,* 1971). This could account in part for *Salmonella* isolations from occasional samples of pasteurized egg products (Peel, 1976).

Recommended pasteurization temperatures for liquid eggs that have received no chemical additives vary from 55.6° to 69°C, and the times of exposure vary from 1½ to 10 min (Table 19.19). Lower temperatures and shorter times increase the risk of survival of salmonellae, and higher temperatures and longer times increase the damage to functional properties (whipping, emulsifying, binding, coagulation, flavor, texture, color, and nutrition) (Forsythe, 1970).

The pasteurization processes required in the United States are listed separately in Table 19.20. These figures take into consideration the heat susceptibility of various *Salmonella* strains in the egg product, and the flow properties of the egg material within the pasteurizer. In the case of egg white, for instance, the average particle will stay at 60°C for 3.5 min, and the most rapidly moving particle will stay at 60°C for 1.75 min.

The British require pasteurization at 57.2° to 57.8°C for 2½ to 3 min for egg white or 64.4°C for 2½ min for liquid whole egg (Corry and Barnes, 1968; Hobbs and Gilbert, 1978).

The Dutch industry uses the following:

Liquid whole egg:	64.5°C for 2½ min
Plain yolk:	64°C for 6 min
10% sugared yolk:	68°C for 10 min
10% salted yolk:	68°C for 10 min
Egg white:	55°C for 20 min

TABLE 19.16

Reduction of Numbers of Various Bacteria during Pasteurization of Liquid Egg White [a, b]

°C	Holding time (min)	pH	Aerobic plate count	Salmonella [a]	Coliform group	Streptococcus faecalis	Reference
57.2	1	9.0	4	—	>3	—	Ayres and Slosberg (1949)
57.2	10	9.0	5	—	>3	—	
60	"flash"	9.0	2	—	>3	—	
55.6	4	—	2.6	>4	—	—	Kline et al. (1966)
56.7	2	—	2.6	>4	—	—	
57.2	2½	9.2	2.0	—	>5.4	0.6	Barnes and Corry (1969)
		9.3	2.7	—	>5.9	>4	

Pasteurization scheme (column group over °C, Holding time (min), pH)

[a] *Salmonella senftenberg* 775W not included.
[b] Reduction of numbers, decrease $\log_{10}$.

TABLE 19.17

Reduction of Numbers of Various Bacteria during Pasteurization of Liquid Whole Egg

Reference	Pasteurization scheme		Reduction in bacterial level, $\log_{10}$			
	°C	Holding time (min)	Aerobic plate count	Salmonella [a]	Coliform group	Escherichia coli
Gibbons et al. (1946)	60	10	—	4.3[b]	—	—
		15	—	—	—	3.6[b]
		20	—	5.6[b]	—	—
		25	—	>7.3	—	—
		30[d]	2	—	—	4.1[b]
Goresline et al. (1951)	62.8	"flash"	0.6–1.3	+[c]	>4.8	—
Murdock et al. (1960)	63.9	2½	1.3–3.3	—	>4	—
Mulder and van der Hulst (1973)	64.4	3[b]	3.5	—	—	—
Murdock et al. (1960)	65.2	9	2.2–2.3	—	>2	—
Heller et al. (1962)	66.1–62.8	2½	>5.1	>6	>3	—
Murdock et al. (1960)	67–68	1¾	1.7–2.8	—	>2	—

[a] Salmonella senftenberg 775W not included.
[b] Counted by use of nonselective media.
[c] Salmonella present after pasteurization.
[d] Streptococcus faecalis was reduced 1.2 logs.
[e] S. faecalis was reduced 3.7 logs, and Enterobacteriaceae, >6 logs.

TABLE 19.18

Level of Salmonellae in 100 Samples of Unpasteurized Egg Products Previously Determined by Qualitative Test to Contain Salmonellae [a]

Most probable numbers (MPN) (per gram)	Number of samples
<1	86
1.4–2.9	10
5.3	1
24	2
110	1

[a] Garibaldi *et al.* (1969b).

Adding 10% salt to egg yolk increases the heat resistance of salmonellae from 2- (Cotterill and Glauert, 1971) to 10-fold (Garibaldi *et al.*, 1969c). See Table 19.21 for more details. Thus, the pasteurization procedure recommended for this product is somewhat more severe (68°–69°C for 3.75 min; Cotterill and Glauert, 1969). Unfortunately, the salt increases the resistance of salmonellae during the period of pasteurization so that a serious "tailing" takes place with such exposure to heat. A temperature of 73°C, therefore, would be preferable; however, damage to the egg product makes such abuse unfeasible, and one of the following four other procedures appears advisable:

1. If the salted yolks are destined for high-acid salad dressing or high-acid mayonnaise manufacture, where the low pH will destroy salmonellae, no pasteurization is necessary.
2. The yolks can be pasteurized first, then the salt mixed in afterward, with precautions to avoid contamination.
3. Acetic or other organic acid can be added to drop the pH to 4.6, whereupon the salmonellae die much more rapidly. Acidified salted yolks can be pasteurized in 1 min at 60°C (Garibaldi, 1968).
4. A batch pasteurization system at 52°C for 72 hr can be used. This temperature is above the growth range of common egg contaminants, including pathogens, and below the egg coagulation temperature. Although preheating to 52°C must be rapid to prevent spoilage, rapid cooling is not necessary because most of the bacteria, including spoilage organisms, salmonellae, and staphylococci, are dead (Cotterill *et al.*, 1974).

Adding 10% sucrose also increases thermal resistance of salmonellae, but to a lesser degree, as shown in Table 19.21. Also note, in this table,

TABLE 19.19

Recommended Pasteurization Temperatures for Liquid Egg Products

Time (min)	Temperature (°C)	Source
Liquid whole eggs		
2	61.1	Winter (1952)
3	60	Goresline et al. (1951)
2.6	60	Winter et al. (1946)
3.5	60	Anellis et al. (1954)
2–4	63.3	Knowles (1953)
2.5	63.9–65	Murdock et al. (1960)
3.5	55.6	Bergquist (1966)
2.5	63.3	China [a]
3	66.1–67.8	Poland [a]
2.5	62.5	Australia [a]
1.5–3	65–69	Denmark [a]
2.5	63.9–65	England [a]
2.5	64.4	England [b]
9	65.2	Germany [a]
2.8	61.7–65.6	N. Ireland [a]
3.5	60	Lineweaver et al. (1969)
10	57 [c]	Brant et al. (1968)
2.5	64.4	Reinke and Baker (1966)
Yolk		
3	65–69	Denmark [a]
3.5	61.1	Lineweaver et al. (1969)
4	60	Paul et al. (1957b)
2	61.1	Paul et al. (1957b)
1	62.2	Paul et al. (1957b)
Salted yolk		
3.5	63.3	Lineweaver et al. (1969)
3.75	68–69	Cotterill and Glauert (1969)
Sugared yolk		
3.75	63	Cotterill and Glauert (1969)
3.5	63.3	Lineweaver et al. (1969)
Whites		
2	56.7	Winter (1952)
3.5–4	56.7	Kline and Sugihara (1966)
3.5	56.7 (pH 9)	Lineweaver et al. (1969)
3.5	60 (pH 7)	Lineweaver et al. (1969)

[a] From a review by Murdock et al. (1960).
[b] Lineweaver et al. (1969).
[c] Batch process.

TABLE 19.20

Requirements for Pasteurization of Liquid Egg Products in the United States [a]

Liquid egg product	Minimum temperature (°C)	Minimum time (min)
Albumen (without use of chemicals)	56.7	3.5
	55.6	6.2
Whole egg	60	3.5
Whole-egg blends (less than 2% added	61.1	3.5
nonegg ingredients)	60	6.2
Fortified whole egg and blends	62.2	3.5
(24–38% egg solids, 2–12%	61.1	6.2
added nonegg ingredients)		
Salt whole egg (with 2% or more salt added)	63.3	3.5
	62.2	6.2
Sugar whole egg (2–12% sugar added)	61.1	3.5
	60	6.2
Plain yolk	61.1	3.5
	60	6.2
Sugar yolk (2% or more sugar added)	63.3	3.5
	62.2	6.2
Salt yolk (2–12% salt added)	63.3	3.5
	62.2	6.2

[a] U.S. Department of Agriculture (1975b).

the effect of other chemical additives on the thermal resistance of salmonellae.

Pasteurizer temperatures often cause egg material to coagulate on the hot surfaces of the plates (Ling and Lund, 1978) and also adversely affect the functional quality of the product. Therefore, research has centered on means to do the following:

1. repair the damaged quality by adding chemicals such as whipping aids at point of use,
2. increase the sensitivity of *Salmonella* to heat by adding chemicals or altering pH,
3. reduce the adverse effects on quality by adding chemicals before pasteurization.

Of all the egg products, egg white is the most sensitive to heat. Heating unaltered egg white at 62°C for 3.5 min alters 3 to 5% of the ovomucin, 90–100% of the lysozyme, and over 50% of the conalbumin (Lineweaver

TABLE 19.21

Heat-Resistance Characteristics of Salmonellae in Various Egg Products

Test organism	Product	D value (min) at °C							z-value (°C)	Reference
		50	52	54	56	58	60	62		
Salmonella oranienburg	whole egg pH 7.0	—	—	4	1.7	—	—	—	4.5	Cotterill *et al.* (1973)
	whole egg + 10% sucrose	—	—	25	10	3.5	—	—	4.3	
	whole egg + 10% salt	—	—	30	12	6	3	—	7.0	
S. typhimurium	whole egg pH 5.5	—	—	17	5.5	—	—	—	4.2	Lategan and Vaughn (1964)
	whole egg + sorbic acid 100 ppm	—	—	14	4.5	—	—	—	4.1	
	whole egg + β-propiolactone 100 ppm	—	—	11	4.5	—	—	—	4.5	
	whole egg + benzoic acid 500 ppm	—	—	6.5	2.2	—	—	—	4.1	
S. typhimurium	Egg yolk	—	—	9.5	3	1.3	0.4	—	4.4	Garibaldi *et al.* (1969c)
	Egg yolk + 10% sucrose	—	—	75	25	10	4	—	4.8	
	Egg yolk + 10% salt	—	—	110	45	17	6	2	4.6	
S. typhimurium	Egg white pH 9.2	10	3.5	1	—	—	—	—	4.2	Garibaldi *et al.* (1969c)
	Egg white pH 9.2 + 10% sucrose	20	7	2.5	0.8	—	—	—	4.3	
	Egg white pH 7.3 + aluminum salt	—	22	6.5	2.5	—	—	—	4.2	
Mix of 3 strains	Egg white pH 9.5	9	3.5	1.1	—	—	—	—	4.8	Kohl (1971)
	Egg white pH 9.5 + 0.5% polyphosphate	1.1	0.6	0.3	—	—	—	—	4.6	

et al., 1967). Even 3.5 min at the recommended 56.7°C increases markedly the whipping time to prepare meringue. Even a minimal 3 min at 54.4°C doubled the whipping time, but whipping aids such as triethyl citrate or triacetin restored this function to nearly normal (Kline *et al.*, 1965). In this range, a 2°C rise in temperature increases the damage 2½- to 3-fold, whereas it increases the destruction rate of salmonellae only 2-fold. Because of this, and because of the buildup of coagulated material on the pasteurizer plates at higher temperatures, it is inadvisable to pasteurize unaltered egg white above 60°C (Kline *et al.*, 1965; Lineweaver *et al.*, 1967). Authorities in the United Kingdom have recommended 57°C for 2½ to 3 min (Corry and Barnes, 1968; Hobbs and Gilbert, 1978).

Unaltered whole egg is less sensitive to damage, in part because iron from the yolk satisfies the chelating capacity of the conalbumin, and in doing so, stabilizes it (Cunningham, 1966). Egg yolk is likewise relatively stable but is difficult to handle because of its high viscosity.

Lower temperatures are feasible with the use of H_2O_2, which, when added to eggs at a level of 0.1%, increases the heat sensitivity of salmonellae. For example, egg white at its normal pH 9 can then be pasteurized at 51.7°C in 2 min. A catalase treatment after pasteurization breaks down the excess H_2O_2 (Lloyd and Harriman, 1957; Rogers *et al.*, 1966).

An adjustment of the pH of egg white from its normal level near 9 to 6.5–7 increases the stability of both the egg white and *Salmonella* to heat, but *Salmonella* less so than the white (Lineweaver *et al.*, 1967). At pH values below 7, the salmonellae become much more susceptible, especially in the presence of organic acids, because the undissociated organic acids contribute their own antimicrobial action in addition to that of the hydrogen ion itself (Chapter 7, Volume I). By this means, heat required to kill salmonellae in eggs can be reduced by a pH adjustment to 5.5 or 6 with lactic, acetic, formic, or propionic acids (Lategan and Vaughn, 1964).

Unfortunately, merely reducing the pH of egg white to increase bacterial destruction by heat is insufficient because of the adverse effect on the egg white proteins, primarily conalbumin. The addition of a metal salt satisfies the chelating power of the conalbumin, and in doing so, makes it relatively stable to heat. At the same time, adjustment of the pH to 7 with lactic acid enhances the stability of ovalbumin, lysozyme, and ovomucoid (Cunningham and Lineweaver, 1965). Either Fe^{3+} or Al^{3+} will work well, but Fe^{3+} turns the albumen pink, so Al as aluminum sulfate is the metal of choice (Cunningham, 1966). With the pH adjustment to 7 and the addition of Al^{3+}, the egg white can be pasteurized at 60°–62°C for 3½ to 4 min. The total amount of protein altered is < 1%; however, whipping time is still increased and the white requires the addition of a whipping aid (Lineweaver *et al.,* 1967). Aluminum sulfate is not permitted as a food additive in some countries.

Another way to stabilize conalbumin is to add sodium polyphosphate. Experiments have shown that 0.5 to 0.75% sodium polyphosphate in egg white permits effective destruction of salmonellae at 52.2° to 55°C for 3.5 min, without damage to functional properties (Chang *et al.,* 1970; Kohl, 1971).

Further research on chemical additives is even more promising. Disodium ethylenediamine tetra-acetic acid (EDTA) at a level of 7 mg/ml of egg white killed 1 million salmonellae in less than 24 hr at 28°C, and 70 mg of sodium polyphosphate per ml of egg white killed 1 million salmonellae in 60 hr at 28°C. When the egg white was adjusted with lactic acid to pH 5.3, and EDTA added at 7 mg/ml, *Salmonella* resistance decreased 100-fold. These sequestrants take up Ca^{2+} and Mg^{2+}, which are then unavailable to the microorganisms. Microorganisms then become susceptible to attack by lysozyme (Garibaldi *et al.,* 1969a).

Pasteurization reduces the aerobic plate count in liquid eggs by 100- to 1000-fold, usually to about 100/g. See Tables 19.15, 19.16, and 19.17. Survivors are mostly *Micrococcus, Staphylococcus, Bacillus,* and a few gram-negative rods (Shafi *et al.,* 1970). Table 19.22 lists the genera iso-

TABLE 19.22

Principal Bacterial Genera in Whole Egg [a]

	Unpasteurized		Pasteurized	
Genera	Before freezing	After freezing	Before freezing	After freezing
Acinetobacter–Moraxella (*Achromobacter*)	0.0%	1.5%	—	—
Enterobacter (*Aerobacter*)	3.0	0.0	—	—
Alcaligenes	25.1	20.0	4.0%	8.3%
Bacillus	7.4	2.0	83.0	83.5
Chromobacterium	1.6	1.6	—	—
Escherichia	5.7	6.8	2.5	0.0
Flavobacterium	29.0	26.9	4.0	0.0
Gram + cocci	5.3	3.5	2.5	0.0
Proteus	15.6	18.1	4.0	8.3
Pseudomonas	7.4	16.0	—	—
Salmonella	2.0	0.0	—	—
Streptothrix	0.0	2.3	—	—

[a] Wrinkle *et al.* (1950).

lated before and after pasteurization in one investigation of whole egg.

Pasteurization also reduces the direct microscopic count by 57 to 88%, presumably because lysozyme dissolves the cells during the heating process (Hall *et al.*, 1971).

3. Filling and Chilling

Once the egg product has been pasteurized, it must be handled carefully to prevent recontamination from the unpasteurized egg, insanitary equipment, containers, dust, or human or animal sources.

It must also be chilled quickly, preferably using a heat exchanger, or if not, then filled into cans and cooled within 1½ hr to 7°C or below to prevent growth of microorganisms. It is most important that the temperature of the egg product pass quickly through the range of most rapid microbial growth—approximately 50° to approximately 7°C.

4. Salt and Sugar

Unpasteurized 10% salted yolks shipped and used promptly could contain salmonellae at point of use. But this amount of salt decreases the water activity (a_w) of the egg yolk to about 0.90 (i.e., 20.3 g salt in 100 g of water phase). Salmonellae will not grow at this a_w, regardless of temperature and will gradually die off in a matter of weeks (Banwart, 1964; Ijichi *et al.*, 1973; Cotterill and Glauert, 1972). Staphylococci could grow during storage without refrigeration (see Section IV,C).

5. Freezing and Thawing

Containers of liquid eggs to be frozen should be placed in the freezer at −23° to −40°C immediately after chilling and should be stacked in such a manner that they will freeze promptly. Temperatures for prolonged storage should be at or below −18°C; some microorganisms can grow at slow rates at or above −10°C (Michener and Elliott, 1964).

Freezing and frozen storage reduce the numbers of microorganisms but usually not to the point of extinction of a given strain, particularly not in the protective proteinaceous egg magma. For example, freezing and frozen storage will not free an egg product of salmonellae. Table 19.22, however, shows that in one investigation, strains of *Escherichia, Flavobacterium,* and gram-positive cocci died out on freezing, after exposure to pasteurization.

Improper thawing often permits unwarranted increases in microbial content (Forsythe, 1970). When a can of eggs is allowed to thaw completely in a warm room, the temperature of the outside portion will remain in the growth range for many hours. Frozen products should be thawed under conditions that permit the temperature of the thawed material to rise to more than 4°C for only short periods. Some manufacturers thaw in a refrigerator at about 4°C; others immerse the cans in cold running water; others have installed crushers to break up the partly frozen mass; still others remove the liquid-egg material as it thaws by centrifuge or otherwise, (Lawler, 1965). See Chapter 29 for additional information on thawing procedures.

B. Spoilage

Shell eggs, for the most part, are efficiently protected against bacterial entry by the cuticle, shell, and membranes and against spoilage by various factors in the white. When broken out, they lose all this protection, and become highly perishable. One exception is the separated white, which retains its capacity to retard certain types of microbial growth. The yolk is rich in nutrients—for example, nitrogen, minerals, sulfur, B vitamins—and is a relatively favorable medium for microbial growth when handled either separately or mixed with white. In fact, the more objectionable forms of rotting result from attack on yolk constituents. When mixed with the white, the yolk eliminates the various protective devices inherent in the white. For example, its iron content satisfies the chelating power of conalbumin, and thus makes minerals available to spoilage organisms (Cunningham, 1966).

The contaminating organisms at time of breaking are primarily those of the egg shell (Table 19.3) and the occasional spoiled egg. The broken-out eggs must be cooled immediately to 7°C or below, especially if cracked or dirty eggs were used (Brown et al., 1966b), to prevent spoilage.

After the eggs are pasteurized, they should again be cooled promptly. Although organisms that spoil eggs under refrigeration have largely been destroyed (Speck and Tarver, 1967), a delay in cooling will permit mesophilic bacteria to grow and will permit the few psychrotrophs present to build up rapidly. In one study (Steele et al., 1967), unpasteurized whole eggs held at 4°C for 8 to 10 days developed 3 million bacteria per gram. Pasteurized eggs, frozen immediately then placed at 4°C for 45 days yielded fewer than 100 bacteria/g. Pasteurized eggs held at 13°C for 24 hr before freezing, followed by storage at 4°C, developed 3 million organisms/g in 24 days.

The refrigerated shelf-life of pasteurized egg products (i.e., time until spoilage is evident) is remarkably long. Clean eggs, broken, cooled, mixed, pasteurized, and cooled under ideal sanitary conditions remain edible for 20–22 days in the refrigerator (Wilkin and Winter, 1947; Kraft et al., 1967b). If the eggs are dirty, even though the magma is pasteurized, the shelf-life in the refrigerator may be only 2 or 3 days, because more microorganisms will survive from a high original level of contamination (Baker, 1974). Most samples of pasteurized eggs from commercial operations in the United States had a shelf-life of 12–15 days at 2°C (Vadehra. *et al.,* 1969b; York and Dawson, 1973). Before pasteurization came into general use in Europe and North America in the 1960s and early 1970s, the shelf-life of refrigerated whole egg was only 5 to 7 days (Wrinkle *et al.,* 1950; Wilkin and Winter, 1947).

Salted yolks, particularly when batch pasteurized and filled hot into cans, have a long shelf-life, even without rapid cooling, and with storage at room temperature. All bacterial vegetative cells capable of growth in 10% salt have been killed by the heat. Eventually a few spores may germinate and grow (Cotterill *et al.,* 1974).

Pasteurization destroys microorganisms such as *Pseudomonas, Acinetobacter,* and *Enterobacter,* which grow in raw albumen, leaving mesophilic organisms like micrococci, staphylococci, *Bacillus* species, fecal streptococci, and catalase-negative rods that can grow at 23°C but not in the refrigerator (Barnes and Corry, 1969; Shafi *et al.,* 1970).

Table 19.23 lists the genera found in raw liquid egg and the changes they produce in liquid egg in pure culture. Table 19.22 shows that many of the genera listed in Table 19.23 would not exist in pasteurized egg except by postpasteurization recontamination.

The odors of spoilage are much more intense in yolk or whole egg than in white. In white, isolated strains caused no putrid odor or H_2S. Most reduced the pH of the white slightly and produced a small amount of trimethylamine. On the other hand, the yolk developed fishy, moldy, and ammoniacal odors, with high levels of H_2S and trimethylamine and a high total volatile base level (Imai, 1976).

C. Pathogens

Salmonellae, the most important pathogens in liquid egg, are discussed in detail in sections describing the effect of processing procedures on microorganisms (Section IV,A) and methods of control (Section IV,D).

Group A streptococci did not grow in liquid whole egg, but several strains of *Salmonella, Escherichia coli,* and *Staphylococcus aureus* grew well above 15.6°C. Below this temperature, none grew. Thus, liquid eggs should be held refrigerated (Gibbons *et al.,* 1944).

Salmonellae grow rapidly in yolks either aerobically or anaerobically (Lineweaver *et al.,* 1969; Lawler, 1965). When the pH of egg white is at

TABLE 19.23

Changes Produced by Different Genera of Bacteria Isolated from Liquid Egg When Inoculated into Sterile Egg [a]

Genus	Number of strains inoculated	Changes produced
Acinetobacter–Moraxella (Achromobacter)	1	No change in 72 hr
Enterobacter (Aerobacter)	4	Slight acid odor after 72 hr
Alcaligenes	17	12 very sour odor in 60 hr 1 musty odor in 48 hr 4 no detectable change
Bacillus	8	6 coagulation within 18 hr 1 very sour odor after 24 hr 1 no change
Chromobacterium	2	No change in 72 hr
Escherichia	15	4 slight acid odor in 60 hr 8 sour odor after 60 hr 3 no detectable change
Flavobacterium	11	2 coagulation in 120 hr 1 fecal odor in 72 hr 4 slight odor in 120 hr 4 no detectable change in 72 hr
Gram + cocci	3	2 produced much gas in 18 hr 1 no change
Proteus	10	3 very flat-sour odor in 60 hr 2 coagulation in 18 hr 4 very sour odor in 60 hr 1 no change
Pseudomonas	7	3 very sour odor in 60 hr 2 gas within 60 hr 1 sour odor in 60 hr 1 no change

[a] Wrinkle et al. (1950).

or above 9, salmonellae are inhibited, but if the pH is adjusted to pH 6–8, they grow well (Banwart and Ayres, 1957).

Staphylococci are a potential hazard in salted yolks, for they can grow readily at the reduced a_w (0.90) of the product. If they survive pasteurization or if they are introduced after pasteurization, they must grow to levels of 10^6 or more before toxin will form. This would require severe abuse, because the egg would have to be stored for several days at ambient temperatures (Ijichi et al., 1973).

D. Control

1. Control of the Process

To prevent heavy bacterial contamination of liquid egg, eggs for breaking should be washed, candled to sort out rots, and should be examined for odor and appearance at breakout. Equipment that contacts contaminated eggs at breakout should be washed and sanitized before reuse (Forsythe, 1970).

To slow microbial growth, liquid egg products should be chilled to 7°C or below promptly after breakout and after pasteurizing. If the product is to be frozen, cans of eggs should be placed promptly into a freezer at −23° to −40°C and stored at or below −18°C. Liquid egg not to be frozen should be chilled quickly and handled promptly.

All product-contact equipment such as pipes, churns, tanks, and pails should be thoroughly cleaned and sanitized at least daily and preferably each 4 hr during use. Equipment should be designed and installed for easy cleaning, as described in E-3-A Sanitary Standards (IAMFES, 1976a,b,c).

Pasteurizers should have (Murdock *et al.*, 1960; Lineweaver *et al.*, 1969; Forsythe, 1970; Kaufman, 1969, 1972) the following characteristics:

1. Flow diversion valves for imperfectly heated egg.
2. Higher pressure on the pasteurized side than on the unpasteurized to prevent raw melange from breaking into the pasteurized material.
3. Recording thermometers at entrance and exit of the pasteurizer and the cooler.
4. Automatic control of temperature.
5. Automatic flow control.
6. Leak detection devices.

See Appendix 1 of Lineweaver *et al.* (1969) for details of pasteurizer operation.

2. Laboratory Tests

Because of the marginal nature of pasteurization, its efficacy must be checked constantly by the laboratory, testing for *Salmonella* as well as for the coliform group or Enterobacteriaceae. These organisms should not be found in the pasteurized material. The reason for testing for salmonellae is obvious, i.e., the pasteurization is designed to kill these organisms. On the other hand, most of the egg entering the pasteurizer will not contain salmonellae, so that one must also test for the coliform group or the Enterobacteriaceae as a measure of heating sufficiency; these indicator organisms are always present in the raw egg melange. Further, their heat resistance is similar to that of the salmonellae. Pasteurization reduces both salmonellae and Enterobacteriaceae 10^3- to 10^5-fold. Egg products commercially produced in the Netherlands sometimes (but rarely) contained salmonellae in 25-g portions when other Enterobacteriaceae were not found in 0.1- or 1-g samples (van Schothorst and van Leusden, 1977). A positive finding for Enterobacteriaceae or the coliform group could mean either that pasteurization was inadequate or that postpasteurization contamination occurred.

In some countries, for example the United Kingdom, a test for the enzyme α-amylase determines the efficiency of pasteurization (Brookes, 1962; Murdock *et al.*, 1960; Shrimpton *et al.*, 1962). At the temperatures and times preferred in the United Kingdom (64.4°C for 2½ min for whole egg) the enzyme α-amylase is destroyed. The α-amylase test is rapid, accurate, convenient, and inexpensive, whereas tests for microorganisms are slow, expensive, and often give variable results among workers and among laboratories. However, α-amylase is not destroyed by the lower heat treatment preferred in the United States (60°C for 3.5 min; Lineweaver *et al.*, 1969), therefore, this test is inapplicable in the U.S. industry and in other areas of the world where lower temperatures apply. The α-amylase test likewise does not apply to salted or sugared eggs and does not detect postpasteurization contamination.

There are two methods to detect the presence of incubator reject eggs (Robinson *et al.*, 1975):

1. An enzyme test comparable to the α-amylase test.
2. A test for β-hydroxy butyric acid, formed on the inhibition if embryonic growth within the egg.

3. Regulatory Programs

Carefully planned and well-enforced regulatory programs in the United Kingdom, in Canada, in the United States, and elsewhere have improved the wholesomeness and safety of liquid- and frozen-egg products in international commerce. Extensive surveys in 1955–56 in the United Kingdom showed that Chinese frozen whole egg and other egg products contained *Salmonella paratyphi* B, *S. thompson,* and other serotypes of *Salmonella*. Many outbreaks of salmonellosis including paratyphoid fever were described (Newell, 1955; Newell *et al.*, 1955; Essex-Cater *et al.*, 1963; Parry, 1963). Various bakery products, including cream confections, have been vehicles of infection. The egg was sometimes not an ingredient of the cream but was used for other goods in the preparation area. In the United Kingdom, after the pasteurization of liquid whole egg was enforced (Statutory Instruments, 1963), salmonellosis from egg products declined (Mc-

Coy, 1975; Hobbs and Gilbert, 1978). There was a reduction in the reported incidence of paratyphoid fever and in food poisoning from *S. thompson, S. typhimurium,* and *S. menston,* for example. The fall in the number of outbreaks from *S. typhimurium* in the 2 years after the introduction of pasteurization was about 25% and from *S. thompson* and *S. menston* even more than this (McCoy, 1975).

Processed eggs had been demonstrated to be a vehicle of the spread of salmonellosis for 15 years before large-scale outbreaks occurred in Canada in the early 1960s. In 1962, a regulation was promulgated to prohibit the sale of *Salmonella*-contaminated processed eggs, but it was not enforced. In 1969, 7 years after the regulations were published, 17.5% of processed eggs contained salmonellae (Table 19.24). However, in January 1970, the Canadian authorities began a rigid enforcement program, and the incidence dropped rapidly to < 0.2% in 1976 (Pivnick, 1978).

In the United States, pasteurization became mandatory in early 1966, but was not under good regulatory control until 1971 (USDA, 1975b; U.S. Department of Health, Education and Welfare, 1966). A slightly lower incidence of *Salmonella* isolations from eggs was indicated in late 1966 (Elliott, 1966), and by 1975, it was at a low but undetermined level. The number of U.S. outbreaks of salmonellosis traced to eggs has shown a dramatic decline during this period. In the 5-year period from 1963 to 1967, inclusive, there were 133 egg-associated outbreaks in the United States (average 26.6 per year). There were none in 1974 and 1975 (Cohen and Blake, 1977).

The Food and Agriculture Organization of the United Nations (FAO/WHO, 1975) has tabulated the various microbiological criteria imposed on dried and frozen eggs by 19 countries. A summary of these criteria is

TABLE 19.24

Salmonella **Contamination of Processed Eggs in Canada** [a]

Year	Number of lots tested	Percent of lots contaminated
1969	559	17.5
1970	1081	3.9
1971	1644	1.5
1972	1269	0.8
1973	1742	0.7
1974	1787	0.8
1975	2280	1.0
1976	2733	0.2

[a] Pivnick (1978).

presented in Table 19.25. In addition, the FAO has established its own criteria for these two commodities, summarized as follows:

	n	c	m	M
Salmonella (25-g portions)	10	0	0	
(For special diets)	30	0	0	
Standard plate count	5	2	5×10^4	10^6
Coliform group	5	2	10	10^3

in which n samples are analyzed, of which c samples may exceed m, but none may exceed M. For further details, see ICMSF (1974) and FAO/WHO (1975).

E. Choice of Case

For an understanding of "Case" see Table A.1 and the brief description in Appendix IV. For more details, see the second book of this series (ICMSF, 1974).

The Commission has established that whole egg and yolks should be assigned Cases 1, 4, and 10, because they are invariably cooked before consumption or treated with acids (reduced concern). The tests include direct microscopic count, standard plate count, and examination for *Salmonella*.

TABLE 19.25

Range of Recommendations of Microbiological Criteria for Dried and Frozen Eggs Imposed by 19 Countries [a]

	Microbial numbers, per gram	
	Dried eggs	Frozen eggs
Standard plate count	15,000 to 600,000	10,000 to 500,000
Coliform group	10 to 100	10 to 300
Coagulase-positive *Staphylococcus*	ND 0.1 [b] to 1,000	ND 0.1 to 1,000
Salmonella	ND 20 [c] to ND 25 × 30 [d]	ND 20 to ND 25 × 30
Yeasts and molds	20 to 100	50

[a] Summarized from FAO/WHO (1975).
[b] ND 0.1 = not detectable in 0.1 g.
[c] ND 20 = not detectable in 20 g.
[d] ND 25 × 30 = not detectable in 30 samples of 25 g each.

On the other hand, egg whites are frequently not cooked, so that Cases for the above three tests are 2, 5, and 11, respectively. If growth might occur in the thawed product, the Cases for egg whites could be increased to 3, 6, and 11.

V. DRIED EGGS

A. Effects of Processing on Microorganisms

The methods of drying eggs are several, but the microbiology is essentially the same in all. Drying kills many of the bacteria in eggs, but once the egg material is dry, the microbial content remains stable at ambient temperatures, except for a very slow drop in total numbers, rarely to the extinction of a strain. The predominant organisms in the dried product are enterococci and aerobic sporing bacilli, the resistant members of the original flora of the egg pulp. The number of salmonellae may be reduced as much as 4 logs during drying (Gibbons and Moore, 1944), but fermented albumen or temperature-abused whole egg can have a high initial number so that survival of some salmonellae is likely (Ayres and Slosberg, 1949; Gibbons et al., 1944). Salmonellae are the principal microbial problem in dried eggs, and the problem becomes more serious if they grow during fermentation for glucose removal.

1. Glucose Removal (Fermentation)

Dried whites normally contain about 0.6% free carbohydrate, primarily as glucose. On storage, particularly above 15°C, the aldehyde group of the glucose combines with the amino groups of the proteins, reducing their solubility, damaging the flavor, and forming insoluble, brown compounds (the Maillard reaction). Removal of glucose from the liquid egg white before drying prevents these reactions. Glucose removal also improves the stability of dried whole egg and egg yolk but to a lesser extent (Stewart and Kline, 1941; Paul et al., 1957a; Forsythe, 1970; Kilara and Shahini, 1973).

The earliest method used to remove glucose simply permitted the natural egg flora to grow at ambient temperatures of 21°–29°C for 2 to 7 days. The length of time was judged by observations of bubbling, consistency, and clarity of samples. Scums and sediments were discarded and a little ammonia added to clear the liquid. The reactions were not well controlled so that objectionable odors and proteolysis developed. This procedure allowed the growth of group D streptococci, *Enterobacter aerogenes,* and other bacteria. Hence, large numbers of fecal indicators were usually found in the dehydrated material. In addition, salmonellae thrived, presenting a health hazard (Ayres, 1958).

On the other hand, most manufacturers now add rapidly fermenting egg white as a starter and raise the temperature to 35°C so that the glucose is removed within 12 to 24 hr. The group D streptococci are favored, for *Enterobacter* and lactic organisms do not grow well in competition at the natural pH of egg white (> 9.0). The group D streptococci do not cause proteolysis or off-odor, but rather acidify the whites to approximately pH 6. In case a batch of eggs should spoil, for subsequent lots manufacturers generally draw upon frozen starter held available for such an emergency. This fermentation method also may permit the rapid growth of salmonellae after the pH drops below 8.

Various investigators recommend the addition of a massive pure culture starter. The first step is to reduce the pH of the egg from its normal 9.0 to 7.0–7.5 with an organic acid such as lactic, then to add the culture, and ferment 12–24 hr at 30°–33°C. Some claim that bacterial cultures are best because the finished product has high whipping quality, good odor, and solubility (Forsythe, 1970), but others prefer a yeast or enzyme (see Table 19.26).

In many of the fermentation procedures listed in Table 19.26, salmonellae can grow if the pH range of the egg white is 6–8 but not if it is 9 or above (Banwart and Ayres, 1957). After the fermentation period, pasteurization is essential to kill salmonellae (Kline and Sonoda, 1951; Ayres and Stewart, 1947), otherwise salmonellae may carry through into the dried product.

2. Destruction of Salmonellae by Hot Storage

Despite the rather efficient procedures developed to pasteurize liquid egg, salmonellae appear occasionally in the final dried packaged product. After the product is dried, there is still another method to destroy the organisms—hot storage.

Microorganisms are heat sensitive when wet, but heat resistant when dry. Whereas the usual pasteurization time for a wet product is 2 to 5 min, that for dried egg is several days. One author calculated that the resistance of salmonellae in dried egg was 653 times greater than in liquid egg (reviewed in Garibaldi, 1966).

The times and temperatures recommended for "pasteurization" of dried egg white are listed in Table 19.27. These exposures have no serious effect on functional qualities.

The advantage of hot storage over the standard wet pasteurization is that there is no possibility of recontamination, because the containers re-

TABLE 19.26

Methods to Remove Glucose from Liquid Egg

Microorganism or agent	Authors	Comment
Natural flora	Ayres (1958)	*Enterobacter,* fecal streptococci and other bacteria
Coliform group	Stuart and Goresline (1942a, b)	Earliest pure culture studies
Saccharomyces apiculatis	Hawthorne and Brooks (1944)	1% inoculum gave yeasty flavor
Saccharomyces cerevisiae	Ayres and Stewart (1947); Hawthorne (1950); Carlin and Ayres (1953); Kline and Sonoda (1951); Ayres (1958)	Yeasty odor from large inocula can be eliminated by 0.1% yeast extract which stimulates activity of small inocula; also may centrifuge to remove yeast
Streptococcus lactis and *Streptococcus faecalis* subsp. *liquefaciens*	Kaplan *et al.* (1950); Ayres (1958)	Resting cells, 37°C for 3 hr; yeast extract 0.1% inhibits acid production
Glucose oxidase and catalase	Baldwin *et al.* (1953); Carlin and Ayres (1953); D. Scott (1953); Paul *et al.* (1957a); Ayres (1958)	Glucose oxidized to gluconic acid; catalase destroys the H_2O_2 that forms
Enterobacter (Aerobacter) aerogenes	Ayres (1958)	Acetyl methyl carbinol production can be minimized by using small inocula with 0.1% yeast extract for stimulation
Escherichia coli	Mickelson and Flippin (1960); Flippin and Mickelson (1960)	*E. coli* shows antagonism for *Salmonella*
Cell-free yeast extract	Niewiarowicz *et al.* (1967)	Desugars in 4 to 5 hr at 5°C

main closed during and after the treatment period. The bactericidal effect of the hotroom treatment depends on the moisture content (Banwart and Ayres, 1956; Carlson and Snoeyenbos, 1970), on the temperature, and on the nature of treatments preceding the heat treatment—i.e., the method of fermentation, the use of ammonia or citric acid, and the method of drying (Northolt *et al.,* 1978). As the time required for inactivation of all the organisms depends upon the number originally present, a shorter period of treatment can be applied to materials with a low level of contamination.

TABLE 19.27

Times and Temperatures of Hotroom Storage to Destroy Salmonellae in Dried Egg Albumen

Pretreatment	Temperature (°C)	Time (Days)	Author
Fermented, pan dried	48.9	20	Ayres and Slosberg (1949)
	54.4	8	
	57.2	4	
3% moisture	50	9	Banwart and Ayres (1956)
6% moisture	50	6	
Spray dried	54.4	7	USDA (1975b)
Pan dried	51.7	5	
Adjusted to pH 9.8 with ammonia, pan dried	49	14	Northolt et al. (1978)
Treated with citric acid	55	14	Northolt et al. (1978)
Spray dried	49	14	Northolt et al. (1978)

B. Spoilage

In the dried product, microorganisms remain alive indefinitely but cannot grow. On reconstitution with water, or on accidental wetting during storage, they will grow and spoil the product (see Section IV,B).

C. Pathogens

Salmonella-contaminated dried egg has caused many outbreaks by direct ingestion, and additional outbreaks by the way of other foods cross-contaminated by dried egg in the kitchen.

Before 1965, dried eggs were often contaminated. For example, in the United States in 1943, salmonellae were isolated from 1810 (35%) of 5198 samples of dried whole egg. Many of the strains isolated were the same as those most commonly occurring in human illness (Solowey et al., 1947). In the United Kingdom at about the same time, salmonellae were isolated from about 10% of 7584 samples of imported dried egg (Haines et al., 1947). In the early 1960s, of 119 samples of Canadian cake mix containing dried egg, 65 (or 54.6%) were positive and caused much illness. In one outbreak, salmonellae spread throughout a bakery from heavily contaminated dried egg to other goods prepared on the premises, and infected the food handlers who became symptomless excretors

(Thatcher and Montford, 1962; Butler and Josephson, 1962; Skoll and Dillenburg, 1963).

Pasteurization, either wet or dry, has come into general use, and national governments have established and enforced regulations requiring it. The incidence of salmonellae in dried egg products is now at a very low, but largely undetermined level.

D. Control

The procedures described under Section IV,D apply also to dried eggs. In addition, the drier should be made of impervious materials without cracks, crevices, and dead-end pockets where product can remain wet and warm. Dried eggs, in common with other dried foods, must be protected against condensation dripping into the dried product. Frequently tank lids, drier hoods, and conveyor covers are cooler than the stream of dried egg, so that condensate forms on their undersides. Dust from the product contributes nutrients so that bacteria, including salmonellae, can grow in the droplets. When the droplets or clumps of damp egg fall back into the main mass they can contribute localized areas of *Salmonella*-contamination. Sampling and analysis is a difficult means to detect such infrequent and localized contamination. The best way to control the problem is to prevent condensation. Warming the surfaces of equipment where condensation occurs is one answer.

Cleaning with water of an area used for a dried product may introduce more hazards. Instead, a vacuum cleaner, dry brushes, and cloths may be used, followed by 95% ethyl alcohol as a disinfectant. If it seems advisable to use water, the equipment must be thoroughly and rapidly dried before reuse.

E. Choice of Case

For an understanding of "Case" see Table A.1 and the brief description in Appendix IV. For more details, see the second book of this series (ICMSF, 1974).

The Commission has chosen for dried eggs, Cases 2 and 11 for the standard plate count and for salmonellae, respectively. In the normal course of events, there will be neither growth nor death among bacteria in dried eggs, thus the Cases fall into the "no change" category. When dried eggs are reconstituted with water, the Cases revert to those of liquid and frozen egg.

VI. FURTHER PROCESSED EGG PRODUCTS

Table 19.2 lists a wide variety of processed products containing eggs. Those that are not cooked, such as meringue pie, egg nog, or dry diet mixes, remain potential hazards from salmonellae that might have survived or by-passed pasteurization. Cooked foods such as custard, cream cakes, and angel cake could be cross contaminated in the preparation area from a contaminated egg ingredient, particularly from a dried egg product (Williams and Hobbs, 1975). However, baked goods that reach 71°C or above are safe from salmonellae (Beloian and Schlosser, 1963; see Chapter 23).

Salmonellae and staphylococci from eggs will die in a few days in acid salad dressing (pH approximately 3.3) or acid mayonnaise; there have been outbreaks of salmonellosis from nonacid mayonnaise. The growth-limiting factors in mayonnaise in the United States are the low a_w (0.93) where salmonellae will not grow, and the low pH (approximately 4) where staphylococci will not grow (Wethington, and Fabian, 1950; Smittle, 1977; see Chapter 24).

20

Fish and Shellfish
and Their Products

I. INTRODUCTION

Fish and shellfish are second only to meat and poultry as staple animal protein foods in most of the world. In some countries (e.g., Japan), they are the principal source of protein. Though dried and salted fish have been items in international commerce for centuries, until comparatively recently, consumption of fish tended to be localized close to areas of capture. Improvements in engines and ship technology have greatly extended the range of fishing vessels, which now forage over all the oceans of the world. Developments in food technology, particularly freezing and canning, have provided means to overcome the problem of perishability of fish and fish products, so that seafoods are now common items of international trade and are consumed in inland areas remote from the seacoast.

Utility and the maintenance of quality are the principal microbiological problems associated with fish and shellfish. Nevertheless, fish and seafood products share with other foods actively moving in world commerce the potential to act as transporters of pathogenic organisms. Particular seafoods, notably shrimp and tuna, are harvested in almost every region of the world and are subjected to primary handling and processing operations that range from highly sophisticated to quite primitive, and from impeccably hygienic to dirty and potentially dangerous. Risk is related to broader conditions of the environment. Generally, where water and air temperatures are low, public health risks to the ultimate consumer are less than in tropical zones. Areas in which cholera, typhoid fever, infectious hepatitis, poliomyelitis, and similar diseases are endemic pose a greater hazard than areas free from such diseases. This is particularly important where animals are harvested close inshore in regions of high human population density. These factors must be taken into account in evaluating the need for extensive testing programs.

Molluscs, as sessile filter feeders, concentrate bacteria and viruses from the water and can be dangerous carriers of enteric pathogens. They can be doubly dangerous because many are eaten raw or only lightly cooked. The frequent disposal of human wastes directly into estuarine and inshore marine waters, rivers, and lakes and the continuing increase in human populations in cities must raise the level of concern about these hazards.

Vibrio parahaemolyticus, Clostridium botulinum type E, and some parasites are the only organisms potentially pathogenic for man occurring naturally on seafoods other than molluscan shellfish. Another particular hazard is ichthyosarcotoxicity, but this is restricted to certain types of fish in particular situations, e.g., scombroid poisoning—tuna; tetradon poisoning—puffer fish; ciguatera—reef fish. Among these, only scombroid poisoning is ascribed to a bacterial origin.

Parasite transfer from fish has become of concern because of the problem with anasakia in Europe and Japan, incidents of parasitic disease in the United States, and the suspected importance of fish as a source of infestation in humans with various trematodes in other parts of the world (ICMSF, 1978).

There is a risk of microbe and parasite transfer to human beings from fish and shellfish grown under artificial conditions (aquaculture), particularly if the system requires that animal or human excreta be fed directly to the fish or shellfish. Aquaculture is growing in importance, and its potential for spread of disease should be thoroughly evaluated.

Vibrio parahaemolyticus is widespread in inshore marine waters, and outbreaks of foodborne illness due to the vibrio are also widespread. Although the vast majority of individuals who become sick from *V. parahaemolyticus* recover quickly and uneventfully, there have been a small number of deaths. Japanese investigators have recently identified a cardiotoxin from the organism.

So many different species of aquatic animals in such a diversity of forms are consumed by man that their processing includes most of the preservative and preparative techniques developed for human foods. The traditional processes of drying, salting, pickling, smoking, and fermenting continue to be applied to fish products by a range of technologies from extremely primitive to highly sophisticated; and such products are widely distributed in world markets. Large quantities of fish, shellfish, and seafood products are canned, mostly by sterilizing processes (commercial sterility), although some are semipreserved. An increasing proportion of the world fish and shellfish catch is frozen either in the raw, unprocessed, or butchered state or as prepared products for consumer use. The variety of prepared consumer products has increased enormously in recent years. The widespread availability of minced fish from deboning machines has, in recent years, greatly accelerated the production of new compounded fish products sold mostly as frozen prepared items.

The great diversity of fish products and the extreme variation in both quality and type of processing procedures used throughout the industry establish a very wide range of bacteriological problems.

A. Definitions

Foods to be covered include all products derived from aquatic animals, exclusive of marine mammals, except convenience foods and animal feeds that are dealt with elsewhere in this volume (Chapters 27 and 17, respectively).

The term "fish" is used both with the specific meaning of fin fish—referring to the free swimming members of Pisces and Elasmobranchii—and as a generic term for seafoods to cover invertebrates as well.

Crustaceae include lobsters, crabs, shrimps, prawns, and related animals characterized by possession of a chitinous exoskeleton.

Molluscs include clams, scallops, oysters, and similar animals of a sessile nature.

No attempt has been made to cover the microbiology of cephalopods and other invertebrates not listed above, since there is little relevant published information.

B. Important Properties

Gross composition of the several hundred species of aquatic animals used as human food varies widely, although, in most cases, the edible portions are striated muscle of roughly comparable composition. They are consumed principally as a source of protein that is high in nutritional quality in terms of digestibility and amino acid composition. The greatest compositional variability in edible marine animals is in fat and water content, which are inversely related.

The fat content varies considerably with species and sometimes with season. Since fish are taken all year, in many cases, this means that there may be wide variation in the fat content of market fish of a single species sampled at different times. Fish lipids are typically triglycerides or phospholipids, characterized by highly unsaturated long-chain acids. They are highly labile chemically and oxidize readily in storage. Carbohydrate content is negligible except in molluscs, where up to approximately 3% glycogen may occur. The mineral content of fish muscle is low but comprises a rather complete range of nutritionally essential elements. Representative values for proximate composition of categories of aquatic animals (nonmammalian) are shown in Table 20.1.

Fish contain a high level of nitrogen-containing compounds other than protein in their muscle tissues, mostly small molecules dissolved in the tissue fluids and apparently utilized actively by bacteria during spoilage. Their concentrations vary with the type of animal (Table 20.2), e.g., Elasmobranchs show high levels of urea and trimethylamine oxide.

The muscle of nonmammalian aquatic animals is similar in general structure to that of mammals, but the amount of connective tissue and pigment is very much less in fish than in mammals.

Fresh fish muscle provides an excellent substrate for the growth of microorganisms because of high water activity, neutral pH, and high level of soluble nutrients. During *rigor mortis,* the muscle undergoes the same biochemical changes as are seen in mammalian muscle, but the ultimate pH of 6.2 or more is higher than in beef. The pH of a few species of flatfish (e.g., halibut) may fall to 5.8. Whether postmortem biochemical change has a direct influence on microbial growth is not clear, but it is a common observation that microbial spoilage processes are not evident until after the resolution of *rigor mortis.*

Microbial spoilage of fish usually is described as a "proteolytic" process. While some protein hydrolysis undoubtedly takes place, the process, at least in the early stages, seems to involve utilization of the nonprotein nitrogen material in solution. With the exception of molluscs, there is no major involvement of carbohydrates in the process, though ribose released

TABLE 20.1

Average Composition of Fish and Shellfish [a]

	Water	Protein	Lipid	Ash
	(%)	(%)	(%)	(%)
Fat fish	68.6	20	10	1.4
Semifat fish	77.2	19	2.5	1.3
Lean fish	81.8	16.4	0.5	1.3
Crustaceans	76.0	17.8	2.1	2.1
Molluscs [b]	81.0	13	1.5	1.6

[a] *Jacquot* (1961).
[b] Molluscs may contain 1–3% glycogen.

TABLE 20.2

Nonprotein Nitrogen (NPN) Material in Fish

	Cod (mg %)	Herring (mg %)	Haddock (mg %)	Elasmobranch (mg %)
Total NPN	419.2	437.0	380.0	—
Bases	257.6	269.9	239.4	—
Creatine	163.0	182.7	—	—
Urea	1.9	4.0	14.5	2000.0
TMAO [a]	~95.0	~40.0	~70.0	~275.0

[a] TMAO = trimethylamine oxide.

as a result of ATP degradation may be utilized. The pH, therefore, increases. However, molluscan spoilage may involve glycogen breakdown leading to a lowering of pH as acid accumulates.

C. Principal Methods of Preservation and Products

The most common method for preserving fish in the fresh state is by chilling with ice or refrigerated brine. Freezing is the most common method of preservation for further processing and is rapidly becoming the most common retail distribution method also. Certain species are typically canned (e.g., tuna, Pacific salmon, sardines), and traditionally many fish are smoked, dried, salted, or fermented. Certain crustaceans may be precooked or blanched (lobster, crab), and precooked formulated products are becoming increasingly important in world trade.

A list of products in the international trade that form the basis of the discussion in this chapter follows:

Fresh raw seafoods
    Fin fish of marine origin
    Crustaceae
    Molluscs
    Fin fish of fresh water origin
Prepared and processed products
    Frozen raw seafoods
    Cooked crustaceae
    Canned seafoods
    Cured and smoked seafoods
    Fermented seafoods

II. INITIAL MICROFLORA

A. Numbers of Microorganisms

The flesh and internal organs of freshly caught, healthy fish are normally sterile, but bacteria may be found on the skin and gills and in the intestine (Shewan, 1962). The numbers of bacteria reported to occur are as follows:

Skin, 10^2–10^7/cm^2 Gills, 10^3–10^9/g   Intestine, 10^3–10^9/g

This very wide range reflects the effects of environmental factors. Thus, counts at the lower end of the range are commonly reported on skin and gills of fish taken in clean, cold waters, while higher counts are found in fish taken from tropical or subtropical waters and polluted areas (Shewan, 1977). The count in the intestine relates directly to feeding activity, being high in feeding fish and low in nonfeeding fish (Liston, 1956).

The counts on crustaceans and molluscs are usually given in terms of the whole animal or the shucked flesh. They are generally in the same range as reported for fish skin counts, i.e., 10^3–10^7/g. Again, the low/high relationship between counts on animals from cold and warm waters is evident (Cann, 1977). Molluscs, because of their sedentary mode of life, carry populations of bacteria that reflect the bacterial condition of the surrounding water, and seasonal variations may be seen with higher counts during summer months.

B. Saprophytic Microorganisms

The most important environmental factor influencing composition of the fish microflora is undoubtedly temperature. Typically, bacterial populations on fish and shellfish from temperate waters are dominantly psychrotrophic, reflecting a temperature of 10°C or less on the sea bed, even though in-shore surface waters may seasonally reach much higher temperatures. The bacteria on fish from tropical areas are mesophilic. Since chilling is the most generally applied technique for fish preservation, this temperature difference greatly affects the bacterial changes taking place during storage of fish and shellfish.

It has often been reported that the microflora of marine fish and shellfish is dominantly halophilic. Actually, in most cases, it would appear that the bacteria are predominantly euryhaline (i.e., able to grow over a wide range of salt concentrations), though showing improved growth at salt concentrations between 2 and 3% (as NaCl). The common use of ice to chill fish and shellfish exposes the bacterial population to decreasingly saline conditions during storage, and it is the euryhaline types that survive and grow out (Table 20.3). There is no consistent difference between the counts on agar with and without sea water, and this is in accord with earlier findings (Bannerjee, 1967). The data also indicate the variability observed in the rate of development of the spoilage flora in fish of different species caught at the same time and held under identical conditions of iced storage.

From a nutritional and biochemical standpoint, the bacteria of fresh fish and shellfish are described as being "proteolytic," rather than "saccharo-lytic." By this, it is meant that they grow more readily in media containing proteins, peptides, or amino acids as the main carbon source rather than polysaccharides or simple sugars. This reflects the nature of the natural substrate(s) on which the bacteria are presumed to be growing: the slimes, exudates, and waste materials available on the animal surfaces. However, the observation may be influenced by the dominant interest of spoilage in-vestigators in protein and muscle tissue component breakdown, leading to an almost exclusive use of protein-based media. Vibrios, which may on occasion dominate the microflora of certain marine animals, actively attack and utilize starch and, often, chitin to provide for their carbon and energy needs. Clearly, chitin may be the most readily available substrate for bac-teria growing on the carapace of crustaceans or in the gut of fish feeding on crustaceans. Nevertheless, it is generally true that most bacteria isolated from marine animals grow best when provided with an ample source of organic nitrogen-containing compounds with or without large amounts of carbohydrates.

TABLE 20.3

Bacterial Counts on Iced Pacific Ocean Fish

Days in ice	Hake (*Merluccius productus*)		Rockfish (*Sebastodes* spp.)		English sole (*Parophrys vetulus*)	
	TSA [a]	TSA-S [b]	TSA	TSA-S	TSA	TSA-S
	(count expressed as $\log_{10}$ per cm^2 skin) [c]					
0	1.18	1.04	2.08	2.15	1.80	1.38
3	2.49	3.16	1.38	1.80	2.30	2.11
7	6.33	6.96	5.54	5.16	4.97	4.77
10	6.27	6.04	4.92	5.30	6.26	6.62
14	7.27	7.76	7.16	6.96	7.18	7.13

[a] TSA: Trypticase Soy Agar made with distilled water.
[b] TSA-S: Trypticase Soy Agar made with aged seawater.
[c] All agar plates were incubated at 10°C for 21 days.

Most investigators have reported that the bacteria from skin and gills of fish and shellfish are predominantly aerobic. This is not universally true, because in situations where vibrios are present in great numbers (Simidu et al., 1969) the population is inevitably facultative in nature. The population of obligate anaerobes on external surfaces is usually negligibly small, but in the intestine, where anaerobic conditions are normal, clostridia may be found in significant numbers (Matches et al., 1974). However, even here, evidence suggests that facultatively anaerobic bacteria predominate.

The majority of reports on the qualitative composition of the bacterial populations of fish and shellfish are based on analyses of animals from temperate and usually northern hemisphere ocean waters. Almost universally, such reports identify gram-negative rod-shaped bacteria of the genera *Pseudomonas, Alteromonas, Moraxella, Acinetobacter, Flavobacterium, Cytophaga,* and *Vibrio* as the dominant types, making up over 80% of the population.

In the case of crustaceans, several workers report a preponderance of *"Achromobacter"** (*Moraxella–Acinetobacter*) and the significant occurrence of coryneform bacteria and micrococci. Reports from subtropical, tropical, and southern hemisphere analyses indicate, at least in some cases, a preponderance of gram-positive bacteria of the *Bacillus, Micrococcus,* and coryneform groups in the bacterial population of fish (Shewan, 1962; Gillespie and Macrae, 1975). A somewhat similar situation appears to exist in the case of freshwater fish, so that cold-water species appear to carry dominantly gram-negative bacteria, and fish in warm freshwater carry a high proportion of gram-positive bacteria (El Mossalami and Sedik, 1973; Shewan, 1977). However, the species distribution on freshwater fish is usually different from that of sea fish, reflecting differences in the environmental bacteria. Even in freshwater fish from cold regions, bacteria of genera such as *Streptococcus, Micrococcus, Bacillus,* and coryneforms can be expected to occur regularly in detectable numbers, though *Pseudomonas, Moraxella, Acinetobacter,* and *Flavobacterium* will normally be dominant. Members of the Enterobacteriaceae and the related genus *Aeromonas* frequently occur on freshwater fish. The varied microfloras of freshwater fish result from the influence of land plants, animals, and soils on the bacterial populations of freshwaters..

Because of their sessile mode of growth close to land and their passive feeding mechanisms, marine molluscs are subject to many of the same influences as lake and river fish. Consequently, the bacteria found on such

---

* In literature prior to 1974, gram-negative rods having the characteristics of *Moraxella* and *Acinetobacter* were usually listed as *Achromobacter* (Buchanan and Gibbons, 1974).

animals, while predominantly similar to those of marine fish, may periodically include disproportionately large numbers of gram-positive species of *Bacillus, Micrococcus,* etc., as well as Enterobacteriaceae and *Streptococcus* (Vasconcelos and Lee, 1972). Vibrios are a consistently large component of the microflora of oysters, at least on the Pacific Coast of the United States and show seasonal increase during summer months. The composition of the bacterial populations on fish and shellfish as reported by various authors is shown in Table 20.4.

In addition to bacteria, yeasts and occasionally fungi are reported to occur on fish and shellfish; however, there is less information available on these microorganisms (Morris, 1975). Yeasts are fairly widely distributed in the sea, though in smaller numbers than bacteria. Fungi are found in inshore and estuarine areas of freshwater. They probably occur adventitiously on aquatic animals, except in the relatively infrequent instances where they are parasitic. There is little or no evidence to suggest an important role for either yeasts or fungi in changes that occur in stored fish and shellfish. However, there have been some reports indicating that a psychrophilic pink yeast may have been responsible for spoilage of stored oysters. It is certainly true that a relatively high proportion of the yeasts from marine environments are "proteolytic."

C. Pathogenic Bacteria and Viruses

The naturally occurring microflora of fish and crustaceans presents less of a public health problem, at least in high-seas species, than the bacteria of land animals, because microorganisms dangerous to man are rarely present. Species caught in inshore waters that may be subject to contamination by human or land-animal waste and industrial or agricultural pollutants are a different matter, of course, and the sessile molluscs are a special case in which hazard may be high.

There are two species of bacteria of public health significance that may occur as part of the normal microflora of fish. These are *Clostridium botulinum* type E and nonproteolytic types B and F, and *Vibrio parahaemolyticus.**

Clostridium botulinum is, of course, a toxigenic heat-resistant, spore-forming organism whose presence in small numbers on fresh fish is harmless. The organism becomes dangerous when conditions of storage or processing permit or even encourage germination of spores, growth, and

---

* For a more extensive discussion of these and other foodborne pathogenic bacteria, see ICMSF (1978) and Chichester and Graham (1973).

TABLE 20.4

Bacteria in Fish and Shellfish, Expressed as a Percentage of the Total Flora for Various Generic Groups

Fish type	Pseudomonas	Vibrio	Acinetobacter–Moraxella (Achromobacter)	Coryneform	Others [a]	Reference
North Sea fish						
(1932)	5	—	56	—	39	Shewan (1971)
(1960)	16	1	23	18	42	Shewan (1971)
(1970)	22	1	41	18	19	Shewan (1971)
Haddock						
(North Atlantic)	26	2	45	4	23	Laycock and Regier (1970)
Flatfish						
(Japan)	20	13	30	17	20	Simidu et al. (1969)
"Pescada"						
(Brazil)	32	—	35	4	29	Watanabe (1965)
Shrimp						
(North Pacific)	10	—	47	3	40	Harrison and Lee (1969)
Shrimp						
(Gulf–ocean)	22	2	15	40	21	Vanderzant et al. (1970)
Shrimp						
(pond culture)	—	5	2	83	13	Vanderzant et al. (1971)
Scampi						
(United Kingdom)	3	—	11	81	5	Walker et al. (1970)
Mullet						
(Queensland)	18	—	9	12	61 [b]	Gillespie and Macrae (1975)

[a] Includes *Flavobacterium*, *Micrococcus*, *Bacillus*, coliforms, and yeast.
[b] Micrococci constitute 49%.

toxin production. The type-E strains have spores of low heat resistance compared to other strains and are unusual mesophiles since they can grow and produce toxin at temperatures as low as 3°C (Schmidt *et al.,* 1961).

Vibrio parahaemolyticus is a mesophilic, gram-negative, mildly halophilic vibrio that can cause gastroenteritis with a characteristic syndrome having some similarities to salmonellosis (Sakazaki, 1969; Fujino *et al.,* 1974). The sickness, which is usually not fatal, occurs only when large numbers (approximately 1 million) of the living organism are ingested and therefore is usually a result of consumption of fish or shellfish products previously held under conditions that permitted multiplication of the bacteria, since the normal level of occurrence of the organisms ($< 10^3/g$) is below an infective dose level. In Japan and other Southeast Asian areas, consumption of uncooked fish is the usual cause of sickness (Okabe, 1974). In the United States and Europe, sickness has occurred from ingestion of precooked crustaceans on which recontamination and growth of *V. parahaemolyticus* have occurred after cooking (Barker *et al.,* 1974). The organism is very sensitive to heat and cold (Liston, 1976).

Other potentially pathogenic bacteria occasionally associated with fish and shellfish include *Clostridium perfringens, Erysipelothrix rhusiopathiae, Edwardsiella tarda, Salmonella* spp., *Shigella* spp., and *Vibrio cholerae.*

These organisms are extraneous contaminants that should not be considered members of the normal bacterial populations of fish and shellfish. *Clostridium perfringens* has been reported to occur in fish in Turkey (Burow, 1974); in the United States (Matches *et al.,* 1974); in Korea (Sohn *et al.,* 1973), and in Japan (Taniguti, 1971). The Japanese studies revealed an incidence of 0 to 83% on fish surfaces and 0 to 15% in the gut. The U.S. work suggested a relationship between sewage (primary-treated) outfall contamination and incidence of *C. perfringens;* the organisms could not be detected in fish from offshore grounds.

Erysipelothrix is carried aboard ship in infected boxes and other wooden surfaces and gains access to the fish by contact. It may cause puncture infections in fish handlers. *Edwardsiella* is likewise a land contaminant. *Salmonella, Shigella,* and the cholera organisms gain access when fish and, particularly, shellfish are held in contaminated water. Because they concentrate bacteria from the surrounding water, molluscs present the greater potential hazard, especially as they are frequently eaten raw.

The potential hazard due to viruses, particularly those excreted in human feces, should be noted. Shellfish, particularly clams and oysters, taken from contaminated waters have been incriminated as the source of a number of outbreaks of infectious hepatitis in recent years (Bryan, 1973) and may serve to transmit other viruses, particularly enteroviruses (ICMSF, 1978; Cliver, 1971).

D. Parasites

Fish are hosts to many parasites, most of which are harmless to man. Freshwater fish serve as intermediate hosts in the life cycle of a number of trematodes that may be transmitted to man if the animals are consumed raw or lightly salted. These include *Cloronchis sinensis, Heterophyes heterophyes, Opisthoreis felineus,* and *O. viverrini.* See the second edition of Book 1 of this series (ICMSF, 1978) for a more complete discussion of them and the crustacean trematode *Paragonimus westermani.* The "fish tapeworm" *Diphyllobothrium latum* and its marine counterpart *D. pacificum* occur in freshwater fish and the marine species "sea wolf" (*Anarhichas orientalis*), respectively, and may be transmitted to man. *Diphyllobothrium* has been reported in humans in the United States and Canada on several occasions in recent years. Among the nematodes, *Anasakis* and probably *Phocanema* are genera transmitted from fish to man. *Anasakis* infections have been reported in Europe and Japan. Recently, worms of the *Phocanema* group have been implicated in human infestations in the United States (U.S. Department of Health, Education and Welfare, 1976a). The *Anasakis* and *Phocanema* worms are present in the flesh of fish and can be readily ingested with raw, lightly pickled, or cold-smoked fish. Cooking destroys the organisms, but there is some evidence that they may survive freezing. *Angiostrongylus cantonensis* has reportedly been passed to man in the Pacific islands by consumption of freshwater shrimp and land crabs (WHO, 1974).

Little is known concerning the possible transfer of parasites by fish raised in ponds fertilized with human or animal manures, since these fish are normally raised and eaten in regions of the world where intestinal parasitization is almost universal, and records of intestinal illness are poor or nonexistent. However, in view of the wide range of products offered by gourmet and specialty stores, the possibility of parasite spread to less highly infested populations through consumption of fish or shellfish raised in manured ponds should not be ignored.

E. Gonyaulax

The other type of intoxication involving microorganisms is the much more serious and frequently lethal syndrome called paralytic shellfish poisoning (PSP) (Schantz, 1973; ICMSF, 1978). This is usually ascribed to saxitoxin, one of several toxins produced by certain dinoflagellates and found in shellfish. The most commonly implicated species are *Gonyaulax catanella* and *G. tamarensis,* but *Pyrodinium phoneus, Gymnodinium breve,* and some other species have also been incriminated. These organisms are consumed by molluscs, including clams, mussels, and oysters. The shellfish are unaffected but retain the saxitoxin and accumulate it in their tissues. The toxin is normally localized in the digestive gland and gills of bivalves, but butter clams tend to concentrate much of the toxin in the siphon. The toxin has no effect on the shellfish but is highly poisonous to man and other warmblooded animals. For a detailed discussion of PSP see the second edition of Book 1 of this series (ICMSF, 1978).

III. CHILLED RAW SEAFOODS

A. Fin Fish of Marine Origin

1. Catching and Processing Procedures

Fish are caught by net, hook and line, or traps in bodies of water more or less remote from the processing plants. Moreover, because of the catching methods, which may extend over several hours, and the unstable and difficult working conditions at sea, there is little control over either the condition of the animals at death or the time of death. This contrasts with the red meat and poultry industry in which animals in good physiological condition are brought live to the processing plant (abbatoir), where they are dispatched quickly and with minimal stress.

Fish after capture must be protected from spoilage to the extent possible during transportation as dead animals to the processing plant. This may require from a few hours to 3 weeks or more. Storage is normally in melting ice or chilled brine (or sea water) solutions at $-2°C$; but increasingly, fishing vessels are equiped with facilities to freeze all or part of the catch at sea, thus effectively halting microbial spoilage.

Fish may be eviscerated on board the vessel prior to packing in ice. This is a common practice on European fishing vessels, at least for larger fish. In other parts of the world, particularly where sailing time between fishing ground and port is short (e.g., Pacific Coast of the United States), fish are stored uneviscerated. Small fish such as herring are not normally eviscerated at the time of capture.

There is controversy concerning the advantages and disadvantages of gutting or not gutting at sea. The process certainly removes a large reservoir of potential spoilage bacteria, but the cutting necessary to achieve this opens flesh surfaces to direct bacterial attack. Products of bacterial growth in the gut and the action of the intestinal enzymes on food and fecal material can cause discoloration in flesh adjacent to the belly cavity, off-flavors, and even digestion of the belly wall in uneviscerated stored fish.

Because of this, it is probably best to gut immediately on capture and wash the carcass thoroughly when this is economically practical.

In general, microbial changes occurring in fish stored in ice or sea water are similar on shipboard and on land. However, conditions are frequently more difficult on board ships and may give rise to undesirable processes. Fish pressed against wooden pen boards may become "bilgy" as a result of anaerobic bacterial growth, and a "bilgy" odor may develop around the gills of fish held in inadequately circulated chilled brine, likewise caused by local anaerobic bacterial growth.

While considerable quantities of fish are sold to consumers whole or eviscerated, the bulk of fish and shellfish undergo further processing in the raw condition. The most common operations for fin fish are filleting and steaking. These operations may be done entirely by hand or by a combination of hand and mechanical processing. A recent development in the separation of flesh from bones and skin is the deboning machine, which yields a minced flesh product quite different from the fillets, steaks, and chunks of meat obtained in more traditional processing.

2. *Saprophytes and Spoilage*

Fish caught by nets or hook and line are commonly killed or die more or less rapidly when brought into the air. There is no substantial evidence that agonal invasion of adjacent tissues or blood vessels by gut bacteria occurs in fish. However, bacteria may gain access to puncture wounds and even to areas bruised during the death struggle and may increase rapidly in these localized areas of tissue (Tretsven, 1964).

Opportunities for bacterial contamination appear at the earliest stages of off-loading the fish from vessels. Modern practice calls for the use of pumps and conveyor systems that may redistribute surface contamination but do not increase it unless polluted water is used in the pumping system. Most fish are probably still off-loaded by transferring them to boxes and baskets that are swung onto the quayside by cranes. Hooks, forks, or pews may be used to facilitate lifting of heavy fish into baskets or onto conveyors. If these penetrate flesh, the puncture area provides an excellent site for bacterial growth, and neighboring tissue may be invaded (Tretsven, 1964). Bacterial buildup in boxes and baskets can be considerable, and this may represent another source of contamination of the fish.

During the operation of capture, the fish comes into contact with nets and ropes, deckboards, human hands and clothing, and this contact continues during the packing and storing operation below decks. It is not surprising, therefore, that freshly caught but handled fish, even from temperate or cold waters, may carry significant numbers of gram-positive bacteria (including coryneforms, *Micrococcus, Bacillus, Staphylococcus*) and even coliform bacteria (in very small numbers). These organisms derive from the shipboard and human environment and are not indigenous to the fish. They are mesophilic or occasionally eurythermic and generally unsuited for competitive growth under the typical conditions of handling and primary processing of fish. Consequently, their occurrence is transient in most instances.

Whether eviscerated or not, most food fish are stored in ice or in refrigerated seawater (RSW) on board the vessel between capture and handling. Clean fresh ice or RSW carry negligible numbers of bacteria, but used ice or RSW may be heavily contaminated with psychrotrophic spoilage organisms (Shewan, 1962). Fortunately, clean ice and RSW are commonly used. The effect of the chilling agents on the bacteria on the fish is a selective repression of growth. Adventitious mesophilic bacteria either fail to grow or slowly die, while the psychrotrophic strains increase in accordance with their capability to compete. For fish caught in temperate waters, the increase in numbers tends to follow a rather characteristic sequence illustrated in Table 20.5.

During this second stage of processing, there is ample opportunity for contamination of products with bacteria from the environment and from the fish plant workers. Under poor quality control conditions where fish temperatures are allowed to rise, bacterial populations may increase (Shewan, 1962). However, under properly controlled conditions this stage of wet fish processing should result in a net reduction in bacterial numbers. Unless the fish have been in shipboard storage for a long time or handling processes have been very poor, the bacteria should be confined to external surfaces and the alimentary canal, and the deeper flesh should be sterile. Of course the numbers of bacteria on the surface can be very large, and transfer of some of these to the exposed flesh surface usually occurs. The microbial level is indicative of good or bad processing practice.

Fish are sold by public auction in many countries. This requires that they be displayed in wooden, metal, or plastic containers on an open or covered quayside or in an auction shed. There are several potential bacteriological dangers in this process. The more or less lengthy exposure to increasing temperatures permits spoilage bacteria to multiply. Containers may be difficult to clean or may be poorly cleaned. Birds, possibly rodents, and flies may gain access to the fish, contaminating them with droppings. Finally, auctioneers, sellers, buyers, and spectators walk among, around, and sometimes across the containers of fish, contributing additional organisms.

At the fish-processing plant, there frequently will be a sorting and washing process prior to temporary iced or refrigerated storage or actual processing. Again, the quality of the water is supremely important. Washing with

TABLE 20.5

Stages in Fish Spoilage

Stage	Tissue changes [a]	Organoleptic changes	Bacterial count	Chemical changes [a]
Stage I (0–5 days in ice)	Rigor mortis ATP → inosine Slight increase, TMA Changes in bacterial types	Eyes bright Flesh firm Color good Gills bright Odor fresh	10^2–10^3/cm^2	TMA $\leqslant$ 1–1.5 mg% VRS $\leqslant$ 2–8 units
Stage II (5–10 days in ice)	Bacterial growth becomes apparent Inosine → hypoxanthine TMAO → TMA NH$_3$ increases	Eyes begin to dull Gill color fades Skin color fades Odor neutral to slightly fishy Texture softening	10^3–10^6/cm^2	TMA < 5 mg% VRS CA 5–10 units TBV $\leqslant$ 15 mg%
Stage III (10–14 days in ice)	Rapid bacterial growth Penetration of tissues Hypoxanthine → xanthine, uric acid, etc. TMA increases rapidly TVB and TVA increase	Eyes sunken Gills discolored and slimy Skin bleached Odor sour and fishy Texture soft	10^6–10^8/cm^2	TMA ~ 10 mg% VRS ~ 10–15 units TVB 20–30 mg% TVA 15–20 mg%
Stage IV (over 14 days in ice)	Bacterial numbers stationary Some change of species General deterioration of flesh TVB and TVA increase rapidly TMA increases or levels off Proteolysis begins H$_2$S and other products formed	Eyes opaque and sunken Gills bleached and slimy Skin very slimy Texture very soft Odor offensive	~10^8/cm^2	TMA > 10 mg% VRS > 20 units TVB > 30 mg% TVB $\geqslant$ 60 mg% H$_2$S indole etc. detectable

[a] TMA: Trimethylamine expressed as mg/100 g fish; TVB: total volatile base expressed as mg/100 g fish; TVA: total volatile acids expressed as mg/100 g fish; VRS: volatile reducing substances expressed as mg/100 g fish.

clean chlorinated water will remove dirt and slime and greatly reduce surface contamination.

During wet-fish processing, contamination most commonly results from
direct handling (coliforms, staphylococci), direct transfer (gut and skin
bacteria to fillet flesh surface), or environmental transfer (contaminated
surfaces, knives, machines, etc). Even under good processing conditions, a
heavy transfer of land- or human-source bacteria to fish is common at this
stage, so that the microflora transiently may show high levels of gram-
positive bacteria even though the total counts in the prepared product may
be considerably lower than in the raw product. The transience results from
the inability of these organisms to survive and grow competitively on fish
flesh as products are held under refrigeration. Thus, within a day or two of
storage after processing, the bacterial populations are once again predominantly gram-negative.

Net increases in count on products undergoing simple processing are indicative of poor manufacturing practices. The potential sources of contamination are many, since surfaces contacting the raw product may become covered with bacteria-laden slime and waste materials. In addition,
there is a very high level of human contact throughout the operation.
Processing is a wet operation, and contamination is minimized in most
plants by copious use of (usually chlorinated) water to wash away product
residues on equipment, tables, floors, and machines during processing. In
addition, workers must be required to wear protective clothing and wash-
able gloves and should be provided with disinfectant dips for hands and
equipment (e.g., filleting knives).

Obviously the extent of contamination with spoilage bacteria depends on
the quality of in-plant housekeeping and sanitation, but it is also a function of the bacteriological quality of the incoming raw material. Bacterial
counts on the surfaces of fish coming into a plant can range from a few
hundreds or thousands to tens of millions per square centimeter. The poor
quality fish with high bacterial level then may contaminate good fish. Also,
as storage proceeds, there is a tendency for more active (spoilage) bacteria
to predominate in the population and to cross-contaminate freshly exposed
surfaces, causing them to spoil more rapidly.

Spoilage bacteria grow almost entirely on the surface of fish. Therefore,
spoilage proceeds slowly when the surface/volume ratio is low, and rapidly
when the ratio is high. Thus, the spoilage rate increases as one passes
from whole fish to gutted fish, to fillets or steaks, and finally to minced
fish. An exception may be fish caught while feeding heavily, for on occasion the intensive enzyme activity digests the gut wall soon after death,
permitting localized microbial entry into the flesh surrounding the belly.

The quantitative changes in bacterial populations are associated with qualitative changes in composition. Typically, gram-negative rod-shaped bacteria rapidly become the dominant type of organism as growth of the population commences (Liston, 1978; see also Table 20.6). While this is due in an overall way to the ability of these organisms to grow at temperatures below 5°C, the capability of the bacteria rapidly and effectively to utilize the nonprotein nitrogen compounds that constitute the available substrate for nutrition is probably the significant determinative factor. This can be seen from Fig. 20.1 in which all group 1 bacteria are spoilage pseudomonads. This is the reason that, in most cases, *Pseudomonas* species (or *Alteromonas* and *Pseudomonas*) become dominant in the bacterial populations of ice-stored fish. Oxidative deamination followed by reincorporation of some of the released ammonia to serve as a nitrogen source seems to be the primary mechanism. This results in an accumulation of volatile amines (mostly ammonia) and free fatty acids. In addition, other amines are produced by the breakdown of nonprotein N compounds, most importantly trimethylamine (TMA) from the reduction of trimethylamine oxide (TMAO), particularly in gadoid fish. In the initial stages of storage, there is little protein breakdown, and organisms that are not particularly proteolytic dominate the spoilage flora. However, in later storage periods when "spoilage" becomes apparent, there is evidence of increased proteolysis (Table 20.7). Sulfur-containing compounds such as methyl mer-

TABLE 20.6

Generic Distribution of the Microbial Flora of English Sole Stored in Ice [a]

	Days in ice					
	3 days		9 days		16 days	
Microflora	Skin (%)	Flesh (%)	Skin (%)	Flesh (%)	Skin (%)	Flesh (%)
Pseudomonas	28	0	72	83	98	87
Acinetobacter–Moraxella (Achromobacter)	18	50	5	3	1	1
Flavobacterium	3	25	3	2	0	0
Alcaligenes	3	0	7	2	1	4
Gram + cocci	33	0	0	0	0	0
Gram + rods	10	25	13	3	0	7
Yeasts	5	0	0	0	0	1
Total %	100	100	100	100	100	100
Total no. of isolates	39	4	64	68	85	82

[a] Bannerjee (1967).

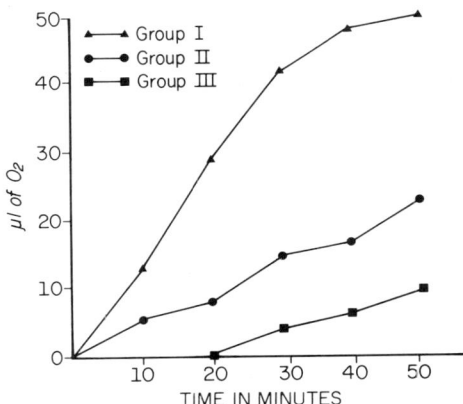

Fig. 20.1. Mean oxygen uptake in the presence of alanine by individual groups of bacteria isolated from fish stored for 3 days in ice (Bannerjee, 1967). Group I: rapid rate of uptake (spoilage *Pseudomonas*). Group II: intermediate rate of uptake (other *Pseudomonas*). Group III: slow rate of uptake (some *Pseudomonas* and all other types of bacteria from fish).

captan, dimethyl sulfide, and H_2S are produced particularly by *Alteromonas* (*Pseudomonas putrefaciens*) in relatively small amounts. Though amines such as ammonia and TMA are major components of spoiling fish aroma, it has been suggested that many of the characteristic odors and flavors of spoilage are ascribed to the compounds listed in Table 20.8, and it has been suggested that the organisms producing them should be considered the "true" spoilage bacteria, even though they may represent only 10% of the total microflora.

Typically, it is reported that in the first few days of storage the surface bacterial populations are composed of a mixture of genera; sometimes *Acinetobacter–Moraxella* (*Achromobacter*) and *Pseudomonas* are dominant, with a scattering of other genera, and sometimes gram-positive bac-

TABLE 20.7

Tyrosine Nitrogen in English Sole Fillets Stored in Polymylar Bags at $0°-2°C$ [a]

Days	Tyrosine	Days	Tyrosine
0	.8.2 [b]	11	14.18
4	10.7	14	19.20
7	12.25	19	24.42

[a] Adapted from Chung (1968).
[b] mg N/100 g muscle.

TABLE 20.8

Partial List of Volatile Compounds Present in Spoiling Fish Muscle in Small Amounts [a]

Ethyl mercaptan	Ethanol
Methyl mercaptan	Methanol
Dimethyl sulfide	Acetone
Dimethyl disulfide	Acetoin
Hydrogen sulfide	Butanal
Diacetyl	Ethanal
Acetaldehyde	Methyl butanal
Propionaldehyde	

[a] Liston et al. (1976).

teria represent a substantial component. As growth begins and numbers increase, *Acinetobacter–Moraxella, Flavobacterium,* and *Pseudomonas* overgrow the other genera. During exponential (logarithmic) growth, *Pseu-domonas* usually becomes the dominant genus, but in a few instances *Acinetobacter* and *Moraxella* have been reported as major or even dominant genera (Walker *et al.,* 1970).

In the latter stages of storage where spoilage is advanced and fish are usually considered inedible, the composition of the bacterial population again may become heterogeneous as species of other genera grow on the large variety of compounds generated by the activity of the pseudomonads.

The bacteria on spoiling fish are quite active biochemically. TMAO is present in essentially all marine fish, and its reduction to TMA is characteristic of the spoilage process in many species. The reduction is believed to be primarily due to bacterial action, and one would expect some correlation between the increase in TMA and the occurrence of TMAO-reducing bacteria. There is indeed an increase of TMAO reducers corresponding in a general way to the increase in TMA, but the failure of these organisms to increase their proportion in the total population suggests that little advantage is to be derived from this property.

Most of the preceding description of the bacteriology of fresh and stored fish is derived from work with fish caught in temperate waters and most commonly with gadoid (codlike; includes cods, haddock, hake, etc.) or other demersal (fish living on or near the sea floor) species stored in ice. There are differences in the bacteriology of fish taken in nontemperate regions, perhaps in pelagic (open water) fish and in fish stored under different conditions.

Coryneform bacteria apparently may be present in large numbers on

African lake fish and may be involved in postmortem spoilage. *Tilapia* from the Nile River were reported to show a change from a gram-positive to a gram-negative microbial population during storage in ice (El Mossalami and Sedik, 1973).

3. Pathogens

Vibrios have been reported on pelagic and on some bottom dwelling fish (Simidu *et al.,* 1969; Sera and Ishida, 1972; Baross and Liston, 1970), and it is not clear whether they survive in significant numbers through the normal handling and processing operations. The frequent occurrence of *Vibrio parahaemolyticus* food poisoning incidents from consumption of raw-fish products in Japan suggests that they do survive and under suitable conditions can increase rapidly in numbers to an "infective" level. It has also been shown by experiment that *V. parahaemolyticus* will outgrow the natural microflora on fish and crustaceae when temperatures are above 20°C (Katoh, 1965; Liston, 1974).

As noted earlier, *Clostridium botulinum,* particularly type E, occurs in small numbers in the gut and sometimes on the skin of fish taken in the northern temperate regions. The spores of this organism may persist through normal handling and can germinate and grow under suitable conditions. However, since these conditions include removal of competing bacteria, anaerobic conditions, and temperature greater than 4°C, it is unusual for *C. botulinum* to be found on fish products except as a result of gross mishandling of smoked or canned products.

Erysipelothrix seems to be associated with dockside soil and is carried aboard fishing vessels in wooden structures from whence it gains access to fish. The organism occasionally grows in deep puncture wounds from fish spines, giving the handler the painful "fish finger." There is no evidence of any problem with this organism beyond the first stages of processing.

Staphylococci are often found on fish and fish fillets after primary processing, but there is no evidence that they will multiply unless there is poor temperature control. Fish are so highly perishable that storage for more than a few hours at temperatures that permit *Staphylococcus* growth will yield a spoiled and repulsive product. The staphylococci are clearly derived in most cases from fish-plant workers, and normal hygienic practices will greatly reduce the level of contamination. This is particularly important where the fish is to be made into a prepared product.

Salmonella and *Shigella* are extremely rare on fresh fish products. The temperatures at which fish are normally held are too low for the growth of these enteric pathogens and, indeed, they tend to die out. Moreover, there are no common sources of contamination in fish plants, since the fish themselves do not normally carry the organisms.

From a public health standpoint, the contamination of the product with potentially pathogenic bacteria from humans or other warm-blooded animals is most important. Such contamination is the most frequent cause of rejection of seafood products by inspection or regulatory agencies. The most common contaminants are coliforms and *Escherichia coli,* enterococci, and staphylococci, and in virtually all cases the cause of contamination is human handling. In some parts of the world, the use of harbor water to wash ice from fish or as a carrying medium in siphon-type pumps is still common. This represents an obviously hazardous situation, since harbors are commonly contaminated with fecal material and may add bacteria of public health significance to the fish.

4. Scombroid Poisoning

Scombroid poisoning is a result of the consumption of tuna (fresh, frozen, or canned) containing high levels of histamine (Kimata, 1961). The symptoms are vomiting, diarrhea, and allergic reactions (including puffiness around the eyes and mouth, tingling, and itching). Antihistamines are helpful in relieving the symptoms.

The normal level of histamine in fresh tuna is less than 20 mg/100 g muscle tissue. Fish involved in scombroid poisoning usually show histamine concentrations in excess of 100 mg/100 g flesh. In an incident in the United States (U.S. Department of Health, Education and Welfare, 1975f) the incriminated tuna had 626 mg histamine/100 g flesh. It is generally accepted that the histamine itself is the cause of scombroid poisoning, but there is some question whether that is so. Some Japanese workers reported the presence of another amine named saurine (Kawabata *et al.,* 1955), which they felt was the toxic agent. Another suggestion is that substance(s) are produced in the tuna flesh which inactivate the human detoxifying mechanisms (ICMSF, 1978).

Tuna and the other scombroid fish have high levels of free histidine in their tissues. After death of the fish, if the temperature of the flesh is not lowered quickly, bacteria may grow and decarboxylate the histidine to histamine. This seems to be the mechanism whereby the compound is produced, and small amounts of histamine are invariably present in dead tuna. Indeed, the measurement of histamine levels was proposed as a spoilage index in scombroid fish at one time (Tomiyasu and Zenitani, 1957). *Proteus morgani, Hafnia alvei,* and more recently, *Klebsiella pneumoniae* have been identified as the probable cause of the high levels of histamine found in tuna implicated in scombroid poisoning (Ferencik, 1970; Omura *et al.,* 1978; Sakabe, 1973; Taylor *et al.,* 1979).These bacteria are mesophiles, and their growth is inhibited at temperatures near 0°C. Thus, rapid chilling is an effective way of preventing toxin development. Nevertheless, there is still some question concerning the whole subject of scombroid poisoning. Enormous tonnages of tuna are caught and processed each year under conditions that might be expected to yield toxic fish in many cases, yet relatively few people become—or report becoming—sick.

5. Interrelations

The organisms that spoil raw seafoods (psychrotrophs) generally grow well from temperatures near 0° to about 30°C. Mesophiles, which include most of the pathogens, generally grow well between about 5° and 40°C. Thus, seafoods refrigerated below 5°C usually will not support the growth of pathogens. Furthermore, in the temperature range where both groups grow well (5° to 30°C), psychrotrophs have a shorter lag period (more rapid onset of growth) and, therefore, tend to spoil the food before the pathogens grow to dangerous levels (reviewed in Elliott and Michener, 1965; Michener and Elliott, 1964).

There are exceptions to these rules: *Clostridium botulinum types E* and *F* can grow at 3.3°C; *Salmonella* species can grow competitively on fish held at or above 8°C (Matches and Liston, 1968a). *Vibrio parahaemolyticus* can also grow on raw fish (Katoh, 1965). Staphylococci are especially poor competitors.

6. Control

To keep microorganisms at satisfactorily low levels in raw fish, handlers should maintain good sanitation on boats and at shore plants, should keep the product under constantly low temperature (< 3°C), and should handle it quickly. In addition, fish (except "feedy" fish) should be kept round for as long as possible before filleting, and should be undamaged by pews or forks. Holes from these tools introduce spoilage bacteria into the clean, edible tissue.

B. Crustaceae

1. Catching and Processing Procedures

Crabs and lobsters are trapped, then transported live to the processing plant. On arrival, any dead animals are discarded.

Shrimp usually are captured by trawlers then iced for transport to the processing plant. Unlike crabs and lobsters, the shrimp die quickly on capture. At the processing plant, they are washed, then shelled by automatic machines. The raw meats may or may not be breaded and are then frozen (see Section IV,A). Some meats are cooked (see Section V,A).

2. *Saprophytes and Spoilage*

Microorganisms on the surfaces of live crabs and lobsters during transport are those of the waters from which they come (often polluted), with the addition of bacteria from the fisherman and his boat. Spoilage organisms are of little importance at this point, because the live flesh remains sterile. The bacterial invasion of the tissues that follows death is of no importance, since the dead carcasses are discarded.

The picture of shrimp spoilage is somewhat different, since the animals die immediately on capture. The trawl picks up a considerable amount of marine mud with the shrimp. Bacteria from this source, the ice, and the boat surfaces have an opportunity to grow during the several days of transport to the shore plant. Thus, most shrimp have a very high count at the time of receipt ashore (Table 20.9).

There have been several reports that suggest that sometimes *Acinetobacter–Moraxella* (*Achromobacter*) and even coryneform bacteria may appear as the dominant group in stored shrimp and molluscs, but other workers report that *Pseudomonas* is usually dominant.

Raw breaded shrimp receives a general microflora—including gram-

TABLE 20.9

Bacterial Counts on Tropical Shrimp before and after Processing [a]

Shrimp type		Coliforms (MPN per 100 g)	Fecal coliforms (MPN per 100 g)	Standard plate count ($\times 10^3$/g) incubated at 20°C	Standard plate count ($\times 10^3$/g) incubated at 35°C
Flower shrimp	Before [b]	7,200	200	5,100	660
	After [c]	16,800	16,800	8,200	39,000
	Before	9.600	480	4,700	140
	After	2,700	80	3,900	670
White shrimp	Before	14,400	90	1,000	250
	After	15,200	15,200	5,500	15,000
	Before	3,300	480	1,100	390
	After	1,500	420	—	760
Pink shrimp	Before	8,400	12	1,100	230
	After	2,100	300	—	310
	Before	14,400	360	1,800	860
	After	14,200	4,300	8,300	1,300

[a] Adapted from Cann (1974).
[b] Raw shrimp, before processing.
[c] Frozen packaged shrimp.

positive and gram-negative species—from the batter mix that overwhelms the marine flora, particularly if the mix is not refrigerated or not changed frequently.

Table 20.9 shows how an unsatisfactory processing operation can introduce contamination or permit growth; more frequently than not, the processed product from such a plant has higher bacterial levels than the unprocessed product.

During spoilage, shrimp flesh seems to undergo many of the same biochemical changes seen in fish. Volatile basic substances increase, and pH rises. Discoloration results from the oxidation of pigments and sometimes from enzyme reactions (melanosis) leading to a condition known as black spot. Shrimp enzymes, not bacteria, are responsible for this. Because of their small size and high surface-to-volume ratio, shrimp spoil more quickly than fish and are usually frozen or otherwise processed quickly on landing.

Crabs and lobsters may undergo endogenous biochemical changes during live storage that may affect their quality, but there are no reports of significant changes due to bacteria.

3. Pathogens

When they are taken in inshore waters that may be polluted, crabs, lobsters, and shrimp may carry adventitious pathogens from sewage or land runoff. If this is so, the bacteria occur in small numbers and are destroyed during processing or die off under the low-temperature storage conditions.

Vibrio parahaemolyticus represents a significant problem in the case of crustaceae. Shrimp and crab have been incriminated in large outbreaks of vibrio food poisoning in the United States (Barker *et al.,* 1974), whereas in Japan such outbreaks usually occur from fin fish (Okabe, 1974). The organism is present apparently on crustaceae taken in warm inshore waters and can grow rapidly (Liston, 1976) when temperatures are allowed to rise. Most cases in the United States have been due to cooked crustaceans that were recontaminated.

Low-temperature strains of *Clostridium botulinum* probably occur on crustaceans. Types A and C have been isolated from shrimp, and C and F from crab and shrimp in the tropics. The hazard from these organisms would seem to be negligible under normal handling and processing conditions.

4. Interrelations

Fresh crab and shrimp from temperate waters carry a microflora dominated by gram-negative rods and (sometimes) coryneforms. Tropical shrimp show micrococci, coryneforms, and gram-negative rods as the principal components of the microflora (Cann, 1977). During the storage of shrimp in ice, psychrotrophic gram-negative bacteria become dominant, and these may be either *Pseudomonas, Acinetobacter–Moraxella (Achromobacter)*, or a mixture of both. Coryneform bacteria may persist at levels in excess of 10%, but micrococci are present in only small numbers.

5. Control

Since crustaceans may be caught close inshore, it is to be expected that some proportion will be from polluted waters and will carry sewage-derived microorganisms. Fortunately, such bacteria are carried on the surfaces of the animals and are usually destroyed by the normal cooking procedure to which crab, lobsters, and most shrimp are subjected. However, cooking does not destroy all bacteria; Cann (1977) reports counts as high as $3.7 \times 10^5/g$ in crab boiled for 30 min and $4.5 \times 10^4/g$ on tropical prawns after boiling for 3 min. Thus, when waters are heavily polluted with human fecal waste or are contaminated with dangerous infective pathogens such as *Vibrio cholerae,* fishing should be prohibited. The problem is more likely to be one of transfer of the pathogen than of foodborne illness of the usual kind.

It is normal practice in commercial fisheries to wash shrimp with clean seawater to eliminate contaminating mud that can carry many bacteria. In some fisheries, larger shrimp are "headed" on board the vessel. This procedure eliminates the head, gills, and thorax containing organs and some of the viscera, thereby removing a large internal source of bacterial contamination (70% of the bacterial load, according to Novak, 1973) but exposing flesh at the broken surface. There seems to be little benefit bacteriologically in this technique, but it does eliminate most of the enzymes responsible for development of black spot (melanosis). Shrimp are normally iced to preserve quality on board ship and also during storage prior to processing on shore. This inhibits growth of mesophilic bacteria.

Picking, the separation of the edible tail meat from the carapace and removal of the "sand vein" (lower intestine) is commonly done by peeling machines that are quite sanitary in larger operations. However, a large amount of shrimp is still hand peeled, often under insanitary conditions.

C. Molluscs

1. Harvesting and Processing Procedures

Clams are usually dug from the sand, and oysters are either raked, hand picked, or trawled from the bottom. They are then transported live in the shell, generally without refrigeration. As with crabs and lobsters, dead animals are discarded.

In many countries, molluscs are delivered to the retail store live "in the shell." But where trade in molluscs is more than local in nature, the animals are frequently removed from the shell (shucked) prior to shipment raw. Shucking is mostly a hand operation, which can lead to contamination of the shucked flesh. Oysters and, to a lesser extent, clams frequently are consumed raw.

After shucking, oyster, clam, and mussel meats that are not to be processed further are packed commonly in glass containers with or without brine (or organic acids in the case of mussels) and stored under refrigeration. Oysters are customarily tumbled in aerated fresh water ("blown") to remove shell fragments and sand and to plump out the body prior to packing.

2. Saprophytes and Spoilage

Molluscs carry a resident population of bacteria that has been found in the case of oysters to fluctuate between 10^4 and 10^6 bacteria/g of tissue. The higher counts occur when water temperatures are high. The microflora is dominated by gram-negative rod-shaped bacteria of the genera *Vibrio, Pseudomonas, Acinetobacter–Moraxella (Achromobacter), Flavobacterium,* and *Cytophaga* (Colwell and Liston, 1960; Vanderzant *et al.,* 1973; Lovelace *et al.,* 1968). This seems to be the resident population of the mollusc. Smaller numbers of gram-positive organisms may also be present.

When molluscs feed in polluted water, they will concentrate contaminating bacteria, including enteric pathogens and also viruses if they are present. Since oysters and other molluscs harvested for human use are normally grown in estuarine areas that receive some waste from land sources, it is not uncommon for small numbers of coliforms to be found in oysters, but they are not part of the normal resident bacterial population.

During spoilage of shucked molluscs, bacterial populations will normally increase to $10^7/g$ or more. Gram-negative proteolytic bacteria, probably *Pseudomonas* and *Vibrio,* are prominent in the spoilage flora. In addition, saccharolytic bacteria are active, fermenting the 3% or more glycogen in the tissues to various organic acids. In at least one, *Lactobacillus* was reported (Shiflett *et al.,* 1966) as a major component of the spoilage microflora and was identified as the fermenting organism.

Biochemically, spoilage seems to include both proteolytic and saccharolytic activity. Ammonia and other amines accumulate, but so do acids. Indeed the pH of molluscs typically falls during spoilage (this contrasts with fish and crustaceae in which pH increases during spoilage). Fresh oysters show pH values from about 6.2 to 6.5, and this decreases to 5.8 or below during spoilage.

3. Pathogens

During feeding, molluscs filter microorganisms from the water. The volume of water pumped by an oyster can be as great as 10 liters/hr. Thus, the ability to concentrate microorganisms is great. Enteric pathogens derived from human or animal sewage, including *Salmonella, Shigella, Vibrio cholerae, Escherichia coli,* and viruses—particularly infectious hepatitis—are of concern because molluscs may be eaten uncooked. However, *Clostridium botulinum* also has been isolated from oysters (Presnell *et al.,* 1967), and presumably *C. perfringens,* staphylococci, and other potentially pathogenic bacteria can occur. Generally, such bacteria are in the alimentary canal of the mollusc and are not removed by simple shucking. *Vibrio parahaemolyticus* is present commonly in molluscs during the warmer summer months in temperate waters (Baross and Liston, 1970) and all year round in warm waters (Vanderzant *et al.,* 1973). Levels of this organism during summer may be high enough to cause infection, but the number of outbreaks due to this is small, indicating that the hazard is low. Enteric pathogens will persist in oysters during storage at low temperatures, but they may be overgrown by the normal spoilage flora.

Control of the bacteriological quality of growing waters, limitation of harvest from polluted areas, and close surveillance of molluscs in transit by public health agencies has kept the number of outbreaks occurring in the industrialized countries remarkably low in recent years, but the potential for explosive outbreaks is still present.

4. Interrelations

The presence of a large natural population of bacteria does not seem to affect the uptake and short-term survival of potentially pathogenic bacteria. However, there is not much information available on the fate of such organisms during the spoilage of oysters, when the pH drops below 6.0. It would be unwise to assume that such conditions would destroy pathogens, however, since some, at least, are known to survive long periods in oysters stored in the shell.

5. Control

Only molluscs harvested from clean, uncontaminated water can be considered bacteriologically safe. The bacteriological definition of clean water is provided in the U.S. National Shellfish Sanitation Program (Houser, 1965; ICMSF, 1974) and Code of Hygienic Practice for Shellfish (Food and Agriculture Organization, 1976). Requirements for purification and secondary processing are listed also. Unfortunately, large quantities of shellfish harvested from uncontrolled areas still enter the human food supply and represent a considerable hazard.

A full description of these control procedures has been published as Appendix 6 of the *second book of this series* (ICMSF, 1974). Samples of water and molluscs should be analyzed for microbial content by methods described in Recommended Procedures for the Examination of Sea Water and Shellfish (American Public Health Association, 1970).

Molluscs are sometimes subjected to a unique cleansing operation prior to shucking or release for sale. The cleansing, which is applied to animals grown in water of suspected bacteriological quality, may be achieved by active depuration or by relaying the animals in clean water areas. In either case, molluscs cleanse themselves of contaminating bacteria by the flushing action of their normal feeding mechanisms. Animals in the depuration systems are subject to a flow of seawater that has been disinfected (potentially harmful organisms removed) by chlorination, ozonization, or by ultraviolet light, and reconditioned by an appropriate treatment to eliminate toxic residues from the water. Harmful bacteria may be eliminated within 24–28 hr in most cases (Presnell *et al.,* 1969). A similar result may be obtained by relaying shellfish in naturally clean water. The process seems to have little effect on the natural microflora of the shellfish. This does, however, raise the question of *Vibrio parahaemolyticus,* which behaves as part of the natural microflora in oysters. Unfortunately, too, depuration may not remove viruses. Experiments using attenuated Polio I and West Coast Oysters (*Ostrea lurida*) have shown residual levels of 10–90 infective units after a 72-hr (simulated) depuration process (Di-Girolamo, 1969), but other experiments, using the Eastern oyster (*Crassostrea virginica*), have shown less than 0.2 plaque-forming units (P.F.U.) after depuration for 48 hr (Hamblett *et al.,* 1969). No cases of infectious hepatitis have been reported from consumption of depurated molluscs, so in practice, standard depuration procedures apparently are effective in dealing with the low levels of virus that may be present. Depuration is essentially a reconditioning mechanism that is less desirable than maintenance of shellstock quality by bacteriological control of the growing waters. The spread of human habitation and activity in the inshore marine and estuarine waters all over the world makes it increasingly unlikely that purity (particularly freedom from virus contamination) can be maintained in all growing areas. Indeed, we should anticipate a continuing encroachment of contamination in all such areas. Therefore, depuration or some similar process may become a mandatory operation in shellfish culture.

D. Freshwater Fish

The processing of raw freshwater fish and shellfish is similar to the processes for marine fish. However, since the animals are commonly captured close to the processing point, the problems of transportation and intermediate handling are greatly reduced. There is a somewhat greater potential for contamination by microorganisms of public health significance because the waters are more likely to be polluted. Indeed, in some countries in which night soil (human feces) is used as a fish pond fertilizer, contamination with enteric pathogens and parasites is a serious hazard. Unfortunately, little work seems to have been done on this. Otherwise, the hazards and expected microbial changes during processing are the same as for marine animals.

IV. FROZEN RAW SEAFOODS

A. Freezing Process

The usual raw seafood destined for freezing is a fillet, or steak, or a shucked oyster, placed into a consumer-sized package weighing from a few decigrams to a kilo or two, or in commercial sizes for reprocessing or repackaging as blocks weighing as much as 8 or 10 kilos. The packages, or blocks, are frozen either by contact with freezing shelves, or more commonly by a cold air blast. In either case, freezing should be rapid. Good manufacturing practice requires that the frozen packages be stored at or below $-18°C$. Thawing is best done in a refrigerator at about $3°C$ or by immersing in cold water.

B. Saprophytes and Spoilage

Bacterial counts on frozen products will reflect the bacteriological quality of the raw material and contamination or physical removal of contamination during processing. The reduction in count resulting from freezing and storage in the frozen state is highly variable, and this makes the assessment of prefreezing quality difficult in some cases (DiGirolamo *et al.,* 1970). The psychrotrophic bacteria in fish are not particularly resistant to freezing stress, but response is so strain-dependent that no general rule can be given.

Spoilage organisms may grow in a raw fish product held too long before freezing, frozen at a grossly slow rate, thawed too slowly, or held thawed too long. The limitation is time and temperature; no microorganisms can grow below $-10°$ or $-12°C$, and the growth rate increases dramatically with increased temperatures above this limit (reviewed in Michener and Elliott, 1964; Elliott and Michener, 1965). Floral changes and biochemical alterations are those described previously (see Section III).

Seafoods held at improperly elevated frozen storage temperatures ($-10°$ to $-5°C$, for example) are likely to support very slow mold growth. Some molds, and possibly yeasts can grow in that range, whereas bacteria grow only at higher temperatures.

Whereas bacteria do not grow in frozen foods, they are able in some degree to survive freezing and frozen storage, so that thawed fish will spoil about as fast as fish that has never been frozen.

Gram-negative bacteria often die more readily than gram-positive, and spores always survive well.

C. Pathogens

Although the gram-negative *Vibrio parahaemolyticus* is quite sensitive to freezing and thawing, it can survive in small numbers (Johnson and Liston, 1973). This is of little concern in Western countries, since these potentially dangerous bacteria would be destroyed by cooking. Only re-contamination of the cooked food, followed by time–temperature abuse would be dangerous. In the Far East (e.g., Japan), where consumption of raw fish is the rule, the presence of *V. parahaemolyticus* in raw fish is an obvious problem. Outbreaks of disease from this organism are common-place in Japan.

Fishborne *Clostridium botulinum* would survive the freezing process, but it presents no hazard unless conditions for its outgrowth and toxi-genesis are provided. Toxins that might be present in the raw product or produced as a result of bacterial growth in seafoods prior to freezing would, of course, not be inactivated by the freezing process. Conditions for bacterial toxigenesis would also permit development of large popula-tions of living bacteria, and this would be detected in most frozen thawed products as an excessively high aerobic plate count and obvious spoilage.

D. Interrelations

It is unlikely that poor temperature control, which is commonly en-countered in distribution of frozen seafoods, will give rise to a bacteri-ologically unsafe product, even if the product is actually thawed and al-lowed to remain at temperatures in excess of $10°C$ for several hours. In most cases, this would result in such rapid multiplication of spoilage bac-teria that the food would be unacceptable to the consumer. The micro-biology of such products would not be different from that of fish handled without freezing (see Section III,A,5).

E. Control

Temperature control is the principal means to stop microbial activity in frozen seafood, i.e., rapid freezing, holding at or below $-18°C$, and rapid thawing.

Microbial analysis gives some information about the quality of the fish before it was frozen. However, all strains of microorganisms are to some degree killed by freezing and frozen storage; the level of bacteria in frozen fish is always lower than it was in the product before freezing, but it is difficult to say how much lower. Fecal streptococci determined on the frozen product may give a more useful measure of the sanitation of processing than *Escherichia coli,* in view of their greater ability to survive freezing (reviewed in Elliott and Michener, 1961). Staphylococci are reasonably resistant to the effects of freezing (Raj and Liston, 1961) and provide some index of the degree of human contact with the product. The PSP toxin in shellfish meats or scombroid poison(s) in tuna can be detected only by laboratory analysis.

V. COOKED CRUSTACEAE (FROZEN OR CHILLED COOKED SHRIMP, LOBSTER MEAT, AND CRABMEAT)

A. Cooking, Picking, and Packaging

The cooking process can be blanching (shrimp, 95–100°C), boiling, or steaming under pressure (lobsters, crabs, 100°C or above). The duration of the cook is generally short, to minimize quality loss.

Meat from cooked crabs and lobsters is picked by hand. Crabmeat is frequently separated from small pieces of shell by flotation in salt brine then washed in fresh water before packaging and chilling or freezing.

To improve shelf-life, crabmeat may be packed into metal containers and pasteurized so that the internal temperature reaches 77°C for at least 1 min. The heated product is then stored under refrigeration ($< 3°C$). This is a method to hold crabmeat in bulk storage for up to 6 months.

Shrimp may be cooked either before or after shelling. The cooked meat is packaged and frozen without further treatment, or after a batter and breading dip.

B. Saprophytes and Spoilage

Cooking of crab and lobster reduces the bacterial count significantly and probably destroys heat labile organisms such as *Salmonella* and *Pseudo-*

monas. However, temperatures below 60°C have been observed after 30 min of boiling in crabs in the middle of a basket (Cann, 1977).

The process of picking and shell separation, which may involve brine flotation, recontaminates the picked meat with a variety of microorganisms. Counts of the order of 10^5/g in picked meat are normal, and organisms present will include gram-positive rods and cocci, gram-negative rods, and yeasts. Storage of picked meat under refrigeration normally results in a progressive increase in count to 10^7/g or more in about 7 days. The microflora is then usually dominated by gram-negative rods, probably *Acinetobacter–Moraxella* (*Achromobacter*), which are presumably responsible for spoilage. Spoilage is apparently predominantly proteolytic in nature, with production of volatile amines.

Shrimp is handled somewhat differently from crab and lobster. It is, of course, landed dead and after varying periods of shipboard storage so that bacterial growth and spoilage can occur prior to the land-based processing. Bacterial counts on shrimp received at shore plants may vary from 10^3 to 10^7/g. Usually higher counts are more characteristic of shrimp from tropical waters, but this is not invariably so. Shrimp cooked by boiling or steaming for a few minutes to release the shell show approximately a 100-fold reduction in count but may rapidly return to the former level after peeling and sorting. Shrimp peeled and packed without cooking generally show little change in count, though freezing may cause a 10-fold drop. Small Pacific pink shrimp typically show a large reduction after peeling. Because of handling, low levels of coliforms, *Escherichia coli,* and staphylococci may be present in finished products. If shrimp are chilled when stored, after cooking and shelling, spoilage proceeds as a result of *Pseudomonas* or *Acinetobacter–Moraxella* (*Achromobacter*) and possibly (in tropical shrimp) coryneform bacteria (Cann, 1977). Spoilage is mostly proteolytic, with amines and ammonia produced.

C. Pathogens

Since the picking process is done by hand, the possibilities for transfer of potentially pathogenic bacteria from humans to crabmeat is high. The flotation in brine provides an opportunity for contamination by staphylococci, which survive well in salt solutions. Fortunately, staphylococci do not compete well with the normal spoilage flora of crabmeat and generally die slowly during storage (Slabyj *et al.,* 1965). However, it is certainly theoretically possible for pathogenic bacteria from human sources to contaminate picked crabmeat and persist for some time.

Vibrio parahaemolyticus occurs naturally on crustaceae, and outbreaks of *Vibrio* food poisoning have occurred as a result of the consumption of contaminated shrimp, crab, and lobster (Barker *et al.*, 1974). The organism is highly sensitive to heat and is readily destroyed by the cooking process.

Cross-contamination to cooked shellfish, followed by storage above 15°C have permitted *V. parahaemolyticus* to grow to high numbers that have caused illness in large numbers of people in the United States (Liston, 1976). Therefore, to minimize the hazard, processors and handlers should (1) heat the meats thoroughly to destroy the organism, (2) introduce physical separation and separate handling of raw and cooked product to avoid cross-contamination, and (3) establish temperature control (below 5°C) of the finished product to prevent growth of *V. parahaemolyticus*.

Strains of *Clostridium botulinum* (usually type E) have been isolated from crabs, but so far there is no evidence of a specific botulism problem. This is peculiar in some ways because the pasteurization process carried out by U.S. East Coast crab processors is insufficient to destroy all *C. botulinum* spores that might be present. Pasteurized crabmeat is stored for several months in cans that can be expected to develop anaerobic conditions. The product is supposed to be held under refrigeration, and certainly storage below 3°C would prevent germination and outgrowth. In the nature of commercial operations, it might be expected that temperature abuse would occur from time to time, but in the 20 years that the pasteurization process has been used there have been no food-poisoning cases attributable to the practice, even though laboratory tests have indicated a possible hazard.

D. Interrelations

As noted earlier, the composition of the microflora during storage of crustaceans changes as spoilage becomes evident. The *Acinetobacter–Moraxella* (*Achromobacter*) group predominates, and *Pseudomonas* or sometimes coryneform bacteria may also be prominent. In most other animal meats, *Pseudomonas* predominates in spoilage under refrigeration. The difference may be that in crustacean meat, the water activity is reduced from the cooking and from the presence of up to 1½% NaCl. Even at 22°C, staphylococci show little capacity to grow to dangerous levels in crabmeat when spoilage bacteria are present and active (Slabyj *et al.*, 1965).

E. Control

The best control of the microbial condition of cooked crustaceae lies in good manufacturing practices. The cooked product should be handled and held at all times in facilities and equipment that does not contact the raw product. Employees should be educated in good hygiene and should wear sanitary gloves or should wash and sanitize hands before handling cooked product. In small plants where separate facilities are not feasible, only good personal hygiene and careful handling can ensure against cross-contamination.

Cooked product should be cooled promptly and held at or below 3°C. Flotation brines should be refrigerated and should be changed at least daily—perhaps oftener in large operations. At each change, the tank should be cleaned and disinfected.

Microbiological line sampling to determine adequacy of sanitation of surfaces and to measure contamination and growth in product will help evaluate postcooking manufacturing operations. Regular tests on final product are advisable. Coliforms, *Escherichia coli, Staphylococcus aureus,* and aerobic plate count are good indicators.

VI. CANNED SEAFOODS

Canned seafoods such as salmon, tuna, sardines, and similar products, which are given a full retort process, should be "commercially sterile" and free from living bacteria that are potentially pathogenic. For such foods, the bacteriological hazards are the same as those for other low-acid canned foods and relate, with one exception, to the problems of improper or inadequate processing or leakage. The one exception relates to scombroid poisoning. The toxin responsible (possibly histamine) is resistant to heat, and incidents of scombroid poisoning have resulted from consumption of canned tuna (U.S. Department of Health, Education and Welfare, 1975f). Sardines from North Africa have been reported to contain up to 200,000 spores of *Clostridium* per g but, though swelled cans sometimes occur, no disease outbreaks have been reported. The microbiologist must depend on good control at the processing level to ensure a safe final product.

Semi-conserved canned products include a wide range of seafoods that present formidable bacteriological problems. The stability and safety of these products depend on a combination of preservatives and pasteurization. Most are pickled products that depend on salt (e.g., anchovies) or a low pH (e.g., mussels) for stability after a heat process which destroys organisms that might be hazardous or might cause spoilage. Yeast (e.g., *Pichia fermentans*) may cause illness after growing in semiconserved seafoods. Some smoked products (e.g., salmon) are also canned, using a minimal heat treatment, and in these cases salt, smoke constituents, and a low a_w are presumed to be the stabilizing factors. In all of these cases, the margin for error is small. There is substantial historic record of food poisoning from semiconserved products (Shewan and Liston, 1955). Increasingly, such products are found in the chilled food display cabinets, where low temperature provides an additional protective effect. Changing public taste, particularly in Western countries, has caused a trend towards milder tasting products. This has led to reduction in the antibacterial action of pickling and curing processes, and it is often this milder type of product that appears in the refrigerated display. These products, when stored at higher temperatures, may support the growth of dangerous bacteria.

VII. CURED AND SMOKED SEAFOODS

Quite large amounts of fish and shellfish are still processed by curing or smoking, often by traditional methods. The most common method of curing involves salt, and sometimes drying. In all cases, the primary preservative effect is from a low a_w. Some of the processes may include fermentation (e.g., pickled herrings), but this is not a factor in bacteriological control. Bacterial counts on fully cured seafoods are generally low, unless there has been extensive surface contamination. Only halophilic bacteria that have no public health significance can grow on such foods. Obviously, bacteria of public health significance can persist on the surfaces of contaminated cured seafoods, and such products can be passive carriers of disease organisms. Generally, however, the microbiological hazards of fully cured or salted seafood products are hardly significant (Graikoski, 1973).

Smoked seafood products vary widely in microbial stability, depending on the nature and degree of severity of processing. Heavily salted, hard-smoked products (e.g., kippered salmon) are similar microbiologically to the fully cured products. Their water content is too low to support bacterial growth, and they represent little or no hazard. Lightly smoked products (kippered herring, finnan haddie, etc.) that are brined only enough to improve flavor, carry a mixed microbial population and are only slightly more stable than unprocessed fish. Characteristically, gram-positive bacteria dominate the microflora of such products after preparation (Shewan, 1962; Lee and Pfeifer, 1973), but gram-negative bacteria gradually become more numerous during refrigerated storage and usually are ultimately responsible for spoilage.

Hot smoked or barbecued products that are not given a long drying process are the most hazardous of smoked seafoods. Internal temperatures during hot smoking are usually sufficient to kill most vegetative bacteria but not spores. Moreover, modern processing techniques involve relatively light salting, so that the smoked flesh provides an excellent medium for the outgrowth of surviving bacteria. There have been several instances of selective outgrowth of *Clostridium botulinum* in hot smoked fish (U.S. Department of Health, Education and Welfare, 1970a) leading to outbreaks of botulism. The Eh of smoked fish flesh is low enough to permit outgrowth of clostridia, and the killing effect of hot smoking eliminates competing bacteria. Since hot smoked fish are frequently eaten with little or no additional heating, this product falls into a high-risk category.*

VIII. FERMENTED FISH PRODUCTS

Fermentation of some of the salted and semiconserved pickled products mentioned earlier is almost incidental to the main heating or preservative process. However, this is not the case with fermented products of the Far East, particularly in Japan, where there is a fairly wide range of fermented fish products prepared for human use.

The product produced in largest quantity is fish sauce (variously called Nuoc Nam, Na Plam, or Patis) prepared by mixing small whole fish, chopped fish, and sometimes plankton or fish viscera with salt to generate a brine mixture with a final salt content in excess of 20%. This mixture is allowed to ferment and digest over a period of several months at temperatures in the range of 25°–35°C. At the end of this time, a clear, golden liquid is obtained, which may be further purified before bottling for sale. When digestion is carried out for only about 4 months, an undigested residue remains (bagoong) that may be separated and sold for consumption. The initial microbial count on the raw fish used in these processes is usually high, and it is possible that bacterial enzymes are involved, together with fish enzymes, in the digestion process. However, during the long total digestion period, the bacterial population drops to very low levels (Saisithi *et al.,* 1966). The presence of 20% NaCl effectively precludes growth by any pathogenic bacteria, and these products appear to present no bacterial problems.

Other products prepared by a fermentation of mixed fish and vegetables or cereals are less safe. At least one such product (izushi), a pickled relish made from raw fish, rice, and vegetables allowed to ferment together for 3–4 weeks, has been the cause of several botulism outbreaks in Japan.

* U.S. regulations require that hot smoked fish should be brought to an internal temperature of 82°C for 30 minutes if brined to contain 3.5% water phase salt or not less than 65°C for 30 minutes if salt content is 5.0% in the water phase (U.S. Department of Health, Education and Welfare, 1976b).

Indian and Eskimo peoples of the Canadian Northern Territories and Alaska have been victims of botulism from a crude proteolytic, essentially anaerobic fermentation of seafood products. There is no intrinsic control, as contributed by salt or acid in more sophisticated processes, against the growth and toxin production by *Clostridium botulinum*. The mixture is simply buried or set near a stove to "ripen."

IX. CHOICE OF CASE

For an understanding of "Case" see Table A.1 and the brief description in Appendix IV.

In *Microorganisms in Foods 2* (ICMSF, 1974), eight categories of fish and crustacean products were assigned Cases, and sampling plans and microbiological specifications were developed for them. The eight categories did not cover all fish products, but only those for which sufficient microbiological data were available to justify specifications that are realistic, attainable, and supportive of the objective of safe, wholesome products.

Fish and frozen fish, which are to be cooked before eating, present minimal microbiological risk. Standard plate count, assigned to Case 1 is an index of quality (utility), and *Staphylococcus aureus* and fecal coliforms at Case 4 will detect gross insanitary handling. However, Japanese experience suggested it would be prudent to assign a *Vibrio parahaemolyticus* test to Case 12 for fish to be eaten raw, since this organism could grow to infective levels during food preparation. Similarly, freshwater fish from warm waters carries a potential *Salmonella* hazard and was assigned Case 10 for this organism.

The hazard from cold-smoked fish to be cooked before consumption is little different from that of fresh fish, but cold-smoked fish to be eaten without further cooking was assigned to Cases 9 and 6 for *Staphylococcus aureus* and fecal coliform counts, respectively. This takes into account the potential for growth of hazardous organisms between smoking and consumption.

With hot-smoked fish products, there is a risk from *Clostridium botulinum* if the fish is brought to an insufficiently high ($< 82°C$) internal temperature during smoking and held unrefrigerated after smoking. This would require assignment to Case 12 or Case 15 for *Clostridium botulinum* (as suggested in ICMSF, 1974, Table 10) depending on the expected use or misuse. Since smoked fish has been implicated in outbreaks due to staphylococci and salmonellae, hot-smoked fish is assigned Case 9 for *S. aureus* and Case 6 for fecal coliforms.

IX. Choice of Case

The hazard to consumers of breaded precooked fish products is on. potential, because such items have not been much implicated in food poisoning. In most cases, such products are held frozen, or to a lesser extent refrigerated, until just before reheating. Although reheating frequently falls short of a full cook, oven heating at the recommended level will inactivate staphylococci (Houghtby and Liston, 1965). For this reason, Case assignments of 2, 5, and 5 for standard plate count, *Staphylococcus aureus,* and fecal coliforms, respectively, were made for these products.

The case assignments suggested in ICMSF (1974) for frozen raw crustacean products, breaded or not, were standard plate count: Case 1; fecal coliforms: Case 4; *Vibrio parahaemolyticus:* Case 10; and *Staphylococcus aureus:* Case 4. *Vibrio parahaemolyticus* is assigned Case 10 because of large-scale outbreaks in the United States due to growth of this organism following recontamination of the crustacean product after cooking.

Precooked crustacean products are more hazardous than the uncooked because they are commonly eaten without further cooking in salads, cocktails, etc., which can be exposed to temperatures above the chilling range before and during serving. Therefore, Cases were set at 3 for standard plate count, 6 for fecal coliforms, 9 for *Staphylococcus aureus,* and 12 for *V. parahaemolyticus.* Products assigned to these Cases included cooked shrimp, prawns, lobster tails, and picked crabmeat.

Dried seafoods, which, in general, are microbiologically stable after processing, were assigned to Case 2 for standard plate count, Case 5 for *Escherichia coli,* and Case 8 for *Staphylococcus aureus* in Table 21 of Book 2 (ICMSF, 1974). These assignments would be appropriate for fish protein concentrates and powdered hydrolysates, as well as the more traditional dried products.

Semiconserved canned seafoods (i.e., fermented, pickled, acidified) are a potentially serious botulism hazard and are, therefore, assigned to Case 15 for *Clostridium botulinum.*

21

Vegetables, Fruits, Nuts, and Their Products

I. INTRODUCTION

Virtually all fruits and vegetables in their natural state are susceptible to spoilage by microorganisms, at a rate that depends on various intrinsic and extrinsic factors. Thus, preservation of plant materials is now achieved by drying, salting, fermentation, freezing, refrigeration, canning, and irradiation. In certain instances, two or more processes may be combined. Sterility rarely is achieved. Rather, the treatment alters the flora or the environment or both in a fashion that extends the stability of the food.

II. VEGETABLES

A. Definitions and Important Properties

A vegetable is the edible component of a plant including leaves, stalks, roots, tubers, bulbs, flowers, and seeds.

Plant tissues, with the exception of certain seeds, are low in protein. Fiber, starch, certain vitamins, minerals and some lipids are the principal components. In general, the pH of vegetable tissue is in the range 5–7. Since the overall composition and pH are very favorable, growth of numerous microbial species can be expected if adequate moisture is present.

B. Initial Microflora

Soil, water, air, insects, and animals all contribute to the microflora of vegetables. The relative importance of these sources differs with the structural entity of the plant; e.g., leaves will have greater exposure to air, whereas root crops will have greater exposure to soil. The activities of man have important effects. For example, the use of pesticides to control insects often limits the spread of microorganisms. Similarly, cultivation, either by hand or mechanically, will introduce and/or distribute micro-organisms into ecological niches from which previously they were absent. Finally, the introduction of human or other animal waste material into the water or soil will have an obvious impact on the flora of vegetables.

Healthy, intact tissues of 10 species of fruits and vegetables contained primarily gram-negative rods (Samish *et al.,* 1961, 1963; Samish and Etinger-Tulczynska, 1963), while healthy undamaged garlic contained up to 25,000 *Leuconostoc mesenteroides*/g, in pure culture (Smith and Niven, 1957).

The means by which microbes penetrate the tissues has not been established clearly, but their presence usually is not deleterious to the growing plant. An equilibrium of coexistence exists, although it can be broken, and spoilage can develop under certain circumstances. External surfaces of vegetables in the field will bear the heaviest microbial load, but the interior tissues will on occasion, contain certain bacteria that by unknown mechanisms have gained entry.

1. Numbers

A recent study of fruits and vegetables at the time of harvest showed that mean counts of molds ranged from $<$1000–67,000/g (or cm²) of tissue. The number of molds on the plant tissue increased when rainfall immediately preceded harvest and when the temperature was below 24°C (Webb and Mundt, 1978). Considerable variation in the microbial load will occur depending on the quantity of soil that adheres to the plant at the time of sampling. Some typical counts are given in Table 21.1.

2. Saprophytic Organisms

Most of the organisms present on fresh vegetables are saprophytes such as coryneforms, lactic acid bacteria, spore-formers, coliforms, micrococci, and pseudomonads derived from the soil, air, and water. Fungi, including *Aureobasidium, Fusarium,* and *Alternaria,* often are present but in relatively lower numbers than bacteria. To date, the strictly anaerobic organisms that also may be present have not been well characterized except

TABLE 21.1

Bacterial Counts on Vegetables upon Arrival at a Processing Plant [a]

Vegetable	Count per gram
Carrots	440,000
Beets	3,200,000
Cabbage	4,000–2,000,000
Lima beans	1,000–150,000
Corn	100,000–10,000,000
Kale	1,200,000–10,000,000
Spinach	2,000,000–23,000,000
Peas	220,000–30,000,000
Snap beans	600,000–3,000,000
Potatoes	75,000–28,000,000

[a] Average data from seven investigations cited in Splittstoesser (1970).

for certain heat-resistant spore-formers important in spoilage of canned vegetables.

3. Pathogens

Generally, animal and human pathogens are absent from fresh vegetables that have not been exposed to human or other animal waste material. Irrigation and fertilization of vegetable and fruit crops with human and animal wastes can contribute the etiologic agents of infectious hepatitis, typhoid fever, shigellosis, salmonellosis, viral gastroenteritis, cholera, amoebiasis, and other enteric as well as parasitic diseases. Illness caused by the consumption of contaminated raw vegetables has been reported (see Section II,D,3).

C. Raw Vegetables

This section will deal with produce that is harvested and shipped directly to the retail market and with those products that are retained in storage for varying periods of time prior to sale.

1. Effects of Harvesting, Transportation, Processing, and Storage on Microorganisms

The microbial load on raw vegetables is influenced by many factors. The hands of personnel involved in picking, trimming, sorting, tying, and packaging, and the equipment used in these operations contribute to the number of microbes and their distribution on the product. Harvesting often injures the produce with the result that (a) nutrients are released to enhance microbial growth, and (b) a portal of entry is afforded to spoilage microorganisms. Containers and vehicles used for transport are an additional source of microorganisms.

There is basically little difference between microbial counts on protected (e.g., peas) and unprotected vegetables [e.g., potatoes (see Table 21.1)]. Similarly, there are few differences in the counts between root and leaf crops, though this may reflect the greater surface area per gram of leaf vegetables. Washing can remove up to 90% of the surface flora, but those organisms trapped in mucilaginous exudate on the vegetable will remain. Washing can be deleterious in that any residual water will allow rapid multiplication of resident microbes.

Additional handling occurs when the produce reaches wholesale and retail markets. It may be unpacked, trimmed, remoistened, repackaged, and displayed for sale. All this may be done hygienically and with adequate refrigeration or in the absence of either or both. Certain vegetables may be cut or chopped for inclusion in delicatessen-type salads. Microorganisms will multiply faster in cut produce, owing to greater availability of nutrients and water. Further handling of produce will afford the opportunity for contamination (including pathogens) originating from either the handler or work surfaces or utensils previously exposed to other food materials (Velaudapillae *et al.,* 1969). Sufficient moisture, appropriate temperature, and adequate time will ensure a continuing increase in the bacterial population, although it is difficult to estimate the magnitude of the change in microbial load during these manipulations.

2. Saprophytes and Spoilage

The predominant microorganisms on healthy raw vegetables are usually bacteria, although significant numbers of molds and yeasts may be present (Koburger and Farhat, 1975; Splitstoesser *et al.,* 1977; Andrews *et al.,* 1978).

Spoilage organisms may be on the plant in the field or may be introduced during harvesting and transport. Most spoilage is caused by fungi, chiefly members of the genera *Penicillium, Sclerotinia, Botrytis,* and *Rhizopus* (Pendergrass and Isenberg, 1974; Wu *et al.,* 1972; Goodliffe and Heale, 1975). Table 21.2 gives a partial listing of the diseases (often manifested by spoilage) of market vegetables and the organisms responsible. It has also been reported that *Sclerotinia sclerotium,* the cause of pink rot of celery, produces phototoxins which cause a blistering cu-

TABLE 21.2

Common Fungal Spoilage Organisms of Market Vegetables

Vegetable	Mold	Type of spoilage
Carrots	*Alternaria*	Black rot
Celery	*Sclerotinia*	Watery soft rot
Lettuce	*Bremia, Phytophthora*	Downy mildew
Onions	*Aspergillus*	Black mold rot
	Colletotrichum	Smudge (anthracnose)
Asparagus	*Fusarium*	Fusarium rot
Green beans	*Rhizopus*	Rhizopus soft rot
	Pythium	Wilt
Potatoes	*Fusarium*	Tuber rot
Cabbages	*Botrytis*	Gray mold rot
Cauliflower	*Alternaria*	Black rot
Spinach	*Phytophthora*	Downy mildew

taneous reaction in field workers who handle the produce (Wu *et al.,* 1972).

Bacteria are responsible for approximately one-third of the total microbial spoilage loss of vegetables. Such spoilage may be due to bacteria that cause soft rots and other rots, spots, blights, and wilts. Soft rots, occurring during transport and storage, are usually caused by coliforms, *Erwinia carotovora,* and certain pseudomonads, e.g., *Pseudomonas fluorescens (marginalis).* Organisms causing spoilage other than soft rot include corynebacteria, xanthomonads, and pseudomonads. Often infection can occur in the field, thereby permitting invasion of the plant tissue by soft-rot organisms as a result of trauma induced during subsequent transport and storage (Lund, 1971). Soft rot of potatoes by *Clostridium* has also been reported (Lund and Nicholls, 1970).

3. Pathogens

In countries where animal wastes are used as fertilizer, it is to be expected that the produce will contain pathogens.

It has been estimated, although with little confirming epidemiologic evidence, that the use of "night soil" on vegetables and fruits in the Orient may cause as much as 20% of recurrent infections of shigellosis, typhoid fever, cholera, and amaebiasis (Worth, 1963; J. C. Scott, 1953). Outbreaks of salmonellosis attributable to contaminated celery, lettuce, cabbage, endive, and watercress have been reported (Geldreich and Bordner,

1971; Bryan, 1977a). Konowalchuk and Speirs (1975), for instance, refer to an outbreak of infectious hepatitis that resulted from ingestion of cress grown in a sewage-polluted stream. Root crops and low-growing leaf and stalk crops are heavily contaminated by using sewage effluent or artifically contaminated irrigation water (Geldreich and Bordner, 1971; Nichols et al., 1971). The efficiency of subsequent washing with chlorinated water to remove pathogens is questionable (Nichols et al., 1971). Survival time of coliforms, pathogens, and enteric viruses on most raw vegetables is moisture- and temperature-dependent and extends significantly beyond the useful life of the product (Geldreich and Bordner, 1971; Nichols et al., 1971; Konowalchuk and Speirs, 1975). The presence of fecal coliforms and pathogens on market produce has been amply documented (Papavassiliou et al., 1967; Velaudapillae et al., 1969; Ercolani, 1976). An outbreak of *Bacillus cereus* food poisoning due to the home-nurtured vegetable sprouts has been reported (Portnoy et al., 1976).

Another consideration is the role of vegetables as carriers of opportunistic pathogens into situations wherein they become dangerous. For example, raw vegetables have served as vehicles for introducing organisms such as *Pseudomonas aeruginosa* and *Klebsiella* into hospital environments. Exposure to such organisms of individuals suffering from burn wounds or recovering from surgery is potentially dangerous.

4. Control

Control of spoilage microorganisms can be accomplished only by strict sanitation of equipment and close control of the environment (i.e., temperature, relative humidity, and atmospheric composition) in which the raw vegetable is maintained. Only in this way can a substantial microbial buildup be prevented.

An additional approach has been to chlorinate the water used for washing or cooling the produce, at levels from 1–2 ppm to 400–500 ppm. Owing to the instability of chlorine in the presence of organic matter, it is doubtful that the lower levels of chlorine do more than ensure that spoilage bacteria do not enter with the water. Furthermore, the addition of moisture in this fashion can contribute to the proliferation of spoilage microorganisms. At the higher levels mentioned, chlorination of wash water has reduced the microbial load on salad vegetables in commissaries. Obviously, the degree of effectiveness depends on pH, temperature and time, and the extent to which the water is recycled during use. Continuous monitoring of chlorine levels is imperative.

Manipulation of both temperature and gaseous atmosphere during storage of vegetables can be very effective in retarding spoilage. In theory, storage at temperatures near 0°C prevents or greatly retards growth of most of the spoilage flora. But not all produce can be stored at such a low temperature. With some (e.g., potatoes) a low-temperature sweating (condensation) occurs if storage is below 7°–10°C.

Simply reducing the temperature is rarely sufficient. Equally important is proper control of the relative humidity (RH) and movement of the air. Many vegetables are maintained at 90–95% RH. Proper air flow prevents the deposition of moisture on the vegetable surface and retards microbial growth. Higher relative humidities are used for some vegetables; for example, a combination of 0°–1°C and 98–100% RH is optimum for long-term storage of carrots, for *Sclerotinia*-induced watery soft rot is minimized (van den Berg and Lentz, 1966). Lower RH causes moisture loss and quality deterioration.

Prevention or retardation of mold spoilage is accomplished primarily by maintaining the vegetable in a physiological condition in which it is more resistant to attack by spoilage organisms rather than by direct action on the fungus itself. Vegetable tissue is living, respiring matter. During storage, oxygen is consumed, and CO_2 is generated. Elevated CO_2 levels and reduced O_2 concentrations are often employed to prevent mold spoilage. To maintain such storage conditions, continuous monitoring and control of the desired levels of each gas are essential (see Chapter 10).

The facultatively anaerobic coliforms that cause soft rot grow quite well under reduced O_2 tension; therefore, controlled gaseous atmospheres have no inhibiting effect on them. In fact, the exclusion or depletion of O_2 will actually enhance the occurrence of soft rot. Low temperature *per se* will not control these organisms since the lower limits for growth of *Erwinia carotovora* and *E. carotovora,* var. *atroseptica* have been reported as 4.0° and 1–2.8°C, respectively. Consequently, control must be achieved by movement and interchanges of air to minimize moisture deposition on product surfaces.

5. Choice of Case

For an understanding of "Case" see Table A.1 and the brief description in Appendix IV. For more details, see Book 2 of this series (ICMSF, 1974).

Vegetables grown with inorganic fertilizers will bear few pathogens of animal origin. Routine testing of these products for either pathogens or indicator organisms is not recommended. However, if human or animal waste materials are used as fertilizer or if contamination from other undesirable sources may have occurred, the product may present a significant health hazard, so that Case 5 for *Escherichia coli* and Case 11 for *Salmonella* would apply.

D. Frozen Vegetables

1. Effects of Processing on Microorganisms

The microflora of vegetables after freezing usually is similar qualitatively to that of the product before blanching and freezing. This is due primarily to organisms introduced from the air. These organisms become airborne during the handling of raw products and eventually settle on various surfaces including those of unprotected postblanch equipment (Mundt et al., 1966; Mundt and Hammer, 1968). A natural selection process ensues, and the microflora is dominated by those organisms best able to utilize the nutrients that become available as the vegetable is processed.

Raw products pick up bacteria during the handling that precedes the blanch. The purpose of the blanch is to inactivate plant enzymes and thereby stabilize the product during subsequent frozen storage. Blanching (usually 86°–98°C for several minutes, depending on the product) often reduces the microbial load 1,000– to 10,000-fold (Splittstoesser, 1970). The microflora of the frozen product is a reflection of the handling the product receives after blanching. Microorganisms will enter the product from slicers, cutters, choppers, conveyor belts, flumes, lifts, hoppers, and fillers. The contribution from each source will depend on the frequency and effectiveness of cleaning these pieces of equipment.

Bacterial growth in or on product prior to freezing is rarely a problem because of the relatively short time elapsed between blanching and freezing; but there are exceptions. For example, to meet weight requirements, it may be necessary to add product manually to packages that have been mechanically filled. Any "weighing reserves" that have been held at room temperature for a substantial period of time, will have high microbial levels that will be reflected in the final product.

The freezing step will kill or injure a portion of the microflora present. Generally, gram-negative organisms are more easily killed or injured by freezing than are gram-positive bacteria. Protracted storage in the frozen state will bring about a further reduction in the numbers, the extent dependent on time, the nature of the organisms, the nature of the food, and the temperature of storage (see Chapter 1).

2. Saprophytes and Spoilage

In most frozen vegetables, the predominant microorganisms are the lactic acid bacteria (Mundt et al., 1967; Splittstoesser, 1970). Many of these are difficult to speciate but significant numbers of Leuconostoc mesenteroides, enterococci, and aerococci are often found. Micrococci and gram-positive and gram-negative rods (including coliforms) also constitute a considerable portion of the total microflora of certain products.

Microbial spoilage of frozen vegetables is rare. In the frozen state, spoilage is essentially precluded by the low temperature and reduced a_w. The few reports of spoilage of thawed products show that the spoilage rate is temperature dependent (Michener et al., 1968; White and White, 1962). There is no evidence that the thawed product will deteriorate more rapidly than material that has never been frozen (see Volume I, Chapter 1).

Aerobic colony counts ranging from 10^1 to over 10^5/g in frozen vegetables are normal and should not be cause for alarm.

3. Pathogens

Frozen vegetables are rarely involved in food poisoning incidents because (a) non-spore-forming pathogenic bacteria do not survive blanching, (b) any pathogens contaminating the product postblanch cannot grow at the temperature of the frozen food, and (c) most are cooked before being consumed.

Because of their excellent public health record, there have been relatively few surveys of the incidence and numbers of salmonellae and staphylococci in frozen vegetables. Salmonellae were not found during a survey of a limited number of samples of frozen peas, green beans, and corn (Splittstoesser and Segen, 1970). Coagulase-positive staphylococci did not exceed 10/g in any of 112 samples of frozen peas, beans, and corn (Splittstoesser et al., 1965). For these reasons, any alleged involvement of frozen vegetables in foodborne illness should be viewed with extreme suspicion and the possibility of contamination and time–temperature abuse during preparation for serving should be thoroughly investigated.

4. Control

Control of microbial levels in frozen vegetables should be relatively simple, because the blanching step eliminates the vegetative microbial cells. Cleaning and sanitizing the equipment between the blancher and freezer frequently and thoroughly will keep the microbial level satisfactorily low; but the conveyor belts are often difficult to clean (Surkiewicz et al., 1967; Splittstoesser et al., 1961).

However, a system that would eliminate all microbial contamination would be undesirable, because the lactic cocci that predominate on processing equipment inhibit the growth of pathogens (specifically staphylococci and salmonellae; Splittstoesser, 1973).

Splittstoesser and Wettergreen (1964) detected coliforms at all stages of frozen-vegetable manufacture but were unable to correlate their presence with any insanitary practice. They concluded that neither the coliform group nor *Escherichia coli* was a suitable indicator of poor sanitation during the processing of frozen vegetables.

5. Choice of Case

For an understanding of "Case" see Table A.1 and the brief description in Appendix IV. For more details, see Book 2 of this series (ICMSF, 1974).

Routine surveillance of product quality by determining standard plate count may be useful. If the vegetable can be used without cooking, Case 2 would apply; if it must be further cooked, Case 1 would apply. No other tests are justifiable in routine surveillance.

E. Canned Vegetables

1. Effects of Processing on Microorganisms

Canning is a process designed to make a variety of products, including vegetables, 'shelf stable.' To achieve this, a given heat treatment commonly called "the cook" or "thermal process" must destroy those microbial forms that are capable of growth under normal conditions of storage. The most heat-resistant forms are the spores of mesophiles and thermophiles. Among the mesophiles, there are numerous spoilage types, as well as three pathogenic types, capable of growing in canned vegetable products (i.e., *Bacillus cereus, Clostridium perfringens,* and *C. botulinum*). The thermophiles are not pathogenic but are generally more heat resistant than mesophiles. However, thermophiles cause spoilage only under very specific conditions, namely, abnormally high storage temperatures. Thus, thermal processes are designed to eliminate those spores capable of growth under normal storage conditions of the canned product.

Most canned vegetables are "low acid," so defined because their pH is above 4.6. All low-acid vegetables are given a heat treatment sufficient to destroy *Clostridium botulinum* and to render the product "commercially sterile" (see Chapter 1). Safe thermal processes for virtually all canned vegetables have been established, based on the rate of heat penetration for each commodity and for each size container. The heat treatment required for low-acid food products is called a "botulinum cook," and is based on the "12D concept" (see Chapter 1). In some parts of the world, the antibiotic nisin is added to certain canned foods to prevent the growth of thermophilic spore-forming bacteria and to obviate the need for a full "botulinum cook" (Jarvis and Morisetti, 1969; see also Chapter 9).

2. *Saprophytes and Spoilage*

Canned vegetables, heat processed to "commercial sterility," may contain viable spores that are unable to grow under normal storage conditions. These spores are thermophiles, from the water used in flumes conveying the product, or from the starch, sugar, or salt in making the brine or syrup, as well as from the raw vegetables themselves.

Canned vegetables may spoil from insufficient processing, container leakage, or high-temperature storage. Insufficient processing will permit the survival of mesophilic spores that will cause off-odors in the product. From a public health standpoint, this form of spoilage is the most important because it indicates that *Clostridium botulinum* may have survived, reproduced, and produced toxin in the product.

Underprocessing, when it occurs, can be attributed to faulty equipment (e.g., inaccurate thermometers, inadequate steam supply, or inaccurate timing devices) or poor operating procedures (e.g., poor fill control, inadequate venting, failure to place product in the retort). In addition, the consistency of a product may vary from batch to batch. An especially viscous batch may not permit as rapid heat penetration as expected, so that the centers of the cans are underprocessed. Rehydration of dried vegetables such as peas must be carefully controlled to avoid such problems.

After the cans have been processed, spoilage organisms can enter through faulty can seams or punctures, or through the hot, still soft seam mastic from cooling water or wet can conveyor tracks.

It is common to find a mixed population of organisms such as micrococci, lactobacilli, and streptococci in cans spoiled due to leakage. If the cooling water has been chlorinated, leaker cans will contain primarily spore-forming organisms (Corlett, 1976).

Prolonged high-storage temperatures can cause three types of spoilage in vegetables:

1. Flat sour spoilage is caused by facultative anaerobic organisms that produce acid but not gas, such as *Bacillus stearothermophilus* and *Bacillus coagulans*. The spores of these organisms germinate only above 40°C, but the vegetative cells can grow in a wider range. The minimum for *B. stearothermophilus* is 30°–45°C and the maximum 65°–75°C. These figures are 15°–25°C and 55°–60°C, respectively, for *B. coagulans* (Buchanan and Gibbons, 1974).

2. Thermophilic anaerobic (TA) spoilage is caused by obligately thermo-

philic spore-forming anaerobes, such as *Clostridium thermosaccharolyti-cum,* which form large quantities of hydrogen and carbon dioxide.

3. "Sulfide stinker" spoilage is caused by the obligately thermophilic spore-forming anaerobe *Desulfotomaculum (Clostridium) nigrificans,* which produces hydrogen sulfide. The cans remain flat but an odor of hydrogen sulfide is detectable, and the food may become blackened if iron is present.

3. Pathogens

In general, commercially canned vegetables have had an excellent safety record. In certain rare instances, however, botulism resulting from underprocessing (Lynt *et al.,* 1975) and staphyloccocal food poisoning due to postprocessing leaker contamination (Bashford *et al.,* 1960) have been directly attributed to canned vegetables.

Home-canned foods are responsible for most of the botulism outbreaks. Whereas canning manuals establish times and temperatures of processes for given recipes and size of container, housewives frequently alter recipes, use larger food containers, use pressure cookers improperly, or taste questionable foods. Sometimes toxic vegetables have normal or near normal odor and appearance.

4. Control

Control of microbial spoilage problems in canned vegetables requires (a) good sanitation in preparing the food for canning, particularly in blanchers where thermophilic organisms grow, (b) choice of ingredients (e.g., spices) containing few thermophilic spores, (c) proper can closure, (d) retorting schedules appropriate to the product and type and size of the container, (e) automatic retort controls and recordings, (f) chlorinated can-cooling water, (g) careful handling of cans after processing, and (h) storage at moderate temperatures.

For additional control information, see Chapter 15 in ICMSF (1974), Chapters 52 and 53 in Speck (1976a), and especially U.S. Department of Health, Education and Welfare (1976g).

One test sometimes performed is a microscopic examination of the finished canned product for the presence of fragments of the mold *Geotrichum candidum,* as a measure of insanitation in the cannery. An abundance of such fragments indicates that the mold grew on the processing equipment (Eisenberg and Cichowicz, 1977). The fragments are not viable in the canned products, and present no health or spoilage hazard.

Routine culturing of canned vegetables for their bacterial content is not recommended.

F. Dried Vegetables

Dried vegetables are inherently stable; they are rarely involved in food-borne illness; and, with the exception of soybeans and other dried beans, they are not important items of commerce. For these reasons, there is a paucity of information on the microbiology of dried vegetables.

1. Effects of Processing on Microorganisms

Some vegetables (e.g., peas and carrots) are blanched before drying; others (e.g., onions, garlic, and salad vegetables) are not. Therefore, the microbial flora of dried vegetables depends in considerable degree on whether the flora of the raw, cleaned product (Section II,D) has been largely destroyed by blanching (Section II,E,1). In addition, vegetables for drying may support the growth of microorganisms if they are held too long at ambient temperatures, or they may become contaminated by unclean equipment (Vaughn, 1951; Goresline, 1963).

Because water evaporates during drying, the product temperature rarely exceeds 35°–45°C, even though the air temperature is 80°–100°C. Thus, drying rarely reduces the number of organisms and may *apparently* increase the number because they are concentrated in a smaller volume of product (Murphy, 1973).

Most vegetables are dried by blowing heated air over trays or through perforated belts in drying tunnels. Loading of the belts or trays is critical because uneven loading can lead to improper air circulation and temporary "wet spots" that will permit microbial growth. Even though the product will eventually dry, the microorganisms will remain viable.

Most organisms on incoming onions reside on the skin, root area, and tops. Many of these organisms can be removed by proper trimming, as is done with onions that will be sliced. On the other hand, diced, chipped, or flaked onions often are not trimmed as extensively, therefore, the bacterial content in the dried product is higher than that of slices. "Pretreating" raw onions in a 2–4% salt brine prior to dehydration reduces microbial numbers in the final product (Firstenberg *et al.,* 1974).

2. Saprophytes and Spoilage

The saprophytic flora of those vegetables that have been blanched prior to drying consists of those organisms that grow well on product-contact surfaces of the equipment. As with frozen vegetables, the lactic acid bacteria predominate. The saprophytes associated with the dried vegetables that have not been blanched will more closely approximate the flora of the raw product. The method of drying also affects the flora. For example, belt-dried onions have lower colony counts than tray-dried onions. The flora of the former consists mainly of bacterial spores; that of the latter has a high "lactic" population (Sheneman, 1973). It is not uncommon to find coliforms, enterococci, and clostridia in dried onions (Vaughn, 1970; Clark *et al.*, 1966) because these organisms are normally associated with raw vegetables. Onion juice is apparently toxic to *Escherichia coli,* which is, therefore, usually absent from this food (Vaughn, 1951; Sheneman, 1973).

The primary microflora of soybeans and other dry beans consists of fungi. These products can be contaminated in the field, during harvest, and while in storage. Fungi contaminate the bean surfaces, and often invade the bean as well (Mislivec *et al.*, 1975; Mislivec and Bruce, 1977). The predominant species are members of the *Aspergillus glaucus* group (see also Chapter 23).

Spoilage of most dried vegetables is rare due to their low a_w. If moisture is inadvertently added to the finished material, spoilage could develop. This rarely occurs.

3. Pathogens

Vegetative cells of bacterial pathogens are rarely present in dried vegetables. However, spores of *Bacillus cereus, Clostridium botulinum,* or *C. perfringens,* if present in the soil, are likely to carry through into the final dried product. But they are harmless unless permitted to grow on reconstitution.

Nonsporulating organisms like *Escherichia coli* or *Salmonella* are destroyed by blanching, but dried vegetables not previously blanched could contain these organisms if they were present in garden soils previously contaminated with animal or human wastes.

Toxigenic fungi have been isolated from dried beans, and some research has been done on the physical, chemical, and biological factors influencing toxin production in beans (Hesseltine, 1976). In one limited survey of soybeans that were heavily molded in the field, 50% of the samples examined contained aflatoxins (Bean *et al.*, 1972); in another study, aflatoxin was found in only 2 of 866 samples (Shotwell *et al.*, 1969b).

Certain further-processed soy protein foods have been contaminated infrequently with salmonellae. In view of the rigorous chemical treatment applied during the manufacture of soy proteins, the presence of salmonellae reflects a recontamination after fabrication (see also Chapter 23).

4. Control

Obviously moldy beans, whether they have developed mold in the field during periods of heavy rainfall, or after the harvest during storage, should be discarded and used for neither animal nor human consumption.

It is reasonable to expect that processing and storage conditions that will provide adequate moisture, temperature, and oxygen levels for a significant period of time will lead to growth of mold and possibly toxin formation. Rapid drying, and the maintenance of low a_w during storage, will prevent this problem.

Control of microbial growth and contamination in dried vegetables is attained by (a) blanching, when applicable, (b) minimizing the time of storage of the cleaned, cut vegetables before drying them, (c) frequent and thorough cleaning of equipment, (d) proper loading of the product into the dryer, to attain even drying, (e) prompt drying to low a_w, (f) hygienic handling of the dried product, and (g) storage of the dried product to preclude entry of moisture.

5. Choice of Case

For an understanding of "Case" see Table A.1 and the brief description in Appendix IV. For more details, see Book 2 of this series (ICMSF, 1974).

Dried vegetables that have been blanched should not contain vegetative cells of pathogens. For those dried vegetables that are not blanched and are usually eaten raw (e.g., as a garnish), Case 5 for *Escherichia coli* and Case 11 for *Salmonella* are recommended. For those that are eaten only after cooking, Case 4 for *E. coli* and Case 10 for *Salmonella* should be used.

G. Fermented and Acidified Vegetables

Vegetables can be preserved or processed by salting, acidifying, a combination of both, or by fermentation. In the Orient, blends of vegetables known as kimchi, safur asin, nukamiso, dua chua, and paw tsay are fermented. Often such blends include fish, nuts, and rice liquid (Pederson, 1971; Orillo *et al.,* 1969). The most commonly fermented vegetable in North America and Western Europe is cabbage; but cauliflower, carrots, radishes, beets, beans, chard, and turnips are also preserved in this way.

1. Effects of Processing on Microorganisms

The flora involved in the fermentation of vegetables is derived from the raw vegetables and from the processing plant equipment used to prepare them for the fermentation tank. The vegetables are not blanched and thus retain the epiphytic lactic acid bacteria that were associated with them in the field (see Section II,D).

The vegetable is first salted with a cover brine or with dry salt crystals. Salting has a pronounced effect in selecting the predominant microflora.

A complex interaction of factors governs the sequence of organisms that develops during fermentation, and during spoilage, if it occurs: pH, salt, organic acids, a_w, temperature, redox potential, oxygen, and carbon dioxide. With minor variations, the fermentation of all plant foods is the same, consisting of a sequential growth of lactic acid bacteria, including *Leuconostoc mesenteroides, Lactobacillus brevis, Pediococcus acidilactici, P. pentosaceus,* and *Lactobacillus plantarum.* Other "lactics," e.g. *Streptococcus faecalis* have been observed, but are not important in the fermentation.

The fermentation of cabbage is typical. It can be fermented intact, i.e., whole heads submerged in a cover brine (Pederson *et al.,* 1962), but more commonly, it is chopped and dry salted. The salt draws out the plant juices containing fermentable carbohydrates and other nutrients, forming brine. Large numbers of soil bacteria and organisms from the harvesting and transport equipment remain. Additional microorganisms enter from spices or condiments. Maintaining the temperature between 20° and 24°C is critical. Higher temperatures often permit the ascendency of undesirable organisms that give an inferior or unmarketable product. Lower temperatures slow acid development and predispose the product to spoilage. The fermentation vats are covered to minimize the development of oxidative microorganisms that utilize the acid generated during the fermentation process.

In the initial stages of fermentation, i.e. before significant quantities of acid have been generated by the lactic acid bacteria, gram-negative facultative anaerobes such as coliforms multiply at a rapid rate. This is a normal occurence and is inconsequential unless the lactic fermentation fails. Under normal circumstances, *Leuconostoc mesenteroides* is the predominant lactic organism in the early stages of fermentation. The pH drops to 4.6–4.9 and *L. mesenteroides* is followed by the more acid-tolerant lactic acid bacteria, i.e., the pediococci and lactobacilli. In a matter of a few weeks the fermentation is complete and the final product has a pH of 3.5–3.8 and a titratable acidity (as lactic acid) of 1.8%. The interaction of salt, acid and the absence of dissolved oxygen preclude the growth of aerobic and many gram-negative bacteria. The finished sauerkraut may be marketed raw, pasteurized, or canned.

2. Saprophytes and Spoilage

Spoilage of fermenting vegetables can occur in a number of ways. One of the principal causes is uneven distribution of salt. If its concentration is overly high in a localized area, certain yeasts (Pederson and Kelly, 1938) or lactobacilli (Stamer *et al.,* 1973) may grow and turn the product pink; if the salt is low, sauerkraut may soften from coliform bacteria.

If oxygen is present, oxidative yeasts can grow rapidly and utilize the developing lactic acid. This will increase the pH and allow the growth of the less acid-tolerant spoilage forms.

The shelf-life of raw sauerkraut is governed by the temperature of storage. The pasteurized or canned sauerkraut will remain microbiologically stable until the container is opened, whereupon oxidative yeasts may enter and spoil the product at a rate that is a function of temperature and the number and types of yeasts entering.

Certain vegetables intended for use in soups may be dry-salted or brined (20% NaCl) and stored at 35°–40°F (Etchells and Bell, 1976). Such products often contain halophiles, cocci, and spores but usually are used before problems develop.

3. Pathogens

It is not surprising that there are no documented cases of food poisoning from the ingestion of commercially pickled foods that were properly manufactured (Etchells and Bell, 1976; Ito *et al.,* 1976). The combination of salt and acidity in raw fermented products precludes the growth of the vegetative cells of any of the foodborne pathogens. Indeed, should these organisms be inadvertently introduced they will not survive but will die at a rate that is influenced by pH, salt concentration, and temperature. Spores of pathogens will survive but are unable to germinate and grow in this environment.

4. Control

Control of spoilage organisms is achieved by (a) proper distribution of salt, (b) maintenance of the appropriate temperature during fermentation, and (c) destruction or inhibition of the activity of oxidative yeasts. Yeast inhibition can be achieved by securely covering the vat surface to exclude oxygen. Alternatively, short-wave ultraviolet light directed to the brine surface during fermentation, or exposure of the surface to oil of mustard are likewise effective.

Shelf stability of the finished product depends on proper refrigeration of raw fermented vegetables, or appropriate pasteurization or canning.

Routine microbiological examination of fermented vegetables is not recommended.

III. FRUITS

The microflora of fruit is somewhat different from that of vegetables. Many fruits also possess natural defense mechanisms such as a thick skin or antimicrobial substances such as essential oils. Also, fruits have an exceptionally good record from a public health standpoint. Most research efforts have been directed toward means of preventing 'field' diseases rather than postharvest spoilage because the heaviest economic losses are sustained prior to harvest.

A. Definition and Important Properties

For purposes here, a fruit is that portion of the plant that contains the seeds. An exception is made for nuts, which are discussed in a separate section.

Fruits contain organic acids in quantities that are generally sufficient to produce a pH value of 4.6 or lower. Though certain fruits have a higher pH, e.g., watermelon, bananas, cucumbers, and figs, the large majority of fruits are considered high-acid products. The low pH and the nature of the organic acid molecule itself are the major influences that select the predominant microflora of fruit.

B. Initial Microflora

Most soilborne, airborne, and insect-vectored bacteria are incapable of growth at the pH level of most fruit tissue. In contrast, molds, yeasts, and certain lactic acid bacteria flourish at such low pH levels. Consequently, with only certain exceptions, the predominant microbial forms on fruits are fungal not bacterial.

As with fresh vegetables, microorganisms on fruit have their origin in the field and on harvesting and handling equipment. All fruits are exposed to microorganisms during growth on the vines and trees prior to harvesting. Insects, soil, and air contaminate fruit with viable fungi. In most cases contamination remains external, but certain molds can penetrate the outer skin or cuticle of fruit (Alderman and Marth, 1974). Some mold species penetrate the flesh of fruit and remain in a latent stage, i.e., cease to multiply for an extended period. This may be because of the inherent resistance of immature tissue or other factors (Eckert, 1975).

C. Raw Fruits

1. Effects of Harvesting and Processing on Microorganisms

Fruit to be eaten raw is often picked while immature and ripened during transport or storage. It is during these operations that spoilage becomes a significant problem. Fruit cannot be harvested or processed (i.e., sorted, graded, and packaged) without some injury. Injured fruit is very susceptible to fungal invasion and spoilage. One of the most important sources of microorganisms in or on fruit is water that is used extensively to wash, rinse, cool, and convey fruit to and within a given process operation. Such water is often recycled and a large buildup of microorganisms can occur unless a biocidal treatment is used. Water can contaminate not only fruit surfaces but also can bring microorganisms into the interior of the fruit where they may remain protected from subsequent treatments.

The relative humidity of the atmosphere during storage and ripening of fruit is very important. Low humidity will cause dessication; high humidity is conducive to microbial growth. It is not likely that relative humidity during storage could be lowered sufficiently to prevent fungal growth and yet yield a suitable product. A proper combination of temperature and humidity can retard, but not eliminate, spoilage.

2. Saprophytes and Spoilage

Because of the acidity of raw fruits, the primary spoilage organisms are fungi, predominantly molds. Spoilage can occur in the field should invasion of the fruit by the appropriate fungi occur. The extent of postharvest spoilage that can be attributed to preharvest invasion and "latency" is difficult to estimate. Invaded fruits pose a serious problem during transport and storage. Because the contamination is internal, biocidal rinses or treatments are ineffective in killing the organisms. These fruits then can serve as foci of infections during subsequent transport or storage.

Handling of fruit at the retail level affords further opportunity for injury. The bins or shelves or hands of employees can inoculate the fruit with spoilage fungi. These fungi grow rapidly causing a breakdown of the structural components of the fruit leading to undesirable textural changes. This is manifested by softening, wilting, or rotting.

Each fruit species, by virtue of its composition and the type of handling it typically receives is subject to spoilage by one or more mold species (see Table 21.3).

3. Pathogens

Bacterial pathogens are not normally associated with fruit due to the acidic nature of the product. In theory, such pathogens could be introduced by direct exposure to fecal material, but if this occurs, they do not seem to persist.

Various molds are able to produce substances (mycotoxins) that are toxic to certain animals. Certain species of these molds can penetrate the skin of fruit and produce mycotoxins in the fruit pre- and postharvest (Alderman and Marth, 1974). For example, patulin, produced by *Peni-*

TABLE 21.3

Common Fungal Spoilage Organisms of Some Market Fruits

Fruit	Molds	Type of spoilage
Oranges	*Penicillium, Alternaria, Trichoderma, Fusarium*	Blue rot, green rot, gray mold rot, blossom end rot
Apples	*Penicillium, Gleosporium, Rhizopus, Alternaria, Physalospora*	Blue rot, black rot, core rot, brown rot
Pears	*Rhizopus, Podosphaera, Penicillium, Alternaria*	Black rot, blue rot, brown rot, powdery mildew
Tomatoes	*Phoma, Alternaria, Sclerotinia, Fusarium*	Anthracnose, phoma rot, watery rot
Eggplant	*Phomopsis*	Blight
Strawberries	*Mucor, Rhizopus*	Gray mold rot, tan rot, leather rot
Cherries	*Cladosporium, Rhizopus, Alternaria*	Green mold rot, black mold rot, brown rot
Grapes	*Botrytis*	Gray mold rot
Lemons, Limes	*Sclerotinia, Oospora, Phytophthora*	Cottony rot, brown rot, sour rot
Peaches	*Trichothecium, Rhizopus*	Pink rot, rhizopus rot
Bananas	*Pestalozzia, Fusarium, Gleosporium*	Finger rot

cillium expansum has been found in moldy apples and in products made from this fruit (Wilson and Nuovo, 1973). It is likely that future investigations will uncover additional examples of mycotoxin production in fruit intended for human consumption.

4. Control

Before harvest, fungicides and insecticides are effective in reducing the incidence of fungal invasion and infected fruit. Beginning with harvesting, control depends on conditions of handling, transport, and storage. Control is made difficult by (a) the variety of spoilage forms that are present, (b) limitations imposed by the physiology of the fruit itself, and (c) conditions favorable to senescence and ripening of fruit that also are conducive to growth and spoilage by fungi (Smith, 1962).

Control of temperature, relative humidity, gaseous atmosphere, and chemical inhibitors are important in preservation of raw fruits. Exposure of citrus fruits and peaches to hot water to remove (or kill) fungi can effectively control the main spoilage flora but also can injure the product, thus predisposing it to invasion by a secondary contaminant. Similarly, many fruits suffer chill injury at low temperatures (Eckert, 1975), which may enhance the rate of spoilage, especially if the fruit is latently infected. For certain fruits, such as strawberries, low temperature, i.e., $<2°C$, is used to extend shelf-life.

Controlled-atmosphere storage, defined as an adjustment of oxygen and carbon dioxide levels, has been successfully applied in storage of certain fruit. The primary factor is the CO_2 level. High levels of CO_2, e.g. $>10\%$, will affect both the fruit and the fungal flora (see Volume 1, Chapter 10). The CO_2 retards the ripening process, thus maintaining the fruit in a more resistant form for a longer period of time. It also directly retards the growth rate of fungi. Lower levels of CO_2 apparently act in a similar fashion on the fruit, but are without effect on the fungi. However, controlled-atmosphere storage is apparently without effect on bacterial or yeast spoilage. Economic considerations, as well as the existence of other practical alternatives, have precluded general application of this process.

Chemical treatments include preharvest application of fungicides and use of certain chemicals in wash water or in waxes used to coat various fruits (Eckert and Sommer, 1967). In addition, chemicals such as biphenyl are impregnated into fruit wraps and into paper sheets used to line shipping containers. These chemicals mainly inhibit sporulation of various fungi, thereby limiting the spread of decay. Thiobendazole and benomyl, sprayed on fruit, are effective against many fungi. Certain genera, e.g. *Alternaria* and *Geotrichum,* are refractory, and some strains may develop resistant forms. This has led to the use of a combination of fungicides (Eckert, 1975).

Although certain varieties of grapes can be held successfully at low temperature, e.g. 1°C, for extended periods, the gray mold *Botrytis cinerea* can grow slowly at this temperature and cause spoilage. Periodic treatment with SO_2 will control this problem (Nelson and Richardson, 1967).

The routine examination of raw fruit for microbial content is not recommended.

D. Frozen Fruits

1. Effects of Processing on Microorganisms

Most fruits to be preserved by freezing are not blanched because heating causes softening and loss of moisture. An exception is fruit destined for bakery products that are to be heated during preparation. Sometimes, chemicals are added to retain quality—for example, ascorbic acid to control oxidation or citric acid to retain color. It is unlikely that such adjuncts have any substantial effect on the microflora of the product. In certain instances thinly sliced sections of fruit are dipped in sulfuring solutions and held for several hours to allow permeation to occur. This treatment does reduce the number of viable microorganisms on the slices.

Microorganisms, chiefly yeasts and molds, proliferate on equipment surfaces and slough off into the product. Freezing kills a portion of the microflora and causes injury to some of the survivors. Frozen storage causes a dimunition in the viable count but the death rate is slow.

2. Saprophytes and Spoilage

The normal microflora of frozen fruit consists mainly of yeasts and molds. Growth of these organisms and consequent spoilage of the product is influenced by the temperature of storage. Conditions which lead to partial or complete thawing will predispose the fruit to spoilage by organisms that are present at the time of freezing and that have survived freezing and storage—notably the yeasts.

3. Pathogens

Properly handled frozen fruit will not contain pathogenic bacteria, because they cannot survive in the acidic environment normal to fruit tissue. Moreover, should gross contamination due to very poor in-plant hygiene occur, the vegetative pathogens will die during and after the freezing. The rate of kill, however, will be lower in frozen fruits than in those held above freezing.

4. Control

Microbial populations on fruits to be frozen are best controlled by adequate washing, removal of obviously diseased fruit, careful handling to prevent bruises, frequent cleaning and sanitation of handling and conveying equipment, and prompt freezing of the prepared fruit. Sulfuring is limited in application, and control depends on good manufacturing practices, including effective sanitation. Frozen storage below $-10°C$ will prevent all microbial growth.

Routine microbiological examination of frozen fruit is not recommended.

E. Canned Fruit

1. Effects of Processing on Microorganisms

The majority of fruits are quite acidic, i.e., pH <4.6. Some of the less-acid fruits—tomatoes and pimientos—are sometimes acidified with citric acid to bring the pH to or below 4.6 prior to heat processing (Powers, 1976). This enables the processor to use a milder heat treatment and, thus, maintain better textural quality. The heat treatment will destroy all vegetative cells, leaving only spores which are unable to germinate and/or proliferate under the acidic conditions (see Chapter 1).

2. Saprophytes and Spoilage

The saprophytic flora of canned fruits is made up of mesophilic and thermophilic spores. Spoilage of fruits can occur due to (a) insufficient processing, (b) thermophilic anaerobes, (c) *Bissochlamys* molds, or (d) leaking containers.

Insufficient processing can permit the survival of butyric anaerobes (e.g., *Clostridium pasteurianum*) that can grow at pH 3.8 in syrups and that produce butyric acid, H_2, and CO_2 in canned fruits (Jakobsen and Jensen, 1975).

Thermophilic anaerobes (*Bacillus stearothermophilus* or *B. coagulans*) can likewise grow at pH 3.8 and can produce "flat sour" spoilage in tomatoes at storage temperatures above 40°C.

Species of the genus *Byssochlamys* produce extremely heat-resistant ascospores that can germinate, grow, and produce moldy tastes and unsightly "buttons" of mycelial growth in certain fruits and fruit juices (King et al., 1969; Splittstoesser et al., 1971). Although the mold is an obligate aerobe, growth does occur within the can until the oxygen level is very low.

Leakers may permit entry and growth of acid-tolerant bacteria, yeasts, or molds. Cans are either flat or swollen depending on the predominant species in the can.

3. Pathogens

Canned fruits are among the safest foods. The few botulism outbreaks that have been reported were almost all from home-canned pimientos, figs, blackberries, peaches, and (rarely), tomatoes. These are fruit products whose pH can lie above the critical pH 4.6–4.8 range; in commercial operations, pH adjustment and control is part of accepted good practice.

However, organisms entering leakage holes could utilize organic acids and raise the pH to a level permitting germination and growth of *Clostridium botulinum*, if it is present. This possibility should be investigated whenever botulism is associated with a food that normally has a pH below 4.6.

4. Control

Spoilage develops chiefly when the spore load in the product is extraordinarily high. This can be prevented by frequent and efficient cleaning of the food handling equipment.

Control of *Clostridium pasteurianum* spoilage of pears was brought about by an interaction of pH, a_w, and heat. The strains of this species most resistant to acidity were least able to grow at reduced water activity levels. Adjusting the pH to 3.8–4.0 and the water activity to 0.97–0.98 permitted a mild heat treatment to be used to prevent spoilage (Jakobsen and Jensen, 1975).

Routine microbiological analyses of canned acid foods are not necessary but maintenance of adequate processing records and appropriate coding systems are mandatory.

F. Dried Fruits

Many fruits lend themselves to drying as a method of preservation. Drying should reduce the a_w to a level at which spoilage organisms cannot grow. Because of the acidity of most fruits, only fungi can cause spoilage, and then only at relatively high a_w. No organisms can grow on fruits whose a_w is 0.60 to 0.70.

1. Effects of Processing on Microorganisms

The drying processes that are applied to fruits are outlined in Volume I, Chapter 4. Many are given an SO_2 treatment either by fumes, spray, or dip (Roberts and McWeeny, 1972). The primary purpose of this treatment is color stabilization, but it also reduces the microbial load and repels insects.

The microbiology and the preservation technology of dried fruits have been influenced in recent years by the consumer's desire for a more moist product. Dried dates, figs, and prunes are often reprocessed to increase the moisture level to satisfy this demand (Nury *et al.*, 1960). Although the moistened product can support the growth of microorganisms, those that can grow in the acid environment are killed by heat before or after filling.

Fruits are often washed before processing, and some are treated with a dilute alkali solution to speed removal of water. Although each process can reduce the microbial load, recirculation of the process water can increase the number of microorganisms on the product (Miller *et al.*, 1963). The means of dehydration used influences the microflora on the dried product. Sun drying, used extensively for certain fruits throughout the world, is subject to the vagaries of climate. Although the *Aspergillus glaucus* group of fungi is most commonly encountered, *A. flavus* has been found in sun-dried raisins (Follstad, 1966), and aflatoxin has been detected in dry figs (Buchanan *et al.,* 1975). Mechanical dehydration can reduce the total number of microorganisms, although the extent of reduction depends on the fruit and process. Natarajan *et al.* (1948) recommended that figs be dried at 54°–60°C. This treatment reduced but did not eliminate the yeast population because of the internal nature of the contamination. On the other hand, prunes are dried at 74°C for 18–24 hr, a process that kills all microorganisms except bacterial spores. Normally, these spores are precluded from germination by the low water activity and low pH of the product.

Freeze drying (described briefly in Chapter 4) is applied to fruits only for special purposes, e.g., military rations. It is an expensive process usually reserved for high cost, animal protein foods. It is a thermally mild process in which the freezing step permits little or no microbial growth but the drying phase permits survival of most contaminating microorganisms.

2. Saprophytes and Spoilage

Although many fruits are microbiologically stable in the dry form, sometimes the water activity is at a level where some species of molds can grow and cause spoilage. Studies by Tanaka and Miller (1963) and Pitt and Christian (1968) listed the species that could spoil dried and "high moisture" prunes. Members of the *Aspergillus glaucus* group were predominant in both products. *Xeromyces bisporis,* a mold that is able to grow at a_w near 0.60 was also encountered. Unless prunes are suitably heat treated or chemically preserved, spoilage even of dried prunes by this organism is inevitable. The relative infrequency of dried-prune spoilage is attributed to the low incidence of contamination by this organism (Pitt and Christian, 1968).

3. Pathogens

Pathogens cannot grow at the pH of most fruits, though some of the less-acid dried fruits such as dates might provide a favorable survival medium.

Aspergillus flavus and aflatoxins have been detected in dried fruit. Since *A. flavus* grows little below a_w 0.80, the presence of aflatoxin indicates slow or inadequate drying of the product.

4. Control

During storage and moisture equilibriation, dehydrated fruits must be protected from insects and rodents because both can serve as vectors of spoilage fungi. The drying process must be rapid enough to preclude significant growth of spoilage forms before the a_w is reduced to a safe level.

For organoleptic reasons, some fruits are not dried to a level of a_w sufficient to prevent microbial growth. Others undergo some rehydration during distribution. Control of spoilage in such products is effected by chemical means, i.e., either treating the fruit with sorbate or packaging under propylene oxide or methyl bromide (Bolin et al., 1972). Alternatively, the product may be hot-filled into suitable containers. The spoilage organisms that may be present die from the heat, and the closure prevents others from entering.

5. Choice of Case

For an understanding of "Case" see Table A.1 and the brief description in Appendix IV. For more details, see Book 2 of this series (ICMSF, 1974).

Although it is unlikely that pathogens will be present on dried fruit, there may be certain instances, e.g., sun drying, in which there is reason to suspect poor hygiene. Case 5 for *Escherichia coli* is suggested.

G. Fermented and Acidified Fruits

The fruits most commonly fermented are cucumbers and olives. Others fermented include green tomatoes and peppers. The processes vary somewhat depending on the type of product desired and its intended use, i.e., direct consumption or garnish.

1. Effects of Processing on Microorganisms

a. Olives. The olive fermentation is quite similar to the bacterial sequence in sauerkraut manufacture (Section II,G), but there are definite features, notably the spoilage patterns, that are distinct. Olives, as harvested, contain a bitter substance (oleuropein) that breaks down into several products (aglycone and elenolic acid) inhibitory to lactic acid bacteria (Fleming et al., 1973). *Leuconostoc,* the initial predominant organism in the fermentation sequence, is the most sensitive of the lactics to these inhibitors. The oleuropein is removed prior to fermentation by treating the olives with dilute alkali that is then leached out with water. Fermentation will not proceed if alkali treatment or leaching is inadequate, and the product may spoil. Most of the normal flora of the "raw" olive is killed by the lye treatment, but organisms are reintroduced during the leaching process and in subsequent steps.

Fermented green olives are covered with 7–8% brine containing 0.6–0.9% lactic acid, then are vacuum-packed in glass jars, and finally pasteurized in the jars at 60°C. Or, in an alternative process, they are filled into the jars, and covered with hot brine at 80°–82°C. Only the first-mentioned process eliminates all lactic acid bacteria and yeasts. The bacterial spores that survive either process are inhibited by the salt and the low pH.

Ripe (black) olives are lye treated, aerated, and given several wash treatments to leach out the residual lye. The process takes 3–4 days, and microbial growth can occur if the olives remain in water at a suitable growth temperature for a protracted time.

b. Cucumbers. Cucumbers are processed into pickles in a variety of ways, including fermentation and direct acidification. The mildest fermentation process begins with raw cucumbers in a low salt (2.6–4.0%) brine. Lactic fermentation takes place for a few days at room temperature and continues during refrigerated storage. Alternatively, "salt stock" pickles are fermented and cured at high-salt concentrations. Normally, the fermentation process is begun at an intermediate salt level (8–10%), and the salt is increased gradually to 15%. However, if the weather is cool when tanks are being filled, the initial salt concentration may be as low as 6% to enhance the rate of fermentation. The initial flora may be largely coliforms, but pediococci and *Lactobacillus plantarum* soon predominate in the fermentation.

Fully cured, salt stock pickles may be made into a variety of products by leaching out much of the salt. Addition of vinegar makes them "sour pickles"; addition of vinegar and sugar makes them "sweet pickles."

Pickles may also be made by direct acidification. Such pickles, termed "fresh pack," are not fermented. They are combined with vinegar and spices and then pasteurized.

Pure culture fermentation of cucumbers yields the best product (Etchells *et al.*, 1964, 1966). If the fermentation is carried out in containers up to 5 gallons in capacity, the cucumbers are heat pasteurized to destroy the natural flora. For large-scale production, pasteurization is impractical, and the natural flora is suppressed by acidifying the cucumber brine to pH 3.3. The use of chlorination to control the flora (Etchells *et al.*, 1973) has largely been discontinued due to production of off-flavors. A pure culture of *L. plantarum,* or a mixture of *L. plantarum* and *Pediococcus cerevisiae,* is then added, and fermentation takes place.

2. Saprophytes and Spoilage

a. Olives. Inadequate leaching of alkali from treated olives will inhibit the normal fermentation process initiated by *Leuconostoc mesenteroides* and will permit spoilage predominantly by yeasts and coliforms (Etchells *et al.,* 1966; West *et al.,* 1941).

Olive fermentation is slower than cabbage fermentation; thus, there is more opportunity for defects to occur. Gassiness and bloating due to heterofermentative lactics, yeasts, and coliforms are the most frequent problems. A "cheesy odor" defect has also been reported (Pederson, 1971). Storage of olives in salt-free acidified brine can lead to gassy defects caused by yeasts of the genera *Hansenula* and *Saccharomyces.* Softening of fermented green olives is attributable to pectinolytic species of *Rhodotorula.* These yeasts are aerobes that grow on the surface of brines. They elaborate enzymes that diffuse into the fruit causing a softening (Vaughn *et al.,* 1969a; 1972). Small white spots commonly occur on the surface of fermented green olives and other fruits. These may be colonies of *Lactobacillus plantarum* (Thompson *et al.,* 1955;Vaughn *et al.,* 1953) or yeasts (Pederson, 1971).

Spoilage of green olives can occur in improperly sealed containers. The predominant initial flora is lactate-oxidizing, obligately aerobic yeasts. Sufficient growth and acid utilization could raise the pH and create favorable conditions for growth of other salt-tolerant microorganisms.

Softening and sloughing of skin and flesh from black olives being prepared for canning has been observed (Vaughn *et al.,* 1969b). Five species of gram-negative bacteria with pectinolytic ability were isolated from the dilute brine solution.

b. Cucumbers. The saprophytic and spoilage flora of pickled cucumbers depends on the method of manufacturing the product. In salt stock pickles with high-salt concentrations (15%), yeasts, obligate halophiles, and coliforms may develop if the acidity generated by the lactic acid bacteria is not sufficient. Dill pickles manufactured by the low-salt brine (<5%) process can show a bloating effect due to yeasts, heterofermentative lactics, and coliforms if the proper fermentation flora do not develop (Etchells *et al.,* 1968). Softening of the fruit flesh also can occur due to enzymic action of yeasts growing in the brine or to enzymes carried into the fermentation on the cucumber. These latter enzymes originate from filamentous fungi that grow on the flower end of the fruit in the field (Etchells *et al.,* 1958). The yeast population in the brine can be altered qualitatively and quantitatively by the addition of sorbic acid. The species of *Brettanomyces, Hansenula, Torulopsis,* and *Saccaromyces* that normally predominate are then repressed, and if the salt content of the brine is low, species of *Candida* will predominate (Etchells *et al.,* 1961). Thus, the benefit derived from use of sorbic acid is somewhat questionable.

Sweet and sour pickles (nonpasteurized) made from salt stock pickles are preserved by vinegar and/or sugar and vinegar. Should either the acid or sugar level be insufficient, spoilage due to lactic acid bacteria or yeasts will develop (Etchells and Bell, 1976).

"Fresh pack" pickles prepared by direct acidification are pasteurized. If pasteurization is inadequate, yeasts and lactic acid bacteria will spoil the product. If the acidity is insufficient in the cover brine, spores of butyric anaerobes will germinate, grow, and spoil the product. High spore counts in products of this kind indicate inadequate washing of the fruit prior to brining but are of no public health significance if the acetic acid content is adequately high and the pH adequately low.

3. Pathogens

There have been no documented reports of food poisoning due to the ingestion of properly fermented and processed fruit. The acidity and salt content preclude survival of the vegetative forms of pathogenic bacteria and prevent the germination of spores that may be on the raw fruit.

4. Control

a. Olives. Control of spoilage in fermented green olive production depends initially on proper lye treatment and complete removal of the lye. This is necessary to ensure sufficient acid generation by lactic acid bacteria in the low-salt brines. Finally, maintenance of anaerobic conditions during the fermentation period is necessary to prevent the growth of oxidative yeasts. Following fermentation, control is achieved by proper pasteurization and container closure.

b. Cucumbers. Control of spoilage depends on the type of pickled product. In all cases, proper brine makeup, proper fermentation sequence (in the fermented pickles), proper acidification and/or pasteurization, and security of jar closure are essential. Pasteurization of nonfermented pickles is accomplished by heating to an internal temperature of 73°C for 15 min followed by a rapid cooling to prevent overcooking and concomitant softening. In some instances, sodium benzoate is used as a preservative if a milder brine or pasteurization treatment is desired. However, even benzoate will not prevent spoilage of pickles prepared under insanitary conditions that introduce large numbers of sugar- or vinegar-tolerant yeasts.

Routine microbiological examination of fermented fruit is not suggested.

IV. NUTS

There are few, if any, public health problems associated with consumption of nuts. The concern, particularly expressed in t literature, is for the consequence of adding contaminated nuts to products (e.g., baked goods or dairy products) that will not be further processed and that, if abused, will support the growth of pathogens. Owing to the paucity of confirmed outbreaks of illness associated with the use of nuts, the concern may be more theoretical than real.

A. Definitions and Important Properties

In the strict botanical sense, nuts are dry, one-seeded fruits that are indehiscent (i.e., do not split open along a definite seam at maturity). However, for purposes of convenience, this section will deal with those seeds commonly known as nuts, e.g., walnuts, peanuts, almonds, etc. Further, these "nuts" will be separated into "groundnuts" (e.g., peanuts) and tree nuts.

All nuts have a more or less rigid outer casing or shell. For "true" nuts, this is an effective barrier against microbial invasion. In others, the integrity of the seam has a direct bearing on the possibility for contamination by microorganisms. Certain types (e.g., coconuts and black walnuts) have a pulpy flesh covering the shell that affords additional protection.

Most nuts have a high oil content and some contain appreciable quantities of protein and carbohydrate. With certain exceptions (e.g., coconut) nuts contain insufficient moisture (i.e., have too low a_w) to support the growth of bacteria. Certain nuts (e.g., peanuts) may be sufficiently "wet" to permit fungal growth before and during harvest.

B. Groundnuts

1. Initial Microflora

The initial microflora of peanuts is derived from the soil, for the nuts grow underground; mold contamination of the pods is unavoidable. The application of fungicides prior to harvest has not been proven effective in reducing the incidence of mold contamination (Pettit *et al.,* 1971). Moreover, the incidence of fungi usually is higher in those nuts harvested from a field in which an identical crop had been planted the previous year (Pettit and Taber, 1968). Species of *Aspergillus, Penicillium, and Fusarium* are dominant in the microflora of nuts in the field.

et al., 1968). They can be detected in peanut products for months after processing, though a gradual loss occurs, presumably due to oxidation (Waltking, 1971). Within a given lot of peanuts, aflatoxin distribution is likely to be heterogeneous, affecting only occasional nuts. Because of this, detection is more difficult in lightly infested lots. However, the toxin level in single infested nuts may be so high that the total toxin in the sample exceeds the legal tolerance.

5. Control

Control of microbial growth is accomplished by control of moisture. At harvest, peanuts contain 40–60% moisture, which must be reduced to <9% to inhibit mold growth (see Table 21.4). This is accomplished by artificial drying in the more highly developed countries. In the lesser developed countries, field drying in windrows frequently is the only means available, so that the moisture content of the crop is subject to the vagaries of the weather.

Individual contaminated nuts usually can be removed during the sorting and grading process by electronic and manual inspection because toxic nuts often appear discolored or damaged. At various stages in processing, peanuts are stored for periods of time. Storage facilities must be secure against insects and must prevent the introduction of moisture that would permit mold growth. If shelled nuts are stored under refrigeration, they must not be exposed to a warm humid atmosphere for transport to processing facilities, else moisture sufficient to support mold growth may condense on them.

TABLE 21.4

The Effect of Moisture Content on the Incidence of Mold Growth on Stock Peanuts [a]

Moisture (%)	Samples analyzed (Number)	Samples molded	
		Number	%
5.0– 7.9	56	0	0
8.0– 8.9	58	0	0
9.0– 9.9	26	8	31
10.0–10.9	16	12	75
11.0–12.6	4	4	100

[a] Golumbic (1965); peanuts are artificially dried farmers' stock peanuts held 23 days at ambient temperature in moisture-proof plastic bags.

6. Choice of Case

For an understanding of "Case" see Table A.1, and the brief description in Appendix IV. For more details, see the second book of this series (ICMSF, 1974).

Peanuts to be eaten raw can be tested for molds (Case 2) and *E. coli* (Case 5), because in the normal course of events, one would expect no process or treatment that would decrease or increase the microbial levels. Peanuts are also tested frequently for aflatoxins. Microbiological examination of roasted nuts is not advocated.

C. Tree Nuts

1. Initial Microflora

Occasionally, it is possible to find low levels of microorganisms in healthy mature nuts taken directly from the tree. However, this is rare; the majority of intact nuts on the tree are sterile. An intact shell is an effective barrier to microbial invasion, and the presence of a hull surrounding the shell further reduces the possibility of microbial contamination of the nutmeat.

Damage to the shell by mishandling or by birds or insects will destroy the natural barrier and predispose the nutmeat to contamination. For example, the pecan weevil larvae burrow through the shell, permitting microorganisms from various sources (e.g., rainwater dripping from the tree, dust, and the insect itself) to contaminate the nutmeat (Wells and Payne, 1975).

The major portion of the microflora of tree nuts arriving at the processing facility originates from the orchards, and in particular, the orchard soil. Nuts shaken from the tree and caught before hitting the ground contain relatively few microorganisms. However, nuts falling to the ground may be caught on canvas spread below the tree, or they may fall directly onto the soil itself, where they are contaminated with soil and/or fecal microorganisms (King *et al.,* 1970; Hyndman, 1963). One very important consideration is whether the land beneath the trees was used for grazing domestic animals or was fertilized with manure. If so, the incidence and level of fecal organisms in nuts are many times greater (Marcus and Amling, 1973). Length of time on the ground and soil moisture levels also bear directly on the incidence of contaminated nuts. Nuts wetted by pools of rainwater followed by drying may crack, thus eliminating the natural barrier to microbial invasion. Such damaged nuts will reflect the microflora of the soil on which they lie, e.g., *Pseudomonas,*

Xanthomonas, Bacillus, Enterobacter, Clostridium, Corynebacterium, Penicillium, Aspergillus, and *Tricothecium* (Smith and Arends, 1976).

2. Effects of Processing on Microorganisms

Handling during transportation from the field to the processing facility may damage the nut shell and distribute contamination from infected to healthy nuts. In addition, contamination from an improperly cleaned and sanitized transport vehicle may be significant. Often, nuts must be stored prior to further processing. Unless properly protected from insects and animals, additional contamination will occur. It is not uncommon for storage bins to be treated with methyl bromide (for insect control) but this has little effect on the microorganisms in or on the nuts themselves. Also, increased a_w levels during storage enhance the growth of fungi, and if the increase is large enough, bacterial multiplication will occur.

Changes in the microflora of nutmeats as they are processed depend on the process applied. Processes can be separated on the basis of whether water is involved at some stage.

Pecans, while still in the shell, are dry cleaned, then washed with water, and redried prior to storage. While the shell is wet, microorganisms enter with water through fine cracks and contaminate the nutmeats. Some of the nutmeats may remain wet during and after drying, permitting molds to grow. The nuts, still in the shell, may then be rewet and either heated at 82°–93°C for 2 min, or immersed in water containing 1000 ppm of chlorine. Though this tempering process reduces the microbial level, it is not sufficient to sterilize the nutmeats (Beuchat and Heaton, 1975). The tempered nuts are cracked and shelled, and the nutmeats are redried to a_w 0.58–0.64, and stored at −18°C. Microorganisms are dispersed during the procedures following tempering; but microbial growth will not occur under these conditions of storage.

During the dehulling of black walnuts, microorganisms in the wash water enter along the suture line of the shell and contaminate the nutmeats. Chlorine in the wash water is ineffective against the microorganisms because of the high level of organic matter (Meyer and Vaughn, 1969).

Almonds are processed for market without the use of water. The nuts (with husks) begin to dry on the tree and are knocked to the ground where they continue to dry. They are swept to collectors where they become comingled with soil and vegetation. The husks break away from some nuts so that the shells are exposed to soil and the microorganisms contained in it. Fumigation to kill insects with low levels of methyl bromide prior to husking has little impact on the microbial population. However, experimental evidence indicates that it could be used at a higher level to

utmeats without leaving an objectionably high methyl
(Schade and King, 1977). Further manipulation during
rses the organisms from "dirty" to "clean" nuts. Reduction
microbial load is a function of the efficiency of separation
om the more heavily contaminated hull material. The majority of
ds are packaged in shell and distributed to the consumer without
further processing. A small percentage may be shelled, blanched, and
sliced. The blanching process reduces the microbial level considerably.

Like other nuts, intact coconuts on the tree contain few viable microorganisms. However, after harvest, the nuts are stored on the ground in
contact with soil and perhaps manure. The husking of coconuts invariably
causes damage to some shells and leakage of the "milk" may occur.
Microorganisms grow profusely in the milk as it leaks to the outside of the
nuts. Loading, transport, and unloading increase the extent of damage and
spread the contamination. Grinding the meat prior to drying also permits
rapid microbial reproduction, but drying the ground product at 93°–121°C
for 30 min reduces the microbial load. However, because of the cooling
effect of water evaporation, the drying step does not completely eliminate
even heat-sensitive groups such as the gram-negative rods.

3. Saprophytes and Spoilage

The saprophytic flora of nutmeats normally reflects the environment
from which they were harvested, particularly those nuts harvested from
the ground and not given a bactericidal treatment during processing (e.g.,
almonds). The predominant bacterial genera on almonds are *Bacillus,
Pseudomonas, Acinetobacter–Moraxella (Achromobacter), Brevibacterium, Xanthomonas,* and *Micrococcus.* Almonds arriving at the processing facility averaged 3000–7000 bacteria and 1500–47,000 yeasts and
molds per gram. The totally dry processing reduced these levels to 580–
1200 bacteria and 170–2700 yeasts and molds per gram (King *et al.,*
1970).

Studies have shown (Meyer and Vaughn, 1969; Kokal, 1965) that
market samples of black walnuts had standard plate counts of 31,000 to
2,000,000/g. There was no correlation between the standard plate count
and the size of the pieces. Nor could a correlation be made between the
standard plate count and the *Escherichia coli* population. However, there
did seem to be an increased level of *E. coli* in smaller pieces of nutmeat.

A study of market pecans revealed that the microbial forms most predominant were soil and plant organisms (*Leuconostoc, Proteus, Enterobacter* [*Aerobacter*], *Pseudomonas, Clostridium, Corynebacterium, Penicillium, Aspergillus, Fusidium,* and *Trichothecium*). These organisms
presumably survived the tempering process or were reintroduced during

the subsequent cracking, shelling, and separating procedures (Chipley and Heaton, 1971). While the penicillia appear to predominate in freshly harvested pecans, the mycoflora shifts during processing to dominance by aspergilli in market pecans (Huang and Hanlin, 1975). The standard plate count in market samples of pecans often is several thousand per gram (Andrews *et al.*, 1978).

Spoilage of most nutmeats is due to fungi owing to the low a_w of the product. Spoilage of coconut is an exception. Bacteria can proliferate rapidly in coconut meat and milk, achieving populations of millions per gram in several hours in the warm climate in which the nuts are grown (Schaffner *et al.*, 1967; Kajs *et al.*, 1976). *Enterobacter aerogenes* often dominates the flora, producing off-odors and gas.

There is some question concerning the value of coliforms or *Escherichia coli* as indicators of insanitary conditions during the processing of nutmeats (Hyndman, 1963; Hall, 1971). The coliform test is invalid since these organisms are part of the normal flora of the raw products. Testing for *E. coli* has limited value since (a) *E. Coli* has been found (albeit rarely) in nuts on the tree (Kokal and Thorpe, 1969) and (b) none of the processes is unequivocally bactericidal. The presence of high numbers of *E. coli* in nuts may indicate that the nuts were harvested in groves in which animals grazed; testing for salmonellae is then to be recommended.

4. Pathogens

The primary pathogens associated with nutmeats are *Salmonella* and mycotoxigenic fungi. Contamination of nutmeats with salmonellae is quite rare (except in coconuts) but can be expected to occur if domestic animals are grazed beneath the nut trees and nuts are recovered from the ground in these groves. Salmonellae will persist on contaminated nutmeats for extensive periods of time under normal conditions of storage, e.g., low-temperature storage of pecans (Beuchat and Heaton, 1975; King *et al.*, 1970).

Mycotoxigenic fungi may be introduced into the nut while it is on the tree [e.g., by the pecan weevil (Wells and Payne, 1975)] or during harvest and storage. This is not a significant problem per se unless sufficient moisture becomes available to permit fungal growth. Lillard *et al.* (1970) found that among pecans on the market, 9% of bakery lots and up to 85% of retail lots contained *Aspergillus flavus*. The majority of the isolates were aflatoxigenic, and substances chromatographically similar to aflatoxins were detected in extracts of the nuts, although the identity of the substances was not confirmed. It is virtually impossible to destroy these toxins by reprocessing. Roasting of pecans at 191°C for 6 min (in oil) or 15 min (dry) resulted, respectively, in 65 and 87% reduction in the aflatoxin content of the samples (Escher *et al.,* 1973). Aflatoxins have been detected very infrequently in California almonds at very low levels (Schade *et al.,* 1975).

5. Control

Domestic animals should be excluded from orchards to minimize the possible introduction of pathogens, and nuts should be prevented from directly contacting the soil.

Vehicles and storage bins used for transport and storage of nuts must be cleaned and protected from vermin and moisture. Dry-processing equipment (e.g., for almonds) must be maintained in a clean condition so that the cracking and separation steps reduce the number of organisms on the nutmeat product.

The water used for husking, washing, or wetting nuts in the shell, (i.e., walnuts and pecans) must be potable. Drying must be rapid and even enough to avoid microbial growth. Heating peeled and cracked coconuts in water at 80°C for 5–10 min prior to grinding significantly reduces the microbial content and destroys salmonellae. This pasteurization step, coupled with strict sanitation of the grinding equipment and proper drying of the ground product will eliminate salmonellae in dried coconut (Schaffner *et al.,* 1967; Kajs *et al.,* 1976).

Heat will reduce the number of viable organisms and the level of mycotoxins on pecans, but it will not destroy either completely. Similarly, propylene dioxide will destroy many bacteria and molds but not all of them (Beuchat, 1973; Blanchard and Hanlin, 1973). Propylene oxide is without effect on mycotoxins.

After nuts have been processed, they must be protected from moisture.

6. Choice of Case

For an understanding of "Case" see Table A.1, and the brief description in Appendix IV. For more details, see the second book of this series (ICMSF, 1974).

Fully processed nuts are too dry to support microbial growth, but microorganisms will remain viable on them for prolonged periods. Another microcidal process usually is not applied before ingestion. Therefore, to evaluate mold level as an indicator of spoilage and general quality, Case 2 would apply. A test for numbers of *Escherichia coli* to determine the possibility of heavy contamination from fecal sources would require Case 5. Tests for *Salmonella* in coconut would require Case 11.

22

Soft Drinks, Fruit Juices, Concentrates, and Fruit Preserves

I. INTRODUCTION

Because of their particular combination of physical and chemical characteristics, soft drinks, fruit juices, and fruit preserves represent a relatively unique ecosystem. This chapter surveys the compositional characteristics of these products and their influence on the microflora of raw materials and end products. It also outlines and discusses the hygienic and preservation requirements for the different product types.

A. Foods Covered

Carbonated soft drinks are beverages made by absorbing carbon dioxide in potable water containing appropriate flavors and colors. The amount of carbon dioxide is not less than that which will be absorbed by the beverage at a pressure of 1 atm and a temperature of 15°C. In addition, the U.S. Food and Drug Administration standard of identity for carbonated soft drinks specifies the ingredients that may be used, such as nutritive sweeteners, flavoring, coloring and acidification agents, foaming and emulsifying agents, stabilizing or viscosity-producing agents, caffeine, quinine, and chemical preservatives. Most countries have similar requirements.

Noncarbonated (still) soft drinks, in addition to the above ingredients, may contain fruit juices and vitamin C.

Fruit juices are the unconcentrated liquids extracted from pure and mature fruit. Because many fruit juices are either too acid or too strongly flavored to be pleasant for consumption, diluting and/or blending is a common practice. The resulting fruit juice drinks usually contain not less than 20% juice, either single strength or reconstituted from frozen or canned concentrates. Artificial flavoring, coloring, and acidifying agents as well as pectins and other additives may be added as permitted by the authority having jurisdiction.

Fruit nectars are the pulpy-liquid soft drinks prepared from one or more fruits, one or more sweeteners, and other ingredients. Many fruit nectars contain not less than 50% pure fruit juice—either single strength or reconstituted from concentrates—together with added essences, vitamin C, pectins, sugar, and acidifying agents.

Concentrated drinks, with or without fruit, are similar to the single-strength types but are to be diluted before consumption.

Fruit preserves or jams are viscous or semisolid products containing one or more fruits, together with permitted sweetening agents and jellying ingredients such as pectin, carrageenan, agar, guar gum, algin, or methyl-cellulose.

Standards of identity for several of the above products either have been established or are being established under the Joint FAO/WHO Food Standards Program of the Codex Alimentarius Commission.

B. Important Properties

1. Acidity

Fruits, which are the main raw material used in fruit juices, soft drinks, and preserves, usually have pH values between 2.0 and 4.5. This results from the high level of organic acids present, i.e., 0.2% in pears to 8.5% in limes (Fig. 22.1). The highest pH values are found in tomatoes (average 4.3, range 3.5 to 5.0; Powers, 1976). Besides citric acid (which may account for 95% of the total acid content in citrus fruit), malic acid, tartaric acid, and quinic acid may also be present (Ingram and Lüthi, 1961). Fruit concentrates, which frequently serve as raw material in soft drinks and fruit preserve manufacture, have the same pH range (Sand, 1973). Tables 22.1–22.3 give examples of the composition and typical pH values of commercial soft drinks, concentrated drinks, and fruit preserves.

2. Water Activity

Water activity (a_w) plays an important role in the preservation of soft drinks and fruit preserves. Many of them—especially jams, marmalades, and concentrated soft drinks—have a high sugar content. In jams, sugar levels of 55 to 65% w/w are fairly common. Concentrated fruit drinks, which from a microbiological point of view can be considered as jams without the specific gel-systems, have sugar levels ranging from 40 to 65%

1.5 2.0 2.5 3.0 3.5 4.0 4.5 5.0

APPLE
PEAR
PLUM
CHERRY
PEACH
APRICOT
BLACK CURRANT
STRAWBERRY
RASPBERRY
GRAPE
ORANGE (bitter)
ORANGE (sweet)
LEMON
LIME
GRAPEFRUIT
PINEAPPLE
PASSION FRUIT
TOMATO

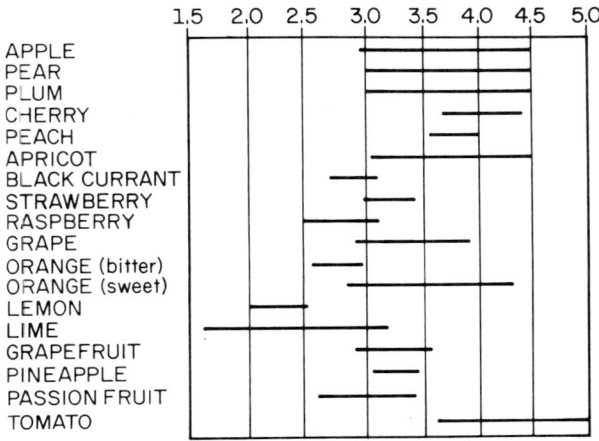

Fig. 22.1. pH ranges of various fruits. (Based on Pollard, 1959; Powers, 1976.)

w/w. As a consequence, the a_w of these products is frequently below 0.90. Soft drinks and fruit juices have sugar levels that range from 5 to 15% w/w, i.e., 5 to 15°Brix. Actual data on the a_w of various types of sugar solutions, i.e., sucrose, glucose, inverted sugar, and glucose syrups are given in Vol I, Chapter 4, Table 4.2. On the realistic assumption that there are no solute–solute interactions (Norrish, 1966), the a_w of mixed solutions containing two or more solutes may be calculated using the following equation (Ross, 1975):

$$a_w = a_{w,1} \times a_{w,2} \times a_{w,3}, \ldots, \text{etc.}$$

where $a_{w,1}$ represents the water activity of the first solution, $a_{w,2}$ that of the second solution, and so forth. By using this equation and the data of Table 4.2, the water activity of a particular mixture containing 40% w/w of sucrose ($a_{w,1} = 0.959$) and, in addition, 40% w/w of glucose ($a_{w,2} = 0.933$) can be calculated as follows:

$$a_w = 0.959 \times 0.933 = 0.895$$

Table 22.4 presents a calculation of a_w in a model concentrated citrus drink.

3. Nutrients

The carbohydrate content of soft drinks and fruit preserves is often high and consists mostly of easily metabolized hexoses (e.g., glucose and fructose), pentoses, and pectins. The organic acid content is usually high but variable and has a strong influence on the pH; many yeasts and molds can

TABLE 22.1

Examples of the Composition, the Optional Additives, and the pH Values of Single-Strength Soft Drinks

Orange drink (with fruit)	g/kg	Cola drink (without fruit)	g/kg
Orange concentrate (60°Brix)	13.20	Cola concentrate	6.4
		Cola essence	0.3
Citric acid	1.4	Sucrose (powder)	84.1
Sucrose syrup (67°Brix)	141.0	Tartaric acid	2.2
		Caffein	0.131
Benzoic acid	0.07	Orthophosphoric acid	0.652
Water	844	Water	906

Orange drink:
Refraction: 10° Brix [a]
Carbonated (2.5 v/v of CO_2) [b] : pH = 2.95
Noncarbonated : pH = 3.2

Cola drink:
Refraction: 8.6° Brix
Carbonated (3.5 v/v of CO_2) : pH = 2.6

Other examples, using lemon, pineapple, cherry, raspberry, etc. as the fruit, show a mean pH of 3.0 when carbonated and 3.3 when not

Other examples like tonic, lemon-lime, champagne soda etc. show a mean pH of 2.8–2.9.

[a] The dry matter content of a particular product is often expressed as "°Brix"; this refers to that sucrose concentration which has the same refraction as the formula concerned: 15°Brix means, for example, that the product has the same refraction as a 15% (w/w) sucrose solution.

[b] The CO_2 content is often expressed in "volumes of CO_2 dissolved in one volume of water": when the CO_2 content is 2.5 v/v, this means that 2.5 volumes of CO_2 are dissolved in 1 volume of water at the specified temperature. The CO_2 pressures, which may be found when various CO_2 volumes are dissolved at various temperatures, are summarized by the American Bottlers of Carbonated Beverages (1957) and by Jacobs (1959).

TABLE 22.2

Examples of the Composition, the Optional Additives, and the pH Values of Concentrated Soft Drinks [a]

Orange squash (with fruit)	g/kg	Grenadine drink (without fruit)	g/kg
Orange concentrate (61°Brix)	55	Grenadine aroma	4
Citric acid	15	Sucrose syrup (67°Brix)	955
Sucrose (powder)	509	Coloring agents	0.4
Sorbic acid	0.5	Tartaric acid	2.5
Water	434	Sorbic acid	0.300
Refraction: 59°Brix; pH 2.5		Water	42
		Refraction: 64°Brix; pH 2.4	
Other examples, with different types of fruit concentrates (apple, cherry, black currant, grapefruit, lemon, pineapple, etc.) have a mean pH of 2.7.		Other examples, like peppermint syrup and other artificially flavored concentrated drinks have a mean pH of 2.4.	

[a] From W. Kooiman and W. Baggerman (unpublished).

TABLE 22.3

Examples of the Composition, the Optional Additives, and the pH Values of Fruit Preserves [a]

Strawberry jam	g/kg	Low-calorie strawberry spread	g/kg
Strawberries	400	Strawberries	422
Sucrose	600	Sucrose	331
Pectin	4	Pectin	7.5
Citric acid	5	Citric acid	2.5
		Coloring agent	0.24
Sorbic acid	0.3	Sorbic acid	0.5
Refraction: 65°Brix; pH 3.3		Refraction: 33°Brix; pH 3.5	
Other types of jam mostly have pH 3.2–3.8.		Most low-calorie spreads have pH 3.4–4.0.	

[a] From W. Kooiman and W. Baggerman (unpublished).

TABLE 22.4

Example of a_w Calculation for a Model Formulation of a Citrus Drink

Ingredient	Weight in formulation (g/100 g)	Water content [a] (g/100 g)	Concentration of individual components		
			g/kg of total water	Molality	a_w
Citrus concentrate (60°Brix)	12.0	4.8	$(12/42.8) \times 1000 = 280$		0.976 [b]
Sucrose	40.0		$(40/42.8) \times 1000 = 935$	2.73	0.940 [c]
Glucose	10.0		$(10/42.8) \times 1000 = 234$	1.30	0.975 [c]
Water	38.0	38.0			

[a] The water content of the final mixture is 38.0 g plus the 4.8 g, which comes from the citrus concentrate, i.e., 42.8/100 g.

[b] The $a_{w,1}$ is calculated, using the experimental datum that the concentrate has an a_w value of 0.900. Because this concentrate is diluted with water over a factor of 12/50 (i.e., 0.24 times), the interpolated a_w value equals 0.976.

[c] Calculated by interpolation in Table 4.2.

use organic acids as carbon sources under aerobic conditions (see Vol. I, Chapter 7). Nitrogen is frequently present but only at very low levels, i.e., 0.05 to 0.15% (Pollard, 1959), and about 60% of it is in the form of free amino acids. In contrast to popular belief, most soft drinks and fruit preserves contain only very low levels of vitamins. Exceptions are black currant juice and rosehips, which contain naturally high levels of vitamin C. Other fruits or fruit concentrates are frequently supplemented with ascorbic acid (vitamin C) as a contributor to flavor and color stability as well as for nutritional reasons. Group B vitamins are practically absent in these products.

4. Oxygen and the Redox Potential

Dissolved oxygen is readily consumed by fruit particles or sulfite, is bound by ascorbic acid, or is removed when the product is pasteurized. As a result, the redox potential is often very low. Replacement of oxygen by CO_2 in a carbonated beverage creates a specific ecosystem. Aseptic filling into oxygen-permeable packaging materials, or any packaging system that permits a large head space, may drastically increase the level of dissolved oxygen or the redox potential. This again has a strong influence upon the ecosystem.

C. Methods of Preservation

The method for preservation of a particular soft drink or fruit preserve is dictated basically by balancing organoleptic quality (flavor, color, texture, and cloud stability), distribution of fruit particles (shreds in marmalades, fruit in jams, etc), processing (especially heat treatment), and the use of preservatives.

1. Natural Antimicrobial Substances

Besides the organic acids, the high sugar content, and the presence of CO_2, many fruits—for example, citrus fruits and cranberries—naturally contain antimicrobial substances such as essential oils, benzoic acid, or sorbic acid (Guenther, 1948). Although an individual compound is rarely or never present in a sufficiently high concentration to assure complete inhibition of microbial growth, the substance may serve as an additional preservative factor.

2. Preservatives

Soft drinks and fruit preserves sometimes require the addition of chemical preservatives to improve their storage stability. These additives should be used judiciously and only when there is a clear need to increase shelf-life, prevent spoilage, or minimize the food-poisoning risk. Many compounds, including antibiotics and various flavoring agents (Chichester and Tanner, 1972; Wallhäuser and Schmidt, 1967), are described in the literature as antimicrobial in soft drinks and fruit preserves. However, not all of these compounds may be used legally. Because of the great variation in types and levels of preservatives permitted in foods (Botma, 1973, 1974; Chichester and Tanner, 1972), the data and recommendations of the Food and Agriculture Organization/World Health Organization (1974) may be used as a reference if a particular country does not list permitted preservatives. For further information on the main physical and chemical data on preservatives, see Table 7.1 in Chapter 7, Vol. I. (See also Section III,D,3.)

II. INITIAL MICROFLORA

Many genera and species of microorganisms, including bacteria, yeasts, and molds, occur in the raw materials that are used in the manufacture of fruit juices, soft drinks, and fruit preserves. Although this group of products rarely presents public health risks, the various ingredients used in their manufacture may contain small numbers of pathogens or adventitious contaminants. (See Section III,C.)

The types of yeasts present—i.e., sporogenous or asporogenous, fermentative or nonfermentative—may vary depending on the type of fruit used, as shown in the following list:

1. Intact apples and freshly pressed juices contain mainly asporogenous genera, e.g., *Candida, Torulopsis, Rhodotorula, Brettanomyces,* and *Cryptococcus.*

2. Fermented apple juices contain mainly sporogenous yeasts, e.g., species of *Saccharomyces, Pichia, Hansenula, Debaryomyces, Lodderomyces,* and *Saccharomycopsis.*

3. Grapes and grape juice are contaminated mainly with sporogenous yeasts, e.g., *Saccharomyces, Zygosaccharomyces, Hanseniaspora, Pichia, Torulaspora, Debaryomyces,* and *Zygopichia.* Asporogenous species like *Kloeckera apiculata,* are also frequently found and comprise together with *Saccharomyces cerevisiae* var. *ellipsoides* about 80% of the yeast flora of grapes and other fruit.

4. Citrus fruits are contaminated mainly with *Candida, Zygosaccharomyces, Hanseniaspora, Saccharomyces,* and *Pichia* species.

Molds frequently isolated from fresh fruit and fruit juices belong to the following genera (Lüthi, 1959; Senser and Rehm, 1965): *Penicillium,*

Byssochlamys, Aspergillus, Paecilomyces, Mucor, Cladosporium, Fusarium, and *Botrytis.*

Many bacterial genera have been isolated from fresh fruit and their juices (Ingram and Lüthi, 1961; Lüthi, 1959; Sand, 1971). Lactic acid bacteria (*Lactobacillus* and *Leuconostoc* spp.), acetic acid bacteria (*Gluconobacter* and *Acetobacter* spp.), and some spore-forming bacteria like *Bacillus coagulans, Clostridium butyricum,* and *C. pasteurianum* (the latter especially in tomato-based products) have been implicated in the spoilage of soft drinks.

III. PROCESSING

A. Effects of Processing on Microorganisms

The processing of soft drinks, fruit juices, and fruit preserves principally involves the mixing together of various ingredients and dispensing into appropriate containers. This may be followed or preceded by a heat treatment; preservatives, including CO_2, may or may not be added.

1. Soft Drinks

Carbonated, single-strength fruit juices may be made stable by combining a heat treatment with a low level of benzoic acid (75 mg/kg). If the CO_2 level is 2.5% v/v, the temperature should not exceed 65°–70°C; if it is 1.5 to 2.0% v/v, the temperature should be 70° to 75°C. Many soft drinks are stable without heat treatment. For example, carbonated cola drink is stable because of its low pH (2.6) in combination with the high CO_2 level (3.5 v/v) and the antimicrobial activity of orthophosphoric acid.

Noncarbonated, single-strength soft drinks may be made microbiologically stable by pasteurization, or by the addition of preservatives. If preservatives are used, it is necessary generally to reduce the pH and/or a_w to achieve satisfactory stability. Pasteurization can be achieved by hot filling followed by additional heat treatment after closure of the container, or by High-Temperature Short-Time (HTST) pasteurization (i.e., 4–10 sec at 90°–95°C) followed by aseptic packaging. Table 22.5 lists typical pasteurization times and temperatures.

2. Tomato Juice

Tomato juice, because of its higher pH (ca 4.3) must be heated more severely to make it microbiologically stable, e.g., a HTST treatment of 118°C for 0.5 to 1:5 min ($F_0 = 0.23$ and 0.7, respectively). Were the juice at pH 4.1 or below, a lesser heating, e.g., 100°C for 1.5 min would probably suffice, and addition of an edible acid to attain this is feasible.

TABLE 22.5

Examples of Pasteurization Conditions for Soft Drinks and Fruit Preserves [a]

Product	Pasteurization treatment	
	Time (min)	Temperature (°C)
Carbonated orange drink (2.5 v/v CO_2)	10–20	65°–70°
Orange juice, noncarbonated	10–15	70°–75°
Concentrated drink (orange syrup) pH = 3.0, 50°Brix	1–5	85°–90°
Strawberry jam (pH = 3.3, 58°Brix)	5–10	80°
Low-calorie strawberry jam (pH = 3.5, 30°Brix)	5–15	80°–85°

[a] These times and temperatures, taken from various sources, should be viewed as examples. They are not recommendations.

However, in large operations in the United States, it is considered impractical to measure the pH of each batch and adjust the heating to fit; therefore, the canning company assumes the pH to be 4.3 and applies the HTST treatment to all lots. After the treatment, the juice is filled into cans or bottles at about 93°C.

3. Concentrated Soft Drinks

Concentrated soft drinks, both with and without fruit, are noncarbonated. To make them stable, depending on composition and required shelf-life, three basic systems are used as follows:

1. By combination of low a_w (65°–70°Brix) and low pH (2–2.5).
2. By hot-filling and/or pasteurization. As the dry matter content is always higher than that of a single-strength drink, the heat treatment should be more severe, i.e., 1–5 min at 85°–90°C. (See Table 22.5.) Preservatives may be added to achieve the required shelf-life after opening.
3. By addition of chemical preservatives. This may be, in many cases, the only option. The amount of sorbic or benzoic acid used depends on the other intrinsic variables and on the policy of food-control agencies but will usually be between 300 and 1500 mg/kg. (See Section III,D,3.) Concentrates should be stored at constant temperature to prevent condensation and possible yeast and mold growth on the surface. This can also be pre-

vented by minimizing the head space or, where permitted, dusting the surface of the concentrate with sorbate.

4. Fruit Preserves

Nearly all fruit preserves, especially jams, marmalades, and low-calorie jams, require heating to 80°–85°C to denature the fruit enzymes and must be hot filled before they gell. A final heating of the product in the closed container gives added security. Inverting the container after hot filling decontaminates the lid. The heat destroys all molds, yeasts, and vegetative cells of bacteria, and the low pH prevents germination and growth of bacterial spores, so the product is stable. However, once the lid is removed, molds and yeasts can enter and grow on the surface. Sorbic or benzoic acid at 200 to 500 mg/kg will inhibit these organisms.

B. Spoilage

Yeasts and molds are the principal organisms that can grow at the low pH and low a_w of soft drinks, juices, and fruit preserves. Among the bacteria, only the lactics and acetic acid organisms can grow.

1. Yeasts

Yeasts predominate in spoilage of acid fruit products because of their high acid tolerance and the ability of many of them to grow anaerobically. Table 22.6 illustrates that most yeasts isolated from fruits can grow at both pH 2 and 1.5, in 45° and 55°Brix sugar solutions, and at 8.5°C. Some are even more tolerant of high Brix and low temperature.

The nutritional requirements of yeasts are minimal; most can synthesize a wide variety of substances essential for growth, including amino acids, carbohydrates, and vitamins. Because of their widespread ability to use simple nitrogen sources, only very low nitrogen levels (0.2 to 0.5 mg/liter) are required for their growth.

As shown in Table 22.7, yeasts that form ascospores are more heat resistant than those that do not. More detailed studies by these workers (Put *et al.,* 1977) established that D values of ascospores of *Saccharomyces cerevisiae* and *S. chevalieri* (the most common spoilage yeasts in canned-fruit products) were ten times that of the vegetative cells. The $D_{60°C}$ values of ascospores of both species were approximately 10 min; z values were approximately 5°C. However, yeast spores are much less heat resistant than are bacterial spores.

Yeasts are "osmophilic," which means that they can grow in high concentration of sugar. Osmophilic organisms may be highly sensitive to salt and, therefore, are not halophilic (Koppensteiner and Windisch, 1971).

TABLE 22.6

Evaluation of the Ability of Six Yeast Genera from Citrus and Other Fruits to Grow at Low pH, in High Sucrose Concentrations, and at Low Incubation Temperatures [a]

Genus	Number of isolates tested	pH		Sucrose concentration (°Brix)			Incubation temperature (°C)			
		2.0	1.5	45	55	65	10°	8.5°	5°	2.5°
Candida	31	4 [b]	3	4	2	1	4	4	2	1
Zygosaccharomyces	14	4	1	4	4	2	4	2	1	0
Hanseniaspora	5	4	4	4	4	0	4	4	0	0
Pichia	5	4	4	4	0	0	4	4	0	0
Saccharomyces	9	4	3	4	4	0	4	4	4	3
Kloeckera	5	4	4	4	4	0	4	4	3	0

[a] From Recca and Mrak (1952).

[b] 4 = 75 to 100% of isolates grew. 3 = 50 to 75% of isolates grew. 2 = 25 to 50% of isolates grew. 1 = 10 to 25% of isolates grew. 0 = no isolates grew.

TABLE 22.7

Thermal Resistance of Asporogenous and Sporogenous Yeasts [a]

Temperature of heating (°C)	Percentage of asporogenous (A) and sporogenous (S) strains [b] showing survival after heating for			
	10 min		20 min	
	A	S	A	S
65°	None	16	None	None
62.5°	25	60	3	30
60°	75	~100	40	~80

[a] From Put *et al.* (1976).

[b] The asporogenous yeasts belonged to the genera *Brettanomyces, Candida, Kloeckera,* and *Torulopsis* (35 strains). The sporogenous yeasts belonged to the genera *Debaromyces, Hansenula, Kluyveromyces, Lodderomyces; Pichia* and *Saccharomyces* (85 strains).

A more readily acceptable approach is to describe all such organisms as "xerophilic," i.e., expressing *the capacity to grow, under at least one set of conditions, at an a_w less than 0.85* (Pitt, 1974).

Growth of yeasts is usually accompanied by formation of CO_2 and alcohol. Yeasts also develop turbidity, flocculation, pellicles, and clumping. If they produce pectinesterase, they can destroy the natural pectin cloud. They can also degrade organic acids (raising the pH) and may form acetaldehyde, which contributes a "fermented flavor."

In spoilage, the following are the most important yeast genera (von Schelhorn, 1951; Sand, 1969).

Asporogenous yeasts (nonascospore forming)	Sporogenous yeasts (ascospore forming)
Rhodotorula	*Debaryomyces*
Candida	*Hansenula*
Brettanomyces	*Lodderomyces*
Torulopsis	*Pichia*
	Saccharomyces
	Saccharomycopsis

For further reading on yeasts from fruits and drinks, see the following: Marshall and Walkley (1951a,b, 1952a,b,c,d), Mossel and Scholts (1964), Mrak and McClung (1940), Sand (1973), and Walker and Ayres (1970).

2. Molds

Mold spoilage of soft drinks and fruit preserves, when it occurs, is frequently (but not exclusively) caused by *Penicillium notatum, Penicillium roquefortii, Cladosporium* spp., or *Byssochlamys* spp. Molds generally have high acid tolerance and are able to grow at low a_w, but these capabilities vary widely among the genera and even among species within a genus. The main xerophilic molds and their lowest reported a_w levels for growth are given in Volume I, Chapter 4. Like yeasts, most molds are not nutritionally exacting, but unlike yeasts, molds, with a few exceptions are strict aerobes. The low redox potential and/or low oxygen tension in many fruit-containing products restricts the development of molds.

Some molds are more resistant to heat than are yeasts, and, therefore, they can predominate in pasteurized products (Marshall and Walkley, 1952c). Mycelial and conidial spores are usually killed at temperatures between 55° and 70°C (King *et al.*, 1969; Lubieniecki-von Schelhorn, 1973). Mold ascospores and sclerotia are much more resistant and some are killed only at temperatures above 80°C. Heat-resistant molds that have been reported to cause the spoilage of soft drinks or fruit preserves are as follows: *Byssochlamys nivea, Byssochlamys fulva, Paecilomyces varioti, Phialophora mustea, Aspergillus fischeri* var. *glaber, Aspergillus fischeri* var. *spinosus,* and *Thermoascus aurantiacum.* Because the occurrence of these molds is restricted to certain geographical areas, types of fruit, conditions of harvesting, and so forth, it is usually not realistic to base a commercial process on their extreme heat resistance (Marshall and Walkley, 1951a,b; Meyrath, 1962). Measures such as the selection of raw materials and cleaning and disinfection effectively control heat-resistant molds.

Molds form colonies and aerial mycelia on the surface, flocculation or floating mycelia within the product, or clarification of cloud by pectin breakdown.

3. Bacteria

Gluconobacter (Acetomonas) species are the main spoilage bacteria because of their ability to grow at relatively low pH (i.e., 3.0 to 3.5) and at low levels of nutrients. They are strict aerobes, and any new developments in packaging techniques and materials that increase oxygen tension in the product will increase the incidence of spoilage by these organisms.

Lactobacillus and *Leuconostoc* species frequently have been isolated from fruit and spoiled soft drinks and sometimes appear as a secondary infection that is introduced during processing. Most lactobacilli that are isolated are heterofermentative, i.e., producing CO_2, acetic acid, and ethanol from sugar. The properties that favor the growth of lactic acid bacteria in soft drinks and fruit concentrates are as follows:

1. Their ability to grow very well at low redox potential.
2. Their acid tolerance (most species can grow at pH 3.5, and 5–10% of them can grow at pH 3.5–3.0).
3. Their high tolerance for CO_2.
4. Their ability to grow at temperatures used for fruit and fruit-juice processing.
5. Their ability to withstand and even grow in the presence of more than 30% w/w sugar.

The growth of lactic acid bacteria causes opalescence of soft drinks and concentrates and sometimes visible bubbles of gas or increased pressure and bursting of containers. In addition, members of the genus *Leuconostoc* and some *Lactobacillus* species produce dextrans and levans, which form a gummy slime or "ropiness." The "buttermilk" off-flavor frequently associated with the growth of *Lactobacillus* species in fruit-containing products is caused by the formation of diacetyl. Growth of acetic acid bacteria also leads to turbidity or opalescence; product flavors are changed by the formation of acids such as acetic and gluconic.

These bacteria are relatively heat sensitive. The heat resistance of *Leuconostoc* and *Lactobacillus* isolates heated in a single-strength orange juice concentrate (42°Brix, pH 3.45) is illustrated in Table 22.8. The high D and z values in the concentrates should be noted. The same type of data has been obtained for acetic acid bacteria (Marshall and Walkley, 1952c).

The principal organism of concern in tomato juice canning is *Bacillus coagulans,* which can produce flat sour spoilage. The HTST treatment at time of canning or bottling is designed to reduce the population of this organism, which requires relatively high temperatures to germinate but after germination can grow in a rather wide range of temperatures (minimum 5°–20°C; maximum 35°–50°C; Buchanan and Gibbons 1974).

C. Pathogens

Soft drinks, fruit juices, concentrates, and fruit preserves are low-risk foods since growth of pathogens does not take place primarily because of the low pH. Foodborne disease from these products does take place occasionally (Wallace and Park, 1933), probably after spoilage organisms have neutralized the natural acids (Powers, 1976; Huhtanen *et al.,* 1976). Tomatoes, tomato juices, and tomato concentrates are more likely to support growth and toxin formation because of their relatively high pH. For

TABLE 22.8

Heat Resistance of Some Lactic Acid Bacteria When Heated in Single-Strength or Concentrated Orange Juice [a]

	Heat resistance when heated in			
	Single-strength juice [b]		Concentrated juice [b]	
Organism	$D_{65.5}$ (min)	z (°C)	$D_{65.5}$ (min)	z (°C)
Leuconostoc: 13 strains, composite suspension	0.04	3.89	0.23	5.56
Lactobacillus: 2 strains, composite suspension	0.28	3.89	1.20	10.0

[a] Adapted from Murdock *et al.*, 1953.

[b] The single strength and the concentrated orange juice had a dry matter content of 9.8° and 42.0°Brix, respectively, and pH values of 3.7 and 3.45, respectively.

this reason, acidification of tomato products to pH 4.1 or 4.2 is common practice.

The minimum pH permitting the growth of salmonellae is 4.05 to 5.50, depending on the nature of the acid (Chung and Goepfert, 1970). Growth on the surface of sliced watermelon caused an outbreak of salmonellosis in the United States in the early 1950s (Gayler *et al.*, 1955). However, in moderately acidic products such as apple juice (U.S. Department of Health, Education, and Welfare, 1975d), the organisms may survive to cause salmonellosis but do not grow, particularly under adverse conditions of temperature, redox potential, or a_w (Chung and Goepfert, 1970). Survival is longer when the pH is high, the storage temperature is low, or the sugar concentration is high (Lüthi, 1959; Mossel, 1963).

The growth of molds on foodstuffs, including fruit and fruit-based products, may be accompanied by the formation of mycotoxins such as patulin, aflatoxin, zeareleone, ochratoxin, sterigmatocystin, and byssochlamic acid (Kadis *et al.*, 1971–1972; Rehm and Meyer, 1967; Riemann, 1969; Rice *et al.*, 1977). Such mycotoxins are usually not destroyed by further processing, e.g., heating.

Some juices destroy viruses and others do not. For example, grape juice and apple juice inactivate polio virus type 1, but pineapple, tomato, grapefruit, and orange juices do not. Ascorbic acid solutions destroy the virus but not in fruit juices (Konowalchuk and Speirs, 1978a,b).

D. Control

Each soft drink or fruit preserve represents an ecosystem characterized by a particular pH value, a specific a_w, ill-defined growth-supporting properties, and more definable growth-inhibiting properties. The growth-supporting potential of the ecosystem should be determined preferably by inoculation of the product with isolates from raw materials, processing plants, and spoiled products of similar composition as well as with strains known to challenge the particular antimicrobial characteristics of the product concerned.

1. Formulation

One means of assessing the relative microbial stability of a product is by use of an ecogram (Mossel and Scholts, 1964). This is a means of expressing the results of inoculation and storage experiments on a particular product using a microflora that represents the potential spoilage flora. It allows an objective assessment to be made of the "sensitivity for spoilage" and uses a single term, the index of growth (I.G.), which is the quotient of the number of isolates showing growth (Ng) in a particular product and the total number of isolated microorganisms used (No) to inoculate the product under investigation, i.e.

$$I.G. = Ng/No$$

When I.G. is 1 or approaches 1, i.e., when all or nearly all the inoculated organisms are able to grow, the product is highly sensitive to spoilage; when I.G. is zero, none of the inoculated microorganisms can grow, and the product is microbiologically stable. The I.G. requires careful interpretation and is only an indicator of spoilage potential. It is important to use realistic inoculum levels and storage conditions when using this procedure.

A model system may serve as a guide for estimating the expected intrinsic stability of various soft drinks and fruit preserves (Kooiman, 1977). The data of Fig. 22.2 and 22.3 were obtained with malt extract broths, having various pH values and sugar contents, and inoculated with a mixture of vegetative yeasts (*Saccharomyces rouxii, S. baillii, S. bispores, Hansenula anomala, Torula mellis*) or mold spores (*Aspergillus ruber, A. niger, Penicillium frequentans, P chrysogenum, P. notatum,* and *Hemispora stellata*), then stored for up to 100 days at 20°C. The figures show that the growth of xerophilic molds is more easily prevented than the growth of xerophilic yeasts. At pH 3.0 with 70% *w/w* sugar added, mold growth is

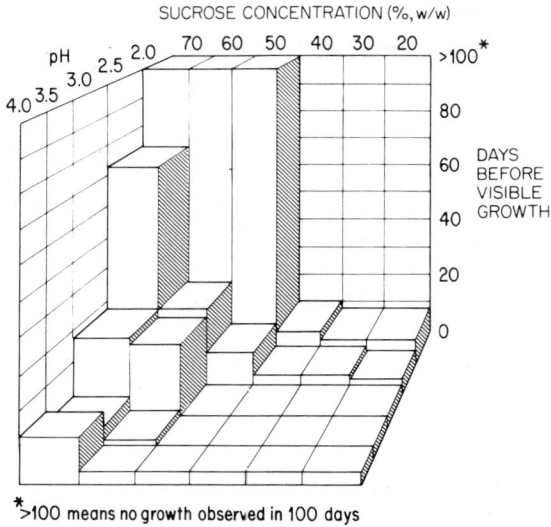

Fig. 22.2. Effect of pH and sucrose levels on the time for visible growth of yeasts to appear in a model system representing soft drinks and preserves. *Saccharomyces rouxii, S. bailii, S. bisporus, Hansenula anomala,* and *Torula mellis* were inoculated into malt extract broths and incubated at 20°C. (From Kooiman, 1977.)

prevented for more than 100 days. However, at the same sugar level, the pH must be lowered to about 2.0 to obtain the same degree of inhibition of xerophilic yeasts. To control the growth of yeasts and molds when the intrinsic stability of the product is insufficient, the required stability may be obtained by heating (pasteurization), chilled storage, or the addition of preservatives.

2. Pasteurization

The objective of heat treatment is the destruction of microorganisms capable of growth in a product during normal conditions of storage. In the case of soft drinks and fruit preserves, a pasteurization treatment, i.e., heating for a specified period of time at temperatures below 100°C, generally will meet this objective.

The pasteurization treatment to be applied depends on the following:

1. *The Microflora* The resistance of spoilage organisms to heat destruction depends on growth conditions and/or the presence of spores (Christopherson and Precht, 1952; Gibson, 1973).

2. *The Product* The a_w, pH, type of product, and presence of antimicrobial substances are very important factors influencing heat resistance and growth of surviving microorganisms (Bell and Etchells, 1952; Dakin,

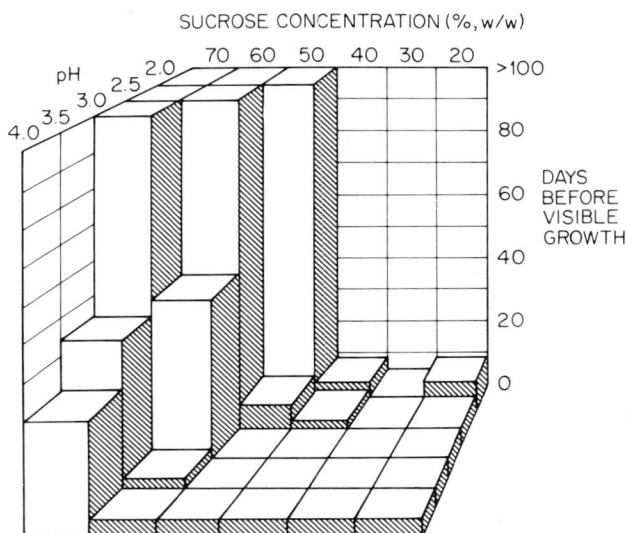

Fig. 22.3. Effect of pH and sucrose levels on the time for visible growth of molds to appear in a model system representing soft drinks and preserves. Spores of *Penicillium chrysogenum, P. frequentans, P. notatum, Aspergillus niger, A. ruber,* and *Hemispora stellata* were inoculated into malt extract broth and incubated at 20°C. (From Kooiman, 1977.)

1962; Lubieniecki-von Schelhorn, 1973; Murdock *et al.,* 1953; Shibasaki and Tsuchido, 1973; Juven *et al.,* 1978). They will result, as might be expected, in a range of pasteurization requirements. Table 22.5 summarizes some of the pasteurization treatments which are applied in practice.

Although the heat resistance of different types of spoilage organisms in these products is often known, there remain many gaps in our knowledge. Pasteurization values (*P* values) may be used for the calculation of pasteurization processes (Shapton *et al.,* 1971). This and other methods for process calculations allow for rapid and easy comparison of time–temperature treatments and allow the producer to establish an optimal balance among heat treatment, organoleptic quality, microbial stability, and processing costs (Del Vecchio *et al.,* 1951; Patashnik, 1953).

3. Preservatives

Many soft drinks, e.g., concentrated fruit drinks, cannot be pasteurized because of damage to their physical and organoleptic properties. When reformulation—e.g., by decreasing the pH or a_w—does not give the required intrinsic stability (see Figs. 22.2 and 22.3), preservatives must be added to obtain the required shelf-life.

In most countries, only benzoic acid, sorbic acid, sulfite, p-hydroxyben-zoic acid and its esters (parabens), and carbon dioxide are permitted as food preservatives to control microorganisms in these products, and only these will be considered here (see Botma, 1973). Their physical and toxicological properties are summarized in Volume I, Chapter 7, Table 7.1, and their effectiveness against a variety of organisms in Table 7.2. These and other data reveal that tolerance varies among genera, species, and even among strains.

Data on the antifungal activity of sorbic acid, benzoic acid, parabens, carbon dioxide and sulfite provide the following information:

1. Sulfite and parabens generally are slightly more effective against molds than against yeasts.

2. Under some circumstances, benzoic acid and sorbic acid do not differ significantly in their effects on molds. However, there are reports (Chichester and Tanner, 1972) indicating that sorbic acid is the more effective against certain molds.

Sorbic acid, sulfite, and parabens are reasonably effective inhibitors of bacterial growth at the pH values of fruit juices and concentrates (see Volume I, Chapter 7, Table 7.2). However, for effective inhibition, *Lactobacillus* species require either significantly higher concentrations of parabens and sorbic acid or lower pH values than do most yeasts and molds (Bell and Etchells, 1952).

The effectiveness of benzoic acid, sorbic acid, and the parabens is greatly enhanced by low a_w and by a high level of undissociated acid molecules. The undissociated molecules are in highest proportion in acid solutions (see Vol. 1, Chapter 7).

The combined effects of a_w, pH, and preservative can be illustrated. First, yeasts (Fig. 22.2) and molds (Fig. 22.3) are inhibited by high sucrose concentration (low a_w) and by low pH; the effect of the two factors is additive. One can then superimpose on these data the effect of sorbic acid (Table 22.9) or benzoic acid (Table 22.10) at various concentrations. The antimicrobial effects of the three factors (pH, a_w, and preservative concentration) are interdependent. Under these conditions, yeasts are more resistant to benzoic and sorbic acids than are molds; this difference is also found for other preservatives, viz. p-hydroxybenzoic acid and its esters (Shibasaki, 1969) and sulfite (Roberts and McWeeny, 1972). Sorbic and benzoic acids are about equally inhibitory to molds, but sorbic acid is more effective against yeasts than is benzoic acid. These data give only an indication of the preservative levels necessary for microbiological stability. Confirmation of the data in an actual product, using the appropriate flora, will generally be necessary.

TABLE 22.9

The Effect of Various Concentrations of Sorbic Acid Superimposed on the Model Systems Illustrated in Figs. 22.2 (Yeasts) and 22.3 (Molds)[a]

		Number of days at 20°C before visible growth is observed at the indicated sucrose concentrations (%, w/w) and pH values											
	Sorbic acid	Yeasts						Molds					
pH	(mg/kg)	20	30	40	50	60	70	20	30	40	50	60	70
2.0	100	•	•	•	•	•	•	62	•	•	•	•	•
2.5		8	14	20	•	•	•	59	59	59	•	•	•
3.0		3	4	8	15	•	•	9	14	22	•	•	•
3.5		3	3	4	9	28	•	9	9	14	•	•	•
4.0		3	3	3	8	24	•	8	9	9	14	•	•
2.0	200	•	•	•	•	•	•	72	•	•	•	•	•
2.5		8	18	42	•	•	•	42	80	•	•	•	•
3.0		6	8	28	24	•	•	12	64	•	•	•	•
3.5		3	6	7	8	•	•	4	59	•	•	•	•
4.0		3	3	6	7	•	•	4	14	22	22	•	•
2.0	300	•	•	•	•	•	•	92	•	•	•	•	•
2.5		36	28	80	•	•	•	67	•	•	•	•	•
3.0		7	13	21	21	•	•	59	•	•	•	•	•
3.5		4	6	17	18	•	•	40	59	•	•	•	•
4.0		3	4	4	9	•	•	14	18	18	•	•	•
2.0	400	•	•	•	•	•	•	•	•	•	•	•	•
2.5		•	•	•	•	•	•	•	•	•	•	•	•
3.0		9	36	41	•	•	•	•	•	•	•	•	•
3.5		8	9	25	•	•	•	•	•	•	•	•	•
4.0		6	6	17	21	•	•	88	82	•	•	•	•
2.0	500	—	•	•	•	•	•	•	•	•	•	•	•
2.5		24	•	•	•	•	•	•	•	•	•	•	•
3.0		13	•	•	•	•	•	88	•	•	•	•	•
3.5		8	21	67	•	•	•	59	•	•	•	•	•
4.0		8	9	28	41	•	•	59	59	•	•	•	•
2.0	600	•	•	•	•	•	•	•	•	•	•	•	•
2.5		•	•	•	•	•	•	•	•	•	•	•	•
3.0		28	•	•	•	•	•	•	•	•	•	•	•
3.5		17	21	•	•	•	•	64	•	•	•	•	•
4.0		8	13	21	•	•	•	42	64	•	•	•	•

[a] From Kooiman (1977). • = more than 100 days.

TABLE 22.10

The Effect of Various Concentrations of Benzoic Acid Superimposed on the Model Systems Illustrated in Figs. 22.2 (Yeasts) and 22.3 (Molds)[a]

| | | Number of days at 20°C before visible growth is observed at the indicated sucrose concentrations (%, w/w) and pH values | | | | | | | | | | | |
| | | Yeasts | | | | | | Molds | | | | | |
pH	Benzoic acid (mg/kg)	20	30	40	50	60	70	20	30	40	50	60	70
2.0	200	48	•	•	•	•	•	48	•	•	•	•	•
2.5		7	7	11	56	•	•	31	56	•	•	•	•
3.0		3	3	7	7	31	56	31	31	•	•	•	•
3.5		3	3	3	7	7	31	11	26	26	56	•	•
4.0		3	3	3	3	3	28	14	7	7	7	28	•
2.0	300	•	•	•	•	•	•	•	•	•	•	•	•
2.5		18	26	48	•	•	•	•	•	•	•	•	•
3.0		7	7	11	•	•	•	60	•	•	•	•	•
3.5		3	3	7	7	31	•	31	•	•	•	•	•
4.0		3	3	3	3	18	•	11	31	7	7	•	•
2.0	400	•	•	•	•	•	•	•	•	•	•	•	•
2.5		•	•	•	•	•	•	•	•	•	•	•	•
3.0		14	11	18	•	•	•	•	•	•	•	•	•
3.5		7	7	7	48	•	•	31	•	•	•	•	•
4.0		3	3	7	11	48	•	26	31	31	•	•	•
2.0	500	•	•	•	•	•	•	•	•	•	•	•	•
2.5		48	•	•	•	•	•	•	•	•	•	•	•
3.0		11	31	18	•	•	•	•	•	•	•	•	•
3.5		7	7	7	14	•	•	84	84	•	•	•	•
4.0		7	3	3	11	31	•	26	31	31	•	•	•
2.0	600	•	•	•	•	•	•	•	•	•	•	•	•
2.5		56	•	•	•	•	•	•	•	•	•	•	•
3.0		14	31	31	•	•	•	•	•	•	•	•	•
3.5		11	11	31	48	•	•	•	•	•	•	•	•
4.0		7	3	3	7	7	•	31	31	77	•	•	•
2.0	800	•	•	•	•	•	•	•	•	•	•	•	•
2.5		•	•	•	•	•	•	•	•	•	•	•	•
3.0		31	31	48	•	•	•	•	•	•	•	•	•
3.5		11	11	31	31	•	•	•	•	•	•	•	•
4.0		7	11	17	7	48	•	•	•	•	•	•	•
2.0	1000	•	•	•	•	•	•	•	•	•	•	•	•
2.5		•	•	•	•	•	•	•	•	•	•	•	•
3.0		•	•	•	•	•	•	•	•	•	•	•	•
3.5		31	48	•	•	•	•	•	•	•	•	•	•
4.0		14	11	18	48	•	•	•	•	•	•	•	•

[a] From Kooiman (1977).   • = more than 100 days.

Many microbial species can adapt to unfavorable growth conditions, i.e., to low pH, or low a_w, or low preservative concentration. The selection of benzoic acid-resistant yeasts under use conditions is well recognized. For example, *Saccharomyces baillii,* at pH 3.0 and in the presence of 40% w/w sucrose, has a benzoic acid tolerance of approximately 700 mg/kg, but the organisms can adapt to benzoic acid concentrations of up to 1200 mg/kg in the same circumstances. The data given in Tables 22.9 and 22.10 relate to nonadapted yeasts; with adapted yeasts the preservative concentrations required will be much higher.

The risk of adaption/selection of resistant types should be reduced as far as possible by the following methods:

1. Good manufacturing practice, i.e., use of hygienic equipment and strict attention to factory hygiene.

2. Avoidance of raw materials that have been treated with preservatives (such as benzoic or sorbic acid) because these may introduce adapted organisms into the factory.

3. Checks on end products and stored batches for the occurrence of adapted yeasts.

Oxidative yeasts (*Pichia membranefaciens, Debaryomyces, Rhodotorula* and *Geotrichum* [*Oospora*]) are more sensitive to inhibition by sulfite than are the fermentative types, e.g., *Hansenula* and *Saccharomyces* (Rehm and Wittmann, 1962). The antimicrobial activity of sulfite is based on its ability to combine with microbial metabolites or enzymes. The reaction rate and equilibrium of some metabolic pathways may be disturbed by the formation of sulfonates and bisulfite addition compounds. Sulfite may also disrupt -S-S bonds in enzymes and form addition compounds with nicotanamide adenine dinucleotide (NAD) (Dupuy, 1959). Only free sulfite is antimicrobial. The sulfite added to a product may decline during storage by oxidation to sulfate in products such as grape juice or by reversible binding in sugar-containing drinks or preserves, particularly those that have been heated (Ingram and Lüthi, 1961). For further information, see Volume I, Chapter 10.

Sulfite is generally more effective at low-pH values against acid-tolerant bacteria than against fungi (Rehm and Wittmann, 1962).

Yeasts are relatively resistant to inhibition by carbon dioxide. Growth of *Saccharomyces chevalieri* can be demonstrated at CO_2 concentrations above 3.9 v/v, corresponding (at 20°C) to a pressure of more than 3.66 kg/cm^2 (Payne and Périgo, 1960). The antimicrobial activity of CO_2 is strongly affected by the following:

1. The sugar concentration. Sugar protects against inactivation by CO_2.

2. The pH value and the presence of organic or inorganic acid. Microbial inactivation increases in proportion to the decrease in pH and/or the increase of undissociated acid.

3. The initial number of contaminants. Although CO_2 is an effective agent when yeast numbers are low, it will not always prevent spoilage when the initial numbers are high.

Carbon dioxide inhibits the development of molds in soft drinks mainly because of the resulting anaerobic atmosphere. Carbonation is frequently applied in combination with benzoic acid (to about 75 mg/kg of the soft drink) and/or a heat treatment.

For further information on CO_2, see Volume I, Chapter 10.

The effectiveness of a preservative, added at a certain concentration to a particular soft drink or preserve, depends not only on the pH and water activity but also on the presence of other growth inhibitors. Such mixtures of preservatives may reinforce or reduce their separate growth-inhibitory activities, i.e., may work synergistically, additively, or antagonistically. The mode of action of such combinations of preservatives is illustrated in an "isobole diagram" (Fig. 22.4), which correlates the inhibitory concentration of one compound with that of a second compound when they are used together. When, for example, 60% of the minimum inhibitory concentration (MIC) of a compound is combined with 10% of the MIC of the second, the compounds act synergistically. When 40% of the MIC of the second compound is used, they act additively. A still higher proportion

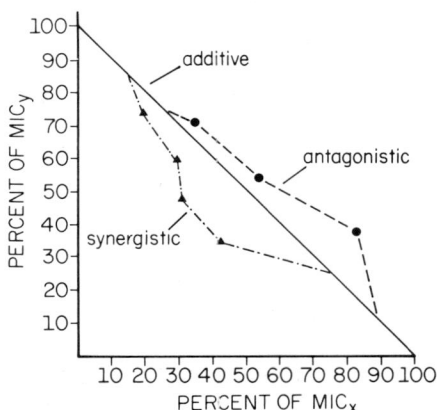

Fig. 22.4. An isobole diagram showing the relation between the Minimum Inhibitory Concentration of compound x (MIC_x) and that of a second preservative (MIC_y), when the two compounds are combined at various concentrations. (From Rehm, 1959.)

of the second compound causes antagonism. Additive activity occurs with some combinations of different parabens or sorbic acid with benzoic acid. Synergistic activity occurs with some combinations of benzoic acid or sorbic acid with sulfite, benzoic acid with formic acid, and sorbic acid with sodium chloride (see also Volume I, Chapter 7). Antagonistic activity has been reported for combinations of benzoic acid and boric acid against *Escherichia coli,* although these compounds are synergistic in inhibiting *Aspergillus* species.

Yeasts are, in most cases, the organisms that can limit shelf-life, and preservatives, therefore, are intended primarily for their control. Pasteurization should, however, be aimed at the inactivation of the most heat resistant spoilage organisms, i.e., the molds.

4. Combination of Pasteurization and Preservatives

Combination of pasteurization with the addition of preservatives allows lower heat treatment and lower preservative concentration. An example of such a combination is a low-calorie fruit spread (Kooiman, 1977). Such a product with a pH of 3.8–4.0 and containing 30% *w/w* sucrose can be made by the following methods:

1. Pasteurization, i.e., heating for 1–5 min at 80°C.
2. Addition of 800–1200 mg/kg sorbic acid if not pasteurized.
3. Application of a combination of a milder heat treatment (i.e., hot filling at 65°C) with 600 mg/kg of sorbic acid.

The exact conditions for such a combination of preserving systems depend strongly on the product concerned.

5. Process Monitoring

After the establishment of formulation and processing parameters the following procedures should be used for process monitoring:

1. Temperature measurements during storage of raw materials, filling, and/or pasteurization.
2. Controls on carbonation by routine measurements of CO_2 pressures in packed products.
3. Measurements of pH, concentration of solids, filling volume, product viscosity, etc. to ensure that product has been correctly formulated.
4. Controls of integrity of closures of bottles, cans, glass jars, or other packaging materials.
5. Good sanitation practices to prevent the establishment and subsequent multiplication of microorganisms on the equipment. Good practices also prevent the adaptation of microorganisms to high concentration of preservatives.

6. Microbiological tests, i.e., the procedures for assessing the microbiological condition of the manufacturing plant and of the raw materials, i.e., by making standard plate counts, counts of yeasts and molds, and direct microscopic examination, as described in Microorganisms in Foods, 1 (ICMSF, 1978), or in Compendium of Methods for the Microbiological Examination of Foods (Speck, 1976a).

7. Applying the diacetyl test, which is an additional tool for checking the sanitary conditions during manufacturing of raw materials (especially citrus and apple juices and concentrates; Murdock, 1967, 1968). This test detects diacetyl and acetylcarbinol, which are end products of bacterial growth.

8. Storage tests on formulated products to establish if organoleptic changes occur on storage and for initial determination of microbiological stability.

End-product analyses will detect only severe processing defects and should be made only when the examining agency does not have access to the manufacturing operation. Control is best based on good manufacturing practices, i.e., properly established processes and formulations together with adequate supervision and monitoring of the entire manufacturing operation.

IV. CHOICE OF CASE

For an understanding of "Case" see Table A.1 and the brief description in Appendix IV. For more details, see the second book of this series (ICMSF, 1974).

Only raw materials (and not the final product) should be analyzed microbiologically on a routine basis. For this purpose only "utility" tests need to be made—i.e., for numbers of yeasts, molds, and lactic bacteria. For this, Case 1 would apply, because the ingredients after being combined will be subjected to conditions that would tend to destroy the microorganisms.

23

Cereals and Cereal Products

I. INTRODUCTION

To a hungry world, direct ingestion of cereals is the most efficient way to obtain nutrients for energy and growth from scarce food resources, and all peoples use more of this commodity than of any other. More than half eat rice as their main food. Therefore, producers, processors, the public, and governmental authorities should be aware of the spoilage, adulteration, and public health problems of these basic foods. This chapter covers only the major grains in commercial production, whose technology is well advanced, and considers only in passing the subsistence farming, hand threshing, and small-scale processing that are so important in the developing nations (see Iizuka, 1957, 1958; Day, 1974; Hoover, 1974). The microbiology of fermented Oriental and Indian cereal products has been reviewed by Beuchat (1978). Cereals are also the basis for most animal feeds, but this chapter covers only human foods.

A. Definitions

Cereal grains include wheat, corn, oats, soybeans, durum, rye, rice, barley, millet, and sorghum.

Propagules are mold cells that are capable of forming colonies on a suitable menstruum (asexual or sexual spores, or fragments of hyphae).

Field fungi are those that grow on grains in the field, where the water activity is in the range 0.95 to nearly 1.00.

Storage fungi are those that can grow at the low water activity levels found in grains during storage (0.7 to 0.9).

Fatty Acid Value (FAV) is the number of milligrams of KOH required to neutralize the acids from 100 g of cereal, dry basis.

Grade is a system for classifying grains by the percentage of abnormal, shrunken, or damaged kernels. There are six grades, in decreasing order of quality, Grades 1 through 5, and "sample." For example, Grade 2 allows 0.2% (wheat or corn) or 0.5% (soybeans) to be "heat" (mold) damaged.

Batter is a mixture of flour and milk or water (usually combined with other ingredients such as sugar, salt, eggs, etc.) whose consistency permits it to pour easily or drop from a spoon. Batters are termed thin or thick, according to their consistency.

Sponge is a batter to which yeast is added. In the sponge process, the yeast is allowed to work in a batterlike mixture before the other ingredients are added.

Dough differs from batter in that it is stiff enough to be handled. In the straight-dough process, all ingredients are added together.

B. Important Properties

The carbohydrate from cereals is our most important energy source, but in addition, cereals contain protein, fat, and fiber (Table 23.1) as well as minerals and vitamins—especially the B group, D, and E. The cereals are, of course, equally nutritious to microorganisms, but microbial activities can be controlled simply by drying to a sufficiently low a_w (i.e., below 0.70). Bacteria are rarely able to grow, and molds can grow to a limited degree, at the marginal moisture levels at which cereal grains and their products are handled in temperate zones. In the humid tropics, mold growth is a serious problem.

TABLE 23.1

Proximate Analysis of Representative Food Grains [a]

Cereal grain	Carbohydrate (%)	Protein (%)	Fat (%)	Fiber (%)	Water (%)
Wheat (whole, hard spring)	69	14	2.2	2.3	13
Corn (field, dry)	72.2	8.9	3.9	2	13.8
Rice (brown, raw)	77.4	7.5	1.9	0.9	12
Oats (rolled dry)	68.2	14.2	7.4	1.2	8.3
Rye (whole grain)	73.4	12.1	1.7	2	11

[a] Adapted from Watt and Merrill (1963).

C. Methods of Handling and Processing Cereal Products

Some cereal grains are eaten after simple cooking, but most are processed in some way. Harvesting in developed countries is nearly all by mowing and threshing machines (combines), whereas in some of the developing countries, mowing is still largely by hand scythe, and threshing by animal or human labor and the wind. Within a few days after harvest, either at the farm or rice paddy, or after transport to a central collecting station, the grains are dried to enhance their keeping quality. Then over a period of up to several years, grains are transported throughout the world and stored in bulk, without further treatment. Most of the grain is milled into flour or meal, then mixed into doughs for baking into breads or pastries or for drying as pasta. Flours also become a basic ingredient of a wide variety of sauces, gravies, sausages, meat loaves, canned foods, and confections. Some grains are precooked and partially milled to make breakfast foods, snacks, and infant foods.

II. INITIAL MICROFLORA (MICROFLORA OF GRAINS IN THE FIELD)

All cereal grains are exposed in the field to a wide variety of organisms from dust, water, diseased plants, insects, soil, fertilizer, and droppings of animals. The numbers and types of microorganisms that are found depend on their resistance as well as on the kind of soil, fertilizers, insects, rodents, and especially weather conditions during and immediately after harvest. Moist, hot weather at this time encourages microbial growth on grains.

The external surfaces of grains at harvest contain hundreds of microbial species (Semeniuk, 1954); only a few of these can invade the grain kernel itself. As a measure of kernel damage, and to evaluate future damage, most investigators sterilize the kernel surface, usually with a solution containing 200 ppm of available chlorine. Then they lift the pericarp (of wheat) or the hull (of barley, oats, or rice) to examine the internal flora. In the field, fungi invade the kernels while they are developing on the plant or after they have matured but before they are harvested. In wheat and barley, the principal invasive field fungi are *Alternaria, Fusarium, Helminthosporium,* and *Cladosporium. Alternaria* is present in nearly 100% of wheat kernels (under the pericarp) and of barley kernels (under the hull). But this minimal invasion is not visible to the naked eye and does no further damage unless the harvested grains are kept moist (Christensen, 1965, 1978).

During cultivation and to the point of husk removal, Southeast Asian rice carries a wide variety of mold species of the genera *Alternaria, Curvularia, Fusarium, Nigrospora, Piricularia, Chaetomium, Sphaeropsis, Aspergillus, Penicillium, Rhizopus, Helminthosporium,* and *Cephalosporium,* as well as bacteria of genera *Pseudomonas, Enterobacter, Micrococcus, Brevibacterium,* and *Bacillus* (Iizuka, 1957, 1958; Majumder, 1974).

Aspergillus flavus may sometimes invade corn kernels in the field after they have been damaged by insect larvae (Christensen and Kaufmann, 1977a).

The fungi that invade starchy cereal grains in the field require kernels of a high moisture level to grow—usually at least 22 to 25% on a wet basis, or 28 to 33% on a dry basis ($a_w = 0.95$ to nearly 1.00) (Christensen and Kaufmann, 1974). When the grains mature, their water content drops, and in any case, they are normally dried within a few days after harvest so that growth by these fungi ceases. If the grain remains moist, growth will continue and discolor the kernel, weaken or kill the embryo, or cause shriveling of the seed; a few field-mold species may form mycotoxins (Christensen, 1965; Christensen and Kaufmann, 1974). Losses from molds are especially high in hot, humid climates (Ito *et al.,* 1971; Majumder, 1974).

During storage of grains containing <13% water, field fungi slowly die, the rate of death depending on the water content and the temperature of the grain (Christensen, 1965).

On grains at harvest time, mold propagules vary from none to several hundred thousand per gram, and bacteria from a few thousand to millions per gram. Bacteria are usually from the families Pseudomonadaceae, Micrococcaceae, Lactobacillaceae, and Bacillaceae (Frazier, 1967).

Fecal indicators are at a low level in field grains, unless there is considerable animal activity in the fields. *Bacillus subtilis* (*mesentericus*) and *B. cereus* are present in small numbers, but *Staphylococcus aureus* or *Salmonella* have not been found in the various surveys that have been made of grains in the fields of Western countries. It is to be expected that occasional spores of *Clostridium perfringens* or *C. botulinum* would be found, since they are present in some soils.

Certain bacterial groups are nearly always present, as shown in the following list:

1. Psychrotrophic bacteria, numbering from 10^4 to $> 10^5$ per g.
2. Actinomycetes, up to $> 10^6$ per g.
3. Aerobic sporulating bacteria, from 10^0 to 10^5 per g. Catalase negative bacteria and aerobic spore formers are irregularly present.

Yeasts occur in great variety on grains, but present no serious problems during storage as long as moisture is limited.

III. HARVESTING, TRANSPORTING, AND STORING GRAINS

A. Effects of Processing on Microorganisms

In great grain-producing areas of the world, the grains are mowed, threshed, and loaded into trucks in one continuous operation. Then they are dried to a moisture level that will keep molds from growing. They are then transported and stored in elevators or bins until trucks, rail, or ships transport them to further storage or mills. In some instances, the grains are heated before use to reduce microbial populations; sometimes they are washed in chlorinated water, or scoured dry then "tempered" in chlorinated water, before the milling operation.

In Asia and Africa, where the bulk of grains is still grown on small farms and harvested by hand (NAS/NRC, 1978), the traditional procedure has been to take advantage of the sun and wind to dry grains. But artificial drying reduces the water content more efficiently and makes the farmer independent of the weather. Drying at 40°–80°C destroys bacteria and molds in proportion to the temperature and time applied (Spicher and Weipfert, 1974). A temperature of 40°C destroys many mold spores and temperatures in the range 65°–80°C destroy all of them (Kopeikovskii and Kostenko, 1963).

In spite of the evident need for driers, they have not received wide acceptance in Asia and Africa because of their high cost, their often unsatisfactory performance in inexperienced hands, and their lack of adaptability to handle several types of grain. Thus, farmers still depend on the wind and the sun and are unwilling, therefore, to plant new types of grain that mature quickly and often must be harvested during the rainy season (NAS/NRC, 1978).

Field molds that survive drying or that recontaminate the dried grain cannot grow in the low-moisture kernels. But during storage, molds remain viable and inactive for many months. Survival is longest at low temperature and low-moisture levels (Christensen, 1978; Andriyevs'kyi and Bondarchuk, 1976).

But the field molds eventually do die (Lutey and Christensen, 1963), so that after long storage and transport in international trade they are at inconsequential levels—a fortunate fact that reduces the hazard of distributing plant pathogens across international boundaries (Wallace and Sinha, 1975).

Molds capable of growth at lower water activities soon take over; the sources of these are the trucks and particularly the storage bins. All grains, therefore, are inoculated to varying degree with fungi capable of growth upon mishandling. Table 23.2 shows the increased incidence of the storage fungus *Aspergillus glaucus* from field to terminal storage bin. The subsequent vulnerability of the grain shipments to spoilage will vary to some degree with the amount of such contamination. The degree of contamination differs among grain-producing countries (Wallace and Sinha, 1975).

Dust, generated each time grain is handled, is a major source of storage fungi spores in the environment of storage facilities and mills. It also contaminates grain directly, for it is frequently collected and returned to the grain. In one study, dust from handling of normal U.S. corn contained up to 33 billion spores for each tonne of grain handled. The number of particles in tailing dust was estimated at 1.7×10^9/g, and the mold spore concentration was 27×10^5/g. This represented 1 spore/62 dust particles. Handling of moderately moldy grain would contribute much higher levels of spores to the dust (Martin and Sauer, 1976).

In Southeast Asia (for example, Thailand and Burma) rice stalks are cut by hand about 30 cm above the soil, then bundled, and laid on the ground to dry. Sometimes the lower grain heads are still submerged in water at the time of harvest. After 7 to 20 days, the imperfectly dried bundles are carried to a nearby area where they are threshed either (1) by hand by striking them against a table or basket, or (2) by the hooves of water buffaloes. The grains are swept up into baskets for storage and

TABLE 23.2

Incidence of *Aspergillus glaucus* **in Wheat at Various Stages of Harvest and Transport** [a]

Source of sample	Number of kernels tested [b]	Percent of kernels yielding *A. glaucus*
Heads of standing shocked and windrowed wheat	2050	4.8
Combines (Mower-thresher)	1000	4.8
Country elevators, new crop	800	50.0
Trucks from country elevators, unloading at the terminal	575	64.0

[a] Adapted from Christensen and Kaufmann (1974).
[b] Cultured without surface disinfection.

further drying in farm houses or crude ventilated storehouses. Storage in the husk (paddy) protects the kernel to some degree from insects, rodents, and fungi (NAS/NRC, 1978). The baskets are then transported to a local godown, where the rice is husked, polished, and packed into 100-kg gunnysacks. (Approximately 60% of the rice is polished in godowns near the paddy field and 40% in centralized mills.) The operator frequently pours various grades of rice onto the floor of the godown to mix them appropriately to meet purchase specifications (Iizuka, 1957, 1958).

In India, the quality of rice is often improved by "parboiling"—that is, generally by treating the paddy (rice in husk) with hot water or steam. This process improves the rice because it (1) makes the husk separate more readily; (2) retains more amino acids, vitamins, and minerals in the grain; (3) hardens the grain, improving its resistance to attack by insects; (4) retains more solids during cooking; (5) destroys lipase, that would otherwise hydrolyze the oil; and (6) improves the yield by 1–2%. The principal disadvantage is that during the treatment the water content rises to 45–50%, so that to avoid rapid mold growth and fermentation, the paddy must be dried promptly to 16–18% moisture, and within a few days to 14–16% (NAS/NRC, 1978).

During storage of rice, the field fungi gradually die, and are replaced by those capable of growing at low a_w. They are, in approximate decreasing order of importance as follows: *Aspergillus* spp., *Penicillium* spp., and *Streptomyces diastaticus* (Iizuka, 1957, 1958). The principal strains are *A. glaucus* and *A restrictus,* both known for their ability to grow at low a_w (Ito *et al.,* 1973). But among the bacteria, *Pseudomonas* is the most frequent in both Asian and Spanish rice (Iizuka and Ito, 1968).

In North America, harvested full-moisture wild rice is often stored in small piles in the open or under refrigeration so that a natural fermentation can proceed. After fermentation, the rice is parched (dried) by the sun or by artificial means until it contains 8–14% moisture. Dried rice is hulled manually or mechanically to remove the outer coat (hull) from the kernel, then scarified to improve water penetration during cooking. Finally, the rice is graded according to kernel size and then packaged (Goel *et al.,* 1970).

Freshly harvested wild rice carries a heavy load of microorganisms: aerobic plate count up to 1.6×10^9/g; molds up to 5×10^5/g. During fermentation, the bacterial level increases markedly, and the molds increase slightly (Goel *et al.,* 1972). From 90 to 99% of the microorganisms are bacteria, primarily *Pseudomonas,* but coliform group and streptococci are present erratically and in smaller numbers. Parching (heating) to dry the grain destroys most microorganisms except *Bacillus* spores; the aerobic plate count drops > 99% (see Table 23.3). Re-

TABLE 23.3

Microbial Content of Raw Fermented Wild Rice, Parched at 65.6°C, and Hulled [a]

	Before hulling		After hulling	
Microbial group	Before parching per g ($\times 10^3$)	After parching per g ($\times 10^3$)	Rice per g ($\times 10^3$)	Hulls per g ($\times 10^3$)
Aerobic plate count	1,800,000	130,000	1,400	34,000
Psychrotrophs	980,000	410	1.6	80
Coliform group	63,000	64	1.7	37
Streptococci	16,000	1,200	3	500
Yeasts	73	1	0.03	1.4
Molds	8,300	3,400	3.9	1,100

[a] From Goel *et al.* (1970).

moval of the hull also removes most of the microorganisms still alive after parching. Cooking, as required in the kitchen, reduces the microbial count further, to about $10^2/g$ (Goel *et al.,* 1970).

In pure culture studies, *Pseudomonas fluorescens, Flavobacterium solare,* and *Acinetobacter/Moraxella* (*Achromobacter*) produced a satisfactory tea-like odor in wild rice, comparable to that of the naturally fermented product (Frank *et al.,* 1976).

B. Spoilage of Grains

1. Water Activity (a_w)

Grain spoilage molds are primarily those capable of growth at minimal a_w levels; they include none of the field fungi. For example, the field penicillia have a minimum a_w for growth near 0.86, whereas the storage penicillia have minima near 0.81–0.83 (Mislivec and Tuite, 1970). Table 23.4 shows the minimum values for a_w permitting growth of the more common storage fungi when incubated near their optimum growth temperatures.

Sporulation also depends on a_w. *Aspergillus glaucus* and other osmophilic species sporulate at a_w as low as 0.64–0.70, whereas *Penicillium* species have a wide a_w range, 0.80–1.00, and some of the Mucorales require high a_w, 0.93–1.00 (Snow, 1949).

These data can be applied to the various cereal products, as shown in Table 23.5. *Aspergillus halophilicus* is halophilic, as the name implies, requiring 10–15% NaCl or 55% sucrose in media for its isolation (i.e., a_w about 0.70). *Aspergillus restrictus, Sporendonema sebi,* and *A. glaucus*

TABLE 23.4

Minimum Water Activity (a_w) for the Growth of Common Storage Fungi at Their Optimum Temperature of Growth [a]

Fungus	Minimum a_w for growth
Aspergillus halophilicus	0.68
A. restrictus	0.70
Sporendonema	0.70
A. glaucus	0.73
A. candidus	0.80
A. ochraceus	0.80
A. flavus	0.85
Pencillium (depending on species)	0.80 to 0.90

[a] Adapted from Christensen and Kaufmann (1974).

grow at a_w 0.78–0.80, where most of the other storage molds cannot grow (Christensen and Kaufmann, 1974). The absolute lower limit of a_w that will permit growth of storage molds over a period of 2 years at a temperature of 21°–27°C is 0.70 (i.e., an interseed Equilibrium Relative Humidity—ERH—of 70%; Christensen, 1956).

In Table 23.5, the percent water must be expressed in an approximate way, because of various factors that affect the relationship between a_w and the percent moisture of grains—for example, their maturity, and therefore the amount of solids and the percentage of oil. See also, Vol. I, Chapter 4.

2. Temperature and Oxygen

Growth of molds depends on temperature, oxygen, and water activity. If two are optimum, growth will occur at extremes of the third; but the growth range in the third is more limited when the other two factors are not optimum (Semeniuk, 1954; Mislivec and Tuite, 1970). Table 23.6 relates both a_w and temperature to the growth of molds in practical laboratory and field tests. Refrigeration of grains (for example, at or below 5°C) permits their storage without damage for prolonged periods even when their water content is too high for safe storage at ambient temperatures (Hodges *et al.*, 1971). For example, soybeans with 14–14.3% moisture will not show mold growth for several years at 5°–6°C but will mold in a few weeks at 30°C. At limiting levels of moisture and temperature, the effect may simply be a prolonged lag period, followed by a slow growth rate. Thus, soybeans with 13% moisture were stored at 5°C for 150–170 days without damage, but at 480–500 days, invasion of

TABLE 23.5

Water Relations of Storage Microorganisms, on Various Cereal Substrates [a]

Microorganism	Minimum water activity	Minimum percent water permitting growth on						
		Wheat, barley, oats, rye, corn, sorghum	Rice		Soybeans	Sunflower		
			Rough	Polished		Seeds	Meats	
Aspergillus halophilicus, A. restrictus	0.65–0.70	12.5–13.5	12.5	14	12–12.5	—	—	
A. glaucus, Sporendonema sebi	0.70–0.75	14.5–15.0	13.5	15	14–14.5	9.5	6	
A. chevalieri, A. candidus,	0.75–0.80	15.0–15.5	14.5	15.5	14.5–15	10.5	7	
A. ochraceus, A. versicolor, A. nidulans								
A. flavus, Penicillium citreoviride, P. citrinum, A. versicolor	0.80–0.85	18.0–18.5	15	16.5	17.0–17.5	11.5	8	
A. oryzae, A. fumigatus, A. niger, P. notatum, P. islandicum, P. urticae	0.85–0.90	19.0–20.0	16.5	17.5	18.5–19.5	13.5	9	
Yeasts, bacteria, many molds	0.95–1.00	22–24	18	—	20–22	—	—	

[a] Adapted from Christensen and Kaufmann (1974, 1977a), Kinoshita and Shikata (1965), Papavizas and Christensen (1957), and Ito et al. (1971).

TABLE 23.6

Moisture Levels and Temperatures Allowing No Growth and Growth of Storage Molds in Grains

Grain	No mold growth occurred			Mold growth occurred			Reference
	Percent water	°C	Time (days)	Percent water	°C	Time (days)	
Wheat	15–15.5	5, 10	365	16–16.5	5, 10	365	Papavizas and Christensen (1958)
	16–18	−5	570				
Corn (maize)	<17.5	35	32	16.5	25–35	32	López and Christensen (1967)
	—	—	—	15.5	25	—	Sauer and Christensen (1969)
	14.2 and 15.5	10	365	14.2 and 15.5	25	365	Sauer and Christensen (1968)
	18.5	5	120	18.5	25–30	7	Christensen and Kaufmann (1977a)
Soybeans	13–14	5	150–170	13	5	480–500	Christensen (1967)
				13–14	25	<480	
	14–14.3	5–6	Several years	14–14.3	30	few weeks	Christensen and Kaufmann (1977a)
Rice	15	4–7	—	15	10–42	—	Kinosita and Shikata (1965)
	13.5	20–25	120	14.5	20–25	120	Fanse and Christensen (1966)
	14	5–15	465	16	25–31	465	Christensen (1969)
	14	25	150	14.5	30	60	Ito et al. (1971)
Sunflower Entire achene	10.5	23.9–35	84	14.7	35	84	Robertson and Thomas (1976)
Meats	9.0–15.7	<5	—	>11	21	—	Christensen (1971)
	6	—	—	>7	—	—	Christensen (1972)

kernels had begun (Christensen, 1967). Conversely, the rate of mold growth increases in approximate proportion to the water content above these minimal levels (Christensen and Gordon, 1948). However, since each mold has its a_w range for growth, there may be a selection of strains that reflects this (Tables 23.4 and 23.5). For short-time storage, higher moisture levels and higher temperatures can be tolerated. However, such treatment makes the lot a poor risk for further storage (Christensen and Kaufmann, 1964).

Based on considerations such as these, Table 23.7 lists the recommended moisture levels for principal grains to be held in short- and long-term storage.

Recent changes in agricultural technology in the United States have resulted in the widespread practice of harvesting high-moisture corn by picker-sheller. The corn frequently has a moisture level in excess of 20% and sustains considerable physical damage in the harvesting. Such conditions are ideal for rapid mold development. After harvest, the corn must either be dried to a moisture level safe for storage or be ensiled and used for animal feed. Several firms in various areas of the world are refrigerating the moist corn, then feeding it directly to animals, or sending it directly to mills for human use. However, even under refrigeration, moist corn will eventually mold so that the product must be handled promptly (Kurtzman and Ciegler, 1970).

Reduced access to oxygen has been somewhat less promising than low

TABLE 23.7

Recommended Maximum Moisture Levels for Short-Term and Long-Term Storage of Principal Grains [a]

	Maximum moisture for storage stability (% water)	
	Short term, 1 year	Long term, 5 years
Wheat	13–14	11–12
Corn	13	11
Barley	13	11
Oats	11	10
Soybeans	11	10
Rapeseed	5–7	—
Rice	14	—

[a] From Golumbic (1965), Sinha and Wallace (1977), and Ito et al. (1971).

water activity and low temperature as a means of inhibiting mold growth in grains. Although molds are primarily aerobes, many can grow at slow rates even without demonstrable oxygen. Anaerobic conditions, especially with preservatives, have been successful in France (Pelhate, 1976) and the United States in preserving moist grains for about 2 weeks at 20°C (Meiering *et al.,* 1966), but the practice has not been applied commercially. This type of storage is equivalent to ensiling of animal feeds, which depends on naturally occurring lactic organisms to reduce the pH by formation of lactic and other organic acids, thus preserving the product against molds and other undesirable microorganisms. For example, moist rice stored under these conditions will sour from lactic bacteria and yeasts which grow to tremendous numbers (Kent-Jones and Amos, 1957).

3. Heating

A moisture content of 10–12%, safe for ambient temperature storage of any grain whose mass is the same temperature throughout, is not safe under the influence of temperature gradients within the bulk. This is particularly true in geographic areas with wide seasonal temperature changes. Temperatures differing within the bulk by only 0.5°–1°C can initiate trouble because moisture evaporates from the grain in a warm section of the bin and condenses in a cool section until the moisture level of this local area is high enough to permit molds to grow (Christensen and Kaufmann, 1977a; Wallace *et al.,* 1976).* For example, if the outside air temperature is much lower than that of the grain, air will pass downward through the grain near the cold outside wall of the bin (or the cold plates of a ship's hold), then pass to the center of the bulk, and rise until it contacts a cool section near the top. An area within 50 cm of the top surface eventually may contain grain whose water content is 20 or even 27%. If the grain is warm when the bin is being filled, the natural insulation of the grain will prevent the center of the bulk from cooling, even in a severe winter (Wallace and Sinha, 1962). As spring approaches with its warmer temperatures, and as the water content slowly increases to 15%, the first molds to grow are *Aspergillus halophilicus,* then *A. restrictus* and *A. glaucus.* Once the grain is at this critical moisture level, the problem begins to be self-perpetuating, for the respiration of the grain and of the mold itself contributes additional moisture to raise the water activity high enough to permit growth of *A. candidus, A. flavus, A.*

---

* Moisture migration is less likely to occur if the grain is in bags, except in the bottom layers. In the primitive conditions of developing countries, grains in bags may continue to dry, whereas bulk stores may deteriorate. Nevertheless, under good commercial conditions, bulk storage is preferred.

ochraceus, and others. By the time these are growing on 10% of the kernels in the moist mass, heating starts. Before this time, growth is so slow that heating is not measurable. As the area warms, additional water forms and evaporates, passing into the grain just above the hot spot, so that the area of decomposition continues to spread within the mass, with no external evidence of trouble (Christensen and Kaufmann, 1964, 1974). The hot spot always spreads outward and upward, never downward through the bulk (Christensen and Kaufmann, 1978), although a hot spot can appear anywhere within the bulk (Wallace and Sinha, 1962). The critical factor is moisture content, so that a pocket of moist grain, from whatever cause, becomes the focal point of the heating. Some operators mix very dry grain with damp grain in order to obtain a safe overall moisture content, but the mixing must be perfect, to void damp pockets.

Eventually penicillia, which require higher water activities, begin to grow, and finally, when the water content exceeds 20%, yeasts and bacteria become active. Inside the mass, conditions are anaerobic or micro-aerophilic, so that lactics grow and ferment the grain. Upon acidification, yeasts are favored, followed by acetic acid bacteria that oxidize alcohols to acids. If this occurs, further fungal growth may be inhibited (Bulla *et al.,* 1978).

Differences in the local environment (inoculum, temperature, type of grain, a_w, redox potential) dictate which microorganisms will grow and which end products will form (Semeniuk, 1954). The size of the inoculum dictates the rate of temperature increase. When the inoculum was 200,000 spores of *Aspergillus flavus* per gram, the temperature rose to 45°C in 2 days; when it was 0.2 per gram, it took 9 days (Christensen and Gordon, 1948).

Each organism raises the temperature to as high a level as it can endure (Christensen and Gordon, 1948). *Aspergillus candidus* and *A. flavus* can raise the temperature to 50°–55°C and keep it there for weeks. Then thermophilic fungi may take over and raise the temperature to 60°–65°C. At this point the grain darkens and may even look charred. Thermophilic bacteria may then grow, raising the temperature to 70°–75°C, producing oxidizable hydrocarbon compounds of low ignition temperature (Milner *et al.,* 1947). In rare instances chemical oxidation then may ignite these compounds. Spontaneous combustion is more common in oily seeds, such as soy, cottonseed, flaxseed, and sunflower than in the starchy cereal seeds (Milner and Geddes, 1946).

Areas directly under the filling spout of the bin may become packed with "fines," which are small pieces of grain and dust. This mass, because of its denseness, is especially vulnerable to heating (Christensen and Kaufmann, 1977b).

4. Insects

Insects such as the granary weevil, *Sitophilus granarius,* aggravate a grain heating problem because they transfer molds and bacteria (Harein, 1969; Majumder, 1974). The insects' own respiration adds to the heating and to the production of moisture. Fumigating for insect control also reduces mold invasion (Table 23.8). However, fumigation has no lasting effect on mold growth or mycotoxin production (Vandegraft *et al.,* 1973). See also Section III,D for interrelations between molds and insects.

5. Kernel Invasion by Molds

Mold invasion can discolor the germ or the whole kernel. Moldy grain produces a musty odor and poor gluten quality in flour and breads made from it. It produces low germinability and inferior malt, so that brewers and distillers will not accept it. Even invisible mold, confined to the inside of the seed will reduce the expected storage time. Some kernels may even contain mycotoxins (Christensen, 1956; Christensen and Kaufmann, 1974, 1977a; Semeniuk, 1954).

Extensive invasion into seeds is by the way of the germ or embryo, generally first visible as hyphae on the surface of the embryo or between the embryo and the surrounding pericarp. These hyphae become visible

TABLE 23.8

Effect of Controlling Insect Infestation in Grains on the Microflora after 3-Months Incubation at 28° to 31°C [a]

Commodity	Treatment	Insects/100 g	Moisture (%)	Microbial count/g ($\times 10^3$)	
				Molds	Bacteria
Green gram	Fumigated	0	11	0	120
	Control	350	40	30	250
Bengal gram	Fumigated	0	11	0	26
	Control	504	45	30	110
Red gram	Fumigated	0	10	0	120
	Control	282	14	6	250
Coriander	Fumigated	0	8	0	70
	Control	400	13	10	120
Wheat	Fumigated	0	9	0	40
	Control	340	14	10	80
Sorghum	Fumigated	0	10	1	26
	Control	525	35	480	64

[a] From Majumder *et al.* (1965).

long before there is any obvious damage to the rest of the kernel. Then sporophores appear on the surface of the germ and in spaces within the germ prior to any externally visible growth. Eventually other tissues of the seed are invaded, whereupon mold and spores are visible externally. Damage such as scratches, cracks, or crushing from harvesting machinery permits rapid invasion into otherwise well-protected surfaces. In beans and soybeans, the pods protect the seeds until they are removed. Cottonseeds can be infected internally through the point of attachment. This type of infection is undetected by external examination (Christensen, 1978; Christensen, *et al.,* 1949).

6. Chemical Changes

During spoilage, certain chemical changes take place. The most important is the production of high fatty acid value (FAV) as a result of mold lipase activity on the glycerides of the oils. Even when molds and their enzymes are destroyed by heat, dry bran forms fatty acids very slowly, indicating that the reaction is not entirely from molds (Loeb and Mayne, 1952). Nevertheless, the FAV can be used as an objective measurement of mold damage (see Section III,E).

Increases in FAV during grain storage are proportional to increased moisture content, increased time of storage, increased temperature, and increased invasion by fungi. Fatty acid value increases much more rapidly and to a higher final level in fine particles of corn than in broken kernels, and much more rapidly and to a higher final level in broken kernels than in sound corn (Sauer and Christensen, 1969).

The respiration process of mold growth is similar to that of the grain itself. Therefore, the production of carbon dioxide occurs at a stable, slow rate in clean grains, but increases above this baseline level with the invasion and growth of molds. The dry-matter loss can be approximated by measuring the carbon dioxide lost to the atmosphere; this correlates well with the amount of mold activity (Steele, 1969). The grain is generally rejected as food by the time mold growth has caused 1 to 2% weight loss, but the point of rejection depends on local habits, and degree of hunger (Saul and Harris, 1979).

7. Microscopic Examination and Culturing

The past history and a prediction of the future keeping time of a lot of grain can be ascertained by studying the mold damage in the laboratory. For example, the growth of *Aspergillus halophilicus* in kernels of stored wheat indicates the grain has been stored for at least 6 months at a moisture content of 13.8–14.3% at a temperature of 20°–30°C. The growth of *A. restrictus* within the kernels of wheat or corn indicates that the grain has been stored for several months at a moisture content of 14–15%, at 20°–30°C. The presence of *A. flavus* in kernels of corn undamaged by insects, indicates that the moisture content must have been at least 18%, because *A. flavus* will not grow at lower moisture levels (Christensen, 1978). If externally visible mold spoilage is already moderately advanced, the grain is a poor risk for further storage (Christensen and Kaufmann, 1977a). If the numbers of mold spores on the external surfaces of a grain are very low, the grain is a good storage risk, even at high moisture levels. In one study (Papavizas and Christensen, 1958), samples of wheat carefully selected as free of storage mold spores did not mold in 19 months at 5° or 10°C even though the moisture content was 16–18%. Samples inoculated with mold spores became moldy in 12 to 19 months.

C. Pathogens and Toxins

1. Mycotoxins

Mycotoxins, due to the growth of molds, are the most important of the microbial health hazards from cereal products. Probably the first mycotoxin to be recognized was ergot, formed by *Claviceps purpurea* in grasses and grains, particularly rye. Ergotism occurred in the Rhine Valley in the year 857 and has been reported a number of times since then (Mirocha, 1969). During the nineteenth century, outbreaks of alimentary toxic aleukia (ATA) from moldy grains were described in Russia, and toxic reactions after consumption of moldy rice were reported in Japan. During World War II, many Russians died from eating moldy millet and wheat that had been permitted to remain in the fields over the winter (Ayres, 1972). Extensive studies by Russian scientists showed the fungi responsible to be *Fusarium* and *Cladosporium* (Hesseltine, 1967). In 1940, Japanese who consumed moldy rice developed nervous disorders that were later attributed primarily to *Penicillium islandicum* (Ayres, 1972). Both this organism and *P. citronum,* also known to produce a mycotoxin, are regularly sought in laboratories of the Food Board and Ministry of Public Health of Japan (Iizuka, 1958). In the 1960s, several investigators reported an association between moldy feeds and animal disease (Lancaster *et al.,* 1961; Forgacs and Carll, 1962; Allcroft and Raymond, 1966).

The events just reported have stimulated extensive investigations into mycotoxins in feeds and foods. It is evident that many of the molds on cereal grains have the ability to produce mycotoxins. In one study of 109 isolates of *Alternaria*—the principal field fungus in grains—79, or 72%

of the total, were toxic to rats. Of 170 isolates of *Fusarium* (the imperfect stage of *Gibberella*), 100, or 59% of the total, were also toxic. Many strains of *Helminthosporium* and *Aspergillus* also have been shown to be toxic, and about 50% of more than 80 isolates of grainborne *Penicillium* were toxic (Christensen *et al.,* 1969). These studies have been repeated many times, with consistent results; namely, that many isolates from grains when inoculated into media or damp grain, then incubated, produce mycotoxins. There have been more than 60 mycotoxins described to date and many more "toxic" antibiotics discarded without study (Hesseltine, 1967). The majority of mycotoxins studied to date can be produced by molds isolated from cereal products. Furthermore, as shown in Table 23.9, the principal mycotoxins found in nature come from cereal products. A comprehensive list of molds and their mycotoxins was published in the second edition of the first book of this series (ICMSF, 1978). Such lists are also available in many of the reviews listed at the end of this discussion of mycotoxins. These reviews also cover the history of mycotoxins, their chemical structure, and the symptoms of mycotoxicoses in animals and man.

TABLE 23.9

Selected Mycotoxins Found in Various Foods [a]

Mycotoxin	Foods found naturally contaminated	Foods supporting experimental production
Aflatoxins	Peanuts, corn, tree nuts, copra, cottonseed, rice Milk (transfer from animal feed)	Grains, fruits, meats, spices, cheese
Sterigmatocystin	Wheat, green coffee	Corn meal, shredded wheat
Ochratoxin	Barley, corn	Grains, peanuts, legumes
Patulin	Rice, apple juice	Rice
Penicillic acid	Corn, dried beans	Corn
Rubratoxins	Corn	Grains
Citrinin	Rice	Grains, rice
Tremorgenic toxin	Corn	Grains
Yellow rice toxins	Rice	Rice
Zearalenone	Corn, other grains	Grains, rice
12,13-epoxy-Δ^9-trichothecenes	Grains, rice	Grains, rice
Stachybotrys toxins	Oats, hay	Grains

[a] From Rodricks and Lovett (1976).

Consumption of any amount of mold growth on grains is a risk to human or animal consumers; the greater risk is to the animals because of the likelihood that they will eat larger quantities of moldy material. The incidence of mycotoxin-positive samples in grains destined for human consumption (i.e., Grades 1 and 2) is low; that in grains intended for animals (Grades 3, 4, 5, or Sample) is higher (Shotwell *et al.*, 1969a, 1970, 1974; Watson and Yahl, 1971; Lillehoj *et al.*, 1975; Ayres, 1972).

The temperature range for mold growth is not always the same as for mycotoxin production. For example, *Aspergillus flavus* grows between 6° and 46°C, but aflatoxins B1 and G1 are produced only between 8° and 37°C (Shotwell *et al.*, 1970). Furthermore, growth at 13°C is relatively rapid, but aflatoxin production is slow, so that there can be good growth without formation of aflatoxin (Schindler *et al.*, 1967). There is little comfort in this, however, since *Fusarium* produces a toxin at temperatures below freezing (Ayres, 1972), and a strain of *Penicillium martensii* produces penicillic acid between —4° and 32°C in corn with 25% moisture.

Nearly all the mycotoxin is to be found in obviously deteriorated kernels (dark color, visible mold, broken pieces). The broken pieces can be sifted out (Kinosita and Shikata, 1965); moldy and discolored kernels can be sorted out for discard by specially constructed machines (Ashworth *et al.*, 1968), or in countries where labor is very cheap, by hand.

Wild rice, improperly prepared, could develop mycotoxins. During the fermentation (Section III,A) the moisture level must remain in the range 35–45% to encourage bacteria, which grow more rapidly than molds. Watering and mixing will keep local areas from drying. A dry crust will support mold growth and development of mycotoxins, which are destroyed neither by the parching nor by the eventual boiling in the kitchen (Goel *et al.*, 1972; Frank *et al.*, 1975).

For further reading on mycotoxins in grains, see the reviews listed on page 78 in the second edition of the first book in this series (ICMSF, 1978), and the following: Bamburg *et al.* (1969); Wogan (1965); Wilson (1966, 1978); Marth (1967); Rodricks and Lovett (1976); Goldblatt (1969).

2. Bacterial Pathogens

Properly handled cereals are so dry that bacteria cannot grow on them. However, they can mechanically carry the viable cells of many pathogens if the grain is exposed to animal or human contamination. Insects, rodents, birds, or human beings may introduce *Salmonella, Escherichia, Shigella,* or *Klebsiella* (Brooks, 1969). Of these, *Salmonella* is the genus of principal concern. The sources may be animal contact in the field; trucks and railroad cars whose previous cargo was animals, meat scraps, tankage, fish meal, poultry or poultry products; insects, mice, rats, and birds in the mill; and human carriers. Once the microorganisms are dried to the low water activity of the cereal, they can remain inactive but viable almost indefinitely even though their numbers will decrease with time. Then, if the cereal is used in a moist food, growth of *Salmonella* occurs. Cooking the cereal will destroy *Salmonella,* but dust from raw, contaminated, dry cereal product may cross-contaminate another more sensitive food under process of manufacture or preparation.

We have seen no comprehensive surveys of mill-grade grains for salmonellae, but one survey of feed ingredients in the fiscal year 1966 in the United States revealed that 0.66% of the grain samples and 2.28% of the oilseed meal samples contained the organisms. In view of the large volume of cereals consumed compared with other foods, even a low incidence of salmonellae is a serious matter. This is particularly true in developing countries, where the direct consumption of cereals is much higher than it is in technically more advanced countries (Hobbs, 1968). In the middle 1960s samples from 10 mills in 10 geographic areas of the United States were analyzed, and no salmonellae were found (Hobbs, 1968). However, sampling to prove a microorganism to be absent is a very difficult approach, because a low-level, infrequent contamination would be missed (see ICMSF, 1974). It is likely that an occasional lot of cereal grain is contaminated.

Bacillus cereus food poisoning is commonly traced to rice dishes, primarily in Chinese restaurants (Schiemann, 1978). Spores survive the cooking, and germinate and grow if the rice is then held between 10° and 49°C for several hours. Rice should be prepared in small quantities, and refrigerated or held above 63°C before consumption (Gilbert *et al.,* 1974). Adding other foods to the rice speeds the growth of *B. cereus,* and, therefore, increases the hazard (Morita and Woodburn, 1977). But *B. cereus* poisoning is not, as many believe, entirely a cereal problem, for many outbreaks occur from other foods (Goepfert *et al.,* 1972).

D. Interrelations

The kind of mold that grows in a given area of a grain storage bin depends on its moisture, temperature, and oxygen requirements. Therefore, one can find different molds in various areas of the same bin (Semeniuk, 1954). For a discussion of the sequence of organisms found during spoilage of grains, see Section III,B,3.

It has been believed generally that the death of field fungi in stored grains depends only on moisture content, temperature, and length of storage. However, laboratory studies have shown that many storage fungi are antagonistic to field fungi. For example, some species of *Aspergillus, Chaetomium, Penicillium,* and *Actinomyces* are antagonistic to each other and especially to field strains such as *Cochliobolus satimus* or *Alternaria tenius* (Sinha *et al.,* 1969a,b; Wallace and Sinha, 1962). Conversely, *Gonatobotrytis* cannot grow in culture except in the presence of *Alternaria.* They generally disappear simultaneously as the grain ages (Sinha *et al.,* 1969a). In addition, on the one hand, aflatoxins behave as antibiotics against certain bacteria, especially *Bacillus* species. On the other hand, fungi and yeasts are not sensitive to the aflatoxins (Hesseltine, 1967).

Insects have a microflora that reflects the kinds of microorganisms in the surrounding environment. Those insects that enter a grain bulk from the outside serve as an inoculum source and transmit microbes to the grain. If the insects develop within the grain mass, they will contain many of the microbes originally associated with the grain. Thus, insects can be vectors of grain disease and can inoculate grains with spoilage organisms.

Microorganisms also may serve as food for insects. The foreign grain beetle *Ahasverus advena* and the hairy fungus beetle *Typhaea stercorea* are nearly always found in moldy grain but never in the absence of mold (Bulla *et al.,* 1978).

Insects are attracted by the warmth and biological products produced by the microorganisms (Golumbic, 1965; Christensen and Kaufmann, 1964), and their respiration, in turn, adds moisture and warmth to the grain (reviewed in Majumder, 1974). High populations of mites often occur together with fungi. One such, *Acarus immobilis,* feeds on *Cladosporium* and certain species of *Aspergillus.* The long-haired mite, *Glycyphagus destructor,* apparently feeds on various seedborne fungi, particularly *Penicillium cyclopium* (Bulla *et al.,* 1978). *Aspergillus flavus, A. ochraceus, A. niger,* and *A. candidus* inhibit egg deposition by the foreign grain beetle. When given a choice of several fungi, the female preferentially oviposits on *Penicillium citrinum* and *A. amstelodami.* Conversely, certain species of *Aspergillus, Alternaria,* and *Penicillium* elaborate antibiotics that can reduce insect growth and reproduction and can kill some insects (Bulla *et al.,* 1978).

E. Control

The best way to avoid microbial growth in grains is to keep them dry. Once they are dry and in storage bins, proper ventilation will remove moisture, prevent condensation, lower and equilibrate temperatures, and prevent heating. The air should be forced by fan, for natural aeration systems using the force of the wind are less satisfactory (Chohan *et al.,*

1972). The fan should be placed to pull air out of the bottom of the bin and should draw cool air into the grain at the top in such a way that the entire bulk is ventilated evenly at a rate of 0.1 ft^3 of air/min/bushel (1m^3 of air/min/12.44m^3 of grain) (Christensen and Kaufmann, 1977b). It is not enough to have a flow of air; the air must be dry ($< 75\%$ relative humidity) and cooler than the grain. Even in areas of high heat and humidity, granary operators in temperate zones can sometimes choose cool and dry days or nights for ventilation. This may not be so in the tropics, for moist warm air drawn through the grain will allow an increase in moisture content so that the grain will mold (Majumder, 1974; Christensen and Kaufmann, 1978). However, if the grain is dried thoroughly by artificial means, then placed in a tightly sealed container or bin, it can keep without spoilage (Majumder, 1974). In moist tropical areas, bins made of metal are less satisfactory than those made of concrete or other insulating materials such as mud, because the metal allows more rapid temperature changes and therefore more condensation of moisture (Majumder, 1974).

As a means of detecting a hot spot where spoilage has become rapid, operators should place thermocouples in strategic positions throughout the bin, not more than 2 m apart, including the area of the spoutline where the "fines" collect. Grain is a good heat insulator, so that even a minor rise in temperature at the position of a thermocouple may signal a hot spot nearby (Christensen and Kaufmann, 1977a,b).

Fumigation also helps control mold growth on grains by killing insects that damage kernels (See Table 23.8) but the effect is not long-lasting (Vandegraft et al., 1973). However, some gases (ammonia, sulfur dioxide, methyl iodide, chloropicrin, methyl bromide, and ethylene dibromide) destroy both molds and insects (Majumder, 1974).

Several investigators have reported experimental success with the use of chemical preservatives to control spoilage of moist grains. Preservatives for this purpose have not yet proved acceptable to the industry. Also, many proposed preservatives would be unacceptable to health authorities. Some that might eventually prove acceptable are propionic acid or calcium propionate (Majumder, 1974; Sauer and Burroughs, 1974; Bothast et al., 1975; Milward, 1976), sorbate or sorbate with carbon dioxide (Danziger et al., 1973; Pomeranz, 1957), acetic acid (Majumder, 1974; Mitsuda and Nakajima, 1977), ammonia (Bothast et al., 1973, 1975; Nofsinger et al., 1977), and propylene oxide (Matz and Milner, 1951).

Another way to minimize heating is to remove and discard the grains or fractions that are troublesome. The "fines"—dust and pieces of broken kernels—tend to settle beneath the entry spout in the bin. They should be removed and stored separately (Christensen and Kaufmann, 1977b).

Sifting and sizing the product to remove broken kernels, foreign matter, and dockage will help maintain quality and reduce susceptibility to mold because the broken kernels are often those most likely to contain mold (USDA, 1968; Kinosita and Shikata, 1965) and to block the passage of drying air.

For some products, such as rice, a sorting or picking operation will remove discolored, shriveled, or otherwise damaged kernels (USDA, 1968). There are machines available that remove darkened or fluorescent kernels at great speed, using a vacuum system to pick kernels from the mass and a light measuring system to detect abnormal kernels for discard (Ashworth et al., 1968). In areas of the world where labor is cheap, this sorting is a hand operation.

For use in the field there is also a portable ultraviolet (black light) lamp useful to detect fluorescent (damaged) kernels of corn in the ear (Christensen and Kaufmann, 1977a). If the mold growth is not far advanced, removing the hull will also remove the mold damage in the case of some seeds, for example, barley (Flannigan and Dickie, 1972).

Washing the grains in water to remove surface dust will also remove much of the microbial load, particularly if the water contains a chlorine compound at a concentration contributing 100–125 ppm available chlorine, and treatment lasts for 3 to 8 min (Kent-Jones and Amos, 1957). See Table 23.10.

Likewise, water to temper the grain (i.e., to raise the water content for more satisfactory milling characteristics) should contain a disinfectant such as chlorine, bromine, chlorine dioxide, or hydrogen peroxide (Hesseltine, 1968; Vojnovich et al., 1972). See also Volume 1, Chapter 14 for a discussion of chlorine and other disinfectants.

TABLE 23.10

Mold Counts (Propagules/g) on Wheat and Milled Products [a]

Grain fraction	Wheat sample number		
	1	2	3
Wheat before washing	2800	3100	1900
Wheat after washing	300	400	500
Tempered wheat	300	400	500
Bran	2600	1900	1200
Shorts (bran, germ, coarse meal)	3200	4700	2700
Patent (first class) flour	1800	3200	400

[a] From Christensen and Cohen (1950).

TABLE 23.11

Effect of Heating and Chlorinating Blight-Infected Corn (Maize) on the Microbial Level in Products Milled from It [a]

	Meal		Flour	
Treatment of corn sample	Bacteria/g	Molds/g	Bacteria/g	Molds/g
Untreated	60,000	200,000	49,000	150,000
0.1% NaOCl solution,				
1 min at 82.2°C	120	120	1,500	860
Indirect steam, 20 min	150	35	550	230
Direct steam injection, 1 min	120	5	2,600	330

[a] Adapted from Vojnovich et al. (1972).

Many investigators have suggested using heat to destroy microorganisms in grains. After heating, grains can be used to make flour that is acceptable as an ingredient in convenience foods. Table 23.11 shows the efficiency of heat near 100°C. Others have suggested less severe heating, but for a longer period. Table 23.12 shows the effect of treating wheat at 60°C for 1–4 hr.

Figure 23.1 shows how the counts of both bacteria and molds fall with storage of corn at 12% moisture at 26°, 37°, and 45°C for up to 25 weeks. Obviously, only adequately dried grain could be held at these temperatures without encouraging microbial growth instead of death.

A sanitation program to keep trucks and bins clean will do much to reduce the level of fungi in the grain; dust or fines should never be re-

TABLE 23.12

Effect of 60°C Heat on the Microbial Content of Wheat (15% Water) and Flour Made from It [a]

	Aerobic plate count/g	
Heat treatment (hr)	Wheat	Flour
Unheated control	1,800,000	49,000
1	250,000	1,800
2	61,000	330
3	36,000	120
4	6,000	200

[a] From Vojnovich and Pfeifer (1967).

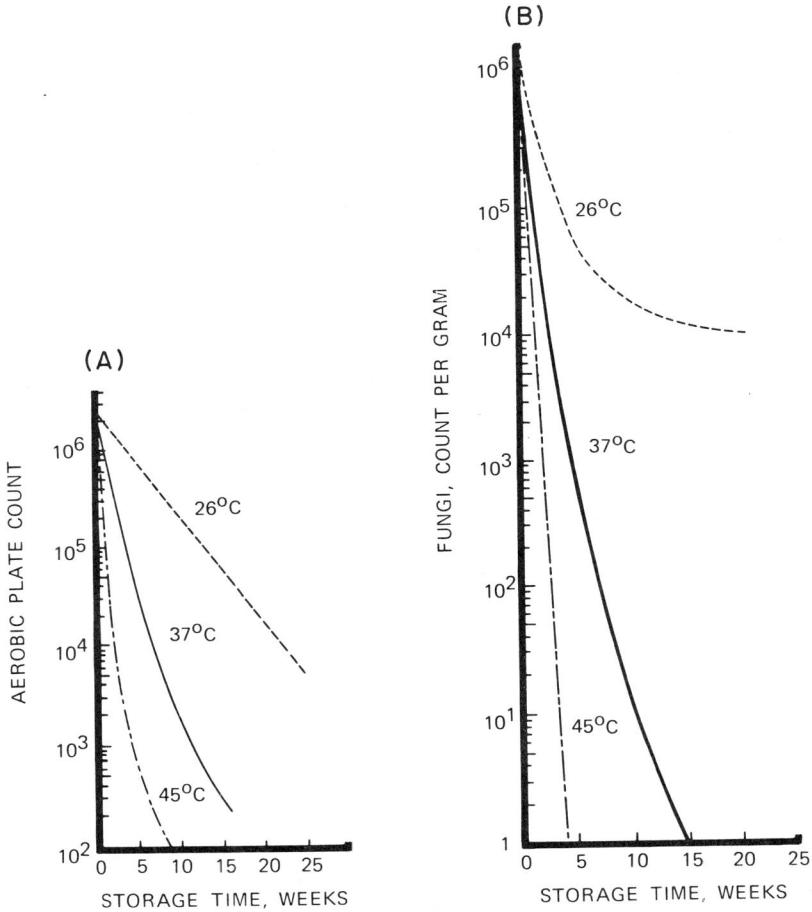

Fig. 23.1. Effect of heated storage on bacterial count (A) and fungal count (B) in blight-infested corn containing 12% moisture. (From Vojnovich *et al.*, 1972.)

turned to the grain. Cooking reduces the level of mycotoxins but generally not to zero. In the United States, the Food and Drug Administration accepts gaseous ammonia to destroy mycotoxins in animal feeds (U.S. Department of Health, Education, and Welfare, 1976c).

Ionizing irradiation at a level of 0.2 to 0.3 Mrad destroys molds that ordinarily spoil rice (Iizuka and Ito, 1968; Ito *et al.*, 1971). However, such a process is illegal in many countries (Vol. I, Chapter 3).

A laboratory for evaluating the condition of grain samples is an essential part of a good operation. Samples of 0.5 kg should be transported to the laboratory in moisture-proof containers. The laboratory should conduct the following tests (Christensen and Kaufmann, 1977a; Christensen, 1956, 1978; Semeniuk, 1954):

1. Determine the types and quantity of shriveled, mummified, or discolored kernels; discolored embryos, scutella, endosperms, or pericarps; bacterial crusts; fungus mycelia, or fungus fruiting bodies.

2. Disinfect the kernel surfaces with hypochlorite solution, and immerse the kernels in agar medium suitable for the growth of molds (see Christensen, 1946, for details on culturing molds from grain).

3. Determine the moisture content [see Christensen (1978) or Okwelogu (1979)].

4. Examine for and identify insects and mites.

5. Determine the Fatty Acid Value (FAV; mg KOH to neutralize the acids from 100 g wheat, dry basis).

Analysis for mycotoxins is not recommended for routine control. A new 4- to 6- hour test for chitin has been recommended to evaluate the amount of mold damage in grains. This test takes the place of mold plate counts (Donald and Mirocha, 1977).

Table 23.13 shows the normal levels of microorganisms in grains and other cereal products, based on industrial experience in the United States.

Grading is an excellent control procedure because it informs the buyer and/or user what the quality is and permits him to select for human consumption the cleanest and safest grains. Table 23.14 shows the percent of heat-damaged (primarily moldy) kernels permitted in various grades of U.S. corn (maize) and how these levels of damage relate to FAV.

IV. FLOURS, MEALS, AND DRY MIXES

A. Processing

Grains for milling are first washed, tempered, screened, and aspirated. The use of chlorinated wash water and tempering water, and the screening and aspirating of broken kernels and "fines" have been described earlier. These steps reduce the microbial level of the grains as they enter the milling operation.

The grains then pass through a milling and sifting sequence that separates the hulls (bran) and the germ and crushes the endosperm into flour. Corn is sometimes ground to flour, meal, or grits without a tempering step.

Soybeans are moistened, dehulled, flaked, extracted with organic solvent

TABLE 23.13

Normal Microbiological Profile of Cereal Grains and Cereal Products [a, b]

Product category	Normal microflora	Range (per gram)	Remarks
I. Raw cereal grains	Molds	10^2–10^4	1. Counts representative of "normal" grains in commercial channels; "mildewed" or "musty" or "spoiled" grain would obviously be beyond these ranges
	Yeasts and yeastlike fungi	10^2–10^4	
	Bacteria		
	Aerobic plate count	10^2–10^6	2. Grains are also routinely tested for salmonellae
	Coliform group	10^2–10^4	
	Escherichia coli	10–10^3	
	Actinomycetes	10^3–10^6	3. Related to amount of soil incorporated in grain sample
II. Flour(s), cornmeal, corn grits, semolina	Molds	10^2–10^4	1. Microbial counts in flour can vary from one storage period to another depending on moisture content and storage conditions; the final observed count is a function of original bioload, proliferation, and die off; counts frequently decrease during storage; however, increases have also been noted
	Yeast and yeastlike fungi	10–10^2	
	Bacteria		
	Aerobic plate count	10^2–10^6	2. Soy flours sometimes contain salmonellae; other flours rarely do
	Coliform group	0–10	
	"Rope" spores	0–10^2	
III. Breakfast cereals and "Snacks"	Molds	0–10^3	1. Cereals are additionally tested for *E. coli* and salmonellae
	Yeasts and yeastlike fungi	0–10^2	
	Bacteria		
	Aerobic plate count	0–10^2	2. Snacks are routinely tested for *Salmonella* and coagulase positive staphylococci
	Coliform group	0–10^2	

(continued)

TABLE 23.13 (continued)

Product category	Normal microflora	Range (per gram)	Remarks
IV. Refrigerated and frozen doughs	Molds	10^2–10^4	1. Yeast counts reflect inoculum intentionally added as part of the product formulation and do not represent "contamination"
	Yeast and yeastlike fungi	10^5–10^6	
	Bacteria		2. Routinely tested for salmonellae.
	Aerobis plate count	10^2–10^6	
	Coliform group	10–10^2	3. The special case of "buttermilk biscuits" that contain 10^2–10^4 lactic acid producers have been omitted from this compilation
	E. coli	0–10	
	Psychrotrophs	10–10^3	
	coagulase-positive staphylococci	10–10^3	
V. Baked goods	Molds	10–10^3	1. Routinely tested for salmonellae and coagulase positive staphylococci
	Yeast and yeastlike fungi	10–10^3	
	Bacteria		
	Aerobic plate count	10–10^3	
	Coliform group	0–10^2	
	E. coli	0–10	
VI. Soy protein	Bacteria		1. Quantitative ranges reflect both original contamination and growth during storage of intermediate moisture products
	Aerobic plate count	10^2–10^5	
	Coliform group	10^2–10^3	2. Routinely tested for molds and yeasts, salmonellae, and staphylococci
	Escherichia coli	0–10^2	
	Psychrotrophs	10^2–10^4	
	Clostridium perfringens	0–10^2	

Category	Test	Count	Remarks
VII. Pasta products	Bacteria		1. Wide ranges in bioloads of these products reflect differences between egg-based and macaroni-type products that are all included in "pasta" category
	Aerobic plate count	10^3–10^5	2. Routinely tested for molds and yeasts, salmonellae, and staphylococci
	Coliform group	10–10^2	3. Soy protein products intended for anaerobic storage (e.g., the canning industry) should be routinely tested for total thermophilic spore formers, flat sour organisms, putrefactive and thermophilic spore formers, and sulfide spoilage organisms
	E. coli	0–10^2	
VIII. Dry cereal mixes	Molds	10^2–10^5	1. Routinely tested for salmonellae and coagulase positive staphylococci
	Yeasts and yeastlike fungi	10^2–10^5	
	Bacteria		
	Aerobic plate count	10^2–10^6	
	Coliform group	0–10^4	
	E. coli	0–10^3	

[a] From Hobbs and Greene (1976).

[b] Table based on "routine" quality control tests normally performed on various items of specified category; data represent industrywide experience; data presented as "orders of magnitude" for illustrative purposes only.

[c] Milled rice produced commercially in the southern area of the United States in 1954 was virtually free from internal infection by fungi and free from internal infection by bacteria, yeast, and actinomycetes. Results from investigations of 1968 relative to the conditions that determine the prevalence of individual kinds of fungi in stored rice were similar to those observed in the 1954 study.

TABLE 23.14

Correlation of USDA Corn Grade with the Proportion of Heat-Damaged
(Primarily Moldy) Kernels and with Fat Acidity Values (FAV) [a]

USDA Grade	Heat-damaged (moldy) kernels (%) [b]	FAV
1	0.1	17
2	0.2	26
3	0.5	36
4	1.0	46
5	3.0	53
Sample	>3.0	125

[a] From Golumbic (1965) and Christensen and Kaufman (1977a).
[b] Other types of damage also contribute to FAV.

to remove oil, caked, and ground to flour. Soy-protein concentrate and
isolated soy protein are subjected to multiple extractions, acid precipita-
tions, and drying. Such procedures destroy the original flora, but also
offer opportunities for further contamination.

Diet foods and dried soups containing cereals sometimes include protein concentrates, dried eggs, yeast, sugar, flavorings, dried dairy products,
and nuts. The microbiology of the mix becomes that of the ingredients
(Karlson and Gunderson, 1965). If all ingredients and the mix remain
dry, there is no growth and little death of the flora (Hobbs and Greene,
1976). See also Chapter 27.

High-grade (patent) flours, which are produced from the endosperms
near the middle of the kernels, are separated from the outer skins (hulls,
pericarps), which are the site of most microbial contamination. Thus, the
flours always have a lower microbial level than the grains from which
they are made (Table 23.12); the bran (hull) has relatively high levels
of microorganisms (Table 23.3). But whole-grain flour or meal, or de-
germinated corn meal or grits, will contain the same amount of fungus-
invaded tissue as the original seeds (Christensen et al., 1969).

Sound clean grains, especially those properly screened, washed, and
tempered, contain few microorganisms. However, contact with mill
machinery introduces contamination of a quantity and variety that is
affected by the degree of cleanliness of the mill equipment. One set of
samples of scrapings and product contained levels of mold shown in Table
23.15. There is a correlation between microbial levels in the product and
the mill sanitation, as shown in Table 23.16. Some investigators have
found even higher mold levels—up to 3.4×10^6/g of flour residues on

TABLE 23.15

Number of Molds Found on Wheat, Static Residues on Milling Machinery, and Flours, Chosen as Typical of Data from 10 U.S. Mills [a]

Grain or milling fraction	Mold propagules (per gram)
Incoming dirty wheat	250
Wet wheat from first temper conveyor	380
Static buildup on first temper conveyor	382,000
Screenings	70,000
Clean tempered wheat from temper bin	250
Aspiration liftings from prebreak system	12,000
First low-grade flour	300
Second clear flour	410
First break flour	470
Fifth break flour	450
Patent flour	210

[a] From Hobbs (1968).

equipment (Christensen and Cohen, 1950) and up to 10^8 in mill dust (Semeniuk, 1954).

Flour with 12% or less water will not support microbial growth (Hesseltine and Graves, 1966). However, water from any source will encourage microbial growth in flour and in residues on mill machinery. Such residues can become damp from high atmospheric humidity, from con-

TABLE 23.16

Effect of Mill Sanitation on the Microbial Content of Flour [a]

Mesophilic plate counts	Number of microorganisms per gram of flour	
	Mills with superior sanitation [b]	Mills with substandard sanitation [b]
Nutrient agar, 37°C	7,300	825,000
Potato dextrose agar, 22°C	3,300	12,000
Mesophilic spores	95	600
Molds	2,200	11,000

[a] From Thatcher et al. (1953).

[b] Average findings from five Canadian mills whose sanitation was judged superior and from five judged substandard, based on the amount of insect and rodent infestation found by inspectors.

densate on cool surfaces, from improper cleanup procedures (Graves *et al.*, 1967; Hesseltine and Graves, 1966), or from insect activity (Thatcher *et al.*, 1953).

The conditions in mill operation that permit activities of insects would most likely contribute to microbial growth. In fact, the level of insect fragments in various areas of the mill correlates with the numbers of microorganisms found (Thatcher *et al.*, 1953). The large numbers of organisms of various microbial groups found in live and dead insects are tabulated in Table 23.17, and the numbers washed from the surface of sieved insect excreta, in Table 23.18. The insects and the static flour from mill machinery introduce many molds not found in the original grain.

Fumigation to kill insects in infested flour raises the level of bacteria dramatically within 24–72 hr, presumably because the dead insects decompose (H. E. Bauman, personal communication).

Wheat flour is usually bleached by oxidizing agents such as nitric oxide, chlorine, nitrosyl chloride, benzoyl peroxide (Frazier, 1967), or more commonly chlorine dioxide (Thatcher *et al.*, 1953). Bleaching reduces microbial numbers to some degree, but spores are little affected (Table 23.19).

Heat and chemicals have been tried specifically to reduce the microbial levels in flour, but heat alone is so effective that chemical treatment seems superfluous. Heating for 45 min in an oven set at 130°C has no adverse effect on functional properties, yet reduces microbial levels markedly (Table 23.20). Table 23.12 shows low microbial levels in flour made from wheat that had been heated at 60°C before milling.

Similarly, heating to 82°C for 1 min or to 66°C for 3 min reduced markedly the bacterial and fungal content of corn grits, corn meal, and corn flour, particularly if the grain was first cleaned and washed. Propylene oxide was most effective in destroying thermophilic bacteria in dry-milled fractions destined for canning (Vojnovich *et al.*, 1970, 1972).

Unheated flours in commerce contain 10^2 to 10^6 bacteria and 10^2 to 10^4 molds per gram (Hesseltine and Graves, 1966; see also Table 23.13). Malts, because they have been incubated with high moisture, have 10^6 or more bacteria and yeasts per gram (Frazier, 1967).

Microorganisms in dry ($< 12\%$ water) flour gradually die during prolonged storage. Among the bacteria, *Flavobacterium* and *Pseudomonas* predominate initially whereas *Citrobacter* (*Paracolobactrum*) later gains ascendancy (Graves *et al.*, 1967). By the time flours have been packaged, transported, and stored, the levels are a few hundred to a few thousand bacteria and 50–100 mold spores per gram. Graham and whole wheat flours at retail contain more—8000 to 12,000 bacteria, and an equivalent number of molds (Frazier, 1967).

TABLE 23.17

Bacteria and Fungi Recovered from Mill Insects by Washing [a]

	Average number per insect							
	Living insects				Dead insects			
Insect species	Nutrient agar plates (37°C)	Potato dextrose agar (22°C)	Mesophilic spores	Fungi	Nutrient agar plates (37°C)	Potato dextrose agar (22°C)	Mesophilic spores	Fungi
Confused flour beetle	2,300	<100	970	70	18,000	34,000	40	700,000
Flat grain beetle [b]	<100	<100	1,600	260	900,000	<100	50	17,000
Black flour beetle	760,000	180,000	3,000	140,000	17,000	1,600	1,300	5,000
Mediterranean flour moth larvae	—	—	—	—	2,000,000	1,500,000	300	46,000

[a] From Thatcher et al. (1953).
[b] This insect is much larger than the others tested.

TABLE 23.18

Numbers of Bacteria and Fungi Washed from Insect Excreta Pellets Obtained from Dead Stock in Flour Mills [a]

| Mill number | Number of microorganisms per gram of sieved excreta concentrates | | | |
	Nutrient agar drop plate count (37°C)	Potato dextrose agar drop plate count (22°C)	Mesophilic bacterial spores	Fungi
3	1,100,000	800,000	43,000	106,000
32	50,000,000	10,000,000	40,000	560,000
27	51,000,000	10,000,000	2,060,000	1,300,000
11	2,100,000	13,000	1,400,000	20,000
14	10,000,000	20,000	1,300,000	10,000

[a] From Thatcher *et al.* (1953).

TABLE 23.19

Effect of Bleaching Agents on Microbial Content of Flours from 32 Canadian Mills [a]

| Microbial group | Average number of bacteria and fungi (per gram) | | Percent of mills showing reduced microbial numbers in bleached flour |
	Unbleached flour	Bleached flour	
Mesophiles and psychrotrophs			
Drop plate counts			
Nutrient agar (37°C)	59,400	24,000	81
Potato dextrose agar (22°C)	28,000	11,000	
Bacterial spores	413	101	66
Molds	3,060	1,570	78
Thermophiles			
Total aerobic spores	9.5	11.4	70
"Flat sour" spores	5.0	6.1	62
Anaerobic spores	9.0	9.8	72

[a] From Thatcher *et al.* (1953).

TABLE 23.20

Effect of Heat on Microorganisms in Flour [a, b]

Organism	Inoculum (per g)	Count after treatment (per g)
Bacillus subtilis (spores)	2,400,000	2,400
	14,300	0
Staphylococcus aureus	1,390,000	250
	860,000	0
Aspergillus flavus	710,000	0
	450,000	0
Escherichia coli	300,000	0

[a] From Wiseblatt (1967).

[b] A layer of flour 6.5-mm-deep was heated 45 min in an oven at 130°C.

The molds in finished flours are mostly *Aspergillus glaucus* and *A. candidus* (Table 23.21). The genera of yeasts found in flours differ markedly from those found in the wheat from which the flour was made, demonstrating again the importance of the mill as a source of the microorganisms. Yeasts become important only when the flour becomes wet.

The bacterial flora of flour is much more diverse than that of the wheat

TABLE 23.21

Molds Isolated from Flour [a]

Prevalent in most samples
 Aspergillus glaucus
 A. candidus
Prevalent in occasional samples
 Penicillium spp.
 Aspergillus flavus
Present in small numbers in some samples
 Aspergillus ochraceus
 A. *niger*
 A. *versicolor*
 A. *terreus*
 Rhizopus nigricans
 Mucor racemosus
 Hormodendron spp.
 Alternaria tenuis
 Helminthosporium spp.
 Fusarium spp.

[a] From Christensen and Cohen (1950).

from which it was made. Both wheat and the flour made from it contain many psychrotrophs, flat sour organisms, and thermophilic spore-forming organisms (Hesseltine, 1968). These are of particular interest to processors of canned and chilled foods in which flour is an ingredient. Soy protein, frequently used for canning, may contribute thermophilic spore formers, flat sour organisms, putrefactive anaerobes, and sulfide spoilage bacteria (Hobbs and Greene, 1976). Rope bacilli come from both insects and insanitary equipment; they are basically soil organisms.

B. Spoilage

For flour and corn meal, 12% water is a critical level, at or below which there will be no microbial growth, and the flour is completely stable. At 15% molds can grow; and at 17%, molds, yeast, and bacteria can grow (Frazier, 1967; Hesseltine and Graves, 1966). The rate of growth is proportional to the water activity and temperature (Kent-Jones and Amos, 1957). If the moisture level is high, as in a flour and water paste, bacteria will predominate because they grow faster than yeasts or molds. Acid-forming lactics will begin acid fermentation, followed by alcoholic fermentation by yeasts, if present, and then oxidation to acetic acid by *Acetobacter,* if present. This sequence is more likely in freshly milled flour than in stored flour because in the latter some of the organisms have died. In the absence of lactics, micrococci may acidify the paste, and in their absence *Bacillus* may grow producing lactic acid, gas, alcohol, acetoin, and small amounts of esters and aromatic compounds. It is characteristic of most flour pastes to develop an odor of acetic acid and esters (Frazier, 1967).

But more commonly, mold growth is predominant because the amount of water is limited. The spoilage sequence of flour whose water content is just above the critical level is similar to that of the grains. Moldy flours have lowered fat, high FAV, and musty odors. Gluten has impaired rheological properties, and yield of both gluten and starch is reduced. Doughs made with moldy flour require longer mix time, and the bread has low loaf volume and poor crumb grain (Pomeranz *et al.,* 1968; Daftary *et al.,* 1970).

C. Pathogens and Toxins

Mycotoxins are the most important health hazard in flours, meals, and dry mixes; salmonellae as a hazard are less important.

Mycotoxins formed in moldy grain can carry through into the flour even surviving a heating step or other procedure designed to kill the molds that formed the mycotoxins. In addition, moist flour and corn meal (> 14% water) will support mold growth in the same way that grains do, and mycotoxins can form (Seeder et al., 1969; Bullerman et al., 1975).

Of 70 molds isolated from flour and bread, 16 were Aspergillus, 48 Penicillium, and six were other genera. Fifteen of the 48 Penicillium species and one strain of Aspergillus ochraceus produced mycotoxins on laboratory media (Bullerman and Hartung, 1973).

Although salmonellae have been detected often in soy flour and sometimes in other oilseed flours, their incidence in wheat flour is very low. In 1958, in 2068 samples of wheat, wheat flour, and dust from several mills, none contained salmonellae (Anonymous, 1958). If salmonellae should enter flour from animal or human carriers, the very large volume of cereals and cereal products will dilute such contamination below the level of detection by a sampling process. However, laboratory experiments have shown that the organisms will remain alive in dry flour for several months (Dack, 1961).

Some cereal products intended for animal feeds are diverted into human food channels. Oilseed proteins are examples of this dual use (Elliott, 1967), and therefore a short discussion of such products is warranted. Table 23.22 shows the relatively high incidence of Salmonella in

TABLE 23.22

Incidence of Salmonella in Oilseed Meals and Alfalfa Imported into Sweden, September 1959 to December 1960 [a, b]

Type of meal	Number of lots analyzed	Salmonella-positive lots		Number of samples analyzed	Salmonella-positive samples	
		Number	Percent		Number	Percent
Cottonseed	10	8	80	525	37	7.1
Soya	18	9	50	914	25	2.7
Peanut	7	3	43	350	6	1.7
Linseed	3	1	33	150	1	0.7
Sunflower	2	1	50	100	1	1.0
Coconut	1	0	0	32	0	0
Alfalfa	8	4	50	391	11	2.8
Totals	49	26	53	2,462	81	3.3

[a] From Rutqvist (1961).

[b] Imports were from Argentina, Denmark, Dominican Republic, French West Africa, India, Indonesia, Iraq, Japan, Mexico, Peru, Poland, Portuguese East Africa, Syria, Turkey, USA, USSR, and West Germany.

oilseed meals imported into Sweden in 1959 and 1960. Similar findings on imported high-protein vegetable meals were reported by Norway, England, West Germany, and Denmark during the 1950s (Elliott, 1967).

In the early 1960s, animal salmonelloses were traced to rape seed meal from a Swedish extraction plant. Although this product was strictly for animal feeding, the finding is important also to human food protection programs. The meal was warm during transport in a closed conveyor from the drier to the bagging area, so that it continued to emit some moisture, which condensed on the cool walls of the conveyor cover. Salmonellae, whose original source was unknown, grew in the condensate, and dropped at intervals into the product flow.

The entire system was cleaned and treated with a chlorine solution; then a cooling system was installed near the drier; the entire conveyor system was heat insulated; and finally, hot air was introduced to keep the conveyor walls warm. No further salmonellae were isolated from the final product of this mill (Rutqvist and Waxberg, 1962).

In 1964, a series of 23 salmonellosis outbreaks occurred in mental institutions in the United States. Epidemiology demonstrated that the source was cottonseed meal and dried yeast that had been mixed with other dry ingredients as a diet supplement. At the mill in Texas from which the cottonseed meal came, 11 of 178 samples (6.2%) taken along the production line and from the final product were positive for *Salmonella*. A hot-air conveying system (82.2°C) to transport the flour to the sacking operation solved what was primarily another condensation problem (McCall *et al.,* 1966; Elliott, 1967).

D. Control

Grains should be fumigated to control insects, then sifted, washed, and tempered in chlorinated water. The grain for patent flour can be heated to reduce microbial levels. The milling machinery should be cleaned regularly to avoid the build-up of static material, where molds and insects thrive. Cleaning of all dry-product areas should ordinarily be done without using water. (See Vol. 1, Chapter 14.) The flour should contain no more than 14%, and preferably no more than 12% water, in order to inhibit microbial growth. The mill should have a program of rodent and insect control and should avoid condensation of moisture where it can drop back into the product or accumulate on the sides of bins, hoppers, conveyors, etc.

After boots and elevators in the flour system have been fumigated, they should be cleaned immediately to remove the dead insects, which otherwise will decompose and contribute large numbers of microorganisms to the flour.

Many processors who use flour as an ingredient for further processed foods have established microbiological specifications for evaluating products as they are received. Canners often set low specifications for sporulating organisms that can survive canning procedures and spoil the product. Table 23.13 lists the normal levels of various microbial groups in flour, dry mixes, and soy proteins.

V. DOUGHS

A. Processing

1. Standard Bread Dough

A typical bread contains wheat flour, water or milk, salt, fat, sugar, and yeast (usually *Saccharomyces cerevisiae*). Ingredients are mixed and the dough permitted to rise (ferment) for several hours at 24° to 29°C, then cut down and kneaded, and permitted to rise again before baking. During this fermentation period, which may last as long as 20 hr, the enzymes of the yeast attack the sugar, to produce carbon dioxide bubbles in the dough.

For economic reasons, the trend has been toward less fermentation time, even to the point of eliminating primary fermentation altogether. In conventional sponge-dough or straight-dough* bread production, the capabilities of bakers yeast are fully utilized, but with the development of high-speed production, some of the functions of the yeast have been taken over by machinery. For example, in the continuous mix process, dough development is accomplished mechanically in an oval chamber with counter rotating impellers. There is insufficient time for metabolic reactions necessary for flavor development. These breads, therefore, lack the flavor that only full microbial action can contribute (Sugihara, 1977).

Some microorganisms are more important than others in development of flavors. Yeast is the principal contributor. Bacteria (usually *Lactobacillus* species) contribute flavor in breads with fermentation periods longer than 8 hr (Kent-Jones and Amos, 1957). The lactic organisms produce short-chain fatty acids, which probably contribute to good flavor in small amounts but which would give undesirable flavors and aromas in large amounts (Robinson *et al.*, 1958).

---

* Bread may be made by the sponge process, in which the yeast is allowed to work in a batterlike mixture before combining with other ingredients, or by the straight dough process in which all ingredients are combined at one time.

Yeast cakes seldom spoil. Sometimes they may mold in moist hot weather or with prolonged refrigerated storage (Bennion, 1967). The industry is seldom bothered by wild yeasts that occur in small numbers in some dough (i.e., *Saccharomyces pastorianus, S. ellipsoideus, Mycoderma cerevisiae, Torula utilis,* etc.) because of the overwhelming abundance of the desirable yeast.

2. Salt-Rising Bread

This product is unique in that the leavening agent is *Clostridium perfringens* rather than yeast (Robinson, 1967). Even though this organism causes foodborne gastroenteritis, it is harmless when used as a leavening agent because it dies during baking. No food poisonings have ever been attributed to salt-rising bread (Dack, 1961). At least one baking company uses *Bacillus cereus* in the same way (Goepfert *et al.*, 1972).

3. Soda Crackers

The formulation for soda crackers is similar to that of bread except that baking soda is added. Yeast *(Saccharomyces cerevisiae)* is a principal ingredient, but since the industry began in 1840, manufacturers have relied on chance contamination by *Lactobacillus* species for acid formation in a 24-hr fermentation period. Recent investigations have revealed the predominance of *L. plantarum* and the secondary role of *L. delbruekii* and *L. leichmannii* in this function. Pure culture starters with these organisms will reduce the fermentation period to about 6 hr, maintain better quality, and permit controlled variations in flavor (Sugihara, 1977, 1978a,b).

4. San Francisco Sourdough French Bread

The traditional method for this sourdough bread is to use a starter built up from the previous batch every 8 hr every day, 7 days a week. It requires a high proportion of active starter sponge for each batch of new starter (about 40%) followed by a fermentation period of 7 to 8 hr at 27°C. The initial pH is 4.4 to 4.5, and the final pH is 3.8 to 3.9. Then the dough requires a starter that makes up about 11% of the final dough mix. This is followed by a similar 7- to 8-hr period of fermentation with a similar pH drop. The very high proportion of starter sponge contributes a massive inoculum and assures a highly acid environment that contains a substantial amount of acetic acid. The low pH and the acetic acid prevent growth of spoilage organisms and loss of the starter (Kline *et al.*, 1970).

The sourdough yeast is *Saccharomyces exiguus (Torulopsis holmii),* and the heterofermentative bacterium is *Lactobacillus sanfrancisco* (Sugihara, 1977). *Saccharomyces exiguus* does not ferment maltose, but *L. sanfrancisco* ferments only maltose. Thus, the two microorganisms do not compete for the same carbohydrate, a fact that probably has contributed to the survival of both organisms in the starter. The yeast is unusually resistant to acetic acid (Sugihara *et al.,* 1970).

5. Sour Rye Bread

Sour rye bread was first mentioned in 800 B.C., and since the fifth century B.C., dried starters made up of cakes of fermented whole grain have been available to bakers. The souring microorganisms are *Lactobacillus plantarum, L. brevis,* and *L. fermenti;* * lactic and acetic acids are the principal compounds formed. The addition of yeast speeds the fermentation process. This bread is generally made with self-perpetuating starter sponges, but pure cultures or the previously mentioned cake are available commercially (Sugihara, 1977).

6. Italian Panettone

In Italy, panettone dough is the basis for Christmas fruit cake, columba (Easter cake), breakfast rolls, and snack cakes. Its production is similarly based on a yeast *(Saccharomyces exiguus)* and one or more species of bacteria *(Lactobacillus brevis, Enterobacter,* and *Citrobacter).* The "madre" or mother sponge has been perpetuated for centuries in very clean surroundings by a special staff consisting in recent years of chemists and microbiologists (Sugihara, 1977). The preparation procedure is strikingly like that of San Francisco sourdough French bread.

7. Idli

Idli, a popular fermented bread in India, is prepared from rice and a legume called black gram mungo *(Phaseolus mungo).* The ingredients are soaked in water, combined and permitted to ferment overnight, then steamed and served hot. The leavening and souring organism is primarily *Leuconostoc mesenteroides,* with *Streptococcus faecalis* and *Pediococcus cerevisiae* playing secondary acidifying roles (Mukherjee *et al.,* 1965). Idli is unique in that the leavening action is solely from the activity of a lactic acid bacterium *(L. mesenteroides).*

8. Frozen and Chilled Dough

In western countries, particularly the United States, frozen or chilled raw or partially baked doughs for baking in the consumer's kitchen have become popular. There are more than 30 different products of which most are breads and rolls, cookies, pizza, and baking powder biscuits (Hessel-

* This organism is not recognized in the 8th edition of Bergey's Manual (Buchanan and Gibbons, 1974).

tine *et al.,* 1969), primarily packaged in hermetically sealed cardboard, plastic film, and metal containers. All refrigerated doughs are chemically leavened, usually by sodium acid pyrophosphate and sodium bicarbonate. Yeast action would otherwise continue during storage and burst the container (Lannuier and Matz, 1967; Lamprech, 1968). Whereas most doughs go promptly into the oven, these special doughs may be held for prolonged periods at temperatures permitting the growth of spoilage organisms. Therefore, the level and types of microorganisms from the ingredients and from the processing area are highly important. Sources of such organisms are the flour, dry milk, eggs, sugar, spices, water, flavors, and the dough-making equipment. Fungi and bacterial counts in the doughs are low unless these ingredients are highly contaminated (Hobbs and Greene, 1976). Usually the fungal flora of the dough is a reflection of that of the flour (Graves and Hesseltine, 1966). Bacterial levels of fresh dough products vary widely, from $10^3/g$ in dinner rolls to $10^8/g$ (lactic acid bacteria) in buttermilk biscuits. All fresh doughs contain wild yeasts (Hesseltine *et al.,* 1969).

B. Spoilage

Doughs in chill storage owe their stability to the following factors (Lannuier and Matz, 1967; Lamprech, 1968):

1. Conditions are anaerobic, therefore inhibiting the growth of molds and other aerobes.
2. Formulations are designed to attain a low a_w.
3. Refrigeration slows microbial and enzyme activity.
4. Manufacturers have strict microbiological specifications for ingredients and maintain faultless sanitation in the manufacturing operation.
5. The pH of most products is relatively unfavorable for most bacterial growth.
6. The leavening agent produces CO_2.

The lactic organisms in doughs, if well controlled, produce delicious sourdough breads. On the other hand, if they are permitted to grow for too long a time, as is often the case in refrigerated doughs, they may cause the package to leak, split open, or explode from gas pressure, revealing a sticky liquid or slimy dough with undesirable odor and flavor (Lannuier and Matz, 1967; Hesseltine *et al.,* 1969). Spoiled doughs reveal flora listed in Table 23.23.

Molds or yeasts are rarely involved in spoilage of refrigerated doughs,

TABLE 23.23

Microorganisms Isolated from Spoiled Doughs [a]

	Isolates	
Microorganisms	Number	Percent
Leuconostoc mesenteroides	390	34.5
Leuconostoc dextranicum	21	1.9
Lactobacillus spp.	597	52.7
Streptococcus spp.	45	3.1
Bacillus spp.	15	1.3
Micrococcus spp.	51	4.5
Gram-negative rods	13	1.1

[a] From Hesseltine *et al.* (1969).

even after 6 to 7 months of storage. The fungus counts are nearly always the same in stored doughs as in freshly prepared doughs (Graves and Hesseltine, 1966).

C. Pathogens and Toxins

Refrigerated doughs may contain *Salmonella* if ingredients such as eggs are contaminated. In a survey in 1962–63, of 145 samples of doughs analyzed, 17 (8.5%) contained *Salmonella*. But among the 17 positive samples, 10 contained eggs, and 4 contained other animal products (Foltz, 1966). The occurrence of *Salmonella* in eggs has been reduced drastically in the United States and other western countries since this 1962–63 survey was made. (See this Volume, Chapter 19.) Dried yeast preparations also sometimes contain *Salmonella*.

D. Interrelations

The San Francisco sourdough French bread process is successful because of a very close relationship between *Saccharomyces exiguus* and *Lactobacillus sanfrancisco*. The yeast cannot use maltose but uses other sugars; the bacterium can use maltose, but cannot use other sugars. Thus, there is no competition for an energy source. The bacterium acidifies the dough to a level that is inhibitory to other yeasts which might compete with *S. exiguus* (Kline and Sugihara, 1971; Sugihara *et al.*, 1971).

E. Control

Normal microbial levels for doughs are listed in Table 23.13. Their microbial quality can be maintained by using ingredients (particularly flour) of good quality, and by maintaining good bakery sanitation. Refrigerated doughs should be dated so that they are removed from the retail case before they spoil. The quality of special sourdough breads can be maintained by careful adherence to ancient traditional procedures or by pure cultures kept under laboratory control.

VI. BREADS

A. Effects of Processing on Microorganisms

At the end of the fermentation process, the dough contains massive numbers of yeasts, and a wide variety of other organisms from the ingredients. During baking, the internal temperature of the loaf approaches or slightly exceeds 100°C (Sugihara, 1977), so that all molds, yeasts, and vegetative bacterial cells are destroyed. Subsequent mold problems are exclusively from postbake contamination from the air, the slicing machine, and from cooling and wrapping equipment. Even a single slice of moldy bread may spread spores throughout an entire building in a few minutes. The air in the average bread manufacturing plant has 100 to 2500 spores/ m^3, but that where old bread is stored may have 10^5 spores/m^3 (Knight and Menlove, 1961).

Among the bacteria, only spores survive, of which the "rope" organism, *Bacillus subtilis,* is the most important. If the ingredients of the dough are heavily contaminated with its spores, some survival is to be expected during baking. The sources of the spores are multiple, but the highest inoculum may be from equipment that has been contaminated by previous batches of bread.

Tortillas (flat corn pancakes) are the staple food in all of Central America. Dry maize is cooked with limestone, soaked in water for about 14 hr, ground into a wet pasty flour, rolled and patted into a pancake, and then cooked for 4 to 5 min on a stove top or "comal." The uncooked tortilla often has excessive levels of a general flora (bacteria, yeasts, and molds) that is largely destroyed during cooking.

Tortillas should be consumed immediately after they are cooked, for they are still moist enough to support rapid bacterial growth (Capparelli and Mata, 1975).

B. Spoilage

Bread is sufficiently moist within the loaf to permit the growth of surviving spore-forming bacteria, including the "rope" organisms. However, molds are the more evident problem, because they grow on the surface where they can be seen. In the United Kingdom, as much as 1% of the annual bread production is lost to molds (Seiler, 1964). The molds most frequently found on baked goods are *Rhizopus nigricans, Penicillium expansum, P. stoloniferum, Aspergillus niger, Monilia (Neurospora) sitophila, Mucor* species, and *Geotrichum* species (Frazier, 1967). But there are many more. One investigation in India revealed 11 genera and 33 species of mold growing on bread (Kapooria and Genda, 1974).

The crust is rather dry and if the relative humidity of the atmosphere is <90%, molds will not grow on it (Frazier, 1967). Also, molds are relatively slow to develop, so that in dry climates, the surface of a slice of bread may dry before mold is evident. In a humid atmosphere, however, and especially on a loaf inside a wrapper, molds will grow rapidly. This is true especially if the bread is wrapped hot from the oven so that droplets of water condense on the inside surface of the wrapper (Kent-Jones and Amos, 1957; Frazier, 1967).

Bread, which has an a_w (see Volume I, Chapter 4) of about 0.95, normally has a shorter mold-free shelf-life than cake, whose a_w ranges from 0.75 to 0.90. Laboratory tests on bread show that increasing the salt or sugar concentration decreases the a_w and thus extends the mold-free shelf-life; addition of more water to the recipe has the opposite effect. Figure 23.2 shows how the water activity of cake affects its mold-free shelf-life.

Rope is a spoilage problem of bread, caused by a mucoid variant of *Bacillus subtilis* or *B. licheniformis* (*B. subtilis* has also been called *B. mesentericus* or *B. panis*). Almost any of the ingredients of bread may contribute the organisms, but flour and equipment that previously have been in contact with contaminated dough are the greatest offenders. The spores easily survive baking and germinate and grow within 36 to 48 hr inside the loaf to form the characteristic soft, stringy, brown mass with an odor of ripe cantaloupe. The bacteria are heavily encapsulated, which contributes the mucoid nature of the material. They also produce amylases and proteases that cause the breakdown of the bread structure. Conditions favoring the appearance of rope are (1) a slow cooling period or storage above 25°C, (2) pH above 5, (3) high spore level, and (4) moist loaf. The water activity inside a loaf is marginal for *B. subtilis* so that rope may appear in localized areas where the moisture content is high. However,

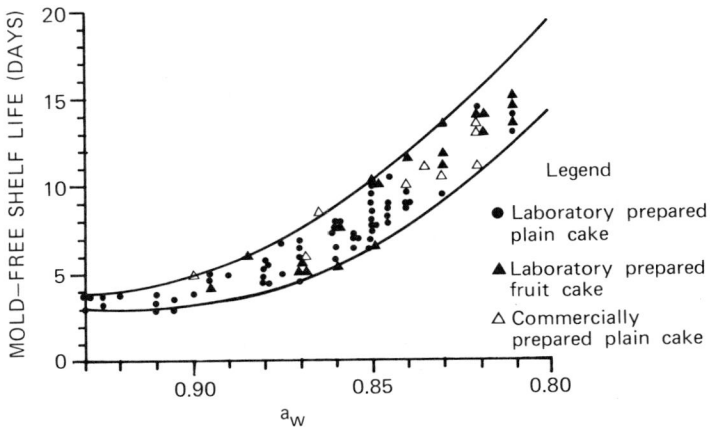

Fig. 23.2. The effect of a_w on the mold-free shelf-life of cake at 27°C. (From Seiler, 1964.)

rope is now rare, because calcium propionate, good sanitation, and good bakery practice keep it under control (Pyler, 1973; Graves *et al.,* 1967; Semeniuk, 1954; Farmiloe *et al.,* 1954).

Red bread is usually caused by Serratia *marcescens,* which may contaminate a product that is baked already and may grow if the moisture level is high enough. Molds such as *Monilia sitophila* may impart a pink color; *Geotrichum (Oidium) aurantiacum* can give the crumb of dark bread a red color (Frazier, 1967).

Chalky bread is characterized by white chalklike spots from the yeast-like fungus *Endomycopsis fibuliger* or a species of the fungus *Trichosporon.*

The chapati, a disc of soft, pliable unleavened bread, is the most common wheat food in India (Hoover, 1974). In tropical areas where the temperature is 40°–45°C, chapaties that are packaged in polyethylene pouches where the equilibrium relative humidity reaches 90 to 95% spoil after approximately 7 days due to the growth of aspergilli (Kameswara Rao *et al.,* 1964, 1966).

Tortillas are a very rich, moist medium for microbial spoilage which will take place within 24 hr under tropical conditions (Capparelli and Mata, 1975).

C. Pathogens and Toxins

Many of the molds that grow on bread are capable of producing myco-toxins either on culture media or on bread (Robinson, 1967; Bullerman and Hartung, 1973). However, people usually will not eat moldy bread knowingly unless they are starving, so the problem is self-controlling, except in impoverished areas of the world.

In canned nonacid bread, previously inoculated with spores of *Clostridium botulinum,* toxin will form on incubation if the a_w of the bread is 0.95 or above. This a_w is equivalent to a moisture content of about 40%. As long as the water content does not exceed 36%, no toxin will form (Wagenaar and Dack, 1954; Denny *et al.,* 1969), but if the can wall is much cooler than the bread, as may occur during freezing or chilling, water may migrate to the can wall, and create a localized area where the a_w rises to the critical level or above. Conversely, when the bread is thawed or warmed, the moisture from the can wall may return to the chilled center of the bread. When the can is held for a week at constant temperature, the moisture level in various portions of the interior equalizes. There have been no reports of botulism, however, from this product.

Acid breads, whose pH is 4.8 or below, are safe from botulism, even when the moisture content is 40% (Wagenaar and Dack, 1954; Weckel *et al.,* 1964; Ingram and Handford, 1957).

Tortillas are often prepared under very primitive conditions, especially in rural areas of Central America. Fortunately, treating the maize with lime and cooking the tortillas destroy nearly all of any aflatoxins that might be present in the original maize, a fact of major importance to millions of people in Latin America (Ulloa-Sosa and Schroeder, 1969). Cooking also destroys bacterial pathogens; however, tortillas are usually immediately recontaminated, and because of their high moisture content and the prevailing warm temperatures, they can support the growth of *Escherichia coli, Staphylococcus aureus, Bacillus cereus,* and possibly other bacterial pathogens. If tortillas are not consumed promptly after they have been cooked, they may be responsible for widespread disease outbreaks (Capparelli and Mata, 1975).

D. Control

The best way to keep dry baked goods from molding is to keep them dry. Cookies (biscuits) require nothing except good sanitation and a low a_w for good stability (Smith, 1972). In bread, low water activity can be maintained on the loaf surface by vapor-proof packaging and by avoiding temperature changes that encourage condensation on the inside of the wrapper. For long storage of bread, cakes, and cookies, freezing is best, but refrigeration or even cool storage will increase the mold-free shelf-life of bread or cake, particularly if the a_w is relatively low (Seiler, 1964).

The number of mold spores on bread has an inverse relation to the mold-free shelf-life (Seiler, 1964). Mold contamination of baked goods can be reduced by separating all handling and storage facilities for returned baked goods, using filtered air in the bread cooling and wrapping areas, frequent cleaning of the slicing machine and other equipment contacting baked goods before packaging, and using ultraviolet light in critical areas such as wrapping machines (Vol. I, Chapter 2). Some bakers first wrap the bread in vapor-permeable wrap, then treat it in a microwave oven for 45 to 90 sec to destroy mold spores, and finally add a second wrap of vapor-impermeable material. In this way, there are no viable mold spores at all on the product (Pyler, 1973).

Most breads and cakes now contain mold-inhibiting chemicals, commonly propionic acid, sorbic acid, or other organic acids, or their salts. For control of bread mold, calcium propionate is popular at a level of 0.1 to 0.32% of the weight of flour used. Generally, 0.2% is recommended (Seiler, 1964), but if this much interferes with yeast fermentation or produces cheeselike flavors, a reduction to 0.1 to 0.15% is advisable. In cake, sorbic acid is effective at a level of 0.03% to 0.125% of the total batter, although it can affect flavor adversely (Pyler, 1973; Seiler, 1964). Chapaties, generally they have a shelf-life of 7 days under tropical conditions. However, with 0.3% sorbic acid and 1.5% salt, they will keep longer than 180 days (Kameswara Rao *et al.,* 1964, 1966).

Sorbic acid inhibits yeast fermentation, however, so that if it is to be used in bread, it must be incorporated into the fat where it stays until the product is baked. Or it can be sprayed onto the surface of the baked loaf (Pyler, 1973). None of the acids or salts mentioned work well with alkaline products like devils food cake (pH 9.0) because antimicrobial activity of these compounds depends on the level of the undissociated acid—usually at its highest level at pH 4 or 5, and at its lowest in the alkaline range. (See Volume I, Chapter 7.) Calcium propionate cannot be used in cake because it reacts with baking powder.

Control of rope depends on good sanitation and selection of ingredients with low spore level, use of chlorine in tempering and washing wheat, quick cooling after baking and prior to packaging, low pH of the bread, and use of organic acid preservatives. The dough should be prepared with the proper amount of yeast, at normal temperature, to obtain a vigorous fermentation that will depress the pH of the dough enough to make it less conducive to rope development (i.e., pH approximately 5). An organic acid can also be used for this purpose.

Baking should be thorough enough to reduce the moisture content of the center of the loaf to the point where rope-forming organisms are slow to grow. The bread should be cooled rapidly and should not be wrapped until its temperature reaches 33°C or below at the center (Pyler, 1973).

Bakery equipment should be kept clean so that it will not contribute spores to the bread.

There are four laboratory procedures for checking the presence and numbers of rope-producing organisms in ingredients: (a) *Practical baking test*—prepare a small batch of bread in the usual way, incubate the loaves, and check for rope after 48 hr; (b) *Qualitative test*—suspend ingredients in nutrient broth, incubate for 48 hr at 37°C, and look for a gray-white pellicle that indicates the presence of *Bacillus;* (c) *Pour plate*—pasteurize a suspension of the ingredient being tested at 80°C for 10 min, then make an agar plate count, and incubate at 37°C. If there are more than 20 organisms/g, the ingredient should be rejected (Frazier, 1967); (d) *Odor qualitative test*—add 75-ml sterile water to 50 g of flour in an Erlenmeyer flask, place it into a water bath at 100°C for 15 min, then incubate it at 30°–35°C for 24 hr, and smell for the typical rope odor (Pyler, 1973).

An even more effective means of controlling rope is to add 0.2% calcium propionate. Some bakers also use acetic, tartaric, citric, or lactic acid or acid phosphate to reduce the pH. But too much acid will interfere with yeast fermentation (Frazier, 1967; Kent-Jones and Amos, 1957).

VII. PASTA

A. Processing

Pastas are made from wheat flour, tap water, seminola, farina, and other ingredients to form a stiff dough of about 30% moisture. The dough is extruded or rolled into a variety of shapes and forms, and then dried at 40°C to a 10–12% moisture level. Drying must be slow, or else the product will crack from the physical strain. Noodles, in the United States, also contain 5.5% egg or egg yolk. Either macaroni or noodles may be enriched with vitamins, minerals, and soy protein to increase their nutritive value.

There is no cooking step in the manufacture, and bacteria grow rapidly during the mixing and drying operations. As a result, the final dried product may have a very high level of microorganisms. In one survey of dried pasta from retail stores in the United States, bacterial levels varied from 10^3 to 7×10^7/g, and 10 out of 23 samples tested contained in excess of 10^5 bacteria/g. *Aspergillus flavus-oryzae* grew from 44 of 47 samples tested, and in most of these it was the predominant fungus. *Penicillium* was present in all samples, and several other genera were represented. The counts of fungi were 75 to 1425 propagules/g (Christensen and Kennedy, 1971). Boiling the pasta kills all molds, yeasts, and vegetative cells of bacteria.

B. Spoilage

There are few reports of spoilage of pasta, since it is almost invariably stored and distributed dry. Some prepared dishes are frozen or canned, but none are held chilled except for local distribution. If drying is too slow, or if the pasta becomes moist or remains moist after manufacture, it might spoil from bacterial or fungal growth. For example, *Enterobacter (Aerobacter) cloacae* has caused gas production in moist macaroni. *Monilia* has caused purple streaks at contact points with paper surfaces during drying (Frazier, 1967). It is possible, of course, that spoiled ingredients, such as liquid egg, may be incorporated into the pasta mix.

C. Pathogens and Toxins

Since pasta products are extruded and dried below pasteurization temperatures, pathogens may survive in the final product. Salmonellae from the egg ingredient have been especially troublesome. Although many *Salmonella* cells may die during these processes, many will survive, especially if they were originally present in high numbers (Hsieh *et al.,* 1976b). During the drying process, the heat resistance of the *Salmonella* cells that survive continually increases (Hsieh *et al.,* 1976a) so that heating the pasta in the dry state would be ineffectual. Once the pasta is fully dry, the organisms remain alive for several months (Walsh *et al.,* 1974; Lee *et al.,* 1975). Of course, boiling will destroy the organisms, but the danger lies in cross-contamination from the dried product to a finished moist food, or in the misguided person who may chew on a dry noodle. In any case, the presence of salmonellae in dried pasta is considered a basis for legal action under the laws of many countries.

The possibility of formation of staphylococcal enterotoxin in pasta has received considerable attention, even though it remains only a potential hazard. In the pasta-mixing equipment (often not cleaned more frequently than once a week), and in the early stages of drying, *Staphylococcus aureus* can grow because conditions are nearly ideal ($35°–40°C$, a_w 0.90–0.95, pH approximately 6). As drying continues, and during storage of the dry product, the cells die out but not necessarily to extinction (Walsh and Funke, 1975). In a survey of dried pasta products in retail stores in the United States, *Staphylococcus aureus* was found in 42 of 1533 packages of macaroni, and in 49 of 1417 packages of noodles; and in related factory investigations, the organisms were found in 179 of 350 samples (Walsh and Funke, 1975). Frozen pasta dishes such as lasagna, ravioli, and pizza frequently contain viable *S. aureus,* and the pasta is one of the sources of the organisms (Ostovar and Ward, 1976). Staphylococcal enterotoxin will not reach a detectable level unless there are at least 10^6 cells/g of dough. This can occur only if the dough is badly abused, for example if it is held for approximately 18 hr at 25°C or approximately 13 hr at 35°C. Any toxin so formed would remain indefinitely in the dried pasta, and would be only partially rinsed out during the boiling, so that it could cause illness when ingested. But the treatment of the dough in these experiments was abusive, and such conditions would not occur in factories working under good manufacturing practice. The noodles would be dry by the time there were three to six generations of growth—quite insufficient for toxin to form at a detectable level. Thus, the hazard remains only potential (Lee *et al.,* 1975; Petrushina, 1976).

Pasta dough, because it contains raw flour, also contains many of the molds from the flour. Many of these strains are capable, in the right circumstances, of producing mycotoxins (Christensen and Kennedy, 1971). But the possibility that mycotoxins would form during processing of pasta is remote; also, the product would be obviously inedible (Stoloff *et al.,* 1978).

Dried spaghetti has been reported to be a source of *Clostridium perfringens* spores (Keoseyan, 1971), and this should be taken into consideration in the handling of soups and other prepared foods that contain pasta, and which could act as a favorable medium for the growth of this organism. However, many other foods are also sources of this ubiquitous organism.

D. Control

Pasta should be dried promptly after mixing, to avoid growth of salmonellae or staphylococci if they are present. Any liquid eggs used for preparation of pasta should be pasteurized before being mixed into the dough. Pasta dough or dried pasta dough should always be boiled before consumption.

Pasta products should be routinely tested for bacterial level, molds, and yeasts, *Staphylococcus aureus,* and (if eggs are an ingredient) for salmonellae.

VIII. BREAKFAST CEREALS AND SNACK FOODS

The breakfast cereal processes (flaking, puffing, or extrusion) require the addition of moisture and, therefore, offer an opportunity for microbial growth. However, heating during the period the product is moist will reduce the microbial load. After such a heat treatment, there are further opportunities of contamination from vitamins, minerals, sweeteners, and colorings that are applied before packaging (Hobbs and Greene, 1976).

IX. PASTRIES

A. Introduction

1. Foods Covered

Pastries usually are cakes or baked shells filled with custard, fruit, creams or imitation creams, honey, nuts, and sauces and may be topped with sugar frostings, fruits, and meringues. They may be fully cooked by baking with the dough (e.g., fruit pies) or cooked separately in bulk and poured into a baked crust or spread onto a baked cake (e.g., cream pies or cream-filled cakes).

2. Important Properties

Many fillings are excellent microbial growth media. Others are only minimally good substrates or they may even be inhibitory to microbial growth because of one or more limiting factors, such as low a_w, low pH, or limited nutrients. Preservation is accomplished by alteration of formulation, refrigeration, or preservatives.

B. Effects of Processing on Microorganisms

There are three basic procedures for manufacture of pastries (Abrahamson et al., 1952).

1. Ingredients of the filling (such as water, sugar, milk powder, cornstarch, egg, salt, possibly citric acid, and flavoring) are combined, cooked, and dispensed into prebaked pastry tubes or shells. Then a chocolate or vanilla icing may be added. Chocolate éclairs, Napoleons, or imitation cream pies are good examples.

2. A preformed (baked or unbaked) pastry shell is filled with combined, uncooked filling (containing ingredients such as water, sugar, milk powder, salt, eggs, cornstarch, flavoring, and possibly shredded coconut). The en-

tire pastry is then baked, cooled, and packaged. Coconut custard pie is a good example.

3. A prebaked pastry shell is filled in two or three steps with ingredients, some of which have been combined and cooked (recipe similar to those used in 1 or 2), but other ingredients such as fruit, flavoring, coconut, chocolate, and whipped cream may be added without cooking. The completed pastry has no final baking. A good example is Nesselrode pie.

Cooking fillings to 76° to 82°C kills all microorganisms except bacterial spores, assuming the entire batch reaches this temperature. But in example 1, there is considerable opportunity for recontamination of the bulk during cooling, conveying, and dispensing (Silliker, 1969). In example 3, there is even more likelihood of contamination, because some ingredients are not cooked at all. Such a practice can be hazardous (Hobbs and Greene, 1976; Silliker, 1969). Cooking or reheating failures sometimes occur if the product is not all brought to an adequately high temperature.

Browning of meringue (heating in an oven at approximately 232°C for about 6 min) offers too little heat to kill bacteria except those in the top layer. Meringue is an excellent insulator because of air bubbles entrapped in the foam. The temperature attained at the meringue–filling interface is often below 44°C during browning (reviewed in Bryan, 1976). Of course, meringue may also be folded into a pastry or dessert with no heat whatsoever.

Freezing and frozen storage of pastries destroy many bacteria, but rarely to the extinction of a given population. Aerobic plate counts and coliform group levels on various custard pies dropped to one-half their original level after 76 days at −20°C, except for the very acid lemon-lime pies whose counts dropped to one-half their original level in only 12 days (Kramer and Farquhar, 1977).

C. Spoilage

Pastries can suffer from the same spoilage problems as bread, but the pastry filling or topping is more susceptible to microbial growth than the cereal product. Most fillings support the growth of spoilage organisms, especially if they contain egg or milk products, and if the a_w is high and the pH near neutral. Cooked fillings spoil from sporulating rods that survive the cooking, unless other organisms are introduced after the cooking step or unless the cooking is inadequate (Castellani et al., 1955). Imitation cream pies spoiled in 48 hr at room temperature (~22°C) with aerobic plate counts of 10^7 to $>10^8$, coliform group levels exceeding 10^6, and Staphylococcus aureus up to 10^6/g (Surkiewicz, 1966).

High-sugar icings or low-pH toppings (i.e., fruit) will not support growth of spoilage bacteria but will eventually permit molds or perhaps yeasts to grow (Silliker, 1969).

The factors in pastries that affect foodborne disease organisms also affect spoilage organisms (Section IX,D).

D. Pathogens

1. Introduction

Cream- and custard-filled pastries can cause foodborne disease. In a 35-year period (1938–1972), there were 439 outbreaks in the United States from ingestion of cream-filled pastries, for which the disease organism could be identified, and 134 for which it could not. Of the former, 85.2% were from *Staphylococcus aureus* and 12.5% were from *Salmonella* (Bryan, 1976). In the period 1945–1947, custard fillings caused 25 outbreaks of which 24 were from *Staphylococcus aureus* (Feig, 1950).

Staphylococcal food poisoning can occur only after the organisms have multiplied under favorable conditions to reach millions per gram. The minimal temperature for enterotoxin production is 10°C (Vol. I, Chapter 13, Table 13.1). It is, therefore, not surprising that foodborne disease outbreaks from cream-filled pastries are attributed primarily to inadequate refrigeration during manufacture or storage (McKinley and Clarke, 1964; Bryan, 1976).

But in the United States, there has been a dramatic drop in foodborne disease from pastries. In 1938 and 1939, 17.8% of all foodborne disease outbreaks implicated pastries. By 1970–1973, the figure had dropped to 2.3%, and by 1976, to 1.8% (Bryan, 1976; U.S. Department of Health, Education and Welfare, 1977a). This remarkable accomplishment has been a result of (a) improved cooking and sanitation procedures, (b) improved refrigeration, (c) altered formulations, and (d) use of preservatives.

2. Cooking and Sanitation

a. Cooking. An adequate cooking process destroys foodborne disease organisms in cream or custard fillings. Both *Staphylococcus aureus* and *Salmonella* are destroyed by merely bringing custards to a second boil after adding the thickening. Baking custards in a pie shell also kills the organisms (Cathcart *et al.*, 1942). However, sometimes the cooking is insufficient so that some organisms survive. For example, the minimum temperature for thickening egg (78°C) may not destroy all staphylococci, but 91°–93°C will do so (Kintner and Mangel, 1953b).

In custard, 60°C required 19 min to kill 10^7 salmonella, or 59 min to kill 10^7 *Staphylococcus aureus.* These times were shortened at $65.7\,^\circ$C, which required 3.5 min to destroy 10^7 salmonella or 6.6 min to destroy 10^7 staphylococci (Angelotti *et al.,* 1961b).

Egg white should be pasteurized by the supplier (and indeed, pasteurization is required in most countries), but sometimes salmonellae survive and appear in the meringue used for pastry topping. Browning the meringue in the oven has little lethal effect on any salmonellae that are present, because heat does not penetrate sufficiently. Adding hot syrup to whipped whites, and whipping again, destroys many salmonellae, but the process is of marginal efficiency.

As a means to destroy any foodborne disease organisms that may be present on the final filled pastry, some have suggested rebaking as a final step in the manufacturing process (Stritar *et al.,* 1936; Gilcreas and Coleman, 1941). This works well for some products, such as custard pies; but for éclairs or Boston cream pies, the quality suffers, so that the process is impractical (Abrahamson *et al.,* 1952).

b. Sanitation. Most contamination occurs while the filling is cooling, or during handling. Any direct contact between human beings and the filling is likely to introduce *Staphylococcus aureus,* for about one-half the human population carries these organisms on the skin and in the mucous membranes (Abrahamson *et al.,* 1952; Surkiewicz, 1966; Silliker, 1969).

Thus, various surveys have demonstrated that *Staphylococcus aureus* is commonly present at low levels in filled pastries of commerce. Investigators in New York City reported *S. aureus* on 10 of 33 éclairs, 3 of 25 Napoleons, 1 of 31 coconut custard pies, and 6 of 31 Nesselrode pies (Abrahamson *et al.,* 1952). Nevertheless, the U.S. Food and Drug Administration has concluded that bakeries operating under the best sanitary conditions can produce cream-filled pastries containing no *S. aureus,* whereas those operating under poor conditions usually cannot (Surkiewicz, 1966). It is apparent that fillings are of such a nature that direct human contact or indirect contact from inadequately sanitized equipment are unnecessary in commercial operations even though such contact occurs with regularity. The level of *S. aureus* in the cooked, cooled filling can be used as a measure of the degree of human contact.

Harmless when present in small numbers, *Staphylococcus aureus* can produce enterotoxin if it is permitted to grow. The sections that follow are devoted to the factors that control such growth in filled pastries (Sections D,3–5).

Salmonellae have been reported in many bakery ingredients, such as flour, milk, eggs, butter, cream, cheese, nuts, coconut, and dried fruit. It is essential, then, that cooked fillings be protected from direct or indirect contact with such "raw" ingredients. Children and the elderly or infirm are especially susceptible to infection from small inocula such as one might find in a pastry in which growth had not occurred.

Only the highest degree of sanitary practice is adequate for handling cooked and cooled cream and custard fillings. They should be protected from dust and droplet contamination from dry foods used as ingredients, and from batters being mixed. Fans should be installed to control air movement from dusty areas. Equipment and personnel in the cooked food area should have no contact with raw ingredients. Equipment should be constructed of impervious metal such as stainless steel and should be disassembled at each cleaning unless it is specifically designed for a clean-in-place program (see Volume I, Chapter 14). Equipment should be cleaned and sanitized every 4 hr during production. Some equipment (e.g., conveyor belts) may be continuously cleaned and sanitized. Employee contact with product should be minimal, and hygienic practices beyond reproach. Personnel should not handle raw foods, insanitary equipment, visit the toilet, or blow their noses, then touch fillings or pastry without first washing and sanitizing their hands. No one with an upper respiratory infection or infected skin should be permitted near the processing area. (See also Volume I, Chapter 14, Section IX.)

3. Refrigeration

To prevent salmonellae from growing, or staphylococci from growing and producing toxin, refrigeration at or below 5°C is essential for most fillings (Segalove and Dack, 1941; Crisley et al., 1964). In fact, most food-protection authorities in countries with advanced technology require such refrigeration (Silliker and McHugh, 1967). It is primarily this factor that has reduced the incidence of foodborne disease from cream-filled pastries in the United States (Bryan, 1976).

The first place where good refrigeration is necessary is during the cooling of the cooked filling. Precooling at room temperature is dangerous (Longrée, 1964). In like manner, cooling in a large mass is inefficient even in a refrigerator, for the center may remain warm for several hours, during which pathogens may grow (McDivitt and Hammer, 1958; Miller and Smull, 1955). Every particle of the mass should pass through the temperature zone in which *Staphylococcus aureus* and *Salmonella* grow (6.7° to 46°C) in 3 hr or less (Hodge, 1960; Angelotti et al., 1961a). To accomplish this, agitation is necessary, preferably with a refrigerated tube agitator or other device to speed the cooling (Longrée, 1964). Or, the hot filling may be dispensed into pastry shells, which are then refrigerated without delay.

The cooled filling should then be handled expeditiously so that no part of it warms to the danger zone for more than a few minutes. Residual food material on equipment may remain several hours at ambient temperature unless it is washed away frequently (Abrahamson *et al.,* 1952). And finally, transport of the finished pastry and storage at retail should be under refrigeration at or below 5°C.

4. Formulations

The traditional cream fillings may contain eggs, milk, shortening, sugar, cornstarch or flour, salt, vanilla and other flavorings, and water (Bryan, 1976). Synthetic cream fillings consist mainly of vegetable oils, stabilizers, emulsifiers, and water. They may also contain sugar, coloring, flavoring, fruit purees, dried milk, and starch (Surkiewicz, 1966). If the a_w is high enough, most will support the growth of various organisms, including *Staphylococcus aureus* (Crisley *et al.,* 1964; Surkiewicz, 1966; Silliker, 1969). Adding milk or small amounts of egg markedly increases the growth of *S. aureus* (Crisley *et al.,* 1964). Cocoa and chocolate inhibit staphylococci (Cathcart and Merz, 1942) or salmonellae (Zapatka *et al.,* 1977) but only in the absence of milk. Formulations that are so devoid of nutrients that *S. aureus* will not grow may also be unpalatable (Cathcart *et al.,* 1947).

The lowest pH for growth of *Staphylococcus aureus* is 4.5 to 4.8, depending on the menstruum; that for *Salmonella* is 4.0 to 4.2 (Volume I, Chapter 13, Table 13.1). The pH of commercial custards used in bakeries is 5.8–6.6 (Bryan, 1976), a range near the optimum for these organisms. Whereas fruit-filled pastries do not support bacterial growth because of the low pH and the presence of organic acids (Cathcart *et al.,* 1947; Ryberg and Cathcart, 1942; see also Volume I, Chaps. 5 and 7 and this volume, Chapter 21), the pH of fruit-flavored cream-filled pastries is well buffered by the milk and eggs so that staphylococci can grow in them readily (Cathcart *et al.,* 1947).

Changes in the formulation can alter the a_w or other factor that will permit a pathogen to grow. High-ratio shortenings, for example, permit incorporation of more water in a cake batter than was possible previously. If this increase is not accompanied by an increase in sugar content, the final product may permit the growth of a pathogen (Bryan, 1976).

Any food whose a_w is below 0.86 cannot support growth of *Staphylococcus aureus* (W.J. Scott, 1953). The surface of a baked custard pie is markedly dehydrated so that *S. aureus* grows only very slowly if at all, whereas it would grow more rapidly in the moist custard (Preonas *et al.,* 1969).

"Butter-cream" fillings (Table 23.24) have a remarkably good safety record despite the fact that they are stored indefinitely at ambient tempera-

TABLE 23.24

Formulation for "Butter-Cream" Filling [a]

Ingredients	Amount (by weight) [b]
Powdered sugar	282.60
Shortening	119.40
Margarine	40.00
Nonfat dry milk	45.00
Salt	3.75
Vanilla extract	1.50
Gelatin	2.80
Water	117.75

[a] From Silliker and McHugh (1967, see for method of preparation).

[b] This formula gives a sugar/water ratio of 2.4/1.

ture. Their a_w, sufficiently low to preclude microbial growth, is adjustable by varying the sugar/water ratio. Growth of *Staphylococcus aureus* sometimes occurs at a ratio of 1.8/1 (or if glucose or invert sugar is used, a ratio of 0.9/1). However, at sucrose/water ratios of 2.1/1 to 3.0/1, no growth generally occurs. Nevertheless, growth has occurred at the interface between a "butter-cream" filling of ratio 2.7/1 and a cake whose a_w was higher than that of the filling (Silliker and McHugh, 1967).

"Obviously, many bakery products present dynamic systems wherein the movement of moisture from one area to another, or the buffering effects on one component or another, or the nutritional contribution of one component or another, may completely nullify what on the surface would appear to be an inherently stable microbiological system. In the study of the microbiology of pastry products, one finds that the whole is not always the sum of its parts" (Silliker, 1969).

5. Preservatives

Growth of *Staphylococcus aureus* in fillings that are otherwise excellent microbial media can be prevented by adding preservatives, either incorporated into the filling or shell or sprayed on the surface. The most popular are the organic acids such as propionic, sorbic, or benzoic, or their salts, usually added at ~0.1% (Preonas *et al.,* 1969; Schmidt *et al.,* 1969; Pyler, 1973).

Their effectiveness is increased by reducing the pH of the filling to the range 3–5, where a high proportion of the acid is undissociated. The level of the undissociated molecule can be calculated for any pH (Preonas *et al.,* 1969; see also Volume I, Chap. 7).

Experiments have shown the value of DL-serine, as opposed to other amino acids, as an antibacterial ingredient in cream pies. Serine was most inhibitory against staphylococci in formulas containing only milk, but much less inhibitory in formulas containing egg, or both milk and egg (Castellani, 1953).

In further experiments, the effectiveness of DL-serine was lessened in the presence of propionate and coconut or cocoa products. Its effectiveness was greatest in a pineapple cream filling (Castellani *et al.,* 1955).

Antibiotics (which are not permitted in foods in many countries) have been proposed as additives. Subtilin controls the growth of food-poisoning enterococci, and a combination of subtilin and terramycin controls the growth of *Salmonella, Bacillus,* and *Staphylococcus* (Godkin and Cathcart, 1953). Hydrogen peroxide also inhibits bacterial growth in synthetic creams (Hobbs and Smith, 1954).

E. Interrelations

The various factors that affect microbial growth in cream- and custard-filled pastries are interrelated. If the conditions are less than optimal in one factor, the range of a second factor that affects growth will be more limited. Thus, temperature (Volume I, Chap. 1), a_w (Volume I, Chap. 4), pH (Volume I, Chap. 5), and organic acids (Vol. I, Chap. 7) all have interrelated effects in the microbiology of pastries.

Staphylococcus aureus does not compete well with spoilage bacteria. However, a properly cooked filling will contain only bacterial spores, so that *S. aureus* contaminating the filling after it has been cooked and cooled will have few competitors. Sometimes *S. aureus* will grow in whipped cream but will not produce toxin, presumably because of competition from other organisms (Post *et al.,* 1961).

F. Control

For good control of the quality and safety of cream- and custard-filled pastries, bakers and handlers should observe the following:

1. Use only pasteurized eggs and dairy products;
2. Cook the pastry filling thoroughly, with appropriate mixing to assure uniformity of temperature;
3. Keep raw ingredients and processes separate from cooked products;

4. Control dusts and aerosols by establishing air movement away from the cooked-product area;

5. Clean and sanitize equipment that contacts cooked fillings every 4 hr during use;

6. Cool cooked fillings rapidly to 5°C or below by refrigeration, while mixing; or fill pastry shells with hot filling and refrigerate immediately. Fillings should be held no more than 3 hr between 6.7° and 46°C;

7. Maintain refrigeration of fillings and filled pastries that are capable of supporting growth of *Staphylococcus aureus* until they are consumed;

8. As an alternative to refrigeration, alter the formulation by reducing the pH, reducing the a_w, or using preservatives to control the growth of pathogens;

9. Wash and sanitize hands before handling cooked product;

10. Minimize hand contact with cooked product, and keep persons with respiratory or skin infections from the cooked-product area.

X. CHOICE OF CASE

For an understanding of "Case" see Table A.1 and the brief description in Appendix IV. For more details, see the second book of this series (ICMSF, 1974).

A. Raw Cereal Products

Raw grains, flours, brans, etc., are invariably cooked before human ingestion, and therefore spoilage tests would fall into Case 1. Storage under dry conditions would neither increase nor decrease a spoilage hazard, so that Case 2 would apply. Storage of inadequately dried grains or cereal products would increase the spoilage hazard, and Case 3 would apply.

However, in case of severe molding, mycotoxins may appear, and the raw products would fall into Case 13—if a sorting and sifting procedure can eliminate toxic kernels—or Case 14 if not. (Cooking would do little to destroy the mycotoxins.) Further storage of a dry grain or cereal product already showing some molding, especially in highly humid conditions, would encourage additional mold growth and a possible mycotoxin problem, which would warrant Case 15. However, Cases 13–15 would dictate elaborate and expensive sampling plans. Routine tests for mycotoxins are not recommended. For this problem, sampling plans dictated by spoilage (Cases 1–3) are more practical.

A test of *Salmonella* in raw grains would require Case 10, since heating would destroy the organism, yet a flour could be airborne and contaminate other foods in the kitchen. Tests for *Bacillus cereus* or *Clostridium perfringens* would be considered Case 8 because the spores would remain dormant in dry storage and would survive cooking.

B. Doughs and Breads

Uncooked doughs may contain *Salmonella* (from ingredients, primarily eggs and yeast), that would die on baking, but that could cross-contaminate other foods in the kitchen (Case 10). *Bacillus* species, if present in the dough, can survive baking and grow inside the baked loaf during handling and storage. Among these species, mucoid strains of *B. subtilis* or *B. licheniformis* may cause the spoilage condition "rope" (Case 3). Sampling to measure mold contamination of bread is impractical.

C. Pasta

During preparation and drying, pasta can spoil from microbial growth, and although during dry storage the condition stabilizes, the microorganisms will die on cooking (Case 1). During this preparation and drying period, there is also a potential hazard from *Staphylococcus*. Preformed enterotoxin would not be destroyed in cooking (Case 8).

A pasta product destined to be cooked and held ready-to-eat in a delicatessen case could spoil (Case 3) or permit the growth of *Staphylococcus* (Case 9) or *Salmonella* (Case 12). Pasta is one of several sources of *Clostridium perfringens* or *Bacillus cereus* in various cooked dishes. The spores would survive cooking and grow in foods held warm for several hours (Case 9).

D. Corn Products

Tortillas and related corn products can support the growth of spoilage organisms before (Case 1) and after (Case 3) the cooking process. Growth of *Staphylococcus aureus* and possible enterotoxin formation can occur during warm storage before (Case 8) or after (Case 9) cooking. Salmonellae or enteropathogenic *Escherichia coli* can grow after cooking if a few cells survive the cook, or are introduced into the cooked product by poor sanitation.

In many instances, education (e.g., improved sanitation, prompt consumption, or use of refrigeration) is a better solution than a sampling plan.

E. Cream- and Custard-Filled Pastries

These products, held warm, spoil primarily from aerobic spore-forming bacteria that survive cooking, or if refrigerated, from molds (Case 3); or they may be frozen for distribution and storage, so that microbial tests measure contamination by indicator organisms such as Standard Plate Count, coliform group, *Escherichia coli,* or *Staphylococcus aureus* (Case 5).

But if mishandled, they can be serious health hazards. They may support the growth of *Staphylococcus aureus* (Case 9) or of *Salmonella* (Case 12) if the organisms survive undercooking or if they enter the product after the cooking step from ingredients, environment, or human contact.

24

Spices

I. INTRODUCTION

Spices and condiments have been used since before recorded history (Parry, 1969a). Spices were used as embalming materials in ancient Egypt, as medicinals in many countries, as sweet scents in religious ceremonies, and to mask the putrid flavor of meat in hot climates lacking refrigeration. A few spices have an antimicrobial effect at the concentrations used in foods and thus serve as preservatives. The main use of spices during the past two millenia, however, has been to add desirable flavors and odors to foods or drinks.

Spices are of interest to microbiologists for four principal reasons. They may (1) become moldy if held at improper humidity and temperature; (2) contain large numbers of microorganisms that occasionally may cause spoilage or more rarely, disease, when introduced into food; (3) exhibit antimicrobial activity, and occasionally aid in preservation; and (4) stimulate microbial metabolism.

A. Definition

Historically, spices were tropical aromatic or pungent vegetable substances used to flavor foods (e.g., pepper, cinnamon, ginger, cloves, and pimento) (Fowler and Fowler, 1964). Subsequently, other materials were included such as leafy parts of plants in temperate zones (e.g., oregano, basil, marjoram, and bay leaves) and seeds of both tropical and nontropical plants (e.g., poppy, mustard, caraway, sesame, anise, and fennel). Today the American Spice Trade Association defines spices as "products of plant origin which are used primarily for the purpose of seasoning food" (Anonymous, 1969). Even dehydrated vegetables such as onion, garlic, celery, green pepper, etc., may be included. Other definitions exist in official and nonofficial sources (e.g., Department of National Health and Welfare, 1977; U.S. Department of Health, Education and Welfare, 1978b; Parry, 1969b; Hardman, 1972; Rosengarten, 1973; Morse, 1977; Merory, 1968; Wohl, 1974). A partial list of spices is given in Table 24.1.

Spices are produced from botanically diverse plants grown in a wide variety of soils and climates. Depending on the spice, the following parts of the plant may be used: bulbs, roots, rhizomes, stems, leaves, bark, berries, buds, arils, and seeds. In some cases, e.g., dill for pickles, the whole plant except the root is used. Histologically and chemically spices

TABLE 24.1

List of Some Frequently Used Spices and Permitted Water Content [a]

Common name [b]	Parts used	Form used	Maximum permitted water content (%) [c]
Allspice (pimento)	Fruit	Whole or ground	12
Anise	Fruit	Whole or ground	10
Basil (sweet)	Leaves	Whole or ground	9
Bay (Laurel)	Leaves	Whole or ground	7
Capsicum (chili)	Fruit		
Caraway seed	Seeds	Whole	11.5
Cardamom seed	Fruit		13
Cassia	Bark	Ground	13
Cayenne	Fruit	Whole or ground	10
Celery seed	Fruit		10
Chervil	Leaves		
Cinnamon	Bark	Ground	
Cloves	Flower bud	Whole or ground	8
Coriander seed	Fruit	Whole or ground	9
Cumin seed	Seed	Whole or ground	9
Dill seed	Fruit	Whole	9
Fennel seed	Fruit	Whole or ground	10
Fenugreek	Fruit	Whole or ground	10
Garlic	Bulb	Chopped, dried, powdered	
Ginger	Rhizome	Dried, ground	12.5
Horseradish	Root	Ground	
Mace	Aril	Whole or ground	8
Marjoram	Leaves (part of tops permitted)	Whole or ground	10

TABLE 24.1 (cont.)

Common name [b]	Parts used	Form used	Maximum permitted water content (%) [c]
Mustard	Seed	Whole or ground	11
Nutmeg	Seed	Whole or ground	8
Oregano	Leaves	Whole or ground	10
Paprika	Fruit	Whole or ground	12
Parsley	Leaves	Dried	
Pepper (black)	Berry	Whole or ground	12
Pepper (white)	Berry	Whole or ground	15
Peppermint	Leaves		
Poppy	Seed	Whole	
Rosemary	Leaves	Whole or ground	8
Sage	Leaves	Whole or ground	10
Savory	Leaves and tops		14
Sesame	Seed		8
Tarragon	Leaves and tops	Whole or ground	10
Thyme	Leaves and tops	Whole or ground	9
Tumeric	Rhizome	Whole or ground	10

[a] From Hardman (1972).
[b] Common names in 11 different languages are listed by Rosengarten (1973).
[c] Department of National Health and Welfare (1977), current or proposed regulations.

and herbs are too diverse to be described here; a concise presentation of these qualities is available (Parry, 1969b).

B. Important Properties

The condition of spices and herbs at harvest and the postharvest processing and storage determine the keeping quality. Spices are resistant to microbial spoilage if they are properly dried and stored. The maximum permitted water content for many spices is listed in Table 24.1, but in practice, the water content is usually much lower. If these moisture contents are exceeded, growth of molds is likely to occur despite the high content of antimicrobial essential oils in some spices.

1. Antimicrobial Activities of Spices

The antimicrobial activity of spices (due to their content of essential oils) is of little significance in preservation of food: the concentrations of essential oils in spiced foods are generally too low (25–250 ppm; Salzer *et al.,* 1977) to prevent microbial growth. Essential oils do, however, sometimes augment other antimicrobial activity (Tanner, 1944) and in exceptional cases, may be the main preservative (Bullerman *et al.,* 1977).

The composition and content of essential oils varies from spice to spice and even within the same spice depending on agricultural practices, geography, and climatic conditions during the growing season (Lawrence, 1978). Also the antimicrobial activity of a given spice may depend only on one or two of the large number of different molecules that make up the spice oil. Some of the antimicrobial molecules may be common to several spices, others unique to a few closely related spices.

Spices that contain the most inhibitory essential oils are cloves, allspice, cinnamon, cassia, and mustard seed. Onion and garlic juice also contain antimicrobial compounds (Abdou *et al.,* 1972), but most herbs and many spices do not possess any antimicrobial activity, or so little that they exert no effect in concentrations used in food. In cloves and allspice, the major antimicrobial components are eugenol, eugenol acetate, or eugenol methyl ether; in cinnamon and cassia, they are cinnamic aldehyde, cinnamic acetate, and eugenol; in mustard seed, it is allyl isothiocynate; and in garlic, it is allylsulfonyl allyl sulfide (Guenther, 1949, 1950). The antimicrobial activity against yeasts and bacteria occurs at concentrations of 10–1000 ppm (Table 24.2). In general, yeasts are more readily inhibited than bacteria, but there are great differences in response among species even in the same genus (Anderson *et al.,* 1953). As the pH is decreased, the inhibitory activity of the oils increases and at pH 4.5 to 5.0, a level of 10 ppm of some oils inhibits certain spoilage organisms (Anderson *et al.,* 1953). The germicidal or inhibitory concentrations given in Table 24.2 must be considered with caution because different investigators have used different methods and test conditions (Koedam, 1977). At best, the values listed can only be approximate.

2. Antimicrobial Effect of Spices in Foods

Cinnamon (about 1%) in raisin bread decreases mold growth and even smaller concentrations prevent production of aflatoxin (Bullerman, 1974): the inhibitory components are eugenol, cinnamic aldehyde, and *o*-methoxy-cinnamaldehyde (Morozumi, 1978). Some essential oils inhibit the development of slime in weiners, especially when sorbate or lysozyme are also added (Mori *et al.,* 1974). Oil of mustard inhibits growth of obligate, aerobic lactate-metabolizing yeasts that form a film of growth on the surface of fermented pickles and sauerkraut (Blue and Fabian, 1943); however, the concentration of the volatile mustard oil must be maintained by a barrier, e.g., a plastic cover. Ground mustard seed (about 1%) and

TABLE 24.2

Concentration of Essential Oils in Some Spices and Antimicrobial Activity of Active Components

Spice	Percent essential oil in whole spice	Antimicrobial compounds in distillate or extract		Antimicrobial concentration [a] (ppm)	Organism
		Compound	Percent		
Allspice (*Pimenta dioica* L.)	3.0– 5.0	Eugenol Methyl eugenol	73–78 9.6	1000(G)	Yeast, *Acetobacter* [b]
Clove (*Syzygium aromaticum* L.)	16.0–19.0	Eugenol Eugenol acetate	72–92	1000(G)	Yeast, *Acetobacter* [b]
Cinnamon bark [c] (*Cinnamomum zeylanicum* Nees in Wall)	0.5– 1.0	Cinnamic aldehyde Eugenol Alcoholic extract of cinnamon	65–76 4–10	100–1000(G) est. 100(I)	Yeast, *Acetobacter* [b] Mold [d]
Cassia (*Cinnamomum cassia* Blume)	1.2	Cinnamic aldehyde Cinnamyl acetate	75–90	100–1000(G) est. 10–100(I)	Yeast, *Acetobacter* [b] Yeast, bacteria [e]
Garlic (*Allium sativum* L.)	0.3– 0.5	Allylsulfonyl Allyl sulfide			
Mustard (*Sinapis nigra* L.)	0.5– 1.0	Allyl isothiocyanate	90	100(G) 22(I)	Yeast, *Acetobacter* [b] Yeast [f] *Vibrio parahaemolyticus* [g]
Oregano (*Origanum vulgare* L.)	0.2– 0.8	Thymol Carvacrol	60–85	100(G)	*Bacillus*
Paprika (*Capsicum annuum* L.)		Capsicidin Capsaicine		est. 100(I) est. 1(I) 50–>100(I)	*Saccharomyces* [h] Various bacteria [h]
Thyme (*Thymus vulgaris* L.)	2.5	Thymol Carvacrol		100(G)	*Vibrio parahaemolyticus* [g]

[a] Est. = estimated; (G) = Germicidal; (I) = Inhibitory. [b] Blum and Fabian (1943); [c] Essential oil from leaves is 70–95% eugenol. [d] Bullerman (1974); [e] Tynecka and Gos (1973); [f] Kosker *et al.* (1949); [g] Beuchat (1976); [h] Gal (1968, 1969).

essential oil of mustard (10–20 ppm) prevent fermentation of fruit juices (Kosker *et al.,* 1949). Apart from these examples, there is little literature on the practical application of antimicrobial activity of spices in foods. In an investigation stemming from an outbreak of botulism from home-canned peppers used in a Mexican style sauce (Glass, 1977), spices added to the sauce at normal levels appeared not to inhibit toxinogenesis (Leonberger and Bernard, 1977).

Spice oils can reduce the heat resistance of microorganisms. As little as 10 ppm of allylisothyocyanate in apple juice decreased the F_{160} (minutes at 71°C for 99.99% destruction) of spores of *Aspergillus niger* from 0.7 to 0.1 min, and 10 ppm in grape juice decreased the F_{160} of *Saccharomyces cerevisiae* from 2.3 to 1.4 min. Somewhat higher concentrations (105 ppm) of oil of cloves and oil of cinnamon accelerated heat destruction of a yeast that caused spoilage of pasteurized pickles (Anderson *et al.,* 1953). Destruction of spores of *Bacillus acidurans* during heating of tomato juice was increased slightly by 20 to 50 ppm of allylisothocyanate (Kosker *et al.,* 1951), but other inhibitory spice oils at organoleptically acceptable concentrations did not accelerate destruction of spores appreciably (Anderson *et al.,* 1949, 1953). However interesting the knowledge obtained from such studies is not generally applied in modern food technology.

The mechanisms of inhibition and destruction by essential oils are varied because the antimicrobial compounds are chemically diverse. Interference with respiratory mechanisms, sulfhydryl enzymes, etc. has been observed (e.g., Tynecka and Gos, 1973). Concentrated spice oils (10–100%) destroy vegetative cells in minutes to hours, the death rate being proportional to the concentration (Ramadan *et al.,* 1972).

Stimulation of microbial metabolism by spices. Only a few reports are available on the stimulation of microbial metabolism by spices. Corran and Edgar (1933) showed that (a) mustard, cloves, and cinnamon were inhibitory to yeasts; (b) several other spices caused little or no inhibition; and (c) black pepper, thyme, bay leaves, marjoram, savory, and rosemary actually increased the rate of metabolism of yeasts. A subsequent study (Wright *et al.,* 1954) indicated that the stimulatory factor for yeast was present in several spices. It was heat stable, not extracted by ether, and lost when the spice was ashed; essential oils were not stimulatory to metabolism. Recently, it has been found (Zaika *et al.,* 1978; Kissinger and Zaika, 1978) that a mixture of spices accelerated the production of lactic acid and drop in pH during production of fermented sausages, an important consideration in preventing the growth of pathogenic bacteria. Salt inhibited the production of lactic acid, but this inhibition was reversed by the spice mixture.

C. Principal Methods of Processing

Most spices, like other raw agricultural materials such as cereals and hay, must be dry at harvest or be dried quickly after harvest to an a_w that will prevent spoilage by molds. The sometimes primitive conditions of culture and harvest, small plots, vagaries of weather, and the humid tropical climates in which some spices are grown can make drying difficult.

The main methods of processing are cleaning, fermentation, drying, grinding, pulverizing, and extraction. Cleaning is required for most spices to remove insects, stones, twigs, and soil. Hand-sorting, winnowing, sifting, and aspiration are applied as required. Despite such procedures, spices frequently are considered adulterated on arrival at the importing country (U.S. Department of Health, Education and Welfare, 1977c).

Fermentation is used for a few products, e.g., for cassia bark to facilitate removal of the outer layer and for allspice berries to develop color and appearance; the microbiology of these fermentations has not been studied. Whether or not the fermentation results in a decreased pH is unknown. The pH of spices is variable, even among lots of the same species (Parry, 1969b).

D. Final Products

Spices may be used whole or ground, but they frequently contain excessive numbers of microorganisms for many uses unless subjected to antimicrobial treatment. For many industrially prepared foods, essential oils obtained by steam distillation (Guenther, 1948) and oleoresins (Cripps, 1972) obtained by extraction with organic solvents, are commonly employed. Essential oils and oleoresins may be used in emulsions or adsorbed on carriers (salt or sugar) to facilitate handling and improve distribution in the food. Oleoresins may be spray-dried and encapsulated (Staniforth, 1972) or adsorbed on cereals to simulate natural ground spices (Norda, 1974). The use of essential oils and oleoresins permits blending from several lots to obtain uniform concentrations of the important ingredients. They have the additional advantage of containing few, if any, microorganisms but have some disadvantages regarding flavor in some food products (Gottschalk, 1977).

II. INITIAL MICROFLORA

The initial microflora of spices is probably the same as that of other agricultural products harvested under similar conditions of soil and climate;

this, however, is speculation, because microbiological studies of unharvested spices are rarely done except to determine the etiological agents of diseases of spice-producing plants.

Dry spices contain those microorganisms indigenous to the spice-bearing plants and the soil in which they are grown. To these are added dust; fecal contamination from birds, rodents, and insects; and contamination from nonpotable water used in some processes, e.g., soaking pepper in the preparation of white pepper. Often, spices contain fungi as incidental contaminants, or as a result of growth during drying, storing, and shipping (Tropical Products Institute, 1972). The microflora that is found is that which has survived drying until the spice is analyzed.

Microbial counts should be the same for whole spice and ground spice unless ground spice is prepared from whole spice with higher bacterial counts. An examination of aerobic plate counts (APCs) for 146 samples of whole spices and 276 samples of ground spices (H. Pivnick, unpublished data) revealed that on the average (geometric mean) only whole mustard seed had a higher APC than its ground counterpart, while whole celery seed, capsicum, cinnamon, and black and white pepper had smaller APCs than their ground counterparts. The differences, however, between whole and ground spices were generally small and did not exceed $\log_{10} = 0.61$. Allspice, whole and ground, had the same counts. In the same samples examined for APC, the yeast and mold counts for whole cinnamon and mustard seed were slightly higher (about $0.5 \log_{10}$) than their ground counterparts, while whole and ground allspice, black and white pepper, celery seed, and capsicum had counts almost identical to each other.

The percentage distribution of aerobic mesophilic bacteria (APC) and molds is given for dry spices that have been sampled in processing establishments or at import (Table 24.3). Most of the APC colonies are *Bacillus* spp., which, in one study were, in order of frequency, *B. subtilis, B. licheniformis, B. megaterium, B. pumilus, B. firmis,* and *B. brevis* (Goto *et al.,* 1971). *Bacillus cereus* is also found frequently. A wide variety of nonsporing bacteria may also be present (El Mossalami and Youssef, 1965; Julseth and Deibel, 1974). Coliforms are often found, but *Escherichia coli* is infrequent (Table 24.4; Kadis *et al.,* 1971; Karlson and Gunderson, 1965), and *Staphylococcus aureus* is rarely found in dry spice (Julseth and Deibel, 1974 and unpublished data). Anaerobes are less numerous than aerobes and are usually *Clostridium* spp. (Inal *et al.,* 1975). Thermophilic bacteria usually constitute 1 to 10% of the bacterial population, although the percentage may be higher occasionally (Kadis *et al.,* 1971); some of the thermophiles are apparently obligate anaerobes (Castell, 1944).

The molds in spices may vary with the spice to some extent, but the

Aspergillus glaucus group is usually most prevalent (Table 24.5) (Christiansen *et al.,* 1967; Hadlock, 1969; Horie *et al.,* 1971).

III. EFFECTS OF PROCESSING ON MICROORGANISMS

A. Harvesting and Initial Processing

The various spices require widely diverse harvest and postharvest methods, which affect the microbial content in different ways. The main consideration is to dry the products as rapidly as possible to prevent spoilage while retaining their desirable characteristics. Examples are given in detail to emphasize the sometimes primitive aspects of harvest and postharvest treatment.

Rhizomes **(Douglas, 1973).** Dried ginger for export is processed as either rough (unbleached) ginger, or as bleached ginger. In Kerala, India, for rough, unbleached dry ginger, the harvested rhizomes are soaked overnight in water. Next day, they are rubbed by hand and cleaned with water. The outer skin is peeled off delicately and carefully with split bamboos with pointed ends, with seashells that have sharp edges, or with glass pieces. The peeled rhizomes are washed in water and then spread out in the sun for 7 to 10 days. After that period, they are further cleaned with pieces of gunny or sacking to remove any loose skin. Finally, they are packed in gunny bags.

In the preparation of bleached ginger, the rhizomes are kept in broad open metal drums filled with water to a height of about 30 cm above the rhizomes. After having been soaked and peeled as above, they are further soaked in 2% filtered lime water. They are then exposed to sulfur dioxide fumes for 12 hr, following which they are dried in sunlight for a day. This process is repeated until the bleaching is satisfactory. Finally, the rhizomes are spread out in the sun for 7 to 10 days for drying. When they are thoroughly dried, they are again checked for any loose skin and then packed in gunny bags.

In Australia much of the ginger is not dried, but is sold in sugar syrup for use in confectionery. Ginger for this purpose must be harvested within a few days to obtain a product with a minimum of fiber. The large amount of product that must be preserved within a short period of time has resulted in experiments to preserve ginger in solutions of salt, acid, metabisulfite, etc. (Brown, 1972; Brown and Lloyd, 1972). Ginger thus preserved in brine can be desalted and sweetened as needed (Seeley, 1975).

TABLE 24.3

Percentage Distribution of Aerobic Plate Counts (APC) and Mold Counts in Untreated Spices [a, b]

Spice	N[c]	Percentage distribution aerobic plate count (APC)							N	Percentage distribution mold count					
		<10	10^2[d] to 10^3	10^3 to 10^4	10^4 to 10^5	10^5 to 10^6	10^6 to 10^7	>10^7		<10^2	10^2 to 10^3	10^3 to 10^4	10^4 to 10^5	10^5 to 10^6	10^6 to 10^7
Allspice	33			3	7	46	42	3	27	37	22	15	18	7	
Anise	22			23	36	36	5		16	56	25	6	13		
Basil	21				14	48	38		17	65	24	6	6		
Bay	41	5	5	46	34	7	3		35	34	29	26	11		
Capsicum (chili)	57		2	9	28	18	31	12	59	44	15	22	7	7	5
Caraway	17		12	35	29	18	6		14	57	7	29	7		
Cardamon	15	7	13	13	27	7	33		15	67	13	20			
Cassia	36	6	66	11	14	3			20	55	5	25	5	10	
Cinnamon	42	2	5	19	48	21	2	2	51	18	33	43	6		
Cloves	28	32	21	25	18	4			26	88		8	4		
Coriander	30		3	3	30	37	13	13	23	4		61	26	9	
Cumin	12				33	42	25		8	25		62		13	
Fennel	16	13	6	13	43	13	13		11	73	27				
Fenugreek	10			20	30	30	20		8	13	37	25		25	

Spice	1	2	3	4	5	6	7	8	9	10	11	12	13	14	15
Garlic	32			16	47	28	9		15	60	40	18	11		
Ginger	33	3		21	15	45	7		28	57	14	23	4		
Mace	28	9		43	50	7			22	59	14	64	29		
Marjoram	21	5			19	43	28	5	14		7	6			
Mustard	67	30	9	33	18	9	1		63	86	6	12	2		
Nutmeg	45	20	11	44	16	4	4		33	52	27	33	9		
Oregano	56	4	2	21	41	23	9		48	35	25	7	9		
Paprika	80			11	11	9	62	18	61	44	44	28	5		
Pepper (black)	108			1	3	5	50	42	82	32	10	34	5	2	23
Pepper (white)	42			12	26	57	5		44	2	2	21	36	23	2
Sage	17			11	41	41	6		14	7	21	33	50		
Savory	10			40	50	10			6		67	33			
Thyme	19			5	11	32	53		16	6		6	81	6	
Turmeric	24			4		21	46	29	32	87	3	6	3		

[a] Spices not treated with ethylene or propylene oxide.

[b] Collated from published and unpublished data for dried spices analyzed in North America, Europe, the Middle East, and Japan.

[c] N: number of samples; total APC = 962; mold count = 808; most of the samples examined for mold count were also examined for APC.

[d] 100 to 999.

TABLE 24.4

Frequency of Occurrence of Coliforms and *Escherichia coli* in Spices Not Treated with Ethylene Oxide or Propylene Oxide [a, b, c, d]

Number of organisms (per g) [e]	Coliforms		*Escherichia coli*	
	Number of samples	Percent	Number of samples	Percent
$<10^{-1}$	110	48	180	79
$10^{-1}-10^1$	29	13	11	5
10^1-10^2	31	14	22	10
10^2-10^3	21	9	10	4
10^3-10^4	20	9	4	2
10^4-10^5	10	4	1	
10^5-10^6	7	3		
Number of samples	228		228	

[a] Whole and ground spices.

[b] *Escherichia coli* was found in the following untreated spices: basil, bay, capsicum, celery seed, coriander, cumin, dill, fennel, garlic, ginger, onion, oregano, parsley, pepper (black), rosemary, sage, and thyme.

[c] *Escherichia coli* is rarely found in spices treated with ethylene oxide or propylene oxide.

[d] Collated from published and unpublished data for dried spices analyzed in North America, Europe, the Middle East, and Japan.

[e] $10^1-10^2 = 10$ to 99; $10^2-10^3 = 100$ to 999, etc.

***Bark* (Rosengarten, 1973).** The first harvest for cassia takes place during the rainy season. Shoots of 2 to 2.5 m are cut close to the ground, and the bark is peeled off in strips and left to ferment in bundles for 24 hr. The corky outer layer of the bark is scraped off, leaving the clean, light-colored bark. Drying—first in the shade, then in the sun—may require 3 to 4 days.

***Berries* (Rosengarten, 1973; Breag et al., 1972).** Allspice (pimento) in Central America and Mexico is harvested almost entirely from the natural virgin rain forests. Branches of large wild pimento trees are chopped off and the berries then picked off the ground. The berries are dried for a few days in the sun on straw mats. At collection centers they are boiled to combat mold growth and then dried and packed for export.

In Jamaica, allspice trees are cultivated. The berries are picked off the trees and fermented for 4 to 5 days in heaps about 60 cm high. The fermentation is accompanied by a temperature rise, and the berries become

TABLE 24.5

Main Components of Mold Flora as Percent of Mold Count [a,b]

Spice	Mold count (per g)	Absidia spp.	A. candidus [c]	A. flavus	A. fumigatus	A. glaucus (group)	A. nidulans	A. niger	A. tamarii	A. terreus	A. versicolor	Mucor pusillus	Penicillium spp.	Rhizopus spp.
Allspice	7.0×10^4	3	+ [d]	1	—	9	—	80	—	—	1	—	6	—
Anise	9.5×10^3	—	—	1	1	55	—	3	—	1	2	—	33	—
Cardamon	1.6×10^3	—	—	3	—	64	12	12	—	1	—	—	—	9
Capsicum (chili)	3.9×10^4	33	—	+	—	69	1	17	+	—	+	1	1	1
Cinnamon	8.7×10^4	4	—	5	1	—	—	62	—	—	10	—	2	—
Coriander	1.3×10^5	—	7	—	—	67	7	1	—	2	—	—	2	—
Cumin	1.5×10^3	—	—	—	—	62	7	7	—	—	—	—	17	—
Fennel	6.7×10^3	2	3	2	2	62	4	6	2	2	5	—	—	—
Fenugreek	2.5×10^3	—	2	—	—	16	6	8	—	2	—	2	60	2
Ginger	1.7×10^3	—	15	3	—	32	—	9	—	—	—	—	35	3
Mace	8×10^2	12	—	—	—	88	—	—	—	—	—	—	—	—
Nutmeg	6.2×10^4	—	—	4	—	70	—	12	8	2	—	—	3	—
Paprika	5.5×10^2	27	—	1	—	27	—	10	—	—	—	—	18	18
Pepper (black)	6.4×10^5	—	2	1	—	92	—	+	+	—	1	—	2	—
Pepper (white)	6.5×10^4	—	16	7	8	2	13	11	2	3	12	—	21	—
Turmeric	0.5×10^2	—	—	—	—	100	—	—	—	—	—	—	—	—

[a] From Flannigan and Hui (1976); see also Pal and Kundu (1972).

[b] Minor components were Thermoascus crustaceus in black pepper; Talaromyces dupontii in fenugreek; Thermomyces lanuginosus in white pepper; Alternaria alternata in red pepper; Fusarium poae in fennel; and Syncephalastrum racemosum in ginger and nutmeg.

[c] A = Aspergillus.

[d] + = <1%; — = absent.

brown. They are then spread out in thin layers on concrete barbecues and turned over several times during sun drying. The time required for drying the berries down to a safe moisture content during dry weather conditions is approximately 5 days. At night and during downfalls of rain that are frequently sudden and heavy, the berries are swept up into heaps and covered; but rewetting is inevitable. This drying process results in an estimated loss of about 10% of the harvested berries due to mechanical damage and mold growth (despite a high content of antimicrobial compounds in the essential oil). Artificial drying reduces this loss (Breag *et al.,* 1972).

Seeds **(Rosengarten, 1973).** Although most seeds come from pods, and may be sterile *in situ,* the processes of harvesting, threshing, and winnowing offer ample opportunity for microbial contamination. Moreover, the smaller the particle, the greater the surface per unit of mass. Thus, large surfaces are available for adherence of microorganisms on most seeds; caraway seeds number about 300,000/kg and poppy seeds, about 2,000,000.

B. Drying

Drying of spice decreases the number of vegetative cells of bacteria. The flora remaining consist mainly of spore-forming bacteria and molds because of their ability to survive long periods on dried foods (Table 24.3). Yeasts may also be present. In general, bacterial counts are from 10^4 to 10^7/g except for cassia, cloves, mustard, and nutmeg, which tend to have smaller bacterial populations. Mold counts rarely exceed 10^5 g. However, extremely large variations occur in the microbial content of different lots of the same spice (Warmbrod and Fry, 1966; Yesair and Williams, 1942).

C. Disinfestation

Spice can be disinfested chemically with methyl bromide or low levels of ethylene oxide to destroy insects. Disinfestation, however, is not highly germicidal. Germicidal treatments require higher concentrations than those used for disinfestation.

D. Gas Treatment for Destroying Microorganisms

Ethylene oxide (Hruby *et al.,* 1961) and to a lesser extent, propylene oxide, are used to destroy microorganisms (see Chapter 10). Spices thus treated are frequently, but erroneously, termed "sterile." Gas treatment destroys 99% or more of the bacteria, most of which are in the spore state, and 99.9% or more of the molds (Eschmann, 1965; Gerhardt, 1969; Lindberg and Nickels, 1976). Bacterial spores are only marginally more resistant than vegetative cells (Blake and Stumbo, 1970; Werner *et al.,* 1970). The effect of ethylene oxide on bacteria (aerobic plate count) under industrial conditions is given in Table 24.6 for ground, powdered, and granulated spices. There was destruction of at least 99.9% in about one-half of the samples. Bacteria in mustard and granulated onion were least susceptible to destruction by ethylene oxide while bacteria in cardamom, black pepper, thyme, and nutmeg were most susceptible. Destruction of molds in the same samples was usually from 99 to 99.9%, occasionally higher.

The moisture content of the spice to be treated should be as high as possible and compatible with keeping quality, at least 6% (Guarino, 1972); the relative humidity in the exposure chamber should be about 35% (Gilbert *et al.,* 1964) and the ethylene oxide concentration about 750 g/m^3 (Coretti and Inal, 1969). The temperature should be elevated

TABLE 24.6

Percentage Decreases in Aerobic Plate Count (APC) of Spices due to Industrial Treatment with Ethylene Oxide [a, b]

Percentage decrease in APC [c]	Percentage of all samples	Spices with median value in category given in column (1) [d]
<80	1.5	None
80–89.9	6.6	Mustard flour, ground mustard seed, granulated onion
90–98.9	22.1	Allspice, basil, chili powder, granulated garlic, mace, oregano, parsley, pepper (white), sage
99–99.9	24.3	Bay leaf, capsicum (chili), celery seed, chili powder, curry, dill seed, paprika
99.9–99.99	24.3	Caraway, cinnamon, coriander, cumin, fennel, fenugreek, ginger, marjoram, rosemary, savory, turmeric
99.99–99.999 [d]	14.7	Cardamon, pepper (black), thyme
>99.999	6.6	Nutmeg

[a] Spices were ground, powdered, or granulated.

[b] Usually 3 to 4 lots of each spice were examined before and after treatment; in all, 136 samples.

[c] Decrease of 99.999% = decrease of 5 $\log_{10}$, e.g. 10^8 per g to 10^3 per g.

[d] Values were calculated and collated from unpublished data; the median of 3 to 4 samples determined placement in categories in column 1.

slightly above ambient, to about 25°–30°C, to speed destruction of micro-organisms (Coretti and Inal, 1969). Chamber temperatures of 50°–60°C may be used; however, it is unlikely that the centers of bags or barrels of spice in the chamber attain these temperatures. The exposure time should not be more than 6 hr, especially for paprika (Szabo, 1968); and 2-hr exposure is used for some spices. A mixture of ethylene oxide and methyl formate is recommended for those few spices (tumeric and mustard seed) whose color and flavor are adversely affected by ethylene oxide (Mayr and Suhr, 1972).

Residues of ethylene oxide and its most common reaction product, ethylene chlorhydrin (Wesley et al., 1965), are limited by regulation in some countries. For instance, in Canada, 1500 ppm of ethylene chlorhydrin is permitted (Department of National Health and Welfare, 1978); in the United States, 50 ppm of residual ethylene oxide is permitted (U.S. Department of Health, Education and Welfare, 1978a).

Propylene oxide is used instead of ethylene oxide in some countries, but not in others where it is prohibited. On a weight basis it is less bactericidal than ethylene oxide, but it is less likely to form toxic by-products. The permitted residue of propylene oxide in the United States is 300 ppm (U.S. Department of Health, Education and Welfare, 1978a).

See also Volume I, Chap. 10 for a more extensive discussion of these gases.

E. Some Novel Treatments

Some novel methods for reducing the microbial content of spices include exposure to hot ethanol vapor (Wistreich et al., 1975) or acidification with hydrochloric acid followed by neutralization. The latter procedure has been examined for decontaminating paprika, and the resultant paste is suitable for use in some meat products (Huszka et al., 1973).* Heating in steam under pressure is not commonly used to destroy microorganisms in spice because it causes a loss of some of the volatile oils (Yesair and Williams, 1942).

F. Irradiation

Irradiation of spices, especially with τ rays, has been found effective in reducing microbial populations (Gerhardt, 1969; Heins and Ulmann, 1971;

* Paprika contributes a red color to meat, and, therefore, unscrupulous persons have sometimes used it to conceal a stale or spoiled condition. It is not legal in some countries.

Goto *et al.,* 1971; Tjaberg *et al.,* 1972; Vajdi and Pereira, 1973; Gotts-chalk, 1977). Depending on the number and type of microorganisms, the dose necessary to reduce the population to a satisfactory level (less than 10,000/g) varies from 0.4–0.7 Mrad (Inal *et al.,* 1975). Irradiated seasoning material used in canned meat products introduces fewer heat resistant bacteria than nonirradiated seasoning, thus enabling reduced thermal processing (Farkas, 1973; Kiss *et al.,* 1974). Gamma irradiation results in less loss of essential oils than treatment with ethylene oxide (Vajdi *et al.,* 1973) but is not legally acceptable in most countries.

Ultraviolet irradiation has little penetrating power and has only limited effectiveness in reducing bacteria on spices (Walkowiak *et al.,* 1971) despite continuous agitation to expose surfaces (Eschmann, 1965) (see Volume I, Chaps. 2 and 3).

IV. SPOILAGE

A. Spoilage of Spices

There is virtually no bacterial spoilage of spices subsequent to harvesting and drying. However, fungal spoilage does occur prior to drying; or it may occur during storage and shipping especially if relative humidity and temperature are high, or localized wetting occurs. Moldy spices introduce viable organisms that may grow in some products, and may introduce moldy off-flavors and enzymes, e.g., pectinase.

Emulsions of mixed essential oils can sometimes support bacterial growth to 10^7–10^8/ml in the aqueous phase when inhibitory molecules (e.g., cinnamic aldehyde in oil of cinnamon) migrate out of the aqueous phase into the oil phase (e.g., oil of nutmeg). Adjustment to pH 4 with lactic acid effectively controls this problem (Pirie and Clayson, 1964).

B. Spoilage of Foods by Microorganisms from Spices

Some spices may contain spores from mesophilic aerobes, mesophilic anaerobes, and flat sour thermophiles. Excessive numbers of such spores introduced with spice into food that is subsequently canned have caused spoilage (Jensen *et al.,* 1934; Castell, 1944; Bean and Salvi, 1970; Fievez and Granville, 1965; Julseth and Deibel, 1974). Thus, a major concern of spice processors is that the microbial load does not contribute to spoilage of food in which the spice is used (Guarino, 1972).

Microbially contaminated spices have been associated with softening of dill pickles and spiced fruit in acidified syrup, but experimental conditions were not adequate to evaluate whether the bacteria or pectolytic enzymes in the spices were responsible (Fabian *et al.,* 1939). The quality of processed meats may be adversely affected by the bacteria and molds introduced with spices (Jensen *et al.,* 1934; Jensen, 1954; Mossel, 1955; Müggenburg, 1956; Walz, 1956; Pohja, 1957; Hruby *et al.,* 1960; Silliker, 1963; Heath, 1967; Hadlock, 1969; Palumbo *et al.,* 1975). However, the potential for spoilage would depend on whether the meat was canned and pasteurized or heated to obtain a shelf-stable product; fermented; or cooked and refrigerated. Psychrotrophic aerobic spore formers are uncommon in spice (Michels and Visser, 1976), and bacteria in spices would probably not cause spoilage in refrigerated products. Even when there is a low probability of contaminated spice causing spoilage (e.g., in dry gravy bases or dehydrated soups) spice may introduce numbers of microorganisms considered undesirable to industrial or regulatory interests (Karlson and Gunderson, 1965; Kadis *et al.,* 1971; Surkiewicz *et al.,* 1972, 1976).

V. PATHOGENS

Spices are not major contributors to foodborne disease. However, they sometimes contain bacteria that can cause foodborne infections; they are frequently contaminated with molds capable of synthesizing mycotoxins; and they sometimes contain mycotoxins. The sources of such pathogens can be soil, fecal material from birds and other animals (Heath, 1964), and nonpotable water used for soaking some spices, e.g., in the production of white pepper and bleached ginger.

A. Bacteria

Spore-forming organisms that are capable of causing gastroenteritis when ingested in large numbers are found in spices in small numbers. To cause illness, they would have to multiply to 10^5–10^6/g of food to which the spice is added. The spore formers of concern are *Clostridium perfringens* and *Bacillus cereus. Bacillus subtilis,* common in spice, has only rarely caused foodborne gastroenteritis.

Clostridium perfringens has been found in up to one-half of samples of black pepper, pimento, coriander, bay leaf, and nutmeg and less frequently in cinnamon and cloves (Nikolaeva, 1967; Powers *et al.,* 1975; Leitao *et al.,* 1973–1974; Krishnaswamy *et al.,* 1971). The number per gram is usually less than 500 and is rarely over 1000.

┖ *Bacillus cereus* was found in 53% of 110 samples of spices, and in each of the seven varieties of spice examined; 89% of the isolates were toxigenic. The number per gram is rarely over 5000 (Powers *et al.,* 1976; Kim and Goepfert, 1971).

Non-spore-forming bacteria, both indicator organisms and those potentially capable of causing disease, are found in spices. Coliforms are often present in some spices and *Escherichia coli* is occasionally present, usually in low numbers (Table 24.4) (Julseth and Deibel, 1974; Powers *et al.,* 1975). *Salmonella* is found infrequently, but in a wide variety of spices (Leitao *et al.,* 1973–1974; Sperber and Deibel, 1969; Guarino, 1972). In one Canadian study (H. Pivnick, unpublished), seven of 1075 samples contained *Salmonella.*

Black pepper contaminated with *Salmonella weltevreden* was responsible for several incidents of human salmonellosis over a wide area (Laidley *et al.,* 1974). The source would probably have remained undetected except for the sudden increase of human salmonellosis due to this serotype, which was unusual in the area where the spice was distributed. In investigations of salmonellosis, the possibility of contaminated spice should not be overlooked.

B. Molds

Molds are frequently present on spices (Table 24.3) (Christensen *et al.,* 1967; Horie *et al.,* 1971; Julseth and Deibel, 1974; Powers *et al.,* 1975), usually fewer than 100,000/g, occasionally more. A few of the molds are mycotoxin producers (Hadlock, 1970; Pal and Kundu, 1972). Spices that are too moldy for use as ground spice, are sometimes steam-distilled to obtain essential oils or extracted with organic solvents to obtain oleoresins; such solvents can also extract mycotoxins.

Mycotoxins have been synthesized by indigenous molds growing on some spices under experimental conditions (Frank, 1966; Schindler and Eisenberg, 1968; Flannigan and Hui, 1976). More importantly, aflatoxins (the only mycotoxins for which spices have been examined) have been found in commercially available spices (J. I. Suzuki *et al.,* 1973; Scott and Kennedy, 1973, 1975; Girgis *et al.,* 1977). The concentrations are usually low (less than 10 μg/kg). Nutmeg appears to be especially prone to aflatoxin production, but levels are usually less than 25 μg/kg (Beljaars *et al.,* 1975). However, as nutmeg is unlikely to be used in food at greater than 1% concentration, the maximum aflatoxin content in food from added nutmeg would be low indeed, about 25 μg/100 kg.

VI. CONTROL

The opportunity for control of microbial spoilage of raw spices is largely limited to adequate drying in the production areas and to prevention of rehydration during shipping and storage.

The decontamination procedures with ethylene oxide or propylene oxide assure destruction of a high percentage of spore-forming bacteria and a high percentage of the less numerous non-spore-forming microorganisms; but such procedures are not used universally in technically advanced countries and not used at all in many parts of the world.

Quality control procedures have been described by many organizations. For instance, the American Spice Trade Association has published specifications for cleanliness (insects, excreta visible mold, etc.); the Association requires that 98–99.5% of the product meet specifications or be reconditioned (American Spice Trade Association, 1971). The American Public Health Association (Guarino and Peppler, 1976) and the American Spice Trade Association (1976) have published methods for microbiological examination of spices. The aerobic plate count and the mold count are the tests most frequently used. Specific types, e.g., spores of thermophiles and *Clostridium perfringens,* are sometimes enumerated, and the presence or absence of *Salmonella* is determined. It is not worthwhile to examine spice for *Staphylococcus aureus,* which is rarely present.

Coliforms frequently sought in spices are usually of dubious value as indicators of fecal contamination (Vaughn, 1970; Firstenberg *et al.,* 1974; Powers *et al.,* 1975) because of the ubiquity of coliform bacteria of non-fecal origin in plant material (Geldreich, 1966; Sheneman, 1973). Examination for *Eschericheria coli* is not a good test to indicate the possible presence of *Salmonella* in spice unless the quantity of spice examined for *E. coli,* usually about 0.1 or 1 g, is increased to the quantity used for determination of *Salmonella* (e.g., 375g, American Spice Trade Association, 1976; 125 grams, H. Pivnick, unpublished Canadian study).

If *Salmonella* is of concern, the spice should be analyzed by methods that include dilution or neutralization of the antimicrobial activity of the spice and pre-enrichment in a nonselective medium (Sheneman, 1973; Julseth and Deibel, 1974; Wilson and Andrews, 1976). Failure to dilute or neutralize antimicrobial compounds when required (e.g., in allspice, cassia, cloves, cinnamon, garlic, oregano, onion) may give false negative results for *E. coli* and *Salmonella.* Some leafy materials, although not inhibitory to *Salmonella,* require a higher ($> 10:1$) broth to spice ratio to wet them adequately. Examination of composited samples for *Salmonella* (Silliker and Gabis, 1973; American Spice Trade Association, 1976) is more economical and more meaningful than examination for coliforms or

E. coli by the MPN method. In the previously mentioned Canadian study, seven of 1075 samples contained *Salmonella* when 125-g samples were examined; five of the seven positive samples contained fewer than 1.8 *E. coli*/g, and the remaining two contained fewer than 23 *E. coli*/g as determined by the MPN procedure.

VII. CHOICE OF CASE

For an understanding of "Case," see Table A.1 and the brief description in Appendix IV. For more details, see book 2 of this series (ICMSF, 1974).

The safety and spoilage hazards associated with spices added to foods depend on the probability of contamination with disease-producing or spoilage organisms and toxins. Spices are used in a wide variety of foods and are thus capable of introducing contaminants that may be of no consequence in some products but highly important in others (Krishnaswamy *et al.,* 1971). For instance, spores of *Clostridium perfringens* introduced into soup mixtures to be canned would be of no consequence because canned soups receive a high thermal process that would kill spores (Case 7). In contrast, these spores introduced into dry gravy base would be of consequence because gravies used in food-service establishments do not receive a high thermal process and are frequently held at temperatures at which the spores could germinate, and at which the vegetative cells could multiply. In this context, Case 9 is indicated. Similarly, *Salmonella* may be of no consequence in chili peppers used in a spaghetti sauce that is to be canned but may introduce some hazard when added directly to cooked spaghetti immediately prior to consumption. In this example, Case 11 would apply, assuming there is, in the normal course of events, no opportunity for salmonellae to grow in the spaghetti. *Salmonella* in black pepper or paprika used in potato salad, or other products that would support growth and are subjected to abusive temperatures, e.g., egg salad in sandwiches taken from home to school, would merit Case 12.

ICMSF (1974) uses Case 2 for the aerobic plate count and molds in spices. For many spices, the microbiological specifications are not achievable unless the spices are decontaminated with ethylene oxide.

In general, the spoilage and health hazards presented by spice must be carefully evaluated in the context of the use of the spice, with the more stringent requirements being reserved for those spices that are added to canned foods that receive a minimal thermal process or to cooked foods that are not reheated before eating.

25

Fats and Oils

I. INTRODUCTION

Fats and oils are subject to attack by a variety of microorganisms, provided that various factors necessary to microbial growth are also favorable, e.g., moisture, nitrogen, and minerals. Microbial enzymes can hydrolyze fats to yield free fatty acids and glycerol; and further, lipoxidases elaborated by microorganisms can produce hydroperoxides. The free fatty acids and glycerol resulting from hydrolytic cleavage are readily metabolized by a wide variety of microbes. A detailed discussion of microbial attack on lipid materials is beyond the scope of this book. For such discussion see Eskin *et al.* (1971).

This chapter addresses itself to a discussion of processed foods in which fat is the primary constituent. Butter, margarine, and peanut butter represent foods that are emulsions comprised of oil (or fat) as the continuous phase and water as the discontinuous phase. Mayonnaise, salad dressing, and related products are examples of emulsions in which water is the continuous phase and fat the discontinuous phase.

The form of the emulsion has a profound relationship to microbial stability. With oil-in-water emulsions such as mayonnaise, the growth rate of microorganisms is not affected by the water dispersion; only the chemical composition of the water phase plays a role. On the other hand, with water-in-oil emulsions such as butter, the water exists as microscopic droplets that are dispersed throughout the oil matrix. The majority of these droplets are sterile, and if microorganisms are present, their growth is restricted by limited water and food supply as well as by inhibitory factors in the aqueous phase such as salt, preservatives, and organic acids.

II. MAYONNAISE AND SALAD DRESSINGS

A. Definitions

Mayonnaise or mayonnaise dressing is defined by the U.S. Food and Drug Administration (U.S. Department of Health, Education and Welfare, 1978c) as an emulsified semisolid food prepared from vegetable oil(s), vinegar, lemon juice and/or lime juice, and egg yolk-containing ingredients. Several optional ingredients may be used including salt, sweeteners, spices, monosodium glutamate, sequestrant(s), citric and/or malic acid, and crystallization inhibitors. If raw eggs or raw egg yolk-containing ingredients are used, the pH must not be above 4.1, and the acidity of the aqueous phase expressed as acetic acid must not be less than 1.4%. Also, the final product must be held for 72 hr before it is made available to consumers. Mayonnaise is creamy, pale yellow, and mild flavored, with a pH range from about 3.6 to 4.0. Acetic acid is the predominant acid and represents 0.29–0.5% of the total product. The aqueous phase contains 9–11% salt and 7–10% sugar. The vegetable oil content must be at least 65%; generally, it is 70–80%.

Salad dressing is defined by the U.S. Food and Drug Administration (U.S. Department of Health, Education and Welfare, 1978d) as an emulsified semisolid food prepared from vegetable oils, vinegar, lemon juice and/or lime juice, egg yolk-containing ingredients, and a cooked or partially cooked starchy paste. Several optional ingredients may be used including salt, sweeteners, spices, monosodium glutamate, stabilizers and thickeners, citric and/or malic acid, sequestrants, and crystallization inhibitors. If raw egg yolk-containing ingredients are used, the pH of the salad dressing must not be above 4.1, and the acidity of the aqueous phase expressed as acetic acid must be not less than 1.4%. In addition, the final product must be held for 72 hr before it is made available to consumers. Salad dressing is creamy, pale yellow, and has a tart flavor. Its pH ranges from about 3.2 to 3.9. Acetic acid is generally the predominant acid and represents 0.9 to 1.2% of the total product. The aqueous phase contains 3.0 to 4.0% salt and 20–30% sugar. The vegetable oil content must be at least 30%. Thus, salad dressing differs from mayonnaise in that it contains (a) a cooked or partially cooked starchy paste, (b) about double the amount of acid, (c) three times the amount of sugar, (d) one-third the amount of salt, and (e) less than one-half the amount of vegetable oil.

"Pourable" dressings of various types generally have a pH in the range 3.5 to 3.9. Many of these contain cheese that may be added either milled or in chunk form. Some of the products are homogenized; others separate into oil and water phases.

"Spoonable" dressings contain less acid than pourable dressings and must be kept refrigerated, in contrast to mayonnaise, salad dressing, and pourable dressings, which are stable at ambient temperature.

B. Important Properties of Mayonnaise and Salad Dressing

The microbial stability of these products is due primarily to the preservative effect of vinegar (acetic acid). Generally, an acetic acid concentration in excess of 1.5% makes the product unpalatable, but a concentration much below this permits spoilage (Kurtzman, 1976). In the pH range found in mayonnaise and salad dressing, acetic acid exists primarily in the undissociated form and thus exerts its maximum antimicrobial activity (see Vol. I, Chapter 7). Sodium chloride in mayonnaise, and sugar and sodium chloride in salad dressing contribute to microbial stability by reducing the a_w to about 0.92 (Smittle, 1977). There are no doubt interactions between reduced a_w and acidity in these products. For example, it has been demonstrated that the sensitivity of *Salmonella typhimurium* to a variety of organic acids is influenced by a_w (Goepfert and Hicks, 1969).

C. Processing Procedure

The manufacture of mayonnaise is generally carried out in relatively small batches using tanks equipped with high-speed beaters. Commonly salt is added to the egg blend, and the mixture is refrigerated. Ultimately the salted egg blend is mixed with spices, sugar, and vinegar to form a premix. Following this, the oil is gradually added. The ingredients are coarsely emulsified and then given a further homogenizing treatment using a colloid mill before packaging.

The production of salad dressings is similar. A cooked starch paste is prepared from starch, water, salt, and sometimes vinegar, then brought to 88°–90°C in either open tanks or in votators (closed systems). Vegetable meats, such as pickles and pimentos are added to some salad dressings, in which case they are introduced into the starch paste before the cooking step. The remaining ingredients are mixed as in the manufacture of mayonnaise and then are blended with the cooled starch paste. The final mixture is passed through a colloid mill or other homogenizing device before being packaged.

D. Initial Microflora

Though the ingredients, particularly spices and eggs, may contribute substantial numbers of microorganisms to mayonnaise and salad dressing, most of them are unable to survive at the low pH of these systems, and accordingly, only small numbers are encountered in the finished product shortly after its production (Kurtzman *et al.,* 1971). The organisms normally encountered include lactobacilli, yeasts, and aerobic spore-forming bacteria. Their occurence in the finished product does not necessarily indicate use of raw materials that are highly contaminated. Rather, it is a reflection of the selective nature of both the processing environment and the finished product itself. Both yeasts and lactobacilli are acid tolerant and thus more capable of surviving in the finished product than are a variety of other organisms that might be expected from raw ingredients. Furthermore, yeasts and lactobacilli comprise a major portion of the contaminating flora from processing equipment for the same reason. The occurrence of aerobic spore-forming bacteria is a direct reflection on the raw materials, particularly spices, which are frequently highly contaminated with members of the genus *Bacillus.* The survival of these organisms in the finished product reflects primarily the resistance of the spores (see also review in Smittle, 1977).

E. Spoilage

Spoilage caused by microbial growth is second in importance to oxidative rancidity that may develop on the surface of the product. Emulsion breakdown due to agitation or freezing occurs on rare occasions.

However, microbial spoilage of mayonnaise, salad dressing, and related products does occur. Though early reports suggested spoilage was caused by the growth of aerobic spore-forming bacteria (Iszard, 1927a,b,c; Pederson, 1930; Appleman *et al.,* 1949; Fabian and Wethington, 1950), more recent reports indicate it to be due almost always to the growth of yeasts and lactobacilli. Because these organisms frequently die after growth in these products, it seems likely that the spoilage problems encountered by earlier workers actually involved growth of yeasts and lactobacilli but that these organisms were overlooked as spoilage agents because they had already died.

Lactobacillus fructivorans was first isolated from spoiled salad dressing (Charlton *et al.,* 1934). Kurtzman *et al.* (1971) later showed it to be a common spoilage organism that required a special isolation medium. Yeasts cause spoilage of a wide variety of dressings. Fabian and Wethington (1950) reported that a species of *Saccharomyces* spoiled French dressings and salad dressings; Williams and Mrak (1949) found that a yeast similar to *Saccharomyces globiformis* was responsible for spoilage of a starch-base dressing. Two-thirds of the spoiled dressing samples examined by Kurtzman *et al.* (1971) contained *Saccharomyces bailii.* This yeast was the only one found, and the results were particularly significant because the samples came from widely separated areas of the United States. The organism ferments glucose rapidly but shows a delayed fermentation of sucrose. This finding explains the long delay between manufacture and spoilage of products contaminated with this yeast, when sucrose is used as a sweetening agent.

Gas production is usually the first evidence of microbial spoilage in these products. This may or may not be accompanied by the development of off-odors and flavors. At times there is a thinning effect on texture. An unusual outbreak of spoilage caused by *Pichia bini* was investigated (J. H. Silliker, unpublished). In this incident, only the surface of the mayonnaise was involved. The spoiled product exhibited noticeable off-flavors and dark buttonlike yeast colonies on the surface. The condition was evident only in jars in which air had entered through faulty closures. The problem was resolved by obtaining better closures, decreasing air flow in the manufacturing area, and instituting an improved sanitation program.

In another instance, growth of a *Geotrichum* species occurred during storage on the surface of mayonnaise in some but not all jars from a particular production line. Investigation revealed that during the 1 or 2 sec that the mayonnaise was exposed before the caps were applied, mold spores in the airstream from a contaminated space heater fell onto the top surface of the mayonnaise in some but not all the jars (R. B. Smittle, personal communication).

Diagnosis of spoilage requires, at times, considerable ingenuity. Because the spoilage organisms are apt to die rapidly, it is not uncommon to find an obviously spoiled product completely devoid of viable microorganisms. Direct microscopic examination of the product will generally reveal the presence of large numbers of yeasts and/or gram-positive rod-shaped bacteria, *i.e.*, lactobacilli. Further complicating the problem is the fact that fastidious lactobacilli, such as *Lactobacillus fructivorans*, do not grow on the laboratory media usually employed but are readily detected with a *Lactobacillus* Selective Agar (LBS). Even on this special medium growth may be quite slow, requiring 7–14 days at 20°–25°C (Charlton *et al.*, 1934).

Contaminated raw ingredients are seldom the primary cause of spoilage outbreaks, though they can be the original source of the spoilage organisms. Faulty sanitation practices, frequently in the mixing and filling equipment, are almost always the direct cause of product failure. Malfunctioning pumps or pistons on filling machines permit product accumulation in locations that are not routinely subject to wet cleaning and disinfecting. When this occurs, foci of microbial growth develop, and because the substrate is acidic, such environments are selective for spoilage organisms.

Investigation usually reveals that the acidity of the finished product is sufficient to prevent the development of small numbers of spoilage organisms, but when the inoculum becomes inordinately high, the same product becomes unstable. To prevent spoilage outbreaks, it is necessary to follow good manufacturing practices. Preventive maintenance programs must be instituted to be certain that portions of equipment not designed for routine cleanup are not contributing to contamination.

F. Pathogens

The fate of foodborne pathogens in mayonnaise and salad dressing has recently been reviewed (Smittle, 1977).

1. Salmonella

Though egg yolks and whole eggs may be contaminated with salmonellae, the organisms die in a matter of days in mayonnaise and salad dressing if the acetic acid concentration is adequate. The pH should be 4.1 or below (0.25% acetic acid), and the product should be held at 18°–22°C for at least 72 hr unless pasteurized eggs are used. (See also this volume, Chapter 19.)

In some mayonnaise manufactured in Europe, the acidifying agent is lemon or lime juice or hydrochloric acid instead of vinegar (acetic acid). Citric and hydrochloric acids are far less active than acetic acid against microorganisms. The U.S. regulations (U.S. Department of Health, Education and Welfare, 1975a,b) also permit acids other than acetic but only within certain limits.

Recently, the World Health Organization cooperated with Spanish authorities in investigating outbreaks of salmonellosis among passengers on four flights between Las Palmas in the Canary Islands and Helsinki. The flights were provisioned with egg mayonnaise. *Salmonella typhimurium* was isolated from several patients, one food handler, and the batch of mayonnaise used for one of the meals. Many of the hundreds of victims were severely ill; there were six deaths (Davies and Wahba, 1976).

The hazards of improperly prepared mayonnaise are further illustrated by the following:

In Denmark, mayonnaise is usually prepared in homes by mixing egg yolks with oil, salt, and lemon juice, however, vinegar (4–7% acetic acid) may be added. In 1955, mayonnaise from a salad factory caused approximately 10,000 cases of salmonellosis, and a circular was prepared by the Danish Administration recommending procedures for preparing mayonnaise so that the pH of the product would be lower than 4.5. These recommendations involved a certain proportion of egg yolk and vinegar, a holding time before using of either room temperature for 4 days or 40°C for at least 2 hr. Since then, no cases of salmonellosis caused by commercial preparations of mayonnaise have occurred in Denmark (B. Simonsen, personal communication).

The following two cases are on record, however, from large kitchens in Denmark:

1. One outbreak was traced to mayonnaise that was mixed with cod. The mayonnaise was prepared by mixing egg yolks with vinegar, holding at room temperature for 20 min, and refrigerating thereafter. The pH of the mayonnaise was found to be 5.1. A sample of the mayonnaise collected 4 days later showed a *Salmonella* count of 180,000/g (Petersen, 1964).

2. In another outbreak, 42 people attended a confirmation party involving lunch, dinner, and sandwiches late in the evening. Forty-one people became ill (Number 42 did not eat lunch) with nausea, vomiting, diarrhea, and abdominal pain. There were two fatalities. The lunch was served with homemade salads, with mayonnaise that was prepared the afternoon before with raw egg yolks, sugar, vinegar, oil, and water. The mayonnaise was left at room temperature until it was mixed with other ingredients before the lunch. Samples of the mayonnaise obtained approximately 2 days after its preparation (with intervening storage at 4°C) showed *Salmonella* counts of 6,000,000/g. Held-over salads showed *Salmonella* counts between 50,000,000 and 120,000,000/g (Meyer and Oxhøj, 1964).

These episodes illustrate the necessity for careful pH control in the preparation of mayonnaise and related products. Mayonnaise is particularly troublesome since it contains eggs.

2. Staphylococci

The a_w of mayonnaise and salad dressing is not sufficiently low to preclude the growth of staphylococci. However, these organisms are destroyed at low pH and die in mayonnaise and salad dressing with a pH of 4.1 or below (Smittle, 1977).

3. Other Pathogens

Mayonnaise and salad dressing will not support the growth of *Clostridium botulinum* and *Clostridium perfringens* since these organisms cannot grow at a pH below 4.7 or at a water activity below 0.95. Thus, independent of the known interactions between water activity and pH, each of these factors separately precludes growth of these organisms. Similarly, low a_w alone is sufficient to preclude spore outgrowth by *Bacillus cereus*. Endospores of these organisms remain viable, but in the event germination occurs, the cells will not grow (Smittle, 1977).

G. Control

The second book of this series (ICMSF, 1974) makes no reference to specifications for mayonnaise, salad dressing, and related products. Indeed, provided that the finished product has a pH of 4.1 or below (0.25% acetic acid), no health hazard exists. Despite this, the public and many public health officials and physicians believe that mayonnaise and salad dressing can cause food poisoning and infection. This belief is perpetuated by the news media. The use of mayonnaise and salad dressing in salads and other convenience foods is discussed in this volume, Chapter 27. The influence of these products, if properly made, on the stability and safety of meat and vegetable salad and sandwich fillings recently has been reviewed extensively (Smittle, 1977). Mayonnaise, salad dressing, and related products increase the stability of foods in which they are incorporated, and the commonly held belief that acid dressings are important vehicles in food-poisoning outbreaks is without merit.

H. Choice of Case

For an explanation of "Case," see Table A.1 and the brief description in Appendix IV. For more details, see the second book of this series (ICMSF, 1974).

The safety of these products rests on control of pH, since no food-poisoning hazard exists if the pH is 4.1 or below. Their stability relates to formulation but also to sanitation within the processing environment. Microorganisms in Foods, 2 (ICMSF, 1974) makes no reference to either mayonnaise or salad dressing, and in view of the foregoing this seems appropriate. The most effective means of preventing disease transmission is through pH control. Furthermore, microbiological tests conducted on the finished product are poor indicators of potential stability.

Mayonnaise, salad dressing, and related products having final pH values above 4.1 offer a potential risk of salmonellosis and *Staphylococcus* food poisoning; the degree of risk depends upon the pH, the acidulant used, and the intended use of the product. If concern for the product stems from the likelihood of contamination with staphylococci, Case 7 would be applicable if the dressing were to be incorporated into another food product and consumed immediately. On the other hand, Case 9 would apply if consumption were to be delayed. The same Cases would apply to low-acid dressings likely to be contaminated with *Clostridium perfringens* or *Bacillus cereus*.

Similarly, Case 10 would apply to low-acid products particularly susceptible to contamination with *Salmonella* if the products were added to foods to be consumed immediately; Case 12 would apply when such dressings were added to foods with a subsequent delay before consumption.

III. PEANUT BUTTER

A. Definitions

Peanut butter is produced from roasted shelled peanuts. The fat content averages 49–52%, and in the United States it may not exceed 55%. Peanut butter consists of 90–93% peanuts. Product containing 90% would contain 26% protein. Generally, salt is added to the product to a final concentration of 1–2%, although dietetic peanut butter contains less.

B. Important Properties

Nut butters, of which peanut butter is the best example, are analogous to margarine and butter in that the oil phase is continuous and the moisture phase discontinuous. The moisture content is less than 1%. Obviously, low water activity accounts for the microbial stability of the product.

C. Processing Procedures

Raw peanuts, generally purchased by the processor in shelled form (see this volume, Chap. 21, Section IV), are first roasted in either continuous or batch type roasters which operate at approximately 170°–180°C. Then they are ground and milled, salt and optional sweetening agents are added, and finally the mix is homogenized and packaged.

No moisture is added at any step in the processing of the roasted nuts to finished product.

D. Initial Microflora

As a raw agricultural commodity grown underground, peanuts are subject to contamination with a wide variety of microorganisms derived from the soil, the air, and the equipment in the shelling plant (see this volume, Chap. 21, Section IV,B,1).

E. Effects of Processing on Microorganisms

The roasting process eliminates all vegetative microorganisms, and the flora of the roasted peanuts is comprised solely of bacterial spores.

During the milling operation, considerable heat is generated, but the process is rapid and at the low a_w of the product virtually no additional microbial destruction occurs. The flora of the finished product will consist solely of small numbers of spore-forming bacteria if good manufacturing practices are followed.

F. Spoilage

The a_w of peanut butter is sufficiently low to preclude the growth of even the most drought-resistant microorganisms, even halophilic molds.

G. Pathogens and Toxins

1. Salmonella

Occasionally, peanut butter has been contaminated with *Salmonella,* but no outbreaks traced to the product have been reported. The organisms come from cross-contamination between raw peanuts and the roasted product, or from the plant environment. When the raw material is contaminated with salmonellae, the numbers present are usually quite low. Under good manufacturing practices, the salmonellae are largely eliminated from the plant environment, along with inert dust, soil, and other foreign material. A *Salmonella* problem can arise from the imprudent use of water during clean-up operations, because any water remaining on the equipment can support the growth of microorganisms, including *Salmonella.* Under such conditions, the whole processing environment can become contaminated with salmonellae. Control requires clear separation between raw peanuts and the processing areas. When water is used for cleaning and sanitizing, it must be removed completely and promptly from all product-contact surfaces (see Vol. I, Chapter 14).

Any salmonellae that contaminate the finished product will survive for long periods of time, just as in a variety of other foods with low a_w. The a_w of the finished product is sufficiently low, however, to preclude *Salmonella* growth.

2. Mycotoxins

Growth of molds can take place on freshly harvested peanuts before they are dried (see this volume, Chapter 21, Sections IV,B,4 and IV,B,5, and Table 21.4). Some of these molds produce mycotoxins, which carry through into finished products, including peanut butter (reviewed in Marth and Calanog, 1976). In the United States, the larger processors routinely conduct tests for mycotoxins on shipments of raw peanuts as received as well as on finished products. Further, after the roasting process the peanuts are electronically sorted in such a fashion that off-condition nuts are eliminated from further processing. This procedure has been effective in reducing the mycotoxin risk presented by peanuts. In the underdeveloped parts of the world, however, poor storage may permit the growth of myco-toxin-producing molds. In these areas, where peanut products often constitute a major part of the diet, the problem is particularly serious because the risk of disease appears to be particularly great in nutritionally deprived people.

3. Other Foodborne Disease Organisms

The low a_w of peanut butter precludes the risk of food poisoning caused by toxigenic microorganisms such as *Staphylococcus aureus, Clostridium perfringens, Clostridium botulinum* and *Bacillus cereus.*

H. Control

Microorganisms in Foods, 2, (ICMSF, 1974) recommends that nut butters be analyzed for *Salmonella.* This seems appropriate, but with good manufacturing practices non-spore-forming bacteria should be absent from the product. The use of coliform group, *Escherichia coli,* and/or Entero-bacteriaceae as indicators of cross-contamination within the plant might be considered.

To reduce mold growth and therefore mycotoxin levels, peanuts should be dried promptly after harvest to a moisture level below 9% (Table 21.4).

I. Choice of Case

For an understanding of Case, see Table A.1 and the brief description in Appendix IV. For more details, see the second book of this series (ICMSF, 1974).

Peanut butter is generally consumed without heating or other treatment that would destroy microorganisms, yet none grow in it. Therefore, for *Salmonella,* Case 11 would apply, unless the peanut butter were to be used as an ingredient in a high-moisture food such as a soup or nonacid salad dressing. Case 12 then would apply because the peanut butter would become moist enough for *Salmonella* to grow. If *Escherichia coli* or Entero-bacteriaceae were to be sought as indicators, Case 5 would apply.

IV. MARGARINE

A. Definitions

Margarine as defined under provisions of the U.S. Food, Drug, and Cosmetic Act embraces any plastic fat composition emulsified with at least 80% fat and with moisture in excess of 1%.

Several types of fats have been used to make margarine; the first was manufactured from oleo oil, prepared by fractional crystallization of beef fat. Subsequently, oleo oil was supplemented with neutral lard as the principal raw material. During and after World War I, deodorized coconut oil was used (Mattil, 1964). Currently, margarine contains soybean oil, cottonseed oil, coconut oil, corn oil, safflower oil, palm oil, peanut oil, or sunflower seed oil. Palm oil, now gaining in popularity (Weihrauch et al., 1977), is used much more extensively in Europe than in the United States. The oils are used in liquid form and they are partially hydrogenated.

Margarine can be made either with or without the addition of fluid skim milk or nonfat dry milk. Special emulsifying agents are added to margarine to confer on it the physical properties of a butter emulsion. Vegetable lecithins, particularly those derived from soybeans, as well as mono- and diglycerides are widely used in Europe and North America.

In the United States, margarine has an average salt content of about 3%, with extremes between 2.25% and 4%. Several unsalted margarines are available for persons on low sodium diets. Regular margarines contain by regulation at least 80% fat. Commonly used preservatives listed in decreasing order of frequency are sodium benzoate, potassium sorbate, calcium disodium ethylenediaminetetraacetate (EDTA), isopropyl citrate, and citric acid (Weihrauch et al., 1977).

Soft- or light-blend margarine contains 60% fat, with an aqueous phase of 40%, including milk solids, salt and other ingredients. Diet margarines contain only 40% fat, with the aqueous phase being correspondingly increased. Such margarines are generally devoid of milk solids.

Margarine spreads have oil contents intermediate between the light-blended and diet margarines. These products contain milk solids.

B. Important Properties

The aqueous phase of margarine contains milk solids and salt. If the salt level in the margarine as a whole is 3%, that in the aqueous phase is 15.8%, a level inhibitory to all foodborne pathogens except Staphylococcus. On the other hand, if it contains only 1% salt, the concentration in the aqueous phase is approximately 5.9%, a level that would permit the growth of most foodborne disease organisms.

The organic acids used as preservatives are not highly effective in non-acidulated margarines and margarinelike products; whereas if the pH is lowered by fermentation of the milk or by addition of acid, then the effectiveness of the preservatives is correspondingly increased (see Volume I, Chap. 7).

The major factor contributing to the stability of margarine is the fine dispersion of the water phase. In margarine with a fine water dispersion, 1 ml of water phase is divided into about 10^{10} droplets with diameters less than 6 μ. Accordingly, most of the water droplets contain no microorganisms. If these droplets contain mold spores, germination may occur, but growth will be at a slow rate into the continuous fat phase because of restricted supplies of water, oxygen, and food. Very coarse emulsions, especially in unpreserved milk margarine, are unsatisfactory because they encourage the growth of molds and yeasts (Muys, 1971). The size of the droplets varies with the type of product, namely, in decreasing order, diet > soft > regular margarines. Regular margarines are sold in sticks or bricks; soft-blended and diet margarines, usually in tubs.

C. Processing Procedures

Milk is received at the plant in cans, tank cars, or tank trucks. It is pasteurized, then refrigerated. Some manufacturers reconstitute nonfat dry milk and use it without pasteurization; others pasteurize and refrigerate it before use. In the past, starter cultures, selected on the basis of their ability to produce lactic acid and biacetyl, were added to the milk, which was then held for flavor development. However, the practice of culturing milk for margarine manufacture has declined. Flavoring agents, including biacetyl and starter distillates, are added directly to the pasteurized milk or water to be used for margarine or margarinelike products.

As indicated in Section IV,A, preservatives may be added to the milk; however, the pH of the aqueous phase (6.5–6.6) is not low enough to ensure the optimal antimicrobial effect from the various organic acids that may be used (see Volume I, Chap. 7). Lowering the pH of the aqueous phase, either by fermentation or by addition of acid increases the antimicrobial effect. However, in North America a sour flavor is considered undesirable, and experience indicates that an acidulated product loses its flavor more rapidly. Acidification is more common in Europe (Muys, 1971).

Salt is added prior to emulsification. Emulsifying agents enter either the oil or the aqueous phase, depending upon their relative solubilities. Emulsifying tanks equipped with efficient agitators are first charged with oil at a temperature somewhat above that desired in the emulsion (35°–45°C), and the cold milk is added slowly so that localized cold spots will not be produced to cause crystallization of the oil. A water-in-oil emulsion is formed.

Three different methods have been used for solidifying the emulsion. The earliest manufacturers poured or sprayed it continuously into a trough or vat of cold running water. The emulsion entered at one end of the vat, and as it passed to the other, it solidified and floated to the surface of the water in the form of a flaky mass that was then skimmed off. The procedure washed out a portion of the milk into the cooling water, which then became a source of microbial contamination.

Subsequently, a procedure involving a chill roll for solidifying margarine was developed. The emulsion was chilled by contact with the exterior surface of a revolving internally refrigerated drum. Under these conditions, the chilling surface and the product were both exposed to the atmosphere, thus affording the opportunity for airborne contamination of the product. With either the cold water or roll method of solidification, it was customary to transfer the solidified emulsion to trucks where the product was held for 24–48 hr at a relatively low temperature before processing. After this tempering, the material was moved through milling rolls or other machines of various designs and finally salted. Then the product was formed into prints, wrapped, and packaged.

After 1937, emulsifying machines came into general use. This has made possible the addition of salt to the mix before the emulsification step. As a result, all the water is present in minute droplets. In the modern votator equipment, emulsification, solidification, tempering, and working are carried out in a continuous flow in an enclosed system, with complete protection of the product from atmospheric contamination.

D. Initial Microflora

The initial microflora of margarine reflects that of the raw materials (primarily the milk and water) and the processing environment. Obviously, cultured milk used in a recipe will contribute large numbers of lactic organisms to the margarine, whereas pasteurized milk will contribute only small numbers of heat-resistant bacterial spores.

E. Spoilage

Margarine constitutes a selective environment for yeasts and molds. Milk-based margarine products are much more susceptible to spoilage than water-based products. The amount of salt, the nature of the packaging, the pH of the aqueous phase, and the presence of preservatives all influence susceptibility to spoilage.

In the United States, high-moisture diet margarines, packaged in tubs, have a tendency to "weep." The moisture, accumulating in pockets, is a good environment for mold or yeast development. Such products are stored under refrigeration.

Milk-based regular margarines occasionally mold on the surface and on the package where the margarine adheres. Frequently, this problem is traced to poor refrigerated storage. Parchment wrappers are more conducive to mold spoilage, because of the tendency of the product to adhere and the permeability of the wrapper to oxygen. Aluminum foil overwraps, because they are impervious to oxygen, inhibit mold spoilage. Cardboard cartons impregnated with potassium sorbate have been of limited value in protecting the surface of the product from mold growth. Packaged margarines containing milk generally carry the instruction that the product should be stored under refrigeration, where mold growth is minimized.

Cladosporium butyri and *Dematium* have been reported to produce ketonic rancidity in margarines; whereas *Cladosporium suaveolens* and *Candida lipolytica* cause fat hydrolysis (Muys, 1971).

F. Pathogens

In theory, milkborne pathogens could be carried into the finished product, but in modern margarine this problem is nil, because the pH, salt concentration, and the size of droplets in the emulsion can be adjusted to kill or inhibit these microorganisms. The greatest hazard lies in the diet- and soft-margarinelike products that contain high levels of moisture and in which the water droplets in the emulsion are larger.

It is incumbent upon the producer of high-moisture, low-salt margarinelike products to evaluate carefully the intrinsic ability of such finished products to destroy *Salmonella* and prevent the growth of staphylococci. Adjustment of pH, combined with effective preservatives, has been a successful avenue (Muys, 1971).

G. Control

The microbial stability and safety of margarine and margarinelike products is controlled by salt level and pH of the aqueous phase, the use of oxygen-impermeable packaging materials, the use of preservatives, and the use of refrigeration. The thorough dispersion of water in minute droplets throughout the oil matrix also inhibits microbial growth.

H. Choice of Case

For an understanding of "Case" see Table A.1, and the brief description in Appendix IV. For more details, see the second book of this series (ICMSF, 1974).

The most effective means of preventing disease transmission by margarine and margarinelike products is through proper formulation, particularly with reference to the aqueous phase. Microbiological tests conducted on the finished product are poor indicators of potential stability.

If a product were to be so formulated that *Staphylococcus* growth was possible, and the product was expected to be stored at ambient temperature, Case 9 would apply. If formulation were such that growth could not occur, or refrigeration was expected to be continuous until consumption, then Case 7 would be applicable, unless the product was incorporated into another food capable of supporting growth and was expected to be stored at ambient temperatures. In this event, Case 9 would apply. Similarly, for *Salmonella,* Case 10 would apply if the composition of the aqueous phase or refrigeration could inhibit growth. Case 12 would apply if the formulation and predicted storage conditions might permit growth of *Salmonella,* or if the product was to be incorporated into another food that permitted growth.

V. BUTTER

A. Definition

Butter consists primarily of milk fat, a small amount of nonfat milk solids, salt (if salted butter), and water. Artificial color and a lactic culture (butter culture) may be used in its manufacture. In the United States, butter is defined by law (statute) (U.S. Department of Health, Education and Welfare, 1976e) and by definition must contain not less than 80% milk fat; the other constituents may vary. The approximate composition of commercial salted butter is 80.5% milk fat, 16.5% water, 2.0% salt, and 1.0% curd (nonfat milk solids). The percentage of water in unsalted butter is somewhat higher.

B. Important Properties

Butter is a water-in-oil emulsion in which the water is dispersed as very small droplets, generally less than 30 μ in diameter (Muller, 1952). The microbial stability of butter depends on proper moisture distribution—

microbiologically, the most important characteristic of properly manufactured butter.

C. Processing Procedures

Butter is made by two distinct methods—a conventional churning process and a continuous process. The following is a brief description of these methods. More detailed descriptions may be found in the texts by Campbell and Marshall (1975) and McDowall (1953) and in the review by Mann (1973).

Both processes begin with cream obtained by centrifugal separation of milk into two fractions: cream containing 30–40% milk fat and skim milk containing from 0.01–0.1% milk fat. The percentage of fat remaining in the skim milk will depend on the test method used and on the efficiency of the separation.

Prior to World War II, cream used for buttermaking was primarily obtained directly from milk-producing farms as farm-separated cream. Cream was transported, usually after a few days storage on farms, to butter-processing plants. Skim milk was retained on farms as animal feed. An increased demand for non-fat dry-milk (dried skim milk) for human food, largely stimulated by the outbreak of World War II and by greater economy of operations, led to a different system of cream procurement for buttermaking. Under this system, currently used almost exclusively in major butter-producing countries, milk is transported by truck directly from the farm to the butter plant where it is separated and the cream made into butter. If receiving stations (milk-gathering centers) are involved in the system, the cream is separated at that point and then transported to butter plants for churning. Skim milk is dried in contiguous or separate plants. Butter is still made from farm-separated cream but the amount is very small.

Conventional or batch-type churning involves vigorous agitation of cream at approximately 10°C in a large barrel-type apparatus called the churn. Almost all batch-type churns in use today are made from stainless steel or aluminum magnesium alloys. As the churn revolves, milk-fat globules coalesce until a separation of the fat in granular form (about the size of a pea or kernel of corn) occurs. The moisture phase (buttermilk fraction) is drained away and is usually condensed and dried at a separate facility. The butter granules may or may not be rinsed ("washed") with water. Salt (if used) is added either by sprinkling it over the surface of the granules or by distributing it in a trench made after the granules have been gathered into a firm contiguous mass. At this point, the system consists of fat granules, moisture residual from the buttermilk or rinse water, and salt. Then the butter is worked, usually by further revolving the churn, which causes the butter mass to be repeatedly lifted upward by internal vanes or shelves and then dropped to the bottom of the churn. This process soon brings the granules into a compact homogenous mass and serves to distribute the water and salt throughout the lipid matrix. The butter may then be placed into large boxes or tubs and stored under refrigeration, or it may be transferred directly into a machine that extrudes, forms, and packages it into retail sized portions—a process called "printing." Butter that has been stored in boxes or tubs also may be printed.

Primarily two processes are used in continuous buttermaking. The Fritz process utilizes cream of about 40% milk fat that is continuously fed into a machine where it is subjected to accelerated churning and working. In the process, the buttermilk is drained away, the granules rinsed, if desired, and salt and added moisture, if either is needed, are thoroughly incorporated. In any event, the butter is thoroughly worked to obtain homogeneity. The butter usually is printed and packaged immediately after churning.

The Alpha process also utilizes cream of about 40% milk fat, which is reseparated to 80% cream, cooled, and "worked" to effect phase inversion, i.e., from an oil-in-water to a water-in-oil emulsion. Working continues until salt and moisture are uniformly incorporated and the butter mass becomes homogeneous.

D. Initial Microflora

The initial microflora of butter is derived partly from the cream used in its manufacture. The nature, sources and control of microorganisms likely to be present in fresh milk and cream, as distinct from farm-separated cream are discussed in this volume, Chapter 18. The microbiology of farm-separated cream differs considerably from that of fresh cream separated at the plant or receiving station. Today, farm-separated cream generally is a minor by-product of the farm. The cream may be held on the farm for a week, often under poor, if any, refrigeration before shipment to the butter plant. Souring usually occurs, often accompanied by undesirable fermentations that lead to extensive fat hydrolysis, protein degradation, and associated off-flavors. Acid production (largely lactic) is caused primarily by lactic streptococci, lactobacilli, yeasts, and coliforms. Mold growth, especially of *Geotrichum candidum,* may be extensive. Gram-negative aerobic bacteria such as members of the genera *Pseudomonas, Alcaligenes, Acinetobacter/Moraxella* (*Achromobacter*), and *Flavobacterium* are largely responsible for the proteolytic and lipolytic changes. For further discussion of farm-separated cream see Foster *et al.* (1957) and Hammer and Babel (1957).

E. Effects of Processing on Microorganisms

1. Neutralization

Neutralization is the process of reducing the acidity of sour cream (primarily farm-separated cream) by the addition of alkaline salts such as calcium oxide, sodium carbonate, or sodium hydroxide. Neutralization prevents excessive loss of fat during churning, and eliminates objectionable oily or fishy flavors, but these problems are not microbial in origin. Neutralization has little, if any, effect on the microbial population of cream.

2. Pasteurization

The objectives of pasteurization of cream for buttermaking are (a) to destroy vegetative forms of pathogens and (b) to enhance the keeping quality (shelf-life) of butter. There are desirable nonbiological effects that result from pasteurization, but these are beyond the scope of this discussion.

Most cream used for buttermaking is pasteurized by a high-temperature, short-time (HTST) process at about 85°C with a holding time at that temperature of 15–16 sec. A low-temperature, holding (LTH) method (vat or batch pasteurization) at about 73°C with a holding time of 30 min was used almost exclusively when butter plants were small in size and procurement of farm-separated cream was practical; its use today is confined to the relatively few plants of that type that still exist in various parts of the world. In any event, either process, properly applied, destroys all but the most heat-resistant vegetative forms of microorganisms, including pathogens and other types which commonly cause flavor- and color-defects in butter. Also, the pasteurization destroys certain naturally occurring enzymes that otherwise would carry over into the butter and cause development of various types of off-flavors. Since subsequent processing steps are not inherently destructive to microorganisms, those which survive pasteurization (almost exclusively spore formers) eventually will be found in the butter, although in smaller numbers since most are lost to the buttermilk.

3. Churning

During churning, whether by the conventional or continuous process, most of the microorganisms in the cream are retained by the aqueous phase, that is, the buttermilk. Generally, the bacterial count of buttermilk is greater than that of the cream or the butter produced from it. This is a reflection of the breakup of bacterial clumps from vigorous agitation of the cream during churning and is another manifestation of the discrepancy often observed between bacterial content and bacterial count (see this volume, Chapter 18).

Over the past 40 years, stainless steel or aluminum-magnesium alloys have almost completely replaced wood for construction of churns. This has effected a marked change in the microbiology of butter made by the conventional process. The wooden churn was the most highly contaminated and most difficult of all dairy equipment to clean and sanitize (disinfect). Microbial contamination and growth (especially that of mold) deep in the porous wood and in crevices between joints occurred commonly. This made routine cleaning and sanitizing less than adequate. As a consequence, the source of molds, yeasts, and bacteria responsible for color and flavor defects of butter often could be traced to the wooden churn. Metal churns, whether of the conventional or continuous type, are easily cleaned and sanitized, often by automated systems. This has essentially eliminated churning equipment as a significant source of microorganisms in butter.

4. Working

The working process causes little quantitative change in the microflora, but the dispersion of water in the form of minute droplets throughout the butter drastically changes the microenvironment. Since the number of water droplets ranging from 10–18 billion/g of butter far outnumber the few thousand microorganisms likely to be present, most of the water droplets are sterile. In those droplets that do contain microorganisms the availability of nutrients is limited (Hammer and Hussong, 1930). Examination of microportions of butter (Long and Hammer, 1938) has shown the irregular distribution of bacteria throughout butter, thus indicating that if growth does occur in butter it is limited to certain areas where water droplets are large and where the usual growth-limiting factors such as pH, salt, and lack of nutrients are not adverse.

During the working process, not only is the moisture dispersed as billions of small droplets within a lipid matrix, but salt, if added, is dissolved in the discontinuous moisture phase. These droplets contain salt in concentrations that are inhibitory to many microorganisms. In butter containing 16% moisture and 2.5% salt, the percentage of salt in the brine is 15.625% if all the salt is uniformly distributed in the moisture phase. Hoecker and Hammer (1945) indicated that indeed the distribution of salt, as well as moisture, varies considerably. It is unlikely that the brine dispersed as droplets throughout the butter has a uniform salt content, and this partially explains the occurrence of microbial deterioration even in highly salted butter. Further, the moisture phase of butter arises from two sources, namely, the buttermilk and the wash water. Hoecker and Hammer (1945) showed that the moisture droplets originating from the buttermilk generally remain free of salt while those arising from the wash water contain most of the salt. These workers found that when organisms causing defects were added to the water, no defects were produced in butter in which salt was either well distributed or poorly distributed; whereas, when organisms were added to the cream, defects often appeared, especially in butter in which salt was poorly distributed. This emphasizes the importance of the proper pasteurization of the cream, the uniform dispersion of moisture in minidroplets, and the uniform distribution of salt.

5. Printing

The extrusion of butter from the printer and subsequent cutting to form retail-size portions may alter the physical structure of butter, affect moisture dispersion, and thus influence the growth of microorganisms. During printing there may be a significant aggregation of moisture droplets that are forced from the butter; and relatively large droplets of moisture may be formed within the mass of the butter. These difficulties will be minimized by printing properly worked butter directly from the churn when it is still soft and pliable. They are maximized if the butter was poorly worked and/or stored under refrigeration prior to being printed. Storage at low temperature causes crystallization of some of the liquid portion of the fat phase. Disturbance of this structure, now modified from that occurring immediately after working, will enhance moisture aggregation during printing. In any event, as a consequence of the above, defects may occur in printed butter that may not develop in unprinted butter. If large moisture droplets are formed during printing, more nutrients become available for microbial growth. Furthermore, if growth has occurred during storage of unprinted butter, subsequent printing may distribute these organisms among droplets in which further microbial activity can occur.

6. Starter Cultures

Bacterial cultures are sometimes used to improve the flavor of butter; the principal flavor compound produced is biacetyl. Cultures are used primarily in making unsalted ("sweet") butter, although the practice varies in different countries; e.g., their use is common in most of Europe, but not in the United States. In earlier times, cultures or "starters" were simply portions of sour milk, cream, or buttermilk that had a desirable flavor. Subsequently, as technology improved, pure cultures of *Streptococcus lactis* or *Streptococcus cremoris* were used and later mixed cultures containing either of the above and *Leuconostoc cremoris* (*citrovorum*) became popular. *Streptococcus lactis* subsp. *diacetylactis* (formerly *Streptococcus diacetilactis*) is used to some extent but reductase activity of some strains reduces the desirable biacetyl to the nonvolatile compounds acetylmethylcarbinol and 2,3-butylene glycol.

Butter cultures may be used in several ways, e.g., added to (a) cream at time of churning, (b) cream followed by holding (ripening) at about $10°–21°C$ for several hours or overnight, or (c) the granule stage and worked into the butter. The amount of flavor contributed to the butter will depend upon the amount of flavor compounds developed during ripening or the amount present in the culture when added directly to the cream at the time of churning or when worked into the butter.

With the continuous churning process, the use of cultures is restricted primarily to ripening the cream at about $20°C$ for about 16 hr before churning. Adding culture at the granule stage is not feasible in the continuous process.

If the starter cultures become contaminated, there is obviously the danger of adversely affecting finished product quality. On the other hand, the culture may mask common off-flavors, and the lowering of pH by the lactic acid formed by the culture may discourage the development of cheesy, putrid, or surface-taint defects. See the review by Babel and Hammer (1944) for further information on the action of starter cultures in butter.

F. Spoilage

Although farm-separated cream may be badly deteriorated, subsequent processing such as neutralization of the acid, vacuum treatment, heat treatment, and use of butter cultures eliminates and/or masks off-flavors. The butter from this cream is acceptable, although much inferior to that made from factory-separated sweet cream.

Spoilage of butter may be of microbial or nonmicrobial origin. Defects of nonmicrobial origin relate primarily to chemical degradation of the fat, i.e., oxidative and/or hydrolytic rancidity. Oxidative rancidity at the surface is probably the most frequent defect of butter made by modern methods. Nevertheless, defects of microbial origin occur occasionally; therefore, some discussion of them may be helpful. See also Foster *et al.* (1957) and Hammer and Babel (1957).

1. Surface Taint

This defect initially involves the surface layers of butter but very quickly may be evident throughout the mass of the product. It involves decomposition of the protein portion of the product and is manifested by a putrid, decomposed, or cheesy flavor, apparently from isovaleric acid or a closely related compound (Dunkley *et al.,* 1942). The causative organism is *Pseudomonas putrefaciens* (species *incertae sedis;* Buchanan and Gibbons, 1974). See the review by Wagenaar (1952) for a detailed discussion of this defect in butter. Prevention of surface taint requires proper pasteurization of the cream, combined with a water supply free of bacteria capable of causing surface taint. Cleanliness of the equipment used in handling pasteurized cream and butter is essential.

2. Rancidity and Fruitiness

The rancid flavor of butter comes mainly from free butyric acid that arises from the hydrolysis of butterfat. This reaction may be catalyzed by naturally occurring lipase in the milk or by enzymes elaborated by numerous bacteria and molds. Since milk lipase is destroyed by pasteurization, proper heat treatment will circumvent this problem, unless lipase activity had occurred in the cream prior to pasteurization, in which case the flavor, if sufficiently intense, may carry over into the butter.

Fruity odor is associated with the activity of lipolytic bacteria, particularly *Pseudomonas fragi* (species *incertae sedis;* Buchanan and Gibbons, 1974) and *Pseudomonas fluorescens*. Both species are proteolytic as well as lipolytic, and the fruity odor is associated with both fat and protein decomposition. The occurrence of these organisms in butter represents postpasteurization contamination from water and equipment used in production and processing.

3. Malty Defect

Malty flavor is produced by certain strains of *Streptococcus lactis* (formerly var. *maltigenes*) that produce 3-methylbutanal (Jackson and Morgan, 1954). The organism causing this flavor, though easily destroyed by pasteurization, may grow extensively in the cream before pasteurization, and the flavor then may carry over into the finished product, even though the organism is destroyed. Furthermore, postpasteurization contamination of the cream may result in growth of the organism and development of the flavor either in the cream or the butter.

4. Discoloration

Black discoloration of butter has been attributed to the growth of *Pseudomonas nigrifaciens* (species *incertae sedis;* Buchanan and Gibbons,

1974), an organism easily destroyed by pasteurization. When this type of discoloration occurs, contamination of the cream or butter necessarily occurred at some point after heat treatment. Meticulous plant sanitation is essential for control of the defect. Mold growth is most responsible for surface discoloration of butter. A variety of genera have been implicated. Contamination occurs most frequently during the packaging and printing operation. High humidity in the churn room and lack of proper ventilation may allow molds to grow on walls and ceilings, thus producing foci of contaminating organisms from which air currents may carry mold spores to butter surfaces. If the butter is moved from a cool area to an area of higher temperature at the time it is being printed and wrapped, moisture is apt to condense upon its surface. If, at the same time, the environment is contaminated with mold spores, the elements favorable for contamination and surface growth of mold are provided.

G. Pathogens

Essentially all commercial butter is made from pasteurized cream, and accordingly butter should present less of a foodborne disease problem than even pasteurized milk, since its physiochemical characteristics are immeasurably more inhibitory to pathogens than are those of pasteurized milk. There have been few recorded outbreaks of disease traced to butter and in almost all of them the butter consumed was made from unpasteurized cream. Two recent outbreaks of staphylococcal food poisoning, caused by butter made by modern practices, are worthy of mention because they illustrate the consequences of carelessness in the handling of cream, either raw or pasteurized, prior to churning and/or failure to properly handle or use butter after its manufacture. The first of these was an outbreak due to whipped butter * and to butter from which the whipped butter was made (U.S. Department of Health, Education and Welfare, 1970b). Enterotoxin A was found in both butters. In this instance, the evidence strongly indicates that the enterotoxin had been present in the cream and was carried over into butter in sufficient amount to cause illness when consumed. Since some whey cream (separated from cheese whey) was used in the butter involved, there is a possibility that the toxin originated in the whey from which the whey cream was obtained (see Chapter 18, Section VI,E). In the second outbreak the vehicle was

---

* Whipped butter is prepared by rapid agitation (beating or whipping) of butter in order to incorporate air into the product. A small amount of milk and/or water may or may not be added during the whipping process. Whipped butter is made commercially and also is commonly prepared at place of use, e.g., in restaurants.

commercially prepared whipped butter (U.S. Department of Health, Education and Welfare, 1977b). *Staphylococcus aureus* counts up to $10^7/g$ indicated that toxin undoubtedly formed in the whipped cream.

When butter containing 1.5% salt was inoculated with *Staphylococcus aureus,* then stored at 10°C for 60 days, the numbers of organisms decreased markedly, whether the original inoculum was high ($10^7/g$) or low ($10^3/g$). In whipped butter, the decrease was less marked; in fact, numbers increased slightly in such butter inoculated at low level. When another set of these two butters was stored at 23°C for 14 days, the numbers of *S. aureus* did not change appreciably. When the experiment was repeated with butter containing 0 to 1% salt, storage at 23°C for 14 days caused a 10-fold increase in numbers (Minor and Marth, 1972a). These experiments show that poor sanitary practice (that might introduce staphylococci) coupled with addition of milk or water during whipping (that might reduce the salt concentration) will materially increase the hazard from *S. aureus* in whipped butter stored above 10°C.

H. Choice of Case

For an understanding of "Case" see Table A.1, and the brief description in Appendix IV. For more details, see book 2 of this series (ICMSF, 1974).

The safety and spoilage hazards associated with butter relate primarily to the effectiveness of pasteurization. (See Volume I, Chapter 1, Section V,C,4) and proper care of cream prior to churning. Recontamination after pasteurization and subsequent opportunity for microbial growth may result in formation of harmful toxins and/or undesirable flavor compounds that may be carried over into butter. Nevertheless, the excellent history of safety and quality of butter made under modern practices would indicate a low degree of concern. Under normal distribution practices, few microbiological changes would be expected to occur, especially in salted butter. The aerobic plate count would give useful information, and Case 5 would apply. However, whipped butter and butter destined to be used in its preparation may be hazardous from *Staphylococcus aureus* (Case 9). The coliform count may be useful for unsalted butter (Case 4); however, the variation in salt sensitivity among coliform species and strains within species makes it less valuable for salted butter (Singh and Nelson, 1948).

I. Summary

Current technology of butter manufacture yields a relatively stable product because the fat is the continuous phase, and within this fat are dispersed literally billions of droplets of moisture. If pasteurization of the cream is effective, then the bulk of the moisture droplets will be free of microorganisms. If the butter is salted, the level of the salt in many of the droplets will be sufficient to inhibit the growth of most of the contaminants present in unsterile droplets. If pasteurization is ineffective, or if plant equipment is contaminated, then the unsterile droplets may contain organisms capable of growth, especially in butter having the water phase poorly dispersed. The consequence of such growth may lead to spoilage, or if pathogens are involved, to outbreaks of illness among those who consume the product. Further, during printing and packaging, the butter may receive surface contamination with organisms capable of growth and development of color defects as well as other defects.

26

Sugar, Cocoa, Chocolate, and Confectioneries

I. INTRODUCTION

A. Foods Covered

This chapter describes the microbiology of sugar, cocoa, chocolate confectioneries, and sugar confectioneries.

Sugar (sucrose) is derived from two sources, sugar cane (*Saccharum officinarum*) and sugar beets (*Beta vulgaris*). *Cane* is a bamboolike perennial growing from 2.5 to 6 m in height in tropical and subtropical temperatures of 18° to 28°C and annual rainfall of about 150 cm. *Beets* are annual vegetables that penetrate 25–40 cm into the ground and are grown mainly in temperate climates. Thus, the initial microflora of the two raw materials differs widely.

The word *sugar* is derived from the Hindu, sarkara, which denotes the hard material obtained by boiling the juice of sugar cane to a thick syrup and allowing the syrup to harden. Pieces of sarkara were called khanda, hence the word candy.

Cocoa beans are obtained from pods growing on cocoa trees (*Theobroma cacao*) in warm humid climates with temperatures of 21°–32°C and annual rainfall of 125–300 cm. *Cacao products* means any form of chocolate, chocolate product, cocoa, or cocoa product (U.S. Department of Health, Education and Welfare, 1975c); additional definitions are available (World Health Organization, 1976, 1978). In some countries, chocolate may legally contain a small percentage of noncocoa lipid (Wolf, 1977). Additionally, many products that simulate chocolate contain a large percentage of noncocoa lipids (Durkee, 1977a,b; Wolf, 1977).

These simulated products do, however, contain significant quantities of cocoa powder and are used widely in confectionery.

Confectionery is a term that has different meanings in different countries (Minifie, 1970). In this chapter, confectionery means candy (sweets) and other food products made with sweeteners, flavorings, milk products, cacao products, nuts, fruits, starches, and other materials (Lindley, 1972; U.S. Department of Health, Education and Welfare, 1975c). Chocolate confectioneries are chocolate in bars and blocks; and nuts, fruits and other materials enrobed in chocolate. Confectionery with simulated chocolate is included. Sugar confectioneries include boiled sweets, toffees, fudge, fondants, jellies, pastilles, and other confectioneries not covered with chocolate. Flour confectioneries (cakes and biscuits) are considered in Chapter 23.

B. Important Properties

Products such as crystalline sugar, cocoa, chocolate, and nonchocolate confectioneries have, with minor exceptions, a water activity sufficiently low to prevent growth of intrinsic microorganisms or extrinsic contaminants. There are, however, stages during the production of sugar and cocoa when spoilage may occur or large numbers of bacteria, yeasts, and fungi may be introduced. These sometimes persist in the final product. The remainder of this chapter is concerned with microbial contaminants, their sources, their detrimental and sometimes beneficial effect on products, their fate during processing and storage, and methods for their control.

II. SUGAR

A. Cane Sugar

1. Initial Microflora

The organisms present on sugar cane come from the soil and decaying plant material. The rhizosphere (soil interacting with root) contains a wide variety of organisms, but it is not known that sugar cane rhizospheres are consistently associated with specific microorganisms. In one study, *Enterobacter* was found in high numbers (10^5/g) in soil close to cane stalk and fewer as the distance from the stalk increased (Mayeux, 1960). *Leuconostoc mesenteroides* may also be found in the rhizosphere, but inconsistently, in numbers up to 5×10^3/g (Tilbury, 1970). Juice exuding from cane stubble in wet, warm weather may contain high numbers of microorganisms, about 10^9 bacteria and 10^6 yeasts and molds per gram (Mayeux, 1960).

The microflora of standing cane is variable, influenced mainly by temperature, humidity, and season (Duncan and Colmer, 1964). Microorganisms may be adventitious, epiphytic, or parasitic. They may grow on the surface or in the liquid in the axils of the leaves, exploit fractures in the epidermis, or invade healthy tissue and cause disease. The leaves and stalks of healthy cane contain bacteria (10^4–10^8/g; Mayeux, 1960; Duncan and Colmer, 1964) and yeasts and molds (10^3–10^4/g). The leaves of cane diseased by bacteria, fungi, and viruses contain higher bacterial populations than healthy cane. The bacteria most commonly found on normal leaves and stalks are species of *Flavobacterium, Lactobacillus, Xanthomonas, Enterobacter* (probably *Klebsiella pneumoniae;* Nunez and Colmer, 1968), *Pseudomonas, Erwinia, Leuconostoc, Bacillus,* and *Corynebacterium;* some of these are potentially pathogenic for plants. Yeasts and molds are also present (Shehata, 1960; Tilbury, 1970).

Some specific ecosystems of sugar cane have been studied. The water in the leaf sheath contains 0.01 to 0.06% sugar and has a pH of 5.7 to 7.7. Bacterial populations in this water are directly proportional to the sugar content, ranging from 10^5 to 10^7/ml, and occasionally much higher (Mayeux, 1960; Duncan and Colmer, 1964). Yeasts belonging to several genera may be present under the sheaths (Bevan and Bond, 1971). The rind of the cane may be breached by insects and by other means. Penetration of cane by the cane borer, *Diatraea saccharalis* results in production of detritus (frass) and invasion and growth of bacteria, yeasts, and molds. Frass has been found to contain per gram, 10^8–10^{10} bacteria, mainly *Enterobacter aerogenes,* 10^3–10^6 yeasts, and 10^3–10^5 molds (Mayeux, 1960). In addition to such insect damage, the rind may be split as a result of growth, freezing, or burning of the leaves (Tilbury, 1969). Exudates that appear on the surface, or tissues that are invaded through the fractures offer unique ecosystems for profuse and concurrent growth of various bacteria, yeasts, and molds (Bevan and Bond, 1971). Whether microbial growth actually develops, however, depends on such factors as the extent of physical damage, the temperature, and the time between damage and harvest; high humidity or persistent rainfall may increase the rate and extent of spoilage by microbes.

2. Effects of Processing on Microorganisms

a. Harvesting. There are many methods of harvesting cane, and each has advantages and disadvantages with respect to microbial destruction of sugar in the harvested cane. Also, temperature, humidity, and time from cutting to crushing greatly influence microbial contamination, growth, and spoilage.

Mature cane may be harvested green (i.e., with leaves on) or it may be burned prior to harvesting to remove the leaves. It may be cut by hand with machetes (cane knives) or by machine; machine-cut cane may be chopped concomitant with harvesting into pieces (billets) of about 30 cm in length, thus exposing many additional cut surfaces to contamination.

Burning to remove leaves may increase the temperature of the stalk to 55°–85°C. Apparently burning does not destroy many heat-sensitive bacteria because a wide variety of asporogenous organisms may be found within minutes of burning (Bevan and Bond, 1971). *Leuconostoc mesenteroides* has been found on stalks with about equal frequency before and after burning and increases markedly with time after burning (Bevan and Bond, 1971). About 25% of swab samples of cane contain *L. mesenteroides* in numbers from 5 to 50,000 per swab, but usually fewer than 1000 (Tilbury, 1970).

As *L. mesenteroides* is the principal cause of deterioration of harvested cane, the vectors of this organism have been studied. Wasps feeding on the sugary exudate of burned cane may contain 10^2 to 10^3 cells per insect. Machetes may be free of *L. mesenteroides* in dry weather, but contain 10^3–10^4 cells per machete in wet weather (Tilbury, 1969). Cutting blades in harvester-choppers and the juice-soaked mud in the chopping boxes always contain *L. mesenteroides*. Blunt and badly adjusted knives in harvester-choppers produce billets with split and bruised ends and fractures along the longitudinal axis (Bevan and Bond, 1971). This facilitates entry of *Leuconostoc* and other mucous- or acid-forming bacteria and hastens deterioration (Egan, 1971). Within 10 min of chopping, *L. mesenteroides* may migrate 7.5 cm into the stalk, possibly by passive transport in the vascular bundles, and thence to the parenchyma (storage tissue) where it multiplies.

Sour cane results when *Leuconostoc* and other acid-forming bacteria grow in harvested cane (Tilbury, 1968). They produce invert sugar, lactic and acetic acids, and frequently dextran. A "sour" odour may result from microbial growth. In hot, humid climates, 1–5% of total sugar may be lost for each day between harvesting and crushing (Tilbury, 1975), but in cooler, dry climates, the loss is usually less than 0.5%. Thus, losses of sucrose may be substantial unless the time between harvest and crushing is minimized.

b. Extraction and Processing. Cane is processed to raw sugar in a sequence of operations (Moroz, 1963), most of which affect the microflora or, *vice versa,* are affected by the microflora. These operations are

(a) cutting, (b) crushing and extraction of raw juice, (c) clarification, (d) evaporation, (e) crystallization, (f) centrifugation, and (g) raw sugar.

The microbiology of raw juice has been reported in numerous studies (e.g., Kopeloff and Kopeloff, 1919; Pederson, 1938; Pederson and Hucker, 1946; Mayeux, 1960; Shehata, 1960; Tilbury, 1970) and reviews (e.g., Moroz, 1963; Owen, 1977). The bacteria most commonly found are species of *Actinomyces, Bacillus, Enterobacter, Escherichia, Erwinia, Flavobacterium, Leuconostoc, Lactobacillus, Micrococcus, Serratia, Streptomyces,* and *Thermoactinomyces;* the yeasts are *Saccharomyces, Torulopsis, Candida* and *Pichia* spp; the molds are *Aspergillus, Penicillium, Citromyces, Cladosporium,* and *Monilia.* An orderly taxonomic presentation of microorganisms found in sugar during production is given by Moroz (1963).

In crushing and extraction, the cane is shredded and passed sequentially through a series of rollers with imbibition (extraction) water passed in the opposite direction. Juice from the first roller may contain up to 19% sugar, and from the last roller, less than 5%. The combined juice is called mixed cane juice or raw juice; the extracted fibrous residue of cane is called bagasse. The crushers and rollers, and the associated systems for transporting fluids constitute a huge continuous culture machine that works for months, frequently with several days between scheduled stops.

The mixed juice is an ideal medium for growth of many organisms, but only a few compete successfully. It has a Brix (percent sucrose *w/v,* or equivalent in soluble solids) of 10–18, pH of 5.0–5.6, abundance of inorganic and organic salts, amino acids and other nutrients, and a temperature usually between 25°–30°C. The bacterial count of juice from the first mill (roller) is between 10^5 and 10^7/ml for normal cane and about 10^8/ml for sour cane. To these incoming bacteria are added microorganisms growing in the crushers, rollers, troughs, and strainers. In this environment, the dextran-forming *Leuconostoc mesenteroides* is especially adapted to compete (Pederson and Hucker, 1946; McCleskey *et al.,* 1947), but other catalase-negative, microaerophilic organisms grow well producing acid, invertase, and dextran. In some mills, *Enterobacter* spp. predominate; in others, alcohol-producing yeasts compete well. Fortunately, the microbial destruction of sugar is minimized by clarification, which is usually prompt but which is delayed sometimes by unscheduled or scheduled stops.

The clarification process normally begins within 10 min of extraction. It involves addition of lime to increase the pH to about 8.0 and rapid heating to at least 80° and often over 100°C. The heating destroys yeasts and vegetative cells of bacteria; and settling and filtration remove scums, precipitates, and suspended solids. Clarification decreases the microbial count by 99.999% (Table 26.1), but dextran and many bacterial spores (Skole

TABLE 26.1

Bacterial Content of Sugar Cane during Processing [a]

Product	Mesophiles/ml (range)	Thermophiles/ml (range)
Raw juice (early) [b]	8×10^6–16×10^6	1×10^1–1×10^2
Raw juice (late) [b]	6×10^8– 8×10^8	—
Clarified effluent	0 –11	0 –8
Press juice	0 – 5×10^4	3×10^3–2×10^5
Evaporator	2×10^2– 3×10^4	2×10^2–2×10^3
Storage tank	1×10^3– 7×10^3	2×10^4–4×10^4
Crystallizer	2×10^3– 4×10^4	3×10^2–2×10^4
Massecuite [c]	1×10^3– 1×10^4 [d]	2×10^3–2×10^4
Raw sugar	3×10^2– 5×10^3 [d]	2×10^2–2×10^3
Molasses	3×10^3– 3×10^5	1×10^3–2×10^4

[a] Adapted from Owen (1977).

[b] Early in season and late in season.

[c] Mixture of sugar crystals and molasses.

[d] Per g.

et al., 1977) remain. The materials removed, collectively termed filter cake mud, contain sugar and other organic material. The mud is an excellent medium for bacterial growth and contains 10^5–10^6 *L. mesenteroides* per gram. Filter cake mud is used as fertilizer, thus returning this species to the cane fields.

Dextran is synthesized from sucrose mainly by *Leuconostoc* in souring of cane, in mixed juice in troughs and pipes, and in debris. In juice, it sometimes resembles frog spawn, or more rarely tapioca granules; in the latter it apparently is produced by *L. mesenteroides* in symbiosis with a yeast and an unidentified bacterium (Bevan and Bond, 1971); a similar apparent symbiosis between a levan-forming bacterium and yeast has also been observed (Scarr, 1949). Dextran may become so thick that it clogs pipes. It may also interfere in other parts of the processing since it increases the time for, and decreases the effectiveness of, the clarification process. During further processing, dextran also interferes with evaporation and crystallization by increasing the viscosity of the massecuite, decreasing the rate of crystallization, and increasing the percentage of malformed crystals. Both dextran and malformed crystals decrease the removal of molasses during centrifugation; this leads either to increased washing in the centrifuge, which decreases yield, or to an undesirably high a_w in the raw sugar, which increases the probability of spoilage of raw sugar by osmophilic yeasts. In some circumstances, the destruction of dextran by dextranase is warranted economically (Tilbury, 1972; Tilbury and French, 1974).

A small proportion of sugar cane (McMaster and Ravnö, 1977) and all sugar beets are extracted by a diffusion process. As the microbiology of the diffusion process for both these materials is similar, it will be described below in the section dealing with sugar from beets.

3. Control of Spoilage during Harvesting and Processing of Cane

The two main opportunities for control are in the field and in the mill. In the field, decreasing the interval between harvesting and processing decreases the opportunity for microbial growth. This interval should be not longer than 24–36 hr for whole stalk cane, and 8–12 hr for chopped cane in hot humid weather or 18 hr in cool dry weather. Cutting with sharp knives decreases ragged cuts and minimizes entry of bacteria through the cut surfaces. In mechanized harvesters, cleanliness and sanitizing of chopping boxes is desirable, but opportunities for such care are minimal. Application of formaldehyde to billets decreases spoilage but is not economical (Egan, 1971). In the mill, hygienic practices vary tremendously from one company to another. An effective program includes cleaning every 8 hr with brushing and with hot water sprayed under high pressure to dissolve sugar and dislodge particulate material. Disinfection with quaternary ammonium compounds, organosulfur compounds, and chlorine is employed but not universally. When used, disinfectants may be sprayed onto cane in the crusher or added to the imbibition water. Some operators do not attempt to control microbial growth and the mills become incredibly filthy with particulate material, pockets of fermenting juice, slime, and foul odors. Only recently have the significant financial benefits of improved hygiene been calculated (Tilbury *et al.,* 1978), although extensive educational and regulatory programs have been in existence for many years (Moroz, 1963).

Methods of measuring the effectiveness of control procedures during processing of cane to raw sugar are available. Two main approaches are (a) loss of sucrose and (b) determination of microbial counts, dye reduction times, titratable acidity, pH, and dextran and other polysaccharide gums. No single method is suitable for all purposes.

4. Microbiology of Raw Cane Sugar

Raw cane sugar is the end product of the cane mill and the raw material for the refinery (Moroz, 1963). It consists of about 98–99.3% sucrose in the crystalline form and about 0.5–2% of molasses, which surrounds the crystals. The molasses consists of sucrose, water, invert sugar, organic acids, amino acids and other nitrogenous compounds, minerals, starch, and dextran and other polysaccharides. The a_w should be at 0.65 or less but is sometimes higher. The pH varies from 5.0 to 6.0.

Raw sugar is obtained from clarified juice by evaporation. The juice containing about 12–15% sucrose is boiled under successively reduced pressures at temperatures up to 110°C (Tilbury, 1967) which precludes microbial growth and destroys all microorganisms except some bacterial spores. When the soluble solids (mainly, sucrose) concentration is about 60°Brix, crystallization is obtained by boiling under vacuum at about 60°C until the solution is supersaturated and crystals are formed. The mixture of crystals and molasses is known as massecuite. Raw sugar is obtained by removing most of the molasses from the sugar crystals in a centrifuge.

The microbiological content of raw sugar consists of bacterial spores, yeasts, and molds. Some bacterial spores are present because they survive the thermal process and are not removed during clarification. Added to these are spores and non-sporing organisms in the press juice. In some sugar factories, large numbers of osmophilic yeasts and molds are found in syrup and molasses in storage tanks and are often not killed in vacuum-pan crystallization. Small numbers of airborne molds and insignificant numbers of yeasts may contaminate products during crystallization, centrifugation, and drying. However, large numbers of osmophilic yeasts may be introduced by physical contact with wet, contaminated sugar in which yeasts are growing, i.e., on surfaces of chutes, conveyors, scales, and storage bins. In some countries, cleaning of inaccessible equipment is difficult because of continuous operation of the mills for a week or more at a time, and wet sugary material may contain 10^4–10^6 osmophilic yeasts per gram (Tilbury, 1967).

The microbes of raw sugar may vary with country of origin, but are generally those that persist at the low a_w of the product. The bacteria are most commonly *Bacillus* spp., which do not grow in the raw sugar but may grow in dilute sugar solutions during the refining process. *Desulfotomaculum (Clostridium) nigrificans, Clostridium butyricum,* and thermophilic *Bacillus* spp. may also be present and are of concern if they persist through the refining process and are present in the refined sugar used in canned food. The most common molds are *Aspergillus* spp., *Penicillium* spp., and *Monilia* spp., while species in several other genera are found less frequently (Tilbury, 1966, 1967).

Yeasts found in raw sugar are mainly osmophilic. Their natural reservoirs appear to be sugar cane, bagasse, filter cake mud, and wet sugary materials in the mill. *Saccharomyces rouxii* and *S. mellis* are most common, but other species of *Saccharomyces* and species of *Hansenula, Pichia, Torulopsis, Candida, Dekkeromyces,* and *Endomycopsis* have been found

(Scarr, 1953a; Tilbury, 1966, 1967; Skole *et al.,* 1977). The minimum a_w permitting growth of some yeasts commonly found in raw sugar is given in Table 26.2.

5. Spoilage of Raw Cane Sugar

Raw sugar may be stored for many months and then shipped long distances. Unless precautions are taken, spoilage may cause large economic losses (Tilbury, 1967). Osmophilic yeasts appear to be most active but xerophilic molds may cause or contribute to spoilage (International Commission for Uniform Methods of Sugar Analysis, 1954). Bacteria, although present, do not cause spoilage because of low a_w, and even yeasts and molds can be prevented from growing if a a_w less than 0.65 is maintained. The reasons for spoilage and the methods to prevent it follow.

The composition of the molasses film surrounding the crystals of sucrose determines whether or not spoilage occurs. The composition varies widely but the following is an approximation: sucrose, 40–50%; water, 25%; invert sugars, 15–25%; and minerals and nonsugar organic compounds, 15–25%. The nonsugar organic compounds contain nitrogenous materials and such hydrocolloids as dextrans, levans, starches, and pectins. Thus, the molasses film is a highly concentrated solution of nutrients readily available for microbial growth. Attraction of water from the atmosphere by the hygroscopic invert sugars and hydrocolloids increases the water content, thus increasing the a_w (Tilbury, 1967).

The a_w of raw sugar may vary widely, from 0.575 to 0.825 (Tilbury, 1967). High a_w values are caused mainly by inadequate centrifugation and

TABLE 26.2

Effect of a_w on Growth of Osmophilic Yeasts from Raw Sugars [a, b]

Yeast	Minimum a_w permitting growth
Saccharomyces rouxii	0.650
Saccharomyces bisporus var. *mellis*	0.700
Torulopsis candida (1)	0.650
Torulopsis candida (2)	0.700
Torulopsis etchellsii	0.700
Torulopsis versatilis	0.700
Hansenula anomala	0.750

[a] From Tilbury (1967).
[b] Tested in sucrose/glycerol syrups for 12 weeks at 27°C.

by high relative humidity. If centrifugation is inadequate, the raw sugar crystals are washed in the centrifuge to decrease the soluble materials in the molasses film; this increases the a_w. During production, the relative humidity of the atmosphere influences the amount of moisture taken up after centrifugation of the raw sugar from the massecuite, and artificial drying may be required. During storage, the water content of the raw sugar increases in direct relationship to both its content of reducing (mainly invert) sugar and the relative humidity of the atmosphere (Table 26.3). Raw sugar with an a_w of 0.70, from which osmophilic yeasts or molds are isolated, should be regarded as unstable (Tilbury, 1966).

Osmophilic yeasts grow in the temperature range 0°–40°C with an optimum of about 27°C, but the optimum and maximum may be increased with increased a_w of the substrate. Spoilage by osmophilic yeasts results from growth in the molasses film with preferential metabolism of the levulose component of the invert sugars. Water and organic acids are produced, which increase the a_w and decrease the pH. Both favor further growth of osmophilic yeasts, and the decreased pH causes hydrolysis of sucrose to produce more invert sugar (Scarr, 1953a; Tilbury, 1966). Inversion may also result from the activity of invertase produced by a few osmophilic yeasts, e.g., some species of *Torulopsis* and *Hansenula* and several xerophilic molds, but most osmophilic yeasts do not produce the enzyme invertase (Scarr and Rose, 1966).

Spoilage will not occur if the a_w remains at less than 0.6. Conversely, it may progress rapidly if the a_w is above 0.7. Shortly after production there is a positive correlation ($+0.83$) between the a_w of raw sugar and counts of osmophilic yeasts. Growth of yeasts continues during bulk storage and transportation, and populations may reach 5×10^7/g of raw

TABLE 26.3

Effect of Reducing Sugar Content in Raw Sugar and Relative Humidity on Water Content of Raw Sugar at Equilibrium [a]

% Reducing sugar	Average % moisture of raw sugar at equilibrium during storage		
	50% (R.H.)	61.7% (R.H.)	78% (R.H.)
0.16	0.33	0.55	1.24
0.24	0.42	0.72	1.43
0.73	0.42	1.19	2.35
1.35	1.00	2.72	5.05

[a] From Tilbury (1966).

sugar if the a_w is high and the temperature favorable (25–35°C). Since growth occurs only in the molasses film, a population of $5 \times 10^7/g$ of raw sugar is actually 50 times larger (2.5×10^9) in the molasses film if it comprises one-fiftieth of the raw sugar. Within a few months after maximum growth has been obtained, the population of viable yeasts may decline by more than 99.99% (Tilbury, 1967).

Although spoilage does not generally become noticeable until yeast populations reach $10^7/g$ of raw sugar, some spoilage probably occurs throughout the growth of yeasts. Attempts to accelerate spoilage for purposes of study in the laboratory have included storage of various samples of naturally contaminated raw sugar at 80% R.H. and 27°C: osmophilic yeast populations increased 1000-fold in 2–6 weeks, water increased to between 0.3 and 5%, and there were losses of sucrose up to 30% with corresponding increases in invert sugar, some of which was metabolized. Molds did not grow. Despite these changes, there were only moderate decreases in pH of the raw sugar. In sharp contrast, *Saccharomyces mellis* can cause a decrease in pH from 5.0 to 2.5 when growing in 60°Brix sucrose with small amounts of invert sugar, thus causing chemical inversion; *S. mellis* does not produce invertase (Tilbury, 1967). A possible explanation exists for differences in pH in sugar solutions and in raw sugar. In raw sugar, mixed populations of osmophilic yeasts growing in the ecosystem surrounding the sugar crystal may cause changes that lead to chemical inversion, e.g., localized production of acid, drop of pH, and subsequent metabolism of the acid with a resulting increase in pH. Such changes in pH would not be detectable by conventional methods of analysis. In contrast, when pure cultures grow in nonaerated solutions of sugar, the acids they produce accumulate.

Spoilage of raw sugar by xerophilic molds is probably not of major concern in modern sugar production. Although a wide variety of xerophilic molds (species of *Aspergillus, Cladosporium, Pullularia,* and *Penicillium*) produce invertase and can invert sucrose when growing in a 67% solution of raw sugar (a_w about 0.83), the a_w of raw sugar is usually less. The lowest a_w permitting growth of xerophilic molds is 0.65, but in practical situations in raw sugar they are not a problem because most isolates from sugar require an a_w of 0.80 for growth (Tilbury, 1967). Moreover, below the surface of raw sugar stored in silos, there may be insufficient oxygen to support growth of molds. It is probable, however, that in the past when raw sugar was commonly stored and transported in jute bags, the surface-to-volume ratio was sufficiently large to permit the a_w to increase as a result of hygroscopic attraction of water and to furnish sufficient oxygen for growth of molds.

6. Control of Spoilage of Raw Cane Sugar

The prevention of spoilage of raw sugar depends on obtaining and maintaining an a_w of less than 0.65. If this is done, the number and types of contaminants is of little consequence—they will not grow. Thus, procedures to minimize contamination by yeast during washing of raw sugar in the centrifuges and by good housekeeping, although desirable, may be less economical than taking steps to obtain and maintain a low a_w. These steps are (a) centrifugation sufficient to obviate the need for washing the raw sugar, thus eliminating wash water as a cause of increased a_w; (b) artificial drying of the raw sugar, if required, to an a_w less than 0.65; (3) and storage in sealed silos at a relative humidity of 65% or less.

Despite the above precautions, conditions during transport to the refinery may not be ideal for maintaining a low a_w. On arrival, examination of raw sugar for yeasts by phase-contrast microscopy rapidly reveals the number of yeasts, and whether they are alive or dead. If a potential spoilage situation exists, the shipment is refined quickly to minimize losses and to prevent mixing with sound material in storage.

Viable counts of osmophilic yeasts may be required both in the raw sugar mill and in the refinery. Dilutions are best carried out in diluent of low a_w to minimize osmotic shock; the agar must have a suitable a_w for osmophiles and contain some invert sugar; and petri dishes must be protected against dehydration during the prolonged (4–6 weeks) incubation that may be necessary to develop visible colonies (Tilbury, 1976).

B. Beet Sugar

1. Initial Microflora

The microflora on beets comes from the soil. The genera identified from beet tissues are *Pseudomonas, Bacillus, Arthrobacter, Erwinia, Flavobacterium, Streptomyces,* yeasts (Bugbee *et al.,* 1975), and *Clostridium.* Thermophilic species of *Bacillus* and *Clostridium* (Allen *et al.,* 1946) are usually present in soil, and the *Bacillus* spp. are especially capable of causing spoilage during processing.

2. Effects of Storage and Processing on Microorganisms

a. Storage, Fluming, and Washing. Beets are grown in temperate climates, harvested prior to freezing, and stored in piles for several days to months. Normally a properly covered and ventilated pile will maintain a temperature of about 5°C despite an external temperature as low as −35°C.

Sugar beets do not spoil readily unless storage is prolonged or they have been frozen and thawed or overheated. Thawed beets become gummy because of microbial synthesis of dextrans or levans. Beets stored without adequate air circulation, mainly due to pockets of trash or earth, may overheat to about 50°C within 2 days, and thereafter show evidence of microbial spoilage (Cole and Bugbee, 1976). Overheating, molding, and waterlogging (immersion in water) result in an increase in invert sugar (Oldfield et al., 1971), but it is not clear that microbial enzymes are the main cause.

Beets are flumed from storage to washers in water at 30°–40°C. In the flume, some soil and accompanying bacteria are washed off, and some sugar is leached from the beets; the flume water becomes a good microbial medium and contains 10^6–10^7 bacteria/ml. The microbial content of flume water may be reduced by using fresh instead of recirculated water. Heavy chlorination may be useful in flumes carrying damaged beets (Moroz, 1963), but chlorination does not destroy spore-bearing thermophiles (Carruthers and Oldfield, 1955). In general, the higher the bacterial population in the flume water, the higher the population in the juice from the beets.

After fluming, the beets are washed to remove residual soil and bacteria. The thermophile content of water after washing beets is about 10^4/ml, but can be reduced by multistage washing (Carruthers and Oldfield, 1955). Chlorination of wash water is only partially effective in reducing the microbial load on beets.

b. Extraction. Beets are cut into cossettes (thin V-shaped strips) and extracted with hot diffusion water in countercurrent diffusers. Two types of diffusion systems are used (a) battery and (b) continuous. Battery diffusers consist of stationary tanks connected in series, each tank containing 2–4 tons of cossettes; at short intervals, one tank with extracted cossettes is disconnected from one end of the system while simultaneously another tank with fresh cossettes is connected at the other end. Continuous diffusers are elongated horizontal or vertical cylinders with rotating components. A continuous diffuser can process several thousand tons of beets per day; the average residence time of cossettes is 60 min and of liquid 20–30 min.

Regardless of the diffuser system, there is a head and tail. At the head end, fresh cossettes enter, and raw juice with about 15% sucrose is drawn off. At the tail end, extraction water is pumped in, and extracted cossettes are removed. Material leaving the tail end contains 0.5–1.0% sucrose.

In addition to the diffusers, there are two ancillary systems that pump, transport, filter, strain, heat, and store liquid associated with the process.

One ancillary system is for collecting and storing diffusion water, the other for collecting and storing raw juice. Only the former will be considered further.

In the diffusers, the sucrose content varies from 0.5 at the tail to 15.0% at the head; invert sugars, minerals, and nitrogenous compounds, although at lower concentrations than those of sucrose, have equally varied gradients; the temperature is from 25° to 75°C; the dissolved oxygen varies from abundant to nil, with nitrate available in variable concentrations as a hydrogen acceptor for some species; and the pH varies from 5.0 to 8.0. Similar variations will be present in the ancillary systems, with some components containing material with a pH as low as 4.0.

The microorganisms that enter the head end with the cossettes are as varied as the microflora of the soil in which the beets were grown. However, only a few genera proliferate. The physicochemical attributes of each ecosystem will determine which organisms will grow; and growth of microorganisms will change some of the attributes, particularly pH, nitrate, and dissolved oxygen. In general, the microflora of fresh cossettes consists mainly of mesophilic bacteria with small numbers of thermophiles (Miyajima and Kamoi, 1960); the thermophiles are of particular significance. Yeasts may grow well in some places, but, in general, molds and yeasts do not grow sufficiently to cause losses of economic consequence.

In the operation of battery diffusers, warm diffusion water enters the tail cell and cold cossettes enter the head cell (Carruthers and Oldfield, 1955). Thus, both of these end cells may be at 35°–50°C and permit growth of mesophiles, while the cells between the head and tail are at 65°–75°C and may permit growth of thermophiles (Allen *et al.,* 1946, 1948b). The mesophiles are mainly *Lactobacillus* and *Leuconostoc* spp., although coliforms may be present. The most common thermophile is *Bacillus stearothermophilus.* However, thermophilic *Clostridium* spp., some producing hydrogen sulphide, may grow.

Since many thermophiles grow rapidly at 65°–70°C, and are not destroyed at the maximum operating temperature of about 75°C, usually there are large numbers of bacteria ready to grow logarithmically with minimum lag when the temperature is suitable. In practice, there are always some cells in a battery at temperatures suitable for thermophilic growth. Although each cell in a battery is emptied in sequence about 1 hr after filling, and there is opportunity for cleaning, some of the ancillary systems are not readily available for cleaning. Thus, there is continuous reinoculation from incoming cossettes from the supply water, and from the diffusion and ancillary systems; bacteria from the last-named source are fully adapted to rapid growth.

Continuous diffusers should be at 75°C for maximum control of thermophilic bacteria. However, 75°C is not practical for all components of the system. In the operation of continuous diffusers, the diffusion water is heated to 71°–74°C prior to entering the tail cell, while fresh cossettes are "scalded" by passing through a tank of hot raw juice, sometimes heavily contaminated with spores of thermophiles, prior to entering the head cell at 71°–74°C. Thus, some cells of a continuous diffuser operate at temperatures of about 70°–75°C, while others are at 68°–70°C; at 68°–70°C a continuous diffuser may approximate a continuous culture system for thermophiles with continuous input of fresh nutrients and removal of wastes. The average retention time of pulp is 60–70 min and of extracting fluid, 20–30 min (Klaushofer et al., 1971).

The facultative aerobic populations, mostly *Bacillus stearothermophilus* may reach 10^6–10^7/ml within a few hours at 70°C (Carruthers and Old-field, 1955; Klaushofer et al., 1971), and sufficient acid is produced to decrease the pH to 5.2–5.4 (Oldfield et al., 1974a). However, other aerobic thermophiles are present at 65°C and above, e.g., *B. coagulans* and *B. thermodenitrificans.** Anaerobic thermophilic *Clostridium* spp. may produce considerable quantities of hydrogen sulfide if the diffusion water is not acidified (Klaushofer et al., 1971).

In continuous cane diffusers, the thermophilic aerobes are qualitatively and quantitatively the same as in beet diffusers, but thermophilic obligate anaerobes have not been found (McMaster, 1975).

The diffusion water used for extracting sugar from cossettes comes from two main sources: fresh water and water obtained from the tail cell, including water pressed from the extracted cossettes. The water from the tail cell contains economically recoverable concentrations of sucrose (0.5–1.0%) but may contain spores of thermophiles if the diffusion cells have permitted abundant thermophilic growth. In addition, diffusion water may carry mesophiles growing in pipes, strainers, filters, settling tanks, etc., and the diffusion-water supply tank.

Diffusion water contains a wide variety of bacteria depending on temperature and application of germicides. Aerobic colony counts are 10^5 to 10^7, thermophilic bacteria are present from 10^3 to 10^4 (Allen et al., 1948a). In some diffusion waters, *Lactobacillus* spp. predominate, especially *L. delbrueckii*. It grows slowly at 55°C, but well at 50°; at 45°C it can decrease the pH of diffusion water from 6.9 to 3.8 in 24 hr. Other lactobacilli are also active at 45°C but not at 55°C (Allen et al., 1946). In modern practice, diffusion water is heated to about 80°C before it enters

* Not listed in the eighth edition of Bergey's Manual (Buchanan and Gibbons, 1974).

the tail cell. This treatment kills vegetative cells and markedly decreases the viable microbial population.

3. Spoilage and Other Adverse Microbial Activities

During the extraction process microbial activities create the following problems:

a. Destruction of Sugar. It is difficult to estimate the amount of sugar destroyed by bacteria in diffusers, but lactic acid production is a useful measure. Thermophiles destroy approximately 2 g of sucrose for every 1 g of lactic acid produced (Oldfield et al., 1974a; McMaster and Ravnö, 1977). On this basis, losses due to microbial growth are estimated to be 0.1–0.2% of sucrose extracted (Moroz, 1963), while other data (Carruthers and Oldfield, 1955; Stark et al., 1953) suggest losses of about 1.0–1.5%. Undoubtedly, the temperature is important in determining the species that grow, the extent of the growth, and the amount of sugar destroyed. Also, destruction is greatly increased in continuous culture (Klaushofer and Parkkinen, 1966). Species vary widely in their ability to destroy sucrose (Table 26.4).

b. Formation of Acids. The pH of sugar beets is 6.2–6.4. Thermophilic bacteria growing at 70°C in diffusion juice may reach viable populations of 10^6–10^7 and reduce the pH to 5.2–5.4. At 65°–70°C, the precursor of lactic acid is sucrose; invert sugar is not fermented to lactic acid (Carruthers and Oldfield, 1955; Oldfield et al., 1974a), although contradictory evidence exists (Klaushofer et al., 1971). In cane diffusers, thermophiles do ferment invert sugar to lactic acid, and this fermentation is more significant in cane because it contains more invert sugar than does beet juice (McMaster, 1975). The main acid producers are *Bacillus stearothermophilus* and *B. coagulans* (McMaster and Ravnö, 1977). The acids are about 95% lactic with the remainder consisting of formic, acetic, and glycollic (Oldfield et al., 1974a; McMaster, 1975). Gas (carbon dioxide and hydrogen) may be produced. In battery diffusers, gas pressure may be sufficient to exert a counter pressure against diffusion juice being pumped into the tanks. In continuous tower diffusers, gas bubbles may decrease the specific weight of cossettes and speed their passage through the diffusion liquid, thus decreasing extraction of sugar. Hydrogen sulfide may also be produced (Klaushofer et al., 1971).

At lower temperatures (40°C) in the diffusion supply water, lactic,

TABLE 26.4

Rate of Destruction of Sucrose by Microorganisms from Beet Juice

Organism	Temperature (°C)	Sucrose destroyed (mg/10^9 cells/hr)
Desulfotomaculum (Clostridium) nigrificans [a]	55	0
Enterobacter (Aerobacter) aerogenes, Flavobacterium, Micrococcus,[b] *Streptococcus* [b]	35	0.1–0.4
Leuconostoc mesenteroides [b] *Lactobacillus* [b]	35	2–8
Clostridium thermohydrosulfuricum [a, c]	66	2–3
Clostridium thermohydrosulfuricum [a, c]	70	7–8
Clostridium thermosaccharolyticum [a]	66	2–3
Bacillus thermophilus [b, c]	55	10–40
Bacillus stearothermophilus [a]	65	108–160
Bacillus subtilus [b]	55	20–60
Saccharomyces [b]	35	1500–3000

[a] From Klaushofer and Parkkinen (1966).

[b] Devillers (1955).

[c] Not listed in the eighth edition of Bergey's Manual (Buchanan and Gibbons, 1974).

acetic and small amounts of butyric acid are formed, and the pH is frequently as low as 4.0–4.5 (Allen *et al.,* 1946, 1948b).

c. Corrosion. Steel in diffusers and ancillary systems corrodes from reaction with lactic acid. The rate of corrosion at 70°C is about twice that at 20°C and increases approximately fourfold for each decrease of one pH unit in the range 6.2–4.2 (Carruthers and Oldfield, 1955; Allen *et al.,* 1948a). Increasing the pH of the diffusion water in the supply tank with lime decreases the overall corrosion but increases the depth of pitting. Inhibition of microbial growth by chlorination of the diffusion water inhibits corrosion in the recirculation system but not in the diffusers.

d. Formation of Slime. The most noticeable manifestations of microbial growth in raw beet juice and related materials is slime. Frequently, it has the appearance of frog spawn but may also appear as gum or jelly. It may become so thick that it clogs pipes, filters, etc. Slime is usually dextran or levan; its precursor is usually sucrose.

Dextran-forming organisms are, in decreasing order of importance, *Leuconostoc mesenteroides, L. dextranicum,* and heterofermentative *Lactobacillus* spp. (Perquin, 1940). Many *Bacillus* species synthesize levan from sucrose and some such as *B. subtilus* grow well at 50°C or less, temperatures common in head and tail cells of battery diffusers and in pipes and tanks of the ancillary systems. In cane diffusers, polysaccharides (mainly levans) are formed by thermophilic *Bacillus* spp. if the temperature drops below 60°C.

    e. Formation of Nitrite. Nitrate is present in sugar beets, usually at 20–200 ppm of nitrate nitrogen; nitrite is absent. In continuous diffusers, *Bacillus stearothermophilus* is the most prevalent thermophile. Some strains reduce nitrate only to nitrite while others reduce it to nitrogen gas. In raw juice, the nitrite level is usually 2–15 ppm nitrite nitrogen, occasionally up to 75 ppm.

    The nitrite, in turn, may combine with other chemicals to reduce the yield and quality of sugar. Sulfur dioxide (added to prevent browning, lower pH, and inhibit thermophiles; Oldfield *et al.*, 1974b; Carruthers *et al.*, 1958) becomes bisulfite in acid solution. Nitrite combines with the bisulfite, reducing its efficacy and also forming imidodisulfonate. The imidodisulfonate cocrystallizes with sucrose, increasing the ash content and causing malformed crystals that impede centrifugation of the massecuite.

    The production of nitrite nitrogen at levels of 25–50 ppm (125–250 ppm as $NaNO_2$) and its persistence over 4–6 hr at pH 5.9–6.3 might be expected to exert a strong antimicrobial effect (see Vol. I, Chapter 8). However, numerous studies (Carruthers *et al.*, 1958) did not show this.

4. Control of Spoilage during Extraction

    Ideally, the diffusers and ancillary systems should be at 75°C throughout to minimize microbial growth. At 70°C, there is abundant growth of *Bacillus stearothermophilus,* while at 80°C there is excessive extraction of pectin that interferes with clarification. Unfortunately, it is impossible to operate all components at 75°C (Carruthers and Oldfield, 1955), and many compartments (cells) of both battery and continuous diffusers are at 68°–73°C. Hence, addition of bactericides is an essential part of the control system.

    In diffusers, formalin (40% formaldehyde) is added to those cells that are most susceptible to microbial growth as evidenced by low pH. Usually shock dosing every 2 hr to obtain localized concentration of 120 ppm formaldehyde in the cells concerned is preferable to continuous dosing. Although 50 ppm of formaldehyde in the juice is adequate to prevent growth, acid production, and nitrate reduction (Carruthers *et al.*, 1958), continuous dosing at low levels is thought to lead to selection of strains less susceptible to inhibition by formaldehyde. A modern beet factory will use about 0.25 kg formalin (40% formaldehyde) per tonne* of beets (Guerin *et al.,* 1972). Several thousand tonnes are processed daily.

Sulfur dioxide at 300 ppm has about the same antimicrobial activity as 50 ppm formaldehyde against thermophiles in diffusers and has the advantage of serving as a controlled supply of acidulant. Chlorine and other disinfectants have been investigated for control of microbial growth in diffusers. However, they are inactivated by organic material or they are too toxic or too expensive (Hucker and Pederson, 1942); iodoacetone may be a useful inhibitor of thermophiles (Guerin *et al.,* 1972).

One aspect of control concerns a choice between microbial production of lactic acid and addition of mineral acid, preferably sulfuric, to diffusion water. Make-up water for continuous diffusers has a pH of 7.0–9.0, and cossettes leaving the tail cell must be at pH 6.0 or less for efficient removal of water by pressing; dried cossettes are used as cattle feed. In some operations, sufficient lactic acid is generated by fermentation to obtain the desired low pH, but lactic acid production during diffusion is not sufficiently controllable or predictable (Oldfield *et al.,* 1974a) to assure that the pH of spent cossettes will be ≤ 6.0.

The efficacy of microbial control in diffusion water and in diffusers themselves may be measured by one or more of the following: pH, titratable acid or lactic acid, and direct microscopic or plate counts. As thermophiles grow and die quickly, both types of microbial counts may be useful. The amount of nitrite in diffusers may be a useful measurement of thermophilic activity as nitrite production precedes decreases in pH (Oldfield *et al.,* 1974b).

C. Refining of Raw Sugar

1. Effect of Processing Raw Sugar on Microorganisms

a. Cane Sugar. The refining of raw cane sugar to the crystalline sugar of commerce is designed to remove impurities and produce crystals of sucrose over 99.9% pure (Kelly, 1967). The process is complex (Figure 26.1). Many of the operations are similar to some in the production of raw cane sugar and may destroy or remove bacteria or they may permit recontamination and growth of the contaminants.

The refining of raw cane sugar requires the following procedures, many of which involve microbiology:

1. *Affination:* Mixing (mingling) raw sugar and syrup to facilitate flow of the mixture into basket centrifuges with perforated sides; centrifugation

* 2240 1b—a long ton.

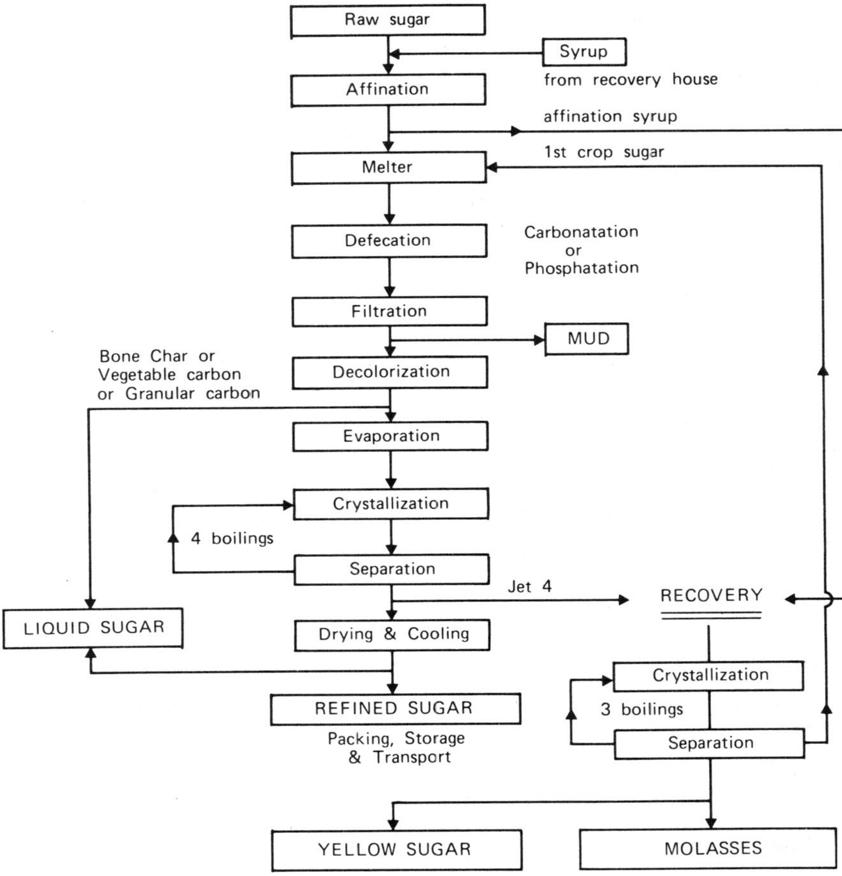

Fig. 26.1. A typical cane sugar refining process (R. Tilbury, Tate and Lyle Research, Reading, England).

to remove the syrup through the perforated sides; washing with water under high pressure while still in the centrifuge to remove most of the molasses film with accompanying microorganisms;

2. *Melting:* Dissolving of the washed sugar with hot water in a tank to obtain a 66°Brix syrup at about 70°C;

3. *Defecation:* Mixing with lime and carbon dioxide, or phosphoric acid and lime, to precipitate impurities, including bacteria;

4. *Filtration:* To remove precipitated impurities and about 99% of bacteria;

5. *Deionization:* To remove ash;

6. *Decolorization:* By filtering solutions at about 90°C through charcoal beds and ion-exchange resins; destruction of some bacteria;

7. *Evaporation:* Crystallization and drying.

The syrup used for affination may contain considerable numbers of spores of flat sour bacteria (Owen, 1977), but the total process of affination, defecation, and filtration removes many microorganisms; others may be destroyed due to "melting" at temperatures of 70°–90°C. Thus, raw sugar, with several thousand bacteria, yeasts, and molds per gram is converted to a solution with few microorganisms. Thereafter, little opportunity exists for microbial growth in the main sequence of operations; temperatures are too high or water activities are too low.

b. Beet Sugar. In North America beets are usually processed to granulated or liquid sugar in a continuous operation; in other areas, raw beet sugar may be stored for subsequent refining (Kelly, 1967). The raw juice is heated to 80°–90°C, adjusted to pH 9.0 or higher with calcium hydroxide and subjected to carbonation ('saturation with carbon dioxide to produce a fine precipitate of calcium carbonate). Impurities and bacteria are precipitated with the calcium carbonate and are removed by filtration. The filtrate is treated with sulfur dioxide to reduce pH and destroy colors, is concentrated by evaporation, then put through an ion exchanger, treated again with sulfur dioxide, and sometimes filtered through kieselguhr to remove most of the remaining bacteria (Scarr, 1968) prior to final evaporation and crystallization. Evaporation is conducted at temperatures and sugar concentrations that preclude microbial spoilage (Miyajima and Kamoi, 1960). Many of the problems associated with refining of beet sugar pertain to cane sugar and are discussed more fully below.

There are three main areas in which microbial growth may occur during refining: in deionization beds, charcoal beds, and sweet waters. Deionization of clarified liquor is carried out at low density (55°Brix) and low temperature (50°C) that permits growth of thermophiles. In charcoal beds, conditions of a_w and temperature prohibit growth during operation, but during "sweetening off" (elution of sugar prior to regenerating charcoal by heating at high temperatures), thermophiles may grow and produce large quantities of gums and slimes (Scarr, 1950). Materials adsorbed on the charcoal may serve as growth factors for *Clostridium thermosaccharolyticum* (Scarr, 1953b).

Sweet waters are sugar-containing waters from several sources, e.g., bag washers, solutions of sugar-containing dust, spillage, and wash water used to recover sugar from filters, charcoal decolorizing beds, and deionizers. The pH may range from 4.5 to 7.5 (usually about 5.5), the Brix from

$0°-60°$ and the temperature from $15°-75°C$. The temperature determines whether microorganisms will grow and if they grow, whether they are mesophiles or thermophiles. Because sweet waters are used in the refinery, e.g., in melting of affinated sugar, dilution of high Brix solutions, etc., the microbial content of sweet waters is important (Tilbury, 1975; Tilbury et al., 1976). If the temperature is favorable, microbial counts may be high (10^4-10^7 per ml).

The principal bacteria in sweet waters are *Leuconostoc, Lactobacillus, Streptococcus,* and *Bacillus* spp. The non-sporing lactic acid-producing rods and cocci frequently compete more successfully than other microorganisms in sweet waters because they have a fermentative metabolism in a sugar-rich environment; they grow over a wide range of pH, temperature, and concentration of sugar; they do not require oxygen; and frequently they are encapsulated, which protects against chemicals and drying.

Spore formers compete well in conditions that are unsuitable for nonsporing lactics. They survive high temperatures and may be the only organisms present when a_w and temperature become suitable for growth. Also, many are facultative or obligate thermophiles (Owen, 1977), and facultative, or more rarely, obligate anaerobes. They are present in all raw sugars (Skole et al., 1977; Rizzuto et al., 1964), and some are removed during the sugar-washing operation in the centrifuge and are circulated in the systems carrying sweet waters. *Bacillus cereus* and *B. subtilis* may survive high temperatures and predominate when conditions are suitable for growth (Skole et al., 1967, 1968; Scarr, 1950). Spores of flat sour bacteria may comprise the majority of thermophilic spores in sweet waters (Owen, 1977).

The yeasts most frequently present are *Torulopsis, Saccharomyces,* and *Hansenula* spp., but other species are also present. Many are osmophilic and some are active producers of invertase. At suitable temperatures there may be 10^5-10^6 yeast/ml.

The main consequences of microbial growth in sweet waters are production of dextran and levan, production of acids, and inversion of sucrose. The total loss of sucrose from such activity in sweet waters may approximate 0.5% of total sugar produced (Tilbury et al., 1976).

Prevention of deterioration in sweet waters is best obtained by maintaining $75°C$ in those components of the system where this is possible and using preservatives where high temperatures cannot be maintained. Several preservatives are effective, e.g., quaternary ammonium compounds, benzoate, formaldehyde, and metabisulfite. Of the legally permissible compounds, metabisulfite appears the most economical. Formaldehyde, although effective, is not legally acceptable in refining, but it is permitted in beet and cane diffusers.

2. Spoilage of Liquid Sugar

Liquid sugar is refined sugar concentrated after the decolorization step (Fig. 26.1) or made by dissolving crystalline refined sugar. It has a sugar content of 66° to 76°Brix depending on the degree of inversion. Osmophilic yeasts (especially *Saccharomyces rouxii, S. cerevisiae,* and *S. mellis)* and molds may grow, although slowly, because lag times and mean generation times generally are inversely proportional to the a_w; bacteria do not grow. Three factors determine growth of yeasts and molds: (a) size of inoculum, (b) availability of nonsucrose nutrients, and (c) gradients with increasing a_w.

Small numbers of fully osmophilic yeasts may fail to grow while some cells in a large population of the same culture may be able to initiate growth. The ash content of liquid sugar may vary from 2.5 to <0.0001%; deionization to a low content of ash may prevent growth of some osmophiles (Scarr, 1963).

Liquid sugar with gradients of a_w are most susceptible to spoilage because strains of yeast with moderate osmophilic characteristics can grow quickly in solutions with high a_w (low concentration of sugar); by adaptation or selection they grow across the gradient to the lowest a_w (highest concentration of sugar). Thus, large populations of highly osmophilic yeasts become established in concentrated solutions of sugar. Gradients of a_w occur because water does not readily mix with highly concentrated solutions of sugar unless agitated. Thus, pockets of water in improperly dried equipment, and condensate that forms on the ceilings and walls of tanks and runs down to the surface of the liquid sugar, provide gradients. Also, improperly washed pipes and valves containing diluted sugar solutions may provide suitable sites for growth to large populations.

Three general methods are used to prevent spoilage of liquid sugar: (a) removal and destruction of microorganisms capable of growth, (b) use of sanitizing agents, and (c) prevention of condensates. Organisms in tanks, pipes, and valves are washed out, destroyed by heat or chlorine (Kelly, 1967), and prevented from growth by long lasting iodophores, which are flushed out before the equipment is used. Liquid sugar may be filtered to remove organisms and heated or passed over ultraviolet lamps to destroy them. Ultraviolet lamps may be placed in the ceilings of tanks to destroy yeasts and molds that settle on the surface of the sugar, but some regulatory agencies prohibit such placement. To prevent condensation, filtered air treated by ultraviolet irradiation is forced over the surface of liquid sugar in otherwise airtight tanks. In preventing spoilage, manufacturers of liquid sugar generally take all necessary precautions. However, food or pharmaceutical manufacturers who purchase shipments too large for immediate use are more likely to suffer spoilage (Scarr, 1963; Moroz, 1963; Tilbury, 1976).

3. Microorganisms in Refined Sugar Capable of Spoiling Other Food

Certain microorganisms that grow during extraction and refining can survive the process or gain entrance after processing. Usually there are fewer than 10^2/g, but if present in sufficient numbers in the refined sugar they cause serious spoilage in foods that have sugar as an ingredient (Scarr, 1968). They may be decreased by irradiation with ultraviolet light (Murphy, 1965). The organisms of most concern are (Cameron and Williams, 1928; Clark and Tanner, 1937) the following:

1. *Bacillus stearothermophilus* and *B. coagulans* that may grow in canned food and can produce acid without gas. As the can of food is not distended by gas, the condition is described as "flat sour," and the two species are designated flat sour organisms. *Bacillus stearothermophilus* is a special nuisance because it forms spores that are very heat resistant and grows at temperatures up to 75°C; it does not grow at pH less than 5.2. In contrast, *B. coagulans* cannot grow at temperatures above 65°C and is less heat resistant than *B. stearothermophilus* but can grow at pH 4.2.

2. *Clostridium thermosaccharolyticum* that grows well at 72°C, but less well at 75°C. In canned food, it may produce sufficient acid to cause hydrogen swells.

3. *Desulfotomaculum (Clostridium) nigrificans* that grows optimally at 55°C. It may cause sulfide stinkers in canned food.

4. Mesophilic bacteria, yeasts, and molds that can grow at the pH of soft drinks.

4. Control of Microorganisms in Refined Sugar

Because sugarborne microorganisms cause spoilage in canned foods and bottled drinks, rigid specifications have evolved (Lynch, 1977). Those for canners sugar apply only to heat-resistant spores: the analytical method stipulates boiling of the sugar solution for 5 min prior to culture (Skole and Newman, 1970). The specifications for bottlers sugar measure heat-susceptible microbes, and samples are not heated. The following are adapted from the references cited:

National Canners Association's Bacterial Standards for Sugar (National Food Processors Association, 1972; Horwitz, 1975): Five samples are examined.

Total Thermophilic Spore Counts—average not more than 125 spores/ 10 g of sugar.

Flat Sour Spores—average not more than 50 spores/10 g of sugar.

Thermophilic Anaerobic (TA) Spores—may be present in up to three of the five samples, but in any one sample not more than four of six tubes inoculated by the standard procedure should contain TA spores.

Sulfide Spoilage Spores—may be present in up to two of the five samples.

Bottlers Standards for Dry, Granulated Sugar (National Soft Drink Association, 1975)

Bacteriological—Bottlers sugar shall not contain more than

200 mesophilic bacteria/10 g
10 yeast/10 g
10 mold/10 g

Bottlers Standards for Liquid Sugar (Sugar Syrup) in 10 g of Dry Sugar Equivalent (DSE)

100 mesophilic bacteria
10 yeasts
10 molds

All microbiological testing is done according to the procedures of the Sugar-Bottling Industries Committee (see also this volume, Chapter 22).

Alternate procedures that have been shown to give equivalent reliability are permitted for testing sugar. Many alternate procedures such as membrane filters, bioluminescence, and fluorescent microscopy have been evaluated, and some are useful (Rizzuto and Jacobson, 1964; Skole et al., 1974; Moroz, 1963; International Commission for Uniform Methods of Sugar Analysis, 1966).

At one time organoleptic tests for objectionable tastes and odors were applied to bottlers sugar, in part because of the presence of butyric acid. The causative organism is probably *Clostridium butyricum,* some strains of which metabolize lactate and produce butyric acid. *Clostridium butyricum* is present in a large percentage of raw sugar and is not readily removed during refining. Butyric acid production is still a problem, but is rare (Rizzuto et al., 1964).

D. Choice of Case

For an understanding of "Case" see Table A.1 and the brief description in Appendix IV. For more details, see the second book of this series (ICMSF, 1974).

The refining of sugar destroys pathogenic organisms, if they are present.

The only refined product that has a history of spoilage is liquid sugar; granulated sugar rarely has been reported to spoil, and then from accidental wetting (Tilbury, 1976). Case 2 is most suitable for granulated sugar that is not to be used in canning, and Case 3 for granulated and liquid sugars that are to be used for canning or bottling.

III. COCOA BEANS, COCOA, AND CHOCOLATE

A. Cocoa Beans

1. Important Properties

Cocoa beans are found in pods surrounded by pulp that is usually sterile. To be converted to cocoa, beans must be removed from the pods, fermented, dried, roasted, ground, and pressed. During fermentation, microbially generated changes decrease pH and increase temperature. These changed conditions destroy germinative capability of the bean and produce desirable color and some flavor. At the same time, nonmicrobial enzymatic activities in the bean effect desirable changes. Drying stabilizes the beans against microbial spoilage, and roasting destroys some microorganisms and develops the full flavor.

The pulp of ripe pods consists of a mass of parenchyma cells derived from the endocarp. Pulp contains 80–90% water; 6–13% sugar of which about one-third is sucrose and most of the remainder invert sugars; 0.5–1% citric acid, and small amounts of aspartic acid, asparagine, and glutamic acid (Forsyth and Quesnel, 1963; Rohan, 1963). The pH is about 3.5. Pulp from unripe pods contains less sugar and is difficult to ferment (Roelofsen, 1958).

The beans consist of testa (shell or skin) and cotyledons. The cotyledons contain about one-third water and one-third fat (cocoa butter), and the remainder consists of phenolic compounds, starch, sugar, theobromine, nonvolatile acids, and many other components in small concentrations (Forsyth and Quesnel, 1963).

2. Fermentation

a. Harvesting. Pods are opened by breaking or cutting, and beans (30–40 per pod) are removed with accompanying pulp. The beans with residual pulp may be fermented in heaps, baskets and barrels, or a system of single or multiple boxes. If fermented in heaps, the mass of beans may vary from 10 to 500 kg. In containers, the mass varies from a few hundred to a few thousand kg. Fermenting beans are covered with leaves, burlap, or boards to retain heat.

In small-scale operations, as practiced primarily in Africa, beans are removed from pods and fermented on plantain leaves on the ground near the trees from which they were harvested. In larger operations, typical of the Americas, beans may be removed from pods close to the trees being harvested and then transported in sacks or boxes to a central location. In some plantations, pods are transported to the fermentation site, and then opened (Ostovar and Keeney, 1973). The size of the fermenting pile, the time between removal of beans and putting into the heap or container for fermentation, and the type of container all may have an effect on the course of the fermentation and, to some extent, on the quality of the final product (Rohan, 1958a; Howell, 1948).

The sources of microflora are varied. The surface of pods, hands, and machetes contain species in at least 17 genera of bacteria as well as unidentified yeasts (Ostovar and Keeney, 1973). At the start of a fermentation, 65% of the microorganisms are from soil or air. However, some selection of microbes may be in effect before the start of fermentation because many microbes active in fermentation are on the walls of boxes, baskets, and sacks that are used sometimes to transport beans to central facilities for fermentation or are left in the sweat boxes from previous fermentations (Ostovar and Keeney, 1973). *Drosophila* spp. also may introduce microorganisms.

b. Flavor Development. During the fermentation, some basic cocoa flavor develops, additional flavor appears during drying and storage, and the full flavor develops on roasting. Enzymes of yeasts may be important in developing flavor (Levanon and Rossetini, 1965), but there is evidence that fermentation is not necessary for the development of basic cocoa flavor (Rombouts, 1952; Wadsworth and Howat, 1954). Possibly there are some limited changes associated with incipient germination that are important (Wadsworth and Howat, 1954), but complete germination is deplored. A comprehensive discussion of chemical and biochemical changes associated with flavor development during fermentation and drying is available (Forsyth and Quesnel, 1963).

c. Microorganisms. The microorganisms found in fermentation and spoilage of cocoa are fermentative, and probably oxidative: yeasts, bacteria that produce acetic and lactic acids, *Bacillus* spp., and molds. Each group will be considered in the context of its capability to participate in the ecosystem of fermentation.

Yeasts. Many of the yeasts found in cocoa fermentations are also found in soil, flowers, and fruits. In fermenting fruits and vegetables, they are often associated with lactic acid- and acetic acid-producing bacteria

(Last and Price, 1969). Possibly they synthesize vitamins important for the lactic acid producers.

The most frequently isolated and most numerous yeasts during fermentation are *Saccharomyces cerevisiae, Saccharomyces* spp., *Candida krusei, Kloeckera apiculata, Pichia fermentans, Hansenula anomola,* and *Schizosaccharomyces pombe.* The numerous synonyms for these species have been critically reviewed (Rombouts, 1953). The fermentative yeasts rapidly dominate and many persist through drying and storage; many samples of stored beans contain 10^5–10^7/g (Hansen and Welty, 1970).

Lactic acid bacteria. Bacteria producing lactic acid have been quantitated by Rombouts (1952) and identified by Ostovar and Keeney (1973). After 1.5 days of fermentation they may constitute 10–20% of the total microflora and number 3×10^8 per bean (Rombouts, 1952). They probably produce substantial amounts of lactic acid. The lactic acid bacteria, and some characteristics important to their role in the fermentation, are listed in Table 26.5.

In one study, during the first 24 hr of fermentation, *Lactobacillus plantarum, Leuconostoc mesenteroides, Streptococcus lactis,* and *Lactobacillus fermentum* were the predominant lactic flora, probably reflecting their ability to grow at the lower temperatures in the fermentation. After 24 hr, as temperatures increased, *Lactobacillus acidophilus, Lactobacillus bulgari-*

TABLE 26.5

Important Characteristics of Bacteria Producing Lactic Acid in Cocoa Fermentations [a]

Species	Acid produced	Gas produced	Maximum temperature (°C) for growth [b]
Lactobacillus plantarum	Lactic		$\leq$45
Leuconostoc mesenteroides	Lactic, acetic	CO_2	<37
Streptococcus lactis	Lactic		<45
Streptococcus thermophilus	Lactic		>50
Lactobacillus bulgaricus	Lactic		>45
Lactobacillus fermentum	Lactic, acetic	CO_2	>45
Lactobacillus lactis	Lactic		>45
Lactobacillus acidophilus	Lactic		>45
Pediococcus cerevisiae	Lactic		<35
Pediococcus acidilactici	Lactic		52

[a] Ostovar and Keeney (1973).
[b] Limiting temperatures (Buchanan and Gibbons, 1974); temperatures given may not precisely describe limits for all strains and conditions.

cus, Lactobacillus lactis, and *Streptococcus thermophilus,* which grow well at 40°–45°C and sometimes higher, were the dominant lactic flora. *Pediococcus acidilactici* and *P. cerevisiae,* although present, never became a large percentage of the lactic flora (Ostovar and Keeney, 1973).

Acetic acid bacteria. Bacteria oxidizing ethanol to acetic acid and oxidizing lactic and acetic acids to carbon dioxide and water are important in fermentation of cocoa beans. Within 2 days they may reach 10^9 per bean (Table 26.6). The acetic acid-forming bacteria may, however, play a less important role in acid production in some types of fermentations than is indicated in Table 26.6 (Ostovar and Keeney, 1973). The acetic acid bacteria are *Acetobacter* and *Gluconobacter* spp. (see Table 26.7). They do not thrive unless there is abundant oxygen and the temperature is <40°C; both of these factors may vary considerably depending on design and operation of the fermentor.

Bacillus. *Bacillus* species are present in low numbers at the beginning of the fermentation, undoubtedly introduced from hands, soil, and the residue of previous fermentations. At the low pH of the fresh pulp (about 3.7), *Bacillus* spp. are unable to grow. However, as the fermentation progresses and the pH and temperature increase, they become the predominating microorganism, numbering about 10^6/g (Ostovar and Keeney, 1973), or 10^7–10^8 per bean (Rombouts, 1952). The role of *Bacillus* spp. as a contributor to the fermentation has not been postulated. Perhaps, they simply grow in an ecosystem that has become exhausted or inimical to other organisms because of decreased sugar, increased oxygen, and temperatures above 45°C for several days.

Bacillus species identified from fermenting cocoa are *B. cereus, B. megaterium, B. subtilis, B. coagulans, B. pumilus,* and *B. stearothermophilus.* Most of these grow well at 45°–50°C, and significantly, the obligate thermophile *B. stearothermophilus* is the predominant microbial species after 72 hr of fermentation (Ostovar and Keeney, 1973). Other species prevalent in dried beans and in cocoa powder are probably also present during fermentation.

Molds. Unlike the yeasts and acid-producing bacteria, most molds are potentially harmful during fermentation, drying, and storage. They vary with the crop, moisture content, and temperature.

Molds are present in fermenting beans, and the low pH favors their growth. If the pile is unturned for 2–3 days, mold growth visible to the unaided eye will appear, but only on the surface of the mass of beans (Roelofsen, 1958). Such growth is not unusual, nor is it considered harmful unless it penetrates the testa and damages the cotyledons (Chatt, 1953). Beans that are obviously moldy, especially if in clumps, are re-

TABLE 26.6

Chemical and Bacterial Changes of Cocoa Pulp and Beans during Fermentation

Duration (days)	Temperature (°C) [a]	pH of pulp [a]	Proportion in pulp (%)			Microbial level average count per bean (×10⁶)				
			Sugar [b] (%)	Ethanol [c] (%)	Acetic acid [c] (%)	Yeasts [a]	Lactic acid bacteria [a]	Acetic acid bacteria [a]	Bacillus spp. [a]	Other bacteria [a]
0	26	3.7	12	0.00	0.04	0.04	<0.0024	0.04	0.08	0.08
0.5	32	3.8	11	0.13	0.12	12.0	<0.13	0.45	0.06	0.06
1.0	38	4.0	4	1.4	0.18	283.0	50.0	280.0	<6.3	<6.3
1.5	42	4.1	0.3	2.8	0.24	320.0	320.0	960.0	<16.0	<16.0
2.0	45	4.2	0.2	2.3	0.48	180.0	30.0	1700.0	30.0	<20.0
3.0	49	4.8	—	1.6	1.20	0.8	<0.1	0.3	9.0	<0.1
5.0	50	6.3	—	0.9	1.40	<0.02	<0.02	<0.02	2.0	<0.02
7.0	46	6.5	—	0.5	1.60	0.15	<0.1	<0.1	9.3	0.6

[a] Adapted from Rombouts (1952).
[b] Adapted from Roelofsen (1958).
[c] Adapted from Forsyth and Quesnel (1963).

TABLE 26.7

Important Characteristics of Bacteria Producing Acetic Acid in
Cocoa Fermentations [a]

Characteristics	Acetobacter [b]	Gluconobacter [c]
Converts ethanol to acetic acid	+	+
Converts acetic acid to carbon dioxide	+	−
Converts lactic acid to carbon dioxide	+	−
pH range of growth	4–7	4–7
Temperature range of growth ($^{\circ}C$)	5–42	7–41

[a] From Buchanan and Gibbons (1974).

[b] *Acetobacter aceti* and *Acetobacter pasteurianus,* each with several subspecies and many synonyms for each subspecies.

[c] *Gluconobacter oxydans,* probably subspecies *suboxydans* and *melanogenes.* These two subspecies have been previously named with over 30 synonyms, frequently in the genus *Acetobacter.*

moved during the postfermentation drying period, but a high percentage of beans may be moldy without appearing to be so (Broadbent and Oyeniran, 1968; Maravalhas, 1966).

The molds commonly found during fermentation are *Aspergillus fumigatus,* especially harmful because it destroys the testa and permits penetration by other molds, *Aspergillus niger, Aspergillus glaucus, Aspergillus flavus, Aspergillus tamarii, Penicillium* spp., and *Mucor* spp. (Chatt, 1953; Roelofsen, 1958). Most molds produce lipases; lipases may contribute to spoilage of chocolate.

d. Sequence of Events during Fermentation. Fermentation will be presented in the following order: sweating, temperature, and normal fermentation.

Sweating. Sweating is the release and drainage of liquid as a consequence of fermentation. In a fresh heap or freshly filled fermentation chamber, the space between the beans is almost completely filled with pulp. The pulp tissue is macerated, probably by enzymes of some yeasts, and liquid escapes through the plantain leaves or through the perforated bottom of the wooden fermentation chamber. The spongy parenchyma that constitutes the pulp (endocarp) is composed of big tubular cells with large intercellular spaces. When these collapse and drain away (pulp is 80–90% water), air can fill the void between the beans. Undoubtedly, much of the ethanol, lactic acid, and acetic acid produced during the fermentation is drained off during the sweating.

Temperature. A temperature of 40°–50°C must be obtained during the fermentation to destroy the capacity of the beans to germinate and to obtain desirable changes in color and flavor. During fermentation, the temperature should be even throughout the mass of beans ($\pm 3\,°C$; Knapp, 1937) and should increase fast enough to destroy the radicle before germination can take place (Chatt, 1953). In some countries, temperatures increase to 48°–50°C, in others rarely over 45°C. The time to reach these temperatures may be 1–3 days depending on the air supplied to the fermenting mass; this supply may depend on mixing of the beans during fermentation and on drainage of sweatings. After about 3–5 days, the temperature begins to decline.

Heaps of beans of 125–300 kg piled on and covered with plantain leaves heat quickly and, uniformly throughout the mass without mixing, while a mass of 1000 kg similarly treated, does not heat quickly and uniformly unless mixed or otherwise provided with air (Rohan, 1958b). In large boxes, mixing is accomplished by turning or by emptying the contents of one box into another at intervals of about 1 day. Additional aeration is obtained as warm air rises, heated by the exothermic microbial activity. This causes cool air to be drawn through perforations in the bottom of the box.

The main heat-generating mechanisms are (Forsyth and Quesnel, 1963; Quesnel, 1968)

Hexose → ethanol	17.8 Kcal/mole
Ethanol → acetic acid	118.2 Kcal/mole
Acetic acid → $CO_2 + H_2O$	209.4 Kcal/mole

Normal fermentation. The wide variety of conditions used for fermenting cocoa beans subjects any description of a normal fermentation to some restraints, especially with respect to time involved. However, several studies (Rombouts, 1952; Roelofsen, 1958; Rohan, 1958a,b; Ostovar and Keeney, 1973) suggest the following sequence of events (Table 26.6). During the first day there are anaerobic conditions, low pH, and high sugar concentration in the pulp. Fermentative yeasts increase 10,000-fold to about 10^8/ml of pulp, and they dominate the fermentation of sugar to ethanol and carbon dioxide. The bacterium, *Zymomonas mobilis,* may also be important in producing alcohol. Also, during the first day *Pediococcus cerevisiae, Leuconostoc mesenteroides,* and *Streptococcus lactis* may produce lactic acid, acetic acid, and carbon dioxide. Production of ethanol and lactic acid provides substrates for a 100,000-fold increase to about 10^9/ml of *Acetobacter* and *Gluconobacter* spp. between 0 and 48 hr.

Maceration and collapse of the pulp and drainage of liquid (sweating) permits entry of air. The oxygen enables both *Acetobacter* and *Gluconobacter* spp. to oxidize ethanol to acetic acid and the *Acetobacter* spp. to oxidize acetic and lactic acids to carbon dioxide and water. These oxidations release sufficient heat to increase the temperature. Above 40°C there is a competitive advantage to several species of *Lactobacillus* (see Tables 26.5 and 26.6) and a disadvantage to yeasts. With the destruction of sugar, almost complete by the end of the second day, the fermentative yeasts— *Zymomonas mobilis*—and lactic acid bacteria are in an ecosystem no longer suitable for growth, and by the third day they decrease to 0.1–1.0% of their maximum populations.

Although the fermentative organisms no longer have an adequate supply of readily available nutrients, the opposite is true for the oxidative organisms. After 1–1.5 days, *Gluconobacter* and *Acetobacter* species and oxidative aerophilic pseudomycelium-forming yeasts have an abundant supply of ethanol, acetic acid, and lactic acid and an increasing supply of oxygen. These two conditions lead to temperatures above 40°C; thus the consequences of their metabolic activities prevent their growth. By the third day, they have decreased to <0.1% of their second-day population, and by the end of the third day, most are killed as the temperature reaches 50°C.

The *Bacillus* spp. generally are unable to grow at the low pH (3.7–4.0) prevailing during the first two days. However, as pH increases, oxygen becomes available, and the temperature increases, conditions become increasingly suitable for their growth. In larger fermentations (Rohan, 1958a,b), temperatures are sustained for several days at about 48°–50°C, and several species of *Bacillus* that grow at 50°C become a significant percentage of the total population, with the thermophiles *B. coagulans* and *B. stearothermophilus* predominating. It is difficult to ascribe a role to the *Bacillus* spp. in the fermentation, but one species, *B. subtilis,* is implicated in synthesis of a component of the flavor of cocoa (Zak *et al.,* 1972).

During normal fermentation, the pH of the pulp, originally at approximately 3.7, increases to about 4.5–5.0, and that of the cotyledons, originally at pH 7.0 decreases to about 5.0 (Rohan, 1958b). The pH of the pulp may increase to 6–7 (Table 26.6; Rohan, 1958a) as the fermentation is prolonged, especially in small heaps of beans fermenting on plantain leaves. It is, however, difficult to reconcile the observed increases in pH of the pulp to 6–7 with the reported concentrations of acetic acid; 1% of titratable acetic acid should result in pH 4 or less. Possibly in small heaps, the greater availability of oxygen results in more complete oxidation of acetic (and possibly lactic) acid than indicated in Table 26.6.

There appears to be a discrepancy between two studies of the fermentation. Rombouts (1952) found maximum microbial counts of about 2×10^9/bean while Ostovar and Keeney (1973) found a maximum of 10^7/g of bean. Since a bean weighs about 1 g one would expect the counts to be about the same. The probable explanation is that more mucilaginous tissue of the disintegrating pulp adhered to the beans sampled by Rombouts (1952) than to those sampled by Ostovar and Keeney (1973), and the pulp contained large bacterial populations. Fermenting plant juice (e.g., when cabbage ferments into sauerkraut) may contain more than 10^9 lactic acid bacteria per milliliter (Pederson, 1971). In contrast, Ostovar and Keeney determined microbial populations per gram of bean disintegrated in a blendor; and the cotyledons during fermentation, unless damaged, contain few if any microorganisms.

Although the normal fermentation just described pertains to the fermentation of large masses of beans (about 250–1000 kg), much of the world crop is fermented in small heaps under primitive conditions, and mixing may be omitted. In addition, pods picked before maturity to prevent invasion by insects contain insufficient sugar, and the pulp does not sweat readily to enable a normal fermentation. Fermentation of cocoa is an art, and many techniques have been adapted to cope with less than ideal conditions (Roelofsen, 1958).

Attempts to improve the product by manipulating the fermentation with added sugar, acids, pure cultures, and mineral nutrients for yeasts generally have been unsuccessful (Roelofsen, 1958).

e. Control of Spoilage during Fermentation. The most important activities for preventing spoilage are (a) mixing that is adequate for the mass of beans being fermented, (b) judicious use of covering materials over the heap to retain the temperature at about 50°C, and (c) adequate drainage to permit escape of sweatings. The shortest period of fermentation which gives the desired changes in color and internal appearance of the bean is the best. Prolonging the fermentation may result in growth of putrefactive bacteria if the pH increases to 6–7. Dominance of the fermentation by *Pseudomonas, Enterobacter,* or *Escherichia* may produce off-odors (Ostovar, 1971).

Care in opening pods with machetes minimizes cutting the beans. Beans with testa penetrated either from cuts or from partial germination due to slow fermentations are more susceptible to damage by molds.

3. Drying

a. Microbiology. At the end of the fermentation, the beans contain about 60% water. They may be washed to remove residual pulp, thus facilitating drying, although washing is not common. Beans are air-dried for 1–2 days in the sun followed by drying by artificial heat for 2–4 days, or more commonly, are air-dried in the sun for 1–4 weeks. At this time the cotyledons contain about 5% water and the testa about 12%. The desirable moisture content for whole dried beans is 6–6.5%, and the critical level is 8% (Hansen, 1975); beans with more than 8% moisture will develop mold in storage.

Molds may grow during drying with *Penicillium* spp., *Aspergillus fumigatus,* and *Geotrichum candidum* (Hansen, 1975) predominating. *Geotrichum candidum* (over 50 synonyms including *G. lactis, Oidium lactis, Oospora lactis*) is abundant, and it is probably active in oxidizing lactic acid; it is unlikely to damage beans (Butler, 1960).

b. Spoilage of Dried Beans. Even properly dried beans will absorb moisture during storage if the relative humidity (R.H.) is too high. Since beans that are partially germinated, or that have broken shells, are more susceptible to spoilage by molds, increased susceptibility must be considered. An R.H. of about 72 (Roelofsen, 1958) to 82% (Rohan, 1963) appears to be maximum for prolonged storage. At 95% R.H., the moisture content of beans increases to 16% (Rohan, 1963). Storage of properly dried beans in polyethylene-lined containers minimizes increase of moisture (Minifie, 1970).

During storage, molds may or may not grow depending on the moisture content. Molds isolated from normal dried beans are *Aspergillus glaucus, A. flavus, A. niger, A. tamarii, Mucor pusillus, Penicillium* spp., *Mucor* spp., and *Geotrichium candidum.* The colony counts are usually 10^4–10^5/g (range 0 to 3×10^5) with the majority of colonies caused by *M. pusillus.* Obviously, moldy beans contain large numbers of *Penicillium* spp., *Aspergillus* spp. (*A. flavus, A. tamarii*), and *Paecilomyces* spp. (5×10^8 to 77×10^9/g; Hansen and Welty, 1970).

Defects may be produced by molds growing in beans. Lipases may cause an increase in free fatty acids from about 1% of the lipid to 7–41%, thus, lowering the melting point and making the chocolate from such beans unfit for coating (Hansen, 1975). Off-flavors are produced (Hansen and Welty, 1970), and there is an increase in total carbonyls (ketoglycerides, keto acids, dicarbonyls, monocarbonyls, and other compounds containing a carbonyl group; Hansen and Keeney, 1970). An excellent objective method for quantitation of mold damage is titration of free fatty acids in extracted lipids (Kavanagh *et al.,* 1970). Mycotoxins may be produced (Scott, 1973; Llwewllyn *et al.,* 1978). Possibly lipases survive processing and cause spoilage in manufactured chocolate. Beans that are obviously moldy are not used for manufacture of chocolate.

Stored beans are subject to infestation by moths, beetles, and weevils. Descriptions of their life cycles and methods for control are available (Rohan, 1963).

4. Processing Beans

a. Cleaning, Roasting, and Winnowing. Beans are cleaned by screening, air currents, and magnets to remove extraneous material. Cocoa beans that are sound and undamaged have few if any microorganisms inside the shell (Meursing and Slot, 1968). Thus, removal of the shell before roasting may, at the same time, remove most of the microorganisms. Two methods of processing are used: (1) roasting and then winnowing to separate shells from cotyledons (nibs) (Lindley, 1972) or, less commonly, (2) winnowing followed by roasting of the nibs. The latter process enables production of cocoa or cocoa liquor with lower microbial counts (Meursing and Slot, 1968).

Roasting heats beans from 15 min to 2 hr at 105° to 150°C. For production of cocoa butter, beans may be warmed only sufficiently to loosen the shell, a method that is unlikely to destroy microorganisms. In some types of roasting, injected moisture increases the lethal activity of heat to microorganisms. During roasting the water content is reduced to 1–2%.

b. Microbiology of Beans before and after Roasting. Prior to roasting, beans contain a wide variety of microorganisms. The bacterial population varies from 5×10^5 to 5×10^7 (Meursing and Slot, 1968; Hansen and Welty, 1970; Barrile et al., 1971). Over 90% of the bacteria are *Bacillus* spp. (Barrile et al., 1971). Non-spore-forming bacteria belonging to the genera *Enterobacter, Escherichia, Flavobacterium, Microbacterium, Micrococcus,* and *Streptococcus* are usually a small percentage of the total population.

Mold and yeast populations in normal beans may vary widely depending on the source, e.g., molds from $<10^3$/g in beans from some countries (Barrile et al., 1971) to 10^5/g in a high percentage of shipments from other countries (Hansen and Welty, 1970); and yeasts from $<10^3$/g to 30×10^5/g. The yeasts most common on dried beans are those that are dominant during fermentation (Knapp, 1937). Species have been listed previously.

Roasting is the main step in the process for destroying microorganisms. If time and temperature of roasting are sufficient, all microorganisms except some bacterial spores are destroyed. During 30 min at 135°C, the aerobic plate count (APC) is reduced to about 10%, and at 165°C to about 1% of the APC of unroasted beans. Roasting at higher temperatures destroys more spores but causes development of undesirable flavor. In one study, many *Bacillus* spp. found on beans before roasting were not found on beans after roasting. After roasting for 40 min at 150°C, the *Bacillus* spp. were, in order of frequency, *B. coagulans, B. stearothermophilus, B.*

circulans, B. licheniformis, B. megaterium, and *B. subtilis* (Ostovar and Keeney, 1973). However, several additional species frequently are found in cocoa powder, suggesting that they too have survived roasting.

c. Processing of Nibs. The nibs are ground and milled to produce chocolate liquor with over 50% fat. Heat generated by milling increases the temperature to 60°–80°C or higher, but at the low a_w of chocolate liquor (usually less than 2% water), microorganisms are not readily destroyed by heat. Chocolate liquor may be pressed to produce cocoa press cake and cocoa butter, mixed with fully or partially dehydrated milk and sucrose to produce chocolate crumb, or stored at temperatures of 60°–80°C, occasionally higher, until used for chocolate, etc. During storage, moisture is reduced to about 1–2% and a slight reduction of asporogenous bacteria may occur. All of the above products may be used in manufacturing chocolate confectionery.

B. Cocoa

1. Important Properties

Cocoa powder contains 9–36% fat and less than 8% moisture; its pH is 5.5–6.2 (natural cocoa) or 7.0–8.0 (alkalized cocoa). As the relative humidity (R.H.) increases, the moisture content increases: at 60, 70, and 80% R.H., the moisture content of cocoa is approximately 8, 11, and 13%, respectively. Storage at 15°–18°C at less than 70% R.H., and with minimum fluctuation of temperature, is recommended (Meursing, 1976).

2. Microbiology

Adequately roasted beans that have been hygienically handled should yield cocoa with few if any yeasts, molds, or asporogenous bacteria. Finding large numbers of organisms other than spore-forming bacteria is a signal that the beans or nibs were not roasted adequately or that the product experienced postprocess contamination and possibly growth during storage because of high a_w.

The microflora of cocoa consists mainly of *Bacillus* spp., with variable numbers of yeasts and molds. A small percentage of samples may contain *Enterobacteriaceae* (Meursing and Slot, 1968) and other non-spore-forming bacteria (Gabis *et al.,* 1970). *Salmonella* spp. have been found (Depew, 1968; Collins-Thompson *et al.,* 1978).

The predominating bacteria are *Bacillus licheniformis, B. cereus, B. subtilis, B. megaterium,* and *B. alvei.* Other species are less frequently present but two of these, the thermophiles *B. coagulans* and *B. stearothermophilus,* may be a potential cause of spoilage in some cocoa-containing products with high a_w (Meursing and Slot, 1968; Gabis et al., 1970; Mossel et al., 1974).

Microbial populations of cocoa vary tremendously, possibly reflecting the microbial quality of the beans and the technology of the country in which the beans were roasted and processed. Many of the bean-producing countries convert a large percentage of the world supply of beans to cocoa, cocoa butter, and chocolate liquor (Wolf, 1977), and this practice is increasing. Some of them have primitive concepts of technology and sanitation as indicated by some recent studies: 36% of industrial cocoa imported into Canada contained over 10^4 Bacillus spp./g, while only 4% of industrial cocoa produced in Canada exceeded 10^4/g (Collins-Thompson et al., 1978). In the Netherlands, at least one manufacturer usually produces cocoa with fewer than 300 Bacillus spp./g (Meursing and Slot, 1968). Yeast and mold counts in cocoa packaged for the retail trade may be high (up to 10^5), although most recommended limits are <50/g. Irradiation of cocoa will destroy microbes, but the flavor is affected adversely (Grünewald and Münzner, 1972). .

Microbial limits are presented as suggestions and recommendations by noncommercial organizations and as specifications in sales contracts by buyers. However, recommended limits may have little impact unless related to economic or regulatory penalties. Some microbial limits are given in Table 26.8.

3. Control

Microbiological methods for the organisms of major interest have been described (Meursing, 1969, 1976). However, the examination for Salmonella spp. requires special mention. Cocoa contains some antibacterial compounds (Busta and Speck, 1968) that express anti-Salmonella activity when cocoa is pre-enriched in the usual lactose or nutrient both at a 1:10 ratio. The use of skim milk as an enrichment medium (Park et al., 1976, 1979) or the addition of casein to lactose or nutrient broths (Zapatka et al., 1977) negates the anti-Salmonella activity. The antibacterial effect is unlikely to compromise determinations for other bacteria (Park et al., 1979) because of lower ratios of cocoa to bacteriological medium.

C. Chocolate

1. Important Properties

Chocolate is of two types, dark chocolate and milk chocolate. Both contain liquor and sugar; the milk chocolate also contains milk supplied either

TABLE 26.8

Recommended Microbial Limits Per Gram of Cocoa

Source	Aerobic plate count	Escherichia coli	Coliform group	Yeasts	Molds	Salmonella	Bacterial spores Flat sour	Other
Minifie (1970)	20,000	<0.1	—	<50	<50	—	—	—
Meursing (1976)	5,000	0	—	<50	<50	<0.1	—	—
I.C.P. Cocoa (1977)	5,000	0	<50	—	—	0	—	—
Van Houten (undated)	5,000	—	<0.1	<50	<50	<0.0025	<5	<1000
Woodward and Dickerson (undated)	10,000	0	0	<50	<50	—	—	—
Meursing and Slot (1968)	20,000	—	0	<50	<50	—	—	—
Collins-Thompson et al. (1978)	100,000	—	<1.8	<2000	<2000	—	—	—
ICMSF (1974)	10,000	—	—	—	<100	—	—	—

as sweetened concentrated block (8% moisture) or crumb (a dry mixture of milk, sugar, and liquor; Howat, 1968). In addition, coatings made with cocoa, sugar, vegetable fats, and emulsifiers may be prepared (Wolf, 1977). The mixtures of ingredients are ground to reduce particle size and then stirred (conching) to further develop chocolate flavor and texture (Cook, 1972).

2. Source and Fate of Pathogens

The only pathogen of importance in chocolate is *Salmonella*. It has been found in cocoa and chocolate on several occasions (Depew, 1968; Goepfert and Foster, 1968; D'Aoust, 1977; Anonymous, 1978) and may survive in chocolate for years (Rieschel and Schenkel, 1971; Dockstader and Groomes, 1971; Tamminga *et al.,* 1976). Although *Salmonella* should be killed by adequate roasting, liquor and cocoa press cake may be produced from inadequately roasted beans in countries with primitive sanitation. Liquor may be heated to 80°–100°C, but after the addition of milk and sugar, chocolate cannot be heated to temperatures that will destroy *Salmonella*. The water content of milk chocolate is 1–1.5%. At this concentration, the conching process (stirring for 10–96 hr) at 46°–70°C destroys few if any *Salmonella* (Foster, 1968; Goepfert and Biggie, 1968; Rieschel and Schenkel, 1971). D_{70} values for *Salmonella* spp. in chocolate are 6–12 hr (Goepfert and Biggie, 1968). *Staphylococcus aureus* (Ostovar, 1973) and *B. cereus* in chocolate and cocoa are probably of little public health significance.

If milk chocolate is contaminated with *Salmonella,* the addition of 2% water to increase the total water to 3.7% raises the a_w sufficiently to enable decontamination at 71°C (Barrile and Cone, 1970). During the heating, the added water evaporates (Barrile *et al.,* 1970).

Salmonella in cocoa (WHO, 1973) and milk chocolate (D'Aoust *et al.,* 1975; P.C. Craven *et al.,* 1975) have caused illness in a large number of consumers. *Salmonella* contamination of milk chocolate, for instance, caused an outbreak of about 200 cases with a 38% hospitalization rate— the infectious product contained 20–90 *Salmonella* cells/100 g. Considering that *Salmonella*-contaminated product was manufactured on several days (D'Aoust *et al.,* 1975) and that a single conch (containing for mixing) contains several thousand kilos, it is probable that several hundred thousand people were at risk (Pivnick, 1974). In this outbreak, the probable source of *Salmonella* was contaminated cocoa beans, but the higher frequency of *Salmonella* contamination in milk chocolate suggests that contaminated milk powder may be a more usual source.

3. Control of Microorganisms

a. Sanitation. Adequate roasting kills *Salmonella*. Thus, physical separation of raw beans from the roasted beans with the roaster as the "germicidal barrier" is a basic means of control. When cleaning equipment, it is important to avoid the use of water, which may permit growth of *Salmonella* in pipes, etc. If water must be used, thorough drying is essential because one of the objectives in manufacturing chocolate is to reduce the water content for technical reasons. Hot cocoa butter is sometimes used to remove chocolate from equipment (see also Volume I, Chapter 14).

b. Laboratory Examination. Laboratory examination of ingredients immediately prior to preparation of chocolate or cocoa is essential as a further control procedure. Various food control agencies require such examinations.

A useful control procedure is the examination of scrap or salvage chocolate from molding machines. Scrap presents a composite sample of material produced over a considerable period of time, and failure to find *Salmonella* or coliforms in such material gives some assurance that the production lot is suitable for shipment. Samples of several other materials (air filters, dust collectors, content of vacuum sweepers, etc.) may be useful indicators of the presence or absence of pathogens (Silliker, 1967).

D. Choice of Case

For an understanding of "Case" see Table A.1 and the brief description in Appendix IV. For more details, see the second book of this series (ICMSF, 1974).

To evaluate the general microbial quality of cocoa, chocolate, and compound coatings containing cocoa, Case 2 would apply (aerobic plate count and mold count) because in the normal course of events the microbial levels will neither increase nor decrease during transport, storage, and consumption. For the same reason, the coliform group and *E. coli* would fall into Case 5 and *Salmonella* into Case 11.

IV. CONFECTIONERY

A. Important Properties

The main ingredients of confectionery are sugar, vegetable oils (mostly hydrogenated), milk and milk products, cocoa and chocolate, egg albumen, gelatins, gums, glazes, pectins, starches, nuts, fruit as jam, candied and dried fruits, colors, spices, flavors, acidulants, and so forth. The microbiology of many of these ingredients is considered in other chapters. This chapter concerns itself with the microbiology of these ingredients only in the context of confectionery; only the most serious faults will be considered.

The microbiology of confectionery reflects the quality of the raw product, the opportunity for destruction of microorganisms during processing, and the opportunity for growth of microorganisms. The opportunity for growth depends primarily on the a_w of the product (see Volume I, Chapter 4), which in general is too low to permit microbial growth.

B. Effects of Processing on Microorganisms

Processing may range from little or no opportunity to destroy microorganisms in vulnerable products to boiling at high temperatures in the preparation of products that are invulnerable to spoilage.

C. Spoilage

Microbial spoilage of confectionery products may best be considered within the framework presented in Table 26.9.

Chocolate-coated fondant (mixture of sugar, egg white, flour, flavoring, and possibly other ingredients) may be especially vulnerable to spoilage as the fondant has an a_w of 0.75–0.84, well within the growth range of

TABLE 26.9

Water Activity of Various Kinds of Candy [a]

Type of product	a_w
Fondant	0.75–0.84
Fruit jellies	0.59–0.76
Marzipan	0.65–0.70
Turkish delight	0.60–0.70
Marshmallow	0.63–0.73
Licorice	0.53–0.66
Gums and pastilles	0.51–0.64
Chocolate	0.37–0.50
Toffee	<0.48
Boiled sweets	<0.30

[a] From Cakebread (1971) and Hilker (1976).

osmophilic yeasts. This product has been fractured and even exploded following the growth of gas-forming *Clostridium sporogenes, Clostridium* spp. (probably *butyricum*), and *Saccharomyces cerevisiae* (Weinzirl, 1922, 1927; Hill, 1925). The source of the *Clostridium* spp. was probably dried egg albumen, which contained 10,000 spores/g (Hill, 1925). Sugar may also introduce *Clostridium* spp. (Weinzirl, 1927).

Preserved fruits, because of their high sugar content, may contain large numbers of osmophilic yeasts due to too long an interval between initiation and completion of the sweetening process as well as failure to heat (82°C for 15–20 min) those products that can be heated successfully. Such contaminated fruits introduce massive inocula into chocolate confectionery; in the confection they may grow and cause bursting. A concentration of 75% sugar is recommended to prevent defects due to osmophilic yeasts (Walker and Ayres, 1970).

"Soapiness" is a defect in "white" chocolate (cocoa butter or other fat with flour, powdered milk, and flavoring). It is most common in products containing coconut and palm oil because of their short and medium chain length fatty acids. Soapiness has been produced by lipolytic enzymes in a *Bacillus* spp. present in the chocolate and powdered milk. Growth in the chocolate was not determined (Witlin and Smyth, 1957) and would be unlikely at the low a_w of the product. Lipolytic enzymes of molds growing in cocoa beans will also adversely affect cocoa butter and other lipids.

D. Pathogens

Confectionery does not support the growth of disease-causing bacteria and, only rarely, supports that of mycotoxigenic molds. Salmonellae will not grow at the a_w of common confections. However, if they enter a confection in the ingredients, they may survive for long periods. At the low a_w of confections, these bacteria may survive a heating step (Goepfert *et al.,* 1970; Gibson, 1973).

Various confection ingredients may contribute salmonellae—primarily coconut (Chapter 21), chocolate (this chapter), milk (Chapter 18), egg albumen (Chapter 19), spices (Chapter 24), and gelatin (Chapter 15). Moldy nuts could introduce mycotoxins (see Chapter 21).

E. Control

As a general principle, a process that will destroy salmonellae offers better protection than laboratory testing does. For example, a solution of sugar syrup, albumen, and gelatin may be pasteurized at 71°–74°C for 15–120 min to eliminate microbiological problems (Minifie, 1970). In the absence of such a procedure, selected ingredients (milk, gelatin, egg albumen, coconut, and chocolate) should be examined regularly for *Salmonella* as a further check on the adequacy of prior treatment.

Nuts should be examined visually for the presence of visible mold. Modern nut processors use photometric equipment to sort out abnormal nuts. Many large processors of confections analyze nut shipments for the presence of mycotoxins. Sweetened fruits should be brought to minimum a_w with minimum lapse of time to prevent growth of osmophilic yeasts.

Additional controls such as reduction of airborne molds (Vakrilov and Murgov, 1970), personal hygiene (Kleinert, 1972; Schenkel, 1972), separation of raw from processed product (Silliker, 1967; Mossel and Sand, 1968), and scheduled examination for microbial content (Ostergren, 1974) are more or less important depending on process or product.

F. Choice of Case

For an understanding of "Case," see Table A.1 and a brief description in Appendix IV. For more details, see the second book of this series (ICMSF, 1974).

Case 3 is appropriate for spoilage organisms in vulnerable products, e.g., osmophilic yeasts in enrobed fruits and Case 11 for *Salmonella* in any confectionery.

27

Miscellaneous Foods

I. INTRODUCTION

The foods discussed in this chapter are either mixtures of ingredients treated separately in the foregoing chapters, or specialty products, such as froglegs that do not fit readily into other chapters.

II. DRY SOUP AND GRAVY MIXES

Dry soup and gravy mixes have many ingredients in common, and some, such as beef-flavor mix can be used as either a soup or gravy (Komarik *et al.*, 1974). These products are made simply by mixing dry commodities, then packaging them in laminated sachets, or other moisture-proof containers. The main ingredients are meats (Chapter 15), poultry (Chapter 16), seafoods (Chapter 20), vegetables (Chapter 21), flours, starches, and thickeners (Chapter 23), fats (Chapter 25), sugars (Chapter 26), milk (Chapter 18), and eggs (Chapter 19) (Binstead and Devey, 1970).

The initial microflora of dry mixes depends on the flora of the dry in- gredients (Karlson and Gunderson, 1965). Mixing and packing under hygienic conditions will have no effect on the flora, although during pro- longed storage many of the nonsporulating organisms will die. Manufac- turers, therefore, can select ingredients with low microbial levels, especially for the preparation of the so-called instant soups, which need not be cooked before consumption. For instance, some firms use spice extracts instead of dry spices, and others use specially processed ingredients (i.e., sterilized spices, or freeze-dried vegetables) (Anema and Michels, 1974; Association Internationale de l'Industrie des Bouillons et Potages [AIIBP],

822

1977). Nevertheless, dry mixes generally contain a wide variety of organisms (Table 27.1). Aerobic plate counts generally lie between 10^3 and 10^5/g (Catsaras *et al.,* 1961; Fanelli *et al.,* 1965; Krugers Dagneaux and Mossel, 1968; AIIBP, 1977; Anema and Michels, 1974; Kadis *et al.,* 1971).

The chemical and microbial stability of dry mixes depends on maintenance of the moisture level at approximately 7% or below (corresponding to a_w 0.1–0.35) by moisture-impervious packaging (see Volume I, Chapter 11).

Salmonellosis outbreaks have occurred from dry mixes containing cottonseed flour and dried yeast (see Chapter 23, Section IV,C).

Staphylococcus aureus or, more frequently, the spore-forming *Clostridium perfringens* or *Bacillus cereus* are often present in dry mixes (Fallesen, 1976; Keoseyan, 1971; Nakamura and Kelly, 1968). If the reconstituted product is held warm, particularly between 30° and 50°C for 6 hr or more, these pathogens, particularly *B. cereus* may grow to levels that will cause illness (Mossel and Krugers Dagneaux, 1963; Coretti and Müggenburg, 1967; Catsaras *et al.,* 1961; Fallesen, 1976; Gilbert and Taylor, 1976).

Control of the microbial quality rests largely on monitoring the microbial levels in the various ingredients. Tests for aerobic plate count and direct tests for pathogens are recommended. Tests for indicator organisms such as Enterobacteriaceae, coliform group, or *Escherichia coli* are less useful because these groups are expected to be present in raw vegetables or because their presence is not a reliable indication of the presence of *Staphylococcus aureus* or *Salmonella* (Silliker and Gabis, 1976).

TABLE 27.1

Occurrence of Different Types of Microorganisms in Dry Soups

Organism	% of samples positive	Numbers (per g)
Enterobacteriaceae	39 [a]	tested in 0.1 g
Coliforms	80 [b]	1×10^1–6.4×10^3; mean 9.2×10^1 [c]
Escherichia coli	3–18 [a, b, d, e]	1–5×10^2 [b]
Aerobic spores	50–71 [b, d]	< 10 [c]–10^4 [d]
Yeasts and molds	—	2.1×10^2–3.9×10^3 [c]

[a] Krugers Dagneaux and Mossel (1968).
[b] Catsaras *et al.* (1961).
[c] Fanelli *et al.* (1965).
[d] Coretti and Müggenburg (1967).
[e] AIIBP (1977).

In the normal course of events, dry mixes *to be cooked* would be re-constituted by boiling in water, which would kill most organisms. Thus, aerobic plate counts would be Case 1; *S. aureus, B. cereus,* and *C. per-fringens,* Case 7; and *Salmonella,* Case 10. On the other hand, *instant* mixes will not always be hot enough after reconstitution to destroy pathogens. In the normal course of events, the reconstituted mixes would be used promptly, so that Case 2 would apply to aerobic plate count, Case 8 to *S. aureus, B. cereus,* and *C. perfringens,* and Case 11 to *Salmonella.* If, after reconstitution, there is expected to be a prolonged holding period at temperatures permitting growth, Cases 3, 9, and 12, would apply, respectively. (See also Appendix IV.)

III. SALADS

Salads are simply mixtures of a variety of foods described in foregoing chapters. Some ingredients are cooked, others not. Usually there are raw vegetables or fruits, which means that salads usually carry a very broad spectrum of microorganisms, often at high levels (Table 27.2). Pathogens are sometimes present.

Most salads are refrigerated, which inhibits microbial growth. They also usually include a sauce or dressing whose low pH and organic acids minimize the growth of most spoilage bacteria as well as that of pathogens (see Vol. I, Chapters 5 and 7). Table 27.3 illustrates the decline in num-bers of pathogens introduced into two types of salad.

Lactobacilli and yeasts are the most frequent cause of spoilage of stored salads because of the low pH. These organisms come primarily from vege-tables and pickles. Although sorbic and benzoic acids inhibit yeasts, they have little effect on lactobacilli (Table 27.4).

Large pieces of meat remain at a relatively high pH within the tissues, since the acids diffuse only slowly, if at all, through solid foods. Eventual equilibrium may be at a less-acid pH value, so that a variety of organisms can grow.

Control of the microbiology of salads rests with the following steps:

1. Whenever feasible use cooked, blanched, or pasteurized ingredients.
2. Clean and wash raw vegetables to remove all adhering dirt.
3. Dice meats and poultry into small pieces, and/or marinate them in acid before mixing them with other ingredients.
4. Keep the pH as low as feasible (while retaining good eating quality) by the use of sauces and dressings containing organic acids (usually acetic acid from vinegar).

TABLE 27.2

Aerobic Plate Counts of Delicatessen Foods [a]

Product	Number of samples	pH range	Proportion of samples with aerobic plate count in following range (%)						
			$<10^3$	10^3-10^4	10^4-10^5	10^5-10^6	10^6-10^7	$>10^7$
Chicken salad	38	4.6–6.2	0	16	31	26	24	3
Egg salad	9	5.1–6.6	0	11	22	11	22	33
Ham salad	24	4.4–5.5	4	8	29	29	13	17
Macaroni salad	119	4.2–5.7	14	31	24	21	10	0
American potato salad	172	4.1–6.1	15	29	27	18	9	2
Sandwich spread	82	4.1–6.0	1	9	16	28	35	11
Shrimp salad	46	4.6–5.8	2	37	35	20	4	2
Tuna salad	27	4.5–5.9	0	48	37	11	4	0

[a] From Pace (1975).

TABLE 27.3

Survival of Food-Poisoning Microorganisms Inoculated into Salads [a]

	Storage temperature (°C)	Days of storage				
		Bacteria/g				
		0	13	21	28	56
Meat and vegetable salad,[b] pH 4.2; acetic acid content 0.5%						
Staphylococcus aureus	20	4×10^9	$<1 \times 10^2$	$<1 \times 10^2$	—	—
	9	1×10^8	9×10^4	$<1 \times 10^2$	—	—
Salmonella	20	2×10^6	<10	<10	—	—
	9	4×10^7	<10	<10	—	—
Clostridium perfringens	20	6×10^5	<10	<10	—	—
	9	5×10^3	5×10^3	5×10^3	3×10^3	<10
Bacillus cereus	20	4×10^3	3×10^3	2×10^3	—	—
	9	3×10^7	6×10^6	6×10^6	4×10^6	3×10^6
Shrimp salad,[b] pH 5.3; acetic acid content 0.3%						
Staphylococcus aureus	20	6×10^9	8×10^3	$<1 \times 10^2$	—	—
	9	2×10^9	4×10^8	3×10^7	6×10^4	1×10^2
Salmonella	20	9×10^8	<10	<10	—	—
	9	3×10^8	4×10^6	2×10^6	1×10^3	<10
Clostridium perfringens	20	6×10^5	<10	<10	—	—
	9	1×10^5	6×10^3	6×10^3	3×10^3	<10
Bacillus cereus	20	3×10^3	2×10^3	2×10^3	—	—
	9	3×10^7	7×10^6	5×10^6	5×10^6	2×10^6

[a] From Holtzapffel and Mossel (1968).
[b] Both types of salads were preserved with a 0.15% mixture of benzoic acid and sorbic acid.

TABLE 27.4

Development of Lactobacillaceae and Yeasts in Salads [a]

	Storage temperature (°C)	Days of storage		
		0	13	21
		(Microorganisms per g)		
Meat and vegetable salad,[b] pH 4.2, acetic acid content 0.5%				
Lactobacillaceae	20	6×10^6	5×10^8	1×10^9
	9	9×10^6	2×10^8	2×10^8
Yeasts	20	6×10^2	$<1 \times 10^2$	$<1 \times 10^2$
	9	8×10^4	6×10^2	7×10^2
Shrimp salad,[b] pH 5.3, acetic acid content 0.3%				
Lactobacillaceae	20	2×10^2	7×10^8	5×10^8
	9	3×10^4	3×10^8	3×10^8
Yeasts	20	6×10^2	4×10^4	6×10^2
	9	2×10^3	4×10^4	5×10^4

[a] From Holtzapffel and Mossel (1968).
[b] Both types of salad were preserved with a 0.15% mixture of benzoic acid and sorbic acid, in a ratio of 1:2.

5. Use strict hygiene in preparation, mixing, and storing the salads.
6. Minimize storage time, and always refrigerate during storage.

Salads prepared according to these guidelines will probably support the growth of lactobacilli during storage; therefore, as a measure of spoilage, a plate count using a medium specifically designed to grow lactobacilli (Speck, 1976a) would be useful, and Case 3 would apply.

However, such salads would gradually kill Enterobacteriaceae, including *Escherichia coli* (Fowler and Clark, 1975), so that tests for these organisms would require Case 4. For the same reason Case 7 would apply to *Salmonella* tests.

Staphylococcus aureus could be used as an indicator of excessive human contact during preparation of the salad. For this, Case 4 would apply. If the salad could support the growth of *S. aureus* (e.g., potato salad at neutral pH), Case 9 would apply. On the other hand, the presence of staphylococcal enterotoxin in an ingredient making up the salad is best sought by direct serological testing (ICMSF, 1978).

IV. PRECOOKED FROZEN FOODS

Innumerable precooked frozen foods are merchandized, their popularity being directly related to consumer convenience. The complexity of these items varies from a single component, such as precooked frozen chicken; to "add-on" products, such as frozen vegetables in sauces; to multicomponent products such as prepared dinners and entrees, pot pies, and pizzas. The consumer is directed, though not always, to heat the product immediately prior to consumption.

The microbiology of the components of such products has been treated in detail in other chapters of this book relating to specific food products. In effect, in producing this type of product, the manufacturer is assuming to varying degree the responsibilities that were previously met by the consumer in preparing similar products in the kitchen. Since some of these products are fully cooked before freezing and others are comprised of mixtures of raw and partially cooked components, no attempt will be made to discuss the full scope of processing conditions involved; these having been discussed elsewhere in this book.

The production of safe products depends upon careful hazard analyses. This begins with the development of a process flow diagram that includes all the pertinent processing factors and details and is specific for the food, process, equipment, and plant involved. It requires the establishment of critical control points in the processing sequence where microbial contamination or growth may occur. Necessary to the establishment of microbiological control is a knowledge of the microflora of the ingredients and the influence of processing and handling on their survival in the finished product. Assessment of acceptable microbial levels in such finished products depends, in part, on the further treatment to be imposed by the consumer. Because of the wide variety of products involved, no attempt will be made either to generalize or to deal with specific products. Factors to be considered in the establishment of microbiological control over this group of products have been concisely discussed (Peterson and Gunnerson, 1974).

Despite the inherent food-poisoning potential of these products, they have enjoyed an excellent public health history attesting to the fact that the food processor has successfully assumed the responsibilities that in former times were met by the ultimate consumer.

Microorganisms in precooked frozen foods can be expected neither to grow nor to die rapidly in the frozen condition. Therefore, aerobic plate counts would require Case 2; fecal indicators (Enterobacteriaceae, coliform group, *Escherichia coli*), Case 5; and *Staphylococcus aureus,* Case 8.

Other tests are not recommended unless special circumstances require them.

V. FROGLEGS

Froglegs have become an important food in international trade because they constitute a valuable export item from developing countries, especially in the Far East. *Salmonella* in the product has been a matter of concern to public health authorities in both exporting and importing countries and many contaminated lots have been rejected (Andrews *et al.,* 1977; Pantaléon and Rosset, 1964; Shrivastava, 1977). Froglegs are mostly packed in pairs and sold frozen.

It is well known that reptiles and amphibia frequently harbor salmonellae (Ang *et al.,* 1973; Bartlett *et al.,* 1977; Sharma *et al.,* 1974) and during transport from the point of capture to the local slaughterhouse, the level of contamination will increase. The processing procedure includes several steps to reduce the number of bacteria and to avoid cross-contamination; but even when the production of froglegs is carried out under the most hygienic conditions, elimination of all salmonellae is difficult to achieve (Nickelson *et al.,* 1975).

To remove soil, feces, slime, and other dirt before slaughter, live frogs are frequently placed in a holding tank containing fresh water. The live frogs are then put into a 10% solution of salt, frequently containing more than 200 ppm of chlorine, to paralyze and anesthetize them. The hindlegs are then cut at the abdomen above the waist in such a manner that the intestines are left intact. The legs are immediately washed to remove blood and all extraneous material. The legs are trimmed by removing bits of membrane, hanging pieces of flesh, and the remaining portion of the cloaca; sometimes they are skinned. Frequently more washings are carried out with chlorine solutions up to 250 ppm, in order to reduce the number of salmonellae on the surface of the skin (Rao *et al.,* 1978; Necker *et al.,* 1978). The legs are finally wrapped in polyethylene films or packed in polyethylene bags and frozen.

Since salmonellae will remain a common organism in the regions of catch, and since contamination during processing can be minimized but not eliminated, salmonellae will continue to be found in froglegs. Irradiation is under consideration in some countries as a solution to this problem (Rao *et al.,* 1978; Necker *et al.,* 1978). Other pathogens are of no concern. Case 10 would be appropriate when froglegs are examined for the presence of salmonellae because they are invariably cooked. A rinsing method to determine the surface contamination by *Salmonella* has been advocated since there is danger of cross-contamination to other foodstuffs.

With respect to the significance to human health, *Salmonella* in frog-legs is not different from that in meats and poultry (see this volume, Chapters 15 and 16).

VI. MEAT PIES

The meat pie is a peculiarly British food, traditionally made with un-cured pork or beef (the steak pie). However, with the development of modern processing techniques and refrigeration, there now exists a wide variety of meat pies that can be differentiated into two categories, the hot-eating and the cold-eating pie.

A. Hot-Eating Pies

Major types of hot-eating pies in Britain are steak, steak and kidney, chicken (with ham or mushroom), meat and potato (Cornish pasty, Forfar bridies, cottage pies), steak and kidney puddings, and sausage rolls.

The steak, steak and kidney, and chicken pies require a two-stage heating during the manufacturing process. It is common to boil the meat filling of these pies in "open-pan" cookers. Once cooked, the meat fill may be cooled before filling into a pastry case, but it is still common to use hot fill, at temperatures ideally above 66°C. The filling operation is then normally followed by baking, sufficient to raise the center tempera-ture above 71°C, but more commonly to above 82°C; however, higher temperatures can cause the meat fill to boil out over the pastry, causing an unsightly product. An innovation in the manufacture of hot-eating pies has been the introduction in the past 10–15 years of pressure cooking and vacuum cooling, which speed the operation.

In the manufacture of Cornish pasties and Forfar bridies, it is usual to prepare the meat filling without heating, then to bake the pie in a slow oven to reach 71° to 82°C at the center.

The cottage pie consists of a meat filling topped with mashed potato, with no pastry. The meat filling is prepared similar to steak pies, filled into aluminum foil dishes, either hot or cold, covered with mashed potato, and baked to a center temperature in excess of 82°C.

The steak and kidney pudding involves the filling of a hot- or cold-cooked filling into a suet pastry, followed by steam cooking to center temperatures in excess of 82°C.

The sausage roll is made by filling cold, uncooked sausagemeat (legally permitted to contain up to 450 ppm SO_2 in Britain) into a pastry case, followed by baking to internal temperatures in excess of 71°C.

It is common to cool the product rapidly under refrigeration; but recently, vacuum cooling has been applied to Cornish pasties, Forfar bridies, and sausage rolls, to bring the temperature to below 21°C in a matter of minutes.

The appearance of hot-eating pies can be enhanced by the application of a glaze to the upper surface of the pastry. These glazes are protein-, starch-, or gum-based and are applied either hot or cold prior to baking, or, as in the case of starch-based glazes, may be applied hot after baking.

The shelf-life of a hot-eating pie is limited by the susceptibility of the gravy to microbial spoilage and absorption of moisture by the pastry. While hot-eating pies are still sold unwrapped, it is common practice for manufacturers to wrap pies in cellulose film or in cardboard cartons. This ensures a shelf-life of about 10 days, allowing up to 7 days for storage and distribution, and 2–3 days refrigerated life at 2°–7°C in the retail outlet. The consumer then restores the optimum organoleptic properties by reheating the pies before consumption.

With the exception of cottage pies and sausage rolls, it is not common practice to freeze baked hot-eating pies. Freezing or deep chill may be used for unbaked pies, commonly where the baking is carried out by other factories, bakeries, retail outlets, or in the home.

Spoilage is most commonly due to molds, introduced usually during the refrigerated cooling of pies, and evident on the inside of the pastry where the moisture content is highest, or on the outside once moisture has migrated. *Mucor* spp. are the fastest growing molds, and it is also common to find *Penicillium, Rhizopus,* and *Aspergillus spp.*

Because of the nature of the production processes, aerobic and anaerobic spore-forming bacteria pose the major pathogen threat. Inadequate cooking and slow cooling provide an ideal situation for the rapid outgrowth of spores, and *Clostridium perfringens* food-poisoning outbreaks frequently occur under such circumstances. *Bacillus spp.* commonly are found in these products. Post-oven glazes are potential sources of bacterial spores but have yet to be incriminated in a food-poisoning outbreak.

B. Cold-Eating Pies

Traditionally, the cold-eating pie was made with uncured pork meat, but currently meat fillings also include cured pork meat (bacon, ham) eggs, poultry meat, and a recent innovation, beef and pickle.

Basically, the cold-eating pie is made by baking an uncooked meat filling in a pastry case to a center temperature in excess of 82°C, followed by cooling under refrigeration or vacuum, and then by injection of a gelatine solution (jelly). The addition of jelly is necessary in order to fill the gap left between the meat filling and the pastry after baking, and to enhance the succulence and flavor of the product.

The jelly traditionally was injected into the pie at or near boiling point, but because gelatin hydrolyzes at sustained high temperatures, it now is more common to inject at temperatures as low as 60°C, especially where in-line heat exchangers are used for pasteurization. The jelly injection can be carried out manually, but automatic machine jellying is coming more into use. The jellying operation is then followed by cooling to enable the jelly to set before further handling.

Glazing of cold-eating pies is common practice, as described under Section VI, A.

Generally, the packing, distribution, and retail storage is similar to that described for hot-eating pies, but the shelf-life is somewhat more limited. The pastry must remain at optimum crispness throughout the shelf-life, for the pie is eaten without further heating.

Like the hot-eating pie, the major spoilage is due to molds, in particular *Mucor, Penicillium, Rhizopus,* and *Aspergillus spp.* Spore-forming bacteria are not so prevalent as in hot-eating pies because of the higher salt levels and the presence of nitrite in cured-meat ingredients.

Food-poisoning incidents involving *Staphylococcus aureus* and *Salmonella spp.* are becoming more frequent in Britain and are invariably associated with the jellying operation. Badly designed equipment, poor hygiene, poor temperature control, and poor handling by the consumer have all been incriminated.

C. Choice of Case

For an understanding of "Case" see Table A.1, and the brief description in Appendix IV. For more details, see the second book of this series (ICMSF, 1974).

Hot-eating pies are to be reheated immediately before consumption. If such heating brings the center of the pie to 70°–80°C, all yeasts, molds, and vegetative bacterial cells will be destroyed. Thus, one would expect all tests to fall in the first category of Table 1 (reduce degree of hazard). However, all too frequently, chilled precooked foods such as meat pies are merely warmed enough to be palatable (or in the instance of cold-eating pies, not warmed at all), so that the first category would not apply.

All meat pies should be stored under refrigeration. Therefore, one must assume that handling after processing will be under conditions that would permit spoilage microorganisms to grow (Third category in Table A1: "May increase hazard"), but not pathogens (Second category).

Aerobic plate count and mold count measure general sanitation, degree of spoilage, and expected shelf-life, for which Case 3 would apply. Direct testing for *Clostridium perfringens, Bacillus cereus,* or *Staphylococcus aureus* would require Case 8. Tests for *Salmonella* would require Case 11.

28

Natural Mineral Waters

I. INTRODUCTION

Neither of the Commission's previous books discussed natural mineral waters. Although these generally have a good health record, their involvement in a recent large epidemic of enteric illness, *vide infra,* has made a discussion of their microbiology important in this book. Natural mineral water, as stated in the Draft European Regional Standard for natural mineral water (FAO/WHO, 1977) is clearly distinguishable from ordinary drinking water because of the following:

1. Obtained directly from a natural or a drilled source from underground water-bearing strata.

2. Collected under conditions that guarantee the original bacteriological purity.

3. Not subjected to treatments other than decantation, gas replacement, or bottling.

4. Bottled close to the point of emergence from the source.

5. Characterized by its content of certain mineral salts and their relative proportions and the presence of trace elements or other constituents.

Natural mineral water should generally contain at least 1000 mg/liter of dissolved salts. These levels may be lower in carbonated waters and in certain types of waters corresponding to the designations in the FAO/WHO document on mineral water (FAO/WHO, 1977).

II. INITIAL MICROFLORA

Directly after emerging from the source and following bottling, natural mineral water contains small numbers of microorganisms, i.e., approximately 10 to 100 per ml. This microflora, known as the autochthonous or indigenous flora, is composed mainly of gram-negative bacteria, i.e., *Flavobacterium* and *Cytophaga* species, together with *Micrococcus* and *Nocardia* species; in addition, species of *Alcaligenes* and *Arthrobacter* may be present (Schmidt-Lorenz, 1976). These autochthonous bacteria are characterized physiologically by their low nitrogen requirements; they need only small amounts of organic compounds for growth but are, nevertheless, chemo-organotrophic rather than autotrophic. The autochthonous microflora, which is as a rule very small in cell-size when present in mineral water, is usually aerobic and psychrotrophic.

III. EFFECTS OF PROCESSING ON MICROORGANISMS

A. Saprophytic Organisms

Shortly after bottling when the open system (the source) is exchanged for a closed system (the filled bottle), multiplication of the autochthonous bacteria begins. This growth is characterized by an alternating increase and decrease in the number of bacteria. Each new population of cells is often composed of another species, using the dead cells of the previous growth as a nutrient, i.e. crytic growth. The characteristic initial increase in bacterial counts, just before bottling, is probably caused by (a) the adsorption of organic compounds to the bottle surface, thus increasing their concentration sufficiently to permit growth and (b) the increase of dissolved oxygen during the filling operation. When tested immediately after bottling, aerobic plate counts at 30° and 37°C should be no more than ten times that of the water taken directly from the source. During the distribution and storage of bottled mineral water, flavobacteria and related members of the autochthonous microflora grow, and counts of up to 10^4–10^5/ml are commonly found. Such growth is more common in mineral waters bottled in plastic bottles and is believed to be due to the characteristic that the plastic permits the entry of air and provides a surface suitable for microbial growth. Such growth is completely normal and is characteristic of all types of low-nutrient water. It in no way affects the suitability of the water for human consumption (Schmidt-Lorenz, 1976).

The allochthonous bacteria, i.e. those which are brought in from outside, e.g. during storage *prior* to bottling, during bottling, or from the environmental air, constitute a second type of bacterial flora. Significant multiplication of such bacteria is rare, mainly because of their nutritional requirements; many of them, e.g., most *Leuconostoc* and *Lactobacillus* species, are inactivated in pure water, physiological saline, or normal drinking water by more than 50% in about 20 min and by more than 90% after 1 hr of storage.

The more persistent types of allochthonous bacteria have physiological characteristics similar to those of the autochthonous species. These include a number of gram-negative mesophilic species such as *Pseudomonas cepacia (P. multivorans), P. fluorescens,* and *P. aeruginosa.*

B. Pathogens

Inoculation of *Escherichia coli, P. aeruginosa,* salmonellae, staphylococci, and *Bacillus* species into natural mineral water and subsequent storage at 20°C usually results after several weeks in a slow but definite decrease in numbers of bacteria (Buttiaux, 1959; Buttiaux and Boudier, 1960).

In the cholera epidemic in Portugal in 1974, bottled, noncarbonated mineral water was implicated as one of the primary vehicles involved; about 3000 persons were affected (Blake *et al.,* 1977). This epidemic illustrated the need for good manufacturing practices and effective bacteriological control in the production of mineral waters.

Obviously, fecal matter in mineral water may contribute virtually any of the enteric pathogens, including viruses and parasites.

C. Control

Carbonation of mineral water to a level of 3–5 volumes of CO_2 has a strong bactericidal effect on most bacteria, including types producing slime. Because of the accompanying pH reduction (with 4–5 volumes of CO_2, the pH will be approximately 3.3) and because of the effect of the carbon dioxide itself, salmonellae, staphylococci, vibrios, and many other bacteria die within a few hours. Thus, carbonation of natural mineral water is an effective means of destroying foodborne disease bacteria; but it will not decontaminate grossly contaminated water (Sharf, 1960).

The same strict hygiene requirements for drinking water should apply to all natural mineral waters (FAO/WHO, 1977). Although carbonation is an effective means of destroying enteric pathogens, it should not be used as a substitute for the maintenance of high standards of hygiene during the processing of carbonated natural mineral waters. Hygienic requirements for the production of mineral waters will be found in the U.S. Code of Federal Regulations (U.S. Department of Health, Education and Welfare, 1976f), and are discussed by Schmidt-Lorenz (1976).

IV. CHOICE OF CASE

For an understanding of "Case" see Table A.1 and the brief description in Appendix IV. For more details, see book 2 of this series (ICMSF, 1974).

Mineral waters should not contain microorganisms that are pathogenic to man and, in general, the microbiological criteria should correspond to those established for drinking water. At present, the FAO/WHO Committee on Natural Mineral Waters is establishing sampling plans and microbiological limits for mineral waters and the International Standards Organization is developing methods of analysis. In view of these activities and the present controversy over the value of using coliform, fecal coliform, and *Escherichia coli* tests as indicators of fecal contamination and possible presence of enteric pathogens in these products no Cases are proposed at the present time. Cases are not proposed for pathogens, as the relevance of tests for these organisms cannot be considered until the controversy over indicator tests is resolved.

29

Preventing Abuse of Foods after Processing

I. INTRODUCTION

After foods leave the processing plant, they are stored and handled in transit, warehouses, retail stores, catering and institutional kitchens, commercial food service establishments, vending machines, and homes. Poor practice can lead to spoilage or foodborne disease. Spoilage problems are treated in detail in this volume, Chapters 15–28; this chapter emphasizes factors that contribute to outbreaks of foodborne disease. For discussions of the significance of the presence of disease organisms in foods, see the second edition of Book 1 of this series (ICMSF, 1978) and appropriate sections of Chapters 25–28.

A. Factors that Contribute to Foodborne Disease

Although foods are sometimes contaminated during production and processing, they are more frequently further mishandled or mistreated during subsequent transit, storage, further processing, and preparation. Most reports of outbreaks of foodborne disease can be traced to malpractices in food service establishments (including catering operations, institutional kitchens, and restaurants) and in homes (U.S. Department of Health, Education and Welfare, 1967–1979; Vernon and Tillet, 1974; Health and Welfare Canada, 1976, 1978; Vernon, 1977; Todd, 1978). Foodborne disease outbreaks result because of a sequence of events, differing somewhat with each etiologic agent, but starting with contamination of the food with an infectious or toxigenic agent. After contamination with toxigenic bacteria or toxigenic molds, the food must be suitable for growth of the organism, and it must be stored at temperatures adequate for such growth for sufficient time for development of large numbers of the organism or elaboration of enough toxin to produce illness. After such growth, the food must not show enough evidence of spoilage so that a person would reject it. Also, subsequent processing, if any, must not kill the organism or denature the toxin. Epidemiologic evidence shows that the faulty procedures listed in Table 29.1 frequently lead to outbreaks of foodborne disease.

B. Typical Outbreaks

The following reports illustrate the postprocess sequence of events that contributed to outbreaks of foodborne disease. Other examples are described by Dack (1956) and Hobbs and Gilbert (1978).

1. Staphylococcal Food Poisoning at a Picnic

Hams frequently have been reported as vehicles of foodborne disease outbreaks, particularly of staphylococcal intoxication. For example, a caterer baked 14 hams in an oven from 9 P.M. until 6:45 A.M. the next day; their internal temperatures were reported to have reached 74°C. While still warm, ten of these hams were boned and sliced by two men. The slices were then stacked into two deep pans, covered with aluminum foil, and put back into the oven at about 8 A.M. with the thermostat control set at 93°C. (Slices at the interior of the mass would not have reached this temperature, however.) At about 2 P.M., the pans of sliced ham were taken by car to a picnic site and put on a serving table where they stayed at ambient outdoor temperatures until served at about 6 P.M. A person who ate some ham when it arrived became ill about one-half hour before the meal was served, but the connection between his illness and the ham was overlooked. Within 4 hr of eating the meal, 244 picknickers showed signs of staphylococcal food poisoning; 120 needed hospital treatment. *Staphylococcus aureus* was isolated from samples of the ham and from swab specimens of acne lesions on the face of one of the men who boned the hams and who was said to have fingered the lesions frequently (Wain and Blackstone, 1956). The food handler no doubt contaminated the ham with small numbers of staphylococci that grew and elaborated toxin during the hours the ham remained in the warm oven, in the car, and at the picnic grounds.

TABLE 29.1

Factors That Contribute to Foodborne Disease Outbreaks [a]

Contributory factors	Outbreaks resulting from foods mishandled in food service establishments (U.S.A., 1973–1976) (%)	Ouotbreaks resulting from foods mishandled in homes (U.S.A., 1973–1976) (%)	Outbreaks resulting from foods mishandled in all places (U.S.A., 1961–1976) (%)	Outbreaks resulting from foods mishandled in all places (U.K., 1969–1976) (%)
Affecting growth				
Improper cooling	63	30	46	77
Improper hot-holding	27	6	16	10
Lapse of day or more before preparing and serving	29	11	21	64
Use of leftovers	7	4	4	—
Faulty fermentations	0	6	—	—
Affecting survival				
Inadequate cooking, heat processing, or canning	5	21	16	21
Inadequate reheating	25	5	12	28
Affecting contamination				
Infected person	26	8	20	16
Contaminated raw ingredients	2	22	11	15
Cross-contamination	6	2	7	9
Inadequate cleaning of equipment	9	1	7	7
Unsafe source	1	11	5	11

[a] From Bryan (1978a).

2. *Clostridium perfringens Gastroenteritis following a School Lunch*

Foodborne disease outbreaks frequently have followed ingestion of turkey meat served in institutional and commercial food service establishments and in homes. In a typical episode, 80% of students and teachers who ate in a school cafeteria experienced a mild gastroenteritis characterized by abdominal cramps and diarrhea (Bryan *et al.,* 1971). *Clostridium perfringens,* Hobbs type 11, was identified from 13 of 14 isolates cultured from stool specimens from five patients. Food history attack rates indicated that the vehicle was either turkey or dressing. Two days before the outbreak, frozen turkeys had been thawed at room temperature in the school kitchen, and the following day some were cooked in a steamer and others in large pots on a range. The cooked turkeys and the liquid in which they were cooked (stock) were cooled on table tops for about an hour before the bones were removed and the stock transferred to 3.8-liter jars. (Spores of *C. perfringens* that were initially present on the raw turkeys would have survived boiling and steaming. Also, *C. perfringens* could have been introduced during boning or transfer of stock into jars.) The stock and meat were then stored in a refrigerator overnight. During this period, the stock (in 3.8-liter containers) and boned meat (in 15-cm-deep pans) did not cool immediately, and thus were within a temperature range suitable for multiplication of *C. perfringens* for over 8 hr. The next morning, the stock was heated and some mixed with cornbread crumbs and other ingredients to make dressing that was baked to a temperature of 100°C. The cold turkey meat was diced, and gravy that had been made from the remainder of the stock was poured over it. Although this mixture was warmed in ovens, it did not reach a temperature lethal for the large numbers of vegetative cells of *C. perfringens* that were present, and therefore, became the vehicle in this outbreak.

3. *Salmonellosis from Barbecued Chicken Prepared in a Retail Store*

Barbecued chicken has been the vehicle for outbreaks of salmonellosis. In one such outbreak involving 25 persons, cooked chickens, which had been purchased from a supermarket, were left at room temperature for several hours before being served.

Investigation of the barbecuing operation at the store revealed that the chickens may have been undercooked and that cross-contamination of cooked chickens from raw chickens and raw meat products was likely. Hot-holding facilities had kept the chickens at 38°C; therefore, salmonellae would have multiplied during this storage. Leftover chickens were refrigerated, and the next day they were reheated in the hot-holding device. Samples of barbecued chicken collected at the store revealed *Salmonella typhimurium,* the serotype responsible for the outbreak. A survey of persons living in the neighborhood of the store and in community hospitals revealed 82 additional cases of S. *typhimurium* infection; two had been fatal. They all reported having eaten barbecued chicken within the week before their illness; most of them had purchased the chicken from the incriminated market. Samples of barbecued chicken collected from the households of the ill were found to contain S. *typhimurium* (Center for Disease Control surveillance data).

4. Bacillus cereus Gastroenteritis from Boiled Rice

Cereal foods are frequent vehicles in outbreaks of *Bacillus cereus* gastroenteritis (see this volume, Chapter 23). In one episode, eight persons began vomiting 1-½ to 2 hr after eating in a Chinese-style restaurant. The only pathogen isolated from their feces was *B. cereus.* Boiled rice had been kept at room temperature throughout the day, and the remaining rice had been refrigerated overnight in a large cooking pot. On the day of the outbreak, this rice was removed from the refrigerator and again kept at room temperature. Samples of rice prepared subsequently contained 3×10^7 B. *cereus*/g (Public Health Laboratory Service, 1973).

In another incident, two persons vomited repeatedly after eating fried rice from a Chinese-style restaurant; *Bacillus cereus* was isolated from the feces of one of these persons. The rice had been boiled the evening before and stored in a large container overnight at room temperature. The next day, portions of the rice were fried for about 3 min upon order. A sample of the boiled rice contained 2×10^9 B. *cereus*/g (Public Health Laboratory Service, 1973).

In these episodes, *B. cereus* spores survived cooking, and after germination the resulting vegetative cells multiplied in large containers during refrigerated storage or during room-temperature storage. Subsequent heating (as in the case of fried rice) failed to kill vegetative cells of the organism.

5. Botulism from a Commercial Pot Pie
Mistreated in a Home

Botulism occurs most often from home-canned foods, but only rarely from commercially prepared foods. Product abuse by consumers can lead to outbreaks. For example, a girl removed a commercially manufactured beef pot pie from the freezer and heated it for 20 min in a gas oven set at 218°C but then decided to eat something else. She turned off the oven but forgot to remove the pie, which remained in the oven for 20 hr with only the pilot light providing heat. The next day, her brother removed the pot pie from the oven with his bare hands, and found it warm, but not hot. He ate two or three bites and then complained that it did not taste right. Two days later he developed clinical symptoms compatible with botulism and was hospitalized. Type A botulinal toxin was identified in his serum, in his stool, and in the remnants of the beef pot pie recovered from the garbage can; *Clostridium botulinum* type A was also recovered from the pie (Hughes *et al.,* 1977).

6. International Outbreaks

Outbreaks of foodborne illness have occurred among airplane passengers on international flights because the caterers or the airlines did not maintain refrigerated foods cold or heated foods hot. Table 29.2 lists some of the outbreaks that have occurred in recent years.

II. INITIAL MICROFLORA

The microflora on foods at the conclusion of processing has been described for each commodity, in this volume, Chapters 15–28. Microbial populations in foods are rarely static and increase or decrease after foods leave processing plants depending on how they are handled (Yeterian *et al.,* 1974). The extent of change depends on properties of the food (e.g., pH, a_w, oxidation reduction potential), on the extent to which the food is protected from subsequent contamination, and on the environment to which the food is exposed during transit, storage, or preparation.

III. EFFECTS OF STORAGE, PREPARATION, AND SERVING ON MICROORGANISMS

Foods arrive at markets, foods service establishments, or homes still in a raw condition, or after having been chilled, dehydrated, heat processed, or frozen. Preparation, and therefore microbiological problems, differ accordingly.

A. Storing

1. Unrefrigerated Storage

Unrefrigerated storage is adequate for most canned foods, dehydrated foods, root vegetables, nuts, highly acidified foods, highly sugared foods,

TABLE 29.2

Some Reported Outbreaks of Foodborne Disease on Civil Air Flights [a]

Year	Route taken			Passengers			Symptoms displayed	Bacteria identified	Vehicle	Contributing factors
	Origin	Via	Destination	Total	Infected No.	Infected %				
1967	London [b] (6 flights)	—	New York	22	—		Diarrhea, abdominal cramps	*Escherichia coli*, various other bacteria	Oysters, mussels	Kept too long at temperature permitting growth of pathogens
				6 crews						
1971	Bangkok [b]	—	Hong Kong	42	23	53	Gastroenteritis	None	Shrimp salad	Not determined
1971	Bermuda [b]	—	Hong Kong	78	34	44	Gastroenteritis	*Shigella sonnei*	Shrimp cocktail	Not determined
1972	Bangkok [b]	Dubai	London	134	12	9	Gastroenteritis	*Vibrio parahaemolyticus*	Cooked crabmeat	Contamination after cooking
1972	London	Bahrein [b]	Sydney	357	47	13	Diarrhea, vomiting (1 death)	*Vibrio cholerae*	Hors d'oeuvres	Infected kitchen staff

Year							Symptoms	Organism	Food	Probable cause
1973	Rome	Lisbon [b]	New York	170	47	28	Diarrhea, vomiting, chills, abdominal cramps, nausea, weakness	*Staphylococcus aureus*	Custard-type dessert	Probably insufficient cooking, and improper refrigeration
	Rome	Lisbon [b]	San José, Costa Rica,	91	50	55				
	Lisbon [b]		Philadelphia	179	150	84				
1975	Tokyo	Anchorage [b]	Copenhagen	343	197	57	Diarrhea, vomiting, nausea, abdominal cramps	*Staphylococcus aureus*	Ham and cheese omelette	Infected kitchen staff, storage at room temperature
1976	Las Palmas [b]	Helsinki	Las Palmas	appr. 2500	appr. 550	appr. 22	Diarrhea, abdominal cramps, fever chills, vomiting, 6 deaths	*Salmonella typhimurium*	Egg salad	Not detected
	Las Palmas [b]	—	Cologne							
	Las Palmas [b]	—	Hannover							
	Las Palmas [b]	—	Hamburg							

[a] F.-P. Gork and D. Grossklaus as cited by World Health Organization, 1977.
[b] Airports where contaminated food was taken aboard.

cheeses, pulses, breadstuffs, dry cereals, pasta, and similar foods. Such foods are stable because they have been heat processed and protected against recontamination (i.e., canned) or because they have low a_w or low pH.

The inherent microbial stability of these foods can be lost unless conditions of storage are maintained, suitable to the food in question. For example, canned foods ordinarily are not processed sufficiently to destroy all sporulating thermophiles. Therefore, they will spoil if they are stored in a very hot warehouse or under tropical conditions (see Volume I, Chapter 1).

An a_w of 0.85 or below will prevent growth of pathogenic bacteria and production of bacterial toxins in foods such as dry grain products (this volume, Chapter 23), nuts and dried vegetables (this volume, Chapter 21), dried eggs (this volume, Chapter 19), dried milk (this volume, Chapter 18), dried feeds (this volume, Chapter 17), and similar products. Relative humidity of storage areas must be controlled to prevent or minimize condensation and absorption of moisture. Temperature fluctuations that occur when foods are transferred from outdoors to indoors and in and out of refrigerators can allow condensation to form on surfaces and cause changes in a_w. When the original container of dried foods is opened or damaged, dried foods (such as flour, sugar, dried milk, dried eggs, rice, or coconut)should be stored in metal cans or bins, glass jars, or similar containers with tight-fitting lids. Sacks of vegetables should rest on pallets to allow circulation of air and stored so that soil and dirt associated with them will not contaminate other foods.

Most fruit preserves (jams, marmalades) and certain concentrated fruit drinks (lemon juice) have a shelf-life after opening of 4 to 8 weeks at room temperature. Molds are more likely to grow than yeasts or bacteria, and readily reach the surfaces of uncovered products. (See also Chapter 22.)

Arrangement of stored foods should permit air to circulate. Bulk foods should be stored preferably on pallets 45 cm above the floor and 30 cm from walls. Individual boxes, cans, or jars, on shelves, should be 5 cm from walls and other stored products. Inventory control and stock rotation are essential to minimize the chances of issuing spoiled products.

Storage rooms should have physical barriers to rodents and insects, such as window screens and metal flashings or cuffs around wires and pipe passages in outside walls. Stored foods must not be exposed to drips from water, sewer, or air conditioning pipes or to flood water, backflow from sewers, or other sources of moisture. Packages must be intact to maintain moisture control. Detergents, disinfectants, other cleaning aids, pesticides, or any poisonous substances should not be stored in the same room as food.

2. Refrigerated Storage

Many prepared foods (e.g., convenience foods, pastes, meat pies, cooked poultry, and luncheon meats) have insufficient intrinsic factors to control growth of microorganisms and must be chilled or frozen to avoid spoilage and proliferation of pathogenic microorganisms during storage. With temperature reduction, bacterial activity declines. (The critical temperatures for growth of microorganisms have been described in Volume I, Chapter 1 and Chapter 13, Table 13.1). Most pathogenic bacteria are unable to multiply below 5°C, although *Yersinia enterocolitica* and *Clostridium botulinum* types E and F can do so. Between 5° and 10°C, however, many foodborne pathogens can multiply slowly (Volume I, Chapter 13, Table 13.1). Published "minimum growth temperatures" are generally determined with pure cultures under ideal conditions in laboratory media near pH 7. In foods, both reduced pH and a competing flora tend to suppress growth of pathogens. Psychrotrophic and psychrophilic bacteria can spoil foods while they are stored in refrigerators, as described in Volume I, Chapter 1 and this volume, Chapters 15–28.

Some cured canned meats and fish products (e.g., large hams, cured pork shoulders, chopped pork, tongue, veal, and fish roe) receive only pasteurization treatment; they must be kept refrigerated, for otherwise anaerobic and facultative bacteria that survived pasteurization can multiply (Chapters 15 and 20).

Temperature control in domestic and commercial refrigerators, as well as those used to transport and display foods, is commonly very poor. A study in the United States, for instance, revealed that 32% of household refrigerators were held at or above 7°C (Jones and Weimer, 1977); chill storage near 0°C will dramatically increase shelf-life over that at 5°–7°C (see Chapter 1, Fig. 1.6; see also Elliott and Michener, 1965). There should be ample refrigerated storage space in every kitchen. Most food service operations need walk-in refrigerators with forced-air circulation to facilitate rapid cooling and metal shelves for ease of cleaning. To minimize microbial growth during cold storage, instructions should be given to (a) keep refrigerator temperatures at 6°C or lower, preferably 1° to 4°C; (b) measure refrigerator temperatures regularly; (c) defrost refrigerators periodically; (d) provide space for air circulation around all items; (e) not overcrowd; and (f) not impede air circulation by putting paper or cardboard on refrigerator shelves.

3. Frozen Storage

During frozen storage, the number of viable microorganisms in most foods decreases, but there is considerable variation depending on the type of substrate and the type of organisms present (see also Volume I, Chapter 1). If (because of power failures or improper temperature control) a frozen food thaws, microorganisms will multiply to an extent related to the suitability of the food. as a microbial medium, the microflora in the frozen product, the temperature above freezing at which the food is held, and the duration of time at this temperature.

B. Thawing

Foods can be thawed either by conduction of heat from the surface to the interior or by heat generated in individual water molecules of the food, as during microwave heating. Conduction thawing is a slower process than freezing, because resistance to heat flow into the food increases as thawing proceeds. The rate of conduction thawing depends on the product's dimensions, proportions (thickness and shape), change in enthalpy, thermal conductivity, initial and final temperatures, surface heat-transfer coefficients, and temperature of the thawing medium. The surface heat-transfer coefficient is the controlling factor in rate of heat transfer for thin portions of solid food; as the thickness is increased, the thermal conductivity of the product becomes the most important rate-controlling factor.

The rate of heat transfer during conduction thawing in air is a function of (a) the difference in temperature between the food and the static boundary of air at the product surface and (b) the air velocity. Foods can be air thawed at refrigerator temperatures, at room temperatures, and at higher temperatures, as in ovens.

Thawing in refrigerators or in rapid thawing devices at temperatures of 10°C or below is recommended because such temperatures minimize microbial growth, weight loss, and discoloration (James et al., 1977). As these products are held in refrigerators after thawing, further deterioration and bacterial growth are delayed.

Thawing foods at room temperature instead of in refrigerators increases weight loss, color change, and the possibility of more rapid growth of microorganisms, particularly the psychrotrophs. Spoilage can result if thawed foods remain at room temperature for several hours. Frozen cooked foods particularly can become hazardous if thawed and held at room temperature because competitive flora (destroyed during cooking) is unavailable to limit growth of pathogenic bacteria.

Thawing time for raw turkeys (wrapped in plastic) can be decreased compared to thawing in refrigerators by putting them in paper bags (sacks) at room temperature. The bag restricts air circulation immediately around the carcass, and the frozen product reduces the temperature of the air in the bag to 5° to 15°. According to Klose et al. (1968), thawing of 1.8- to 10-kg plastic-wrapped turkeys in double-layer paper bags at 12.8°C, 21°C, or 28.9°C, would be completed before a *Pseudomonas* sp. could complete four generations of growth (not including the lag phase). Surface temperatures of turkey skin during thawing by this method remained below 10°C until thawing was completed.

Thawing time can be further decreased by exposing products to high temperatures as in cooking. Beef joints weighing 2.25 kg were cooked while frozen (−20°C) and others were cooked after thawing (20°C). Cooking times required for the frozen joints were up to 2.3 times greater than those required for the thawed meat (James and Rhodes, 1978). Mean roasting time for frozen, whole turkeys weighing 9 to 11 kg was 2.5 hr longer than for nonfrozen, whole turkeys; mean roasting times for frozen, cut-up turkeys were 50% to 75% longer than for nonfrozen, cut-up turkeys (Fulton et al., 1967). Because of the increased time of cooking, frozen whole turkeys are often cooked too briefly to reach internal temperatures lethal to vegetative bacteria such as salmonellae. Several outbreaks of salmonellosis have resulted from frozen, whole turkeys inadequately cooked (Sanders et al., 1963; Rogers et al., 1967; Tucker et al., 1968). On the other hand, cooking from the frozen state is usually satisfactory for small items such as fish sticks, steaks, hamburgers, precooked meals, or small packages of vegetables.

The mechanism of heat transfer during thawing in water is similar to that in air, except that the process is more rapid. Bacterial counts on pork legs thawing in water at 10°C increased slightly; those in water at 20° and 30°C, considerably more. The obvious choice was 10°C. Increasing the velocity of water movement had little effect on thawing time, and increased the bacterial count (Bailey et al., 1974).

Although thawing foods in water can shorten thawing time, there are many practical problems with the procedure:

1. Large tanks or sinks are required.
2. Large carcasses or containers are difficult to handle.
3. Portions of unwrapped product break up and particles disperse during thawing.
4. Products become wet, which makes them vulnerable to microbial growth, and they sometimes develop a bleached appearance.

5. Microorganisms concentrate in the thaw water from unwrapped products, unless there is a substantial rate of water change.

6. The thaw water associated with raw meat and poultry can splash and contaminate the nearby environment.

7. Sinks and tanks used to thaw unwrapped meat or poultry can become contaminated and must be thoroughly cleaned and disinfected after use so as not to contribute to subsequent cross-contamination of other foods or utensils.

Thawing frozen foods by microwaves is faster than by conduction, but certain foods are thermally unstable under such treatment. If, during microwave irradiation, a portion of the food becomes slightly hotter than the surrounding material, proportionately more energy will be absorbed within that region at an ever-increasing rate (Bailey, 1975). Microwave thawing is best suited to small portions of food of uniform composition.

Thawed raw meat and poultry and the associated thaw and drip water are important sources of salmonellae and other microorganisms that can contaminate equipment, utensils, and hands of food workers; these organisms can then be spread to other foods and to the kitchen environment.

C. Rehydrating

Properly packaged dried foods are stable to microbial attack until they are reconstituted. If potable or boiled water is not available to rehydrate dried products, they should be thoroughly reheated before serving. Rehydrated foods should be consumed immediately or stored in refrigerators.

D. Preparing Raw Foods

Foods, as received in the kitchen, harbor microorganisms characteristic of the product and of the processes to which they have been subjected. Checked or cracked eggs and eggshells, for example, have been sources of salmonellae in eggnog (raw eggs in milk), ice cream mixes, custards, and cream-filled pastry, making them vehicles in outbreaks of salmonellosis. *Salmonella, Clostridium perfringens,* and *Staphylococcus aureus* frequently are associated with raw poultry and raw red meat; *Campylobacter* is sometimes present (Bruce *et al.,* 1977). Vegetables frequently have a high level of gram-negative bacteria from soil and water. Vegetables that have been fertilized with human feces, irrigated with wastewater or river water, or washed or freshened in contaminated water, can be contaminated with a variety of enteric bacteria, viruses, or parasites. Saltwater fish and shellfish, particularly in the warmer seasons, frequently harbor *Vibrio parahaemolyticus*. Raw milk may contain staphylococci, streptococci, brucellae, salmonellae, or other milkborne pathogens. These organisms may survive if the foods that harbor them are inadequately cooked or served without cooking; or they may be spread from the raw foods to other foods during handling and preparation. Therefore, workers should wash their hands after handling raw foods, particularly raw meat, poultry, fish, and eggshells. Washing poultry under a stream of water from a tap and wiping the carcass with a cloth will remove few salmonellae or other organisms, but the treatments can be a first step in cross-contamination to other foods.

Salads and other foods served raw can acquire foodborne pathogens from the farm environment; from cross-contamination; contaminated ingredients, utensils, and equipment; and carriers who have poor personal hygiene. Contaminated cutting boards, knives, and workers' hands are common transmitters of microorganisms (Gilbert and Watson, 1971; Jordan *et al.,* 1973; Bryan and McKinley, 1974).

Vegetables and fruits, and salads made exclusively from them, do not contain the several amino acids necessary to support the growth of certain foodborne pathogens (e.g., *S. aureus* and *C. perfringens*), but they can be vehicles of other pathogens such as *Shigella* and hepatitis A virus, which require only small inocula to cause illness. Some vegetables (e.g., grated raw cabbage)are often mixed with sufficient vinegar or mayonnaise to inhibit microorganisms or even kill them.

Grinders, containers, knives, cutting boards, or other items of kitchen equipment that have contacted raw foods, particularly those of animal origin, should not be used for any other food, raw or cooked, which will not subsequently be heated unless they are first thoroughly washed and disinfected, otherwise they may transfer contaminating organisms. Also, separate grinders or other equipment should be used for raw pork and raw beef to prevent the spread of *Trichinella spiralis* from pork (a product that should be cooked thoroughly) to beef (a product that is often eaten rare or even raw). If the same grinder must be used, it should be cleaned thoroughly after it has contacted raw pork. (See Vol. 1, Chapter 14 for a discussion of cleaning aids and methods.)

Rags or sponges used to wipe surfaces or clean up spills (particularly in raw preparation areas) can spread contamination; they can become the source of excessive contamination if they remain for hours at ambient temperature while wet with nutritious food material. Disposable paper towels eliminate the problem. However, if rags or sponges are necessary, they should be washed after use and dried promptly to preclude the opportunity for microbial growth (see also Volume I, Chapter 14).

E. Cooking

Most cooking processes reduce the number of non-spore-forming pathogenic microorganisms on surfaces, and even in internal regions of food, to safe levels, but some do not (Longrée, 1972). Egg dishes, such as omelets and scrambled eggs, which are prepared with dried or processed egg products are sometimes cooked inadequately to kill salmonellae (Vollum and Taylor, 1947). Water added to instant soup concentrate is sometimes not hot enough to kill pathogenic microorganisms. Cooking procedures (heating for 30 min in an oven set at 218°C) as recommended on labels of commercially processed, frozen chicken pot pies, have permitted the survival of foodborne pathogens and indicator organisms (Canale-Parola and Ordal, 1957).

Inadequate cooking of turkey and barbecued chicken has contributed to several outbreaks of salmonellosis. The temperature of dressing (stuffing) in body cavities of turkeys rises very slowly during cooking. Turkey breasts or thighs may reach 74°C an hour or more before the center of dressing in the body cavity comes to this temperature. The surface of the cavity in contact with the dressing also heats very slowly (Bryan, 1977b).

Thorough cooking can kill a large proportion of vegetative foodborne pathogens in food, e.g., 74°C in the geometric center of poultry or dressing kills large numbers of salmonellae, staphylococci, and enterococci (Castellani et al., 1953; Webster and Esselen, 1956; Rogers and Gunderson, 1958; Milone and Watson, 1968). When a quantity of chicken à la king that contained 10^7 cells of salmonellae/g was heated to 65.6°C, *Salmonella manhattan* was killed in 0.3 min, but heat-resistant *Salmonella senftenberg* 775W was killed only after 10 min (Angelotti et al., 1961b). At 60°C, these strains were completely destroyed in 3 and 81.5 min, respectively. Likewise, an inoculum of 10^7 of an enterotoxigenic strain of *Staphylococcus aureus* was destroyed after 5.2 min at 65.6°C or 47 min at 60°C (Angelotti et al., 1961b).

Center temperatures and times for a 10^7 reduction of six serotypes of *Salmonella* in 22 kg or larger cuts of beef were 54.4°C for 121 min, 57.2°C for 37 min, 60°C for 12 min, and 62.2°C for 5 min (Goodfellow and Brown, 1978).

Bacterial spores can survive most cooking procedures. For example, spores of *Clostridium perfringens* survived in turkey rolls cooked to an internal temperature of 84°C (Strong and Ripp, 1967). Those of *Bacillus subtilis* and *C. bifermentans* predominated in poultry cooked to internal temperatures of 79°C or above (Patterson and Gibbs, 1973). Heat kills organisms that compete with spore formers and drives out oxygen, causing foods to become more anaerobic. Heat also activates spore germination. Thus, cooking contributes to outbreaks of *C. perfringens* and *B. cereus* gastroenteritis, if subsequent proliferation of survivors is not prevented either by proper hot holding or by proper refrigeration.

When red meat or poultry is roasted, the skin and other external flesh surfaces reach 74°C usually within 1 hr (Bryan and Kilpatrick, 1971; Bryan and McKinley, 1974, 1979). Internal regions, however, require longer to reach this temperature. Cutting 9-kg turkeys in half speeds the heating of the geometric center to 74°C by only a small margin (Esselen *et al.,* 1956; Wilkinson *et al.,* 1965; Bryan and McKinley, 1974).

The internal temperature continues to rise for an interval ranging from a few minutes to longer than one-half hour after cooked poultry or large cuts of meat are removed from ovens, depending on their size and the temperature of the heating unit. This contributes to the lethal effect of cooking on microorganisms (Bryan and McKinley, 1974, 1979).

Cooking in microwave ovens can kill non-spore-forming microorganisms, e.g., *Salmonella typhimurium* and *Escherichia coli* in soup (Culkin and Fung, 1975); *Salmonella* and *Staphylococcus aureus* in chicken, chicken and broth, and white sauce (Woodburn *et al.,* 1962); but it is not adequate to kill spores of *Clostridium perfringens* (Craven and Lillard, 1974).

The Food Service Sanitation Manual of the U.S. Department of Health, Education and Welfare (1976d) stipulates that center temperatures of stuffings, poultry, and stuffed meat and poultry should be at 74°C or above and that of pork at 65.6°C or above with no interruption of the cooking process. If these recommendations are followed, vegetative forms of bacteria and parasites initially present in these products should be killed. A bayonet-type thermometer should be used to determine temperature in the geometric center regions of turkeys, large cuts of meat, and large casseroles.

Products sampled immediately after cooking will have low aerobic plate counts and no *Salmonella, Escherichia coli,* or *Enterobacteriaceae;* if such microorganisms are present the cooking was inadequate.

There is little or no hazard of foodborne illness if foods are eaten soon after cooking (Roberts, 1972). But, as time between cooking and eating increases, temperature control during the interim becomes of increasing concern. If foods cannot be cooked on the day they are served, they should be cooled rapidly and reheated thoroughly just prior to serving. The greatest emphasis for prevention of foodborne illness must be placed on control of the time and temperature of storage after cooking.

F. Hot Holding

While awaiting sale or serving, foods are sometimes kept warm in a steam table, a bain marie, a hot-air cabinet, or under infrared lamps. If the equipment is improperly designed or operated, the foods may remain for long periods at temperatures permitting growth of pathogens (Bryan and Kilpatrick, 1971; Bryan, 1977b; Bryan et al., 1978; Bryan and McKinley, 1979), and such malpractices have lead to outbreaks of food-borne disease (Hodge, 1960; Bryan, 1978a).

A number of outbreaks of salmonellosis from barbecued chicken in Canada stimulated experiments on growth of salmonellae on the cooked chicken surfaces. In barbecued chicken held at 40°C, 10^3 cells of S. typhimurium increased to 10^6/g in 4 hr and to 10^8/g in 8 hr. A Salmonella inoculum of 10^3/g increased to 10^5/g on chicken skin in 6 hr when stored at 43°C and to 10^4/g in that time at 30°C. In 24 hr, the numbers reached 10^8 and 10^9/g, respectively. During this 24-hr period they increased to 10^7/g at 22°C, but they neither grew nor died at 4° or 55°C (Pivnick et al., 1968).

Escherichia coli, Staphylococcus aureus, Clostridium perfringens, and two types of Salmonella, inoculated onto the surface of meat loaf, survived holding up to 2 hr at an internal temperature of 60°C. This survival was explained by the investigators to be due to the cooling effect of evaporating moisture on the surface of the meat loaf. Another explanation is that organisms at the surface become coated with soluble proteins and dry to a point at which they are more resistant to heat. When peas were contaminated with these organisms and held under infrared lamps or in air in a cabinet at 60°C, a plastic wrap increased the rate of microbial destruction, apparently because it reduced surface evaporation and its attendant cooling effect (Maxcy, 1976).

It is unlikely that any practical arrangement of infrared lamps can maintain safe temperatures of food on open display. For instance, a 200-W infrared bulb, 20 cm from the surface of roast beef, keeps the surface directly exposed at an average temperature of only 46°C (Bryan and Kilpatrick, 1971). Bulbs are seldom located this close to food, and the bottom and shaded areas of food receive little if any heat.

Clostridium perfringens grows well at 45°C, somewhat at 50°C, but not at all at 55°C (Smith, 1963; Rey et al., 1975). Therefore, cooked foods stored in hot-holding devices should be kept at or above 56°C and preferably near or above optimal temperatures for eating (60° to 62.8°C) (Blaker et al., 1961; Thompson and Johnson, 1963). These temperatures, given enough time, will kill all vegetative bacteria (Angelotti et al., 1961a; Goodfellow and Brown, 1978). Warming units must be set above 60°C to maintain foods at that temperature, and foods should ordinarily enter these units at a higher temperature.

Canned foods in vending machines are held at about 66°C so that they are at a suitable temperature for eating (Peterson *et al.*, 1960). At this temperature, a thermophile, such as *Clostridium thermosacharolyticum* can grow and spoil the product. If experiment shows that retorting at higher temperatures or for longer times to kill such organisms produces a low-quality product, one solution is to hold the canned product above 74°C to prevent all thermophilic growth (but quality may suffer) (Townsend, 1966). Another solution is to eliminate the can, maintain the food frozen, and heat it on demand to serving temperature by microwave irradiation (Frank, 1963).

Samples of foods held at or above 60°C in warming units usually reveal low mesophilic, aerobic, or anaerobic plate counts, low indicator organism counts, and no *Salmonella*. If, however, the temperature has been below 50°C for a few hours, plate counts can be high, and pathogens may have multiplied.

G. Cooling

It is commonly believed, unfortunately, that meat and poultry can be kept at room temperature after cooking and that refrigeration is unnecessary. In a national survey in the United States, for instance, 46% of home-makers were not concerned about leaving cooked meat at room temperatures for 2 hr or longer (Jones and Weimer, 1977), yet this practice and any other that will permit susceptible foods to remain between 7° and 55°C for hours, are the most frequent contributing factors in foodborne illness (Bryan, 1978a). In addition, many homemakers and food handlers are unaware that a large volume of food placed into a refrigerator may take hours to cool, during which time growth of mesophilic bacteria, including pathogens, can take place. When gravy, for instance, was subjected to procedures that could occur in the preparation of school meals, bacteriological tests showed that the greatest increase in the number of aerobic bacteria occurred during cooling (Tuomi *et al.*, 1974).

Liquid foods, such as milk or broth, cool by convection; i.e., the heat differential creates movement within the liquid, and heat passes quickly to the outer surfaces. Solid, semisolid, or viscous foods, however, cool by conduction. That is, there is little or no movement within the food mass, and heat must pass through the food to the outer surface. The following discussion applies primarily to conduction cooling.

Several factors affect cooling rates: the state of the food (liquid or solid), the mass (weight) of food, the size and shape of a solid food, the surface-to-volume ratio of a food stored in containers, the coefficient of heat transfer of the food and its container, the initial temperature of the food, the type and temperature of the cooling menstrum, the velocity of the air or water at the food surfaces, and whether the food is agitated. Many of these factors can be controlled to aid cooling efficiency.

Cooling at room temperature is slow because of the small temperature differential between the food and the air. Refrigerated cooling of large food masses is also slow because of low conductivity through the mass, even with a large temperature differential between the food and the refrigerated air.

At any given temperature differential, the rate of cooling decreases as the size of the mass of food increases. This principle has been illustrated, for instance, in chicken salad (Lewis *et al.,* 1953), in custard (McDivitt and Hammer, 1958), in beef carcasses (Bailey and Cox, 1976), and in cooked beef (Bryan and Kilpatrick, 1971; Bryan and McKinley, 1979).

The size and shape of the container and the extent to which it is filled (mass and surface-to-volume ratio) greatly influence cooling times. The internal temperature of a given volume of food falls faster in a shallow pan than the same volume will in a large, deep container. For instance, it took 24 hr for chicken salad in a 30-pound can to cool from 26.1° to 8.3°C (the temperature of the refrigerator), whereas it took only 3 hr when filled to a height of only 10 cm in a shallow pan (Weiser *et al.,* 1954). *Staphylococcus aureus* grew in custard that was stored in 20- and 14-liter pots in a refrigerator at 4.4°C, but no growth occurred in custard that was stored in shallow pans in a refrigerator (McDivitt and Hammer, 1958).

Cooling rate is also affected by the material of the container and its thickness. Foods of equal characteristics and volume stored in containers made of good conductors of heat, such as stainless steel, cool faster than the same type and volume of food stored in containers of crockery, glass, or plastic. For instance, 12 liters of pie filling cooled faster in enamel pans than in earthenware crocks, and 4 liters of potato salad cooled faster in stainless steel pans than in crocks (Miller and Smull, 1955).

In general, a very cold refrigerator cools a food faster than a moderately cold one does (Longrée and White, 1955; Bailey and Cox, 1976). Foods will cool faster in refrigerators than at room temperature, faster in freezers than in refrigerators, and initially faster in ice baths than in refrigerators or freezers (Miller and Smull, 1955; Hodge, 1960; Bryan and McKinley, 1974).

Movement of air around foods dissipates heat faster than still air (Bailey and Cox, 1976). Placing foods so that there is space around each item, the use of wire racks, and the use of fans in refrigerators enhance air circulation (Moragne et al., 1960; Newcomer et al., 1962). Large walk-in refrigerators, because of their greater volume and capacity for cooling, will dissipate heat faster than small reach-in refrigerators, so foods will usually cool faster in the former.

Cooling rates are also speeded by agitating the food. Vertical mixers have been used to reduce temperatures of cornstarch pudding (McDivitt and Hammer, 1958) and turkey stock (Bryan and McKinley, 1974). Rapid cooling has been accomplished by stirring with a scraper-lifter action (Moragne et al., 1963) and stirring with a cold tube agitator (Longrée et al., 1960, 1963; Moragne et al., 1961).

Rapid cooling has also been accomplished by putting slices of cooked meats into pans that are in contact with ice and by putting pots of turkey stock (10 liters in a 20-liter pot) in an ice bath (Bryan and McKinley, 1974).

The number of mesophilic aerobic or anaerobic microorganisms in cooled foods depends on previous contamination and on the efficiency of the cooling process. High counts are frequently indicative of microbial growth and may reflect poor cooling practices.

The rapid chilling of cooked foods (i.e., cooling to 25°C or below within an hour or two) and subsequent storage in shallow containers (not exceeding 10 cm in depth) in a refrigerator at or below 5°C will slow spoilage and prevent pathogenic bacteria from multiplying.

H. Freezing

Sometimes kitchen-prepared foods are frozen for subsequent use. Microorganisms usually decrease during freezing and frozen storage, so microbial counts are usually lower after freezing. (See Vol. I, Chapter 1.) Refreezing has no adverse effect on the safety of the food, if between periods of frozen storage, the thawed food is stored at temperatures that preclude multiplication of pathogens. In fact, refreezing a thawed food lengthens the period during which it will remain edible and safe (Elliott, 1961; Elliott and Michener, 1962).

I. Preparing Previously Heated Foods

Raw foods of animal origin are major sources of pathogens that contaminate cooked foods (see Section III,D). Contamination comes also from food service workers or homemakers who are intestinal carriers of pathogens, such as Shigella, hepatitis A virus, and Salmonella typhi, and who do not wash their hands or ineffectively wash after defecating or urinating, and from such workers who have skin infections or who are nasal carriers of *Staphylococcus aureus*. Such persons may contaminate cooked foods during kitchen operations.

The numbers of *Staphylococcus aureus* entering foods from food workers are considered a measure of the degree of human contact, and in this sense are "indicator" organisms. Foods contaminated with small numbers of this organism are safe to eat. Only if the food is a good substrate and is held for several hours in the temperature zone within which *S. aureus* can grow and produce toxin (see Table 13.1) will it be hazardous to eat. Several outbreaks have stemmed from this abuse (Hodge, 1960; Munch-Peterson, 1961, 1963).

Contamination of cooked foods can be prevented, or at least minimized, if food workers wash their hands thoroughly after handling raw poultry and meat and their wrappers, after going to the toilet, after coughing or sneezing, and again before handling cooked foods following work breaks. For homemakers the same is true, but they should also wash after caring for persons with diarrheal illnesses. (See Volume I, Chapter 14 for hand-washing procedures and other procedures useful in preventing foodborne disease.) Although handwashing can remove many microorganisms that have been acquired, it will not be effective in removing staphylococci that reside in the skin.

Cross-contamination can be prevented or minimized if food workers and homemakers either do not use the same equipment and utensils for raw and cooked foods, or thoroughly wash and disinfect equipment and utensils between use for raw foods of animal origin and for cooked foods. To ensure prevention of cross-contamination, utensils, equipment, and surfaces used for raw foods of animal origin should be washed thoroughly and disinfected as soon as such raw foods are removed. Whenever possible, different areas, equipment, and utensils should be used for raw foods than for cooked foods, and different workers should be assigned tasks involved with handling cooked foods. These policies can be followed in large establishments, but they may not be practical in small establishments. In small establishments, workers must be informed of the hazards, and supervisors must require their staff to wash their hands frequently and to clean and disinfect equipment and utensils immediately after using for raw foods and check to see whether or not they do so.

Cooked ingredients should be chilled rapidly for use in salads (e.g., potato, macaroni, chicken, turkey, and tuna). The salads should be put into shallow containers as soon as they are mixed and stored at 6°C or below until served. Addition of sufficient mayonnaise, vinegar, high-acid (low pH) salad dressing, or other high-acid ingredients to reduce the pH of salads to 4.5 or below will also inhibit the growth of pathogenic bacteria (Kintner and Mangel, 1953a; Wethington and Fabian, 1950; Longrée et al., 1959; Smittle, 1977). (See also this volume, Chapter 25, Section II.)

Microbial levels in salads are affected by ingredients, pH, and method of preparation. Raw vegetables introduce a general flora (this volume, Chapter 21), and spices frequently introduce spores (this volume, Chapter 24). The presence of coliforms, Enterobacteriaceae, fecal coliforms, or *Escherichia coli* in properly cooked foods suggests postcooking contamination.

J. Reheating

Foods served a day or more after preparation, and foods left over from a previous meal then inadequately reheated, often cause outbreaks of foodborne disease (Hodge, 1960; Bryan, 1978a). Just warming foods to make them palatable is usually insufficient to kill either spores or vegetative bacteria that have survived cooking or that have been introduced after cooking.

Such foods can be made safe (except for preformed staphylococcal enterotoxin) by thorough reheating such as wrapping in foil and baking, heating in steamers, heating on ranges with stock or gravy in covered pans, adding slices or small pieces of boiling stock or gravy, and heating in microwave ovens (Bryan and Kilpatrick, 1971; Bryan and McKinley, 1974, 1979). Units of food that weigh less than 1.4 kg (3 pounds) should be reheated to at least 74°C, and food mixtures or units of food of more than 1.4 kg should be reheated to at least 71°C at the geometric center, unless there is evidence that other time–temperature combinations will effectively kill large numbers of foodborne pathogens (Goodfellow and Brown, 1978).

Reheating is often the last line of defense in the prevention of foodborne disease. In this regard, thorough reheating is more crucial than the initial cooking, because if bacteria have survived the process, or if there has been postcooking contamination and inadequate hot storage, chilling, or cold storage, the bacterial contaminants can be killed during reheating. Although staphylococci are killed by recommended reheating practices, their enterotoxins will not be destroyed. In such cases, the product could become a vehicle of staphylococcal food poisoning in the absence of viable *Staphylococcus aureus*.

Samples taken after thorough reheating should yield low aerobic plate counts and no indicator organisms or vegetative forms of pathogenic bacteria; if the heating is inadequate, the samples may yield high microbial levels, including pathogens.

K. Serving

While foods are being prepared for the table or being served, they can be contaminated from hands or equipment. Such improprieties may offend

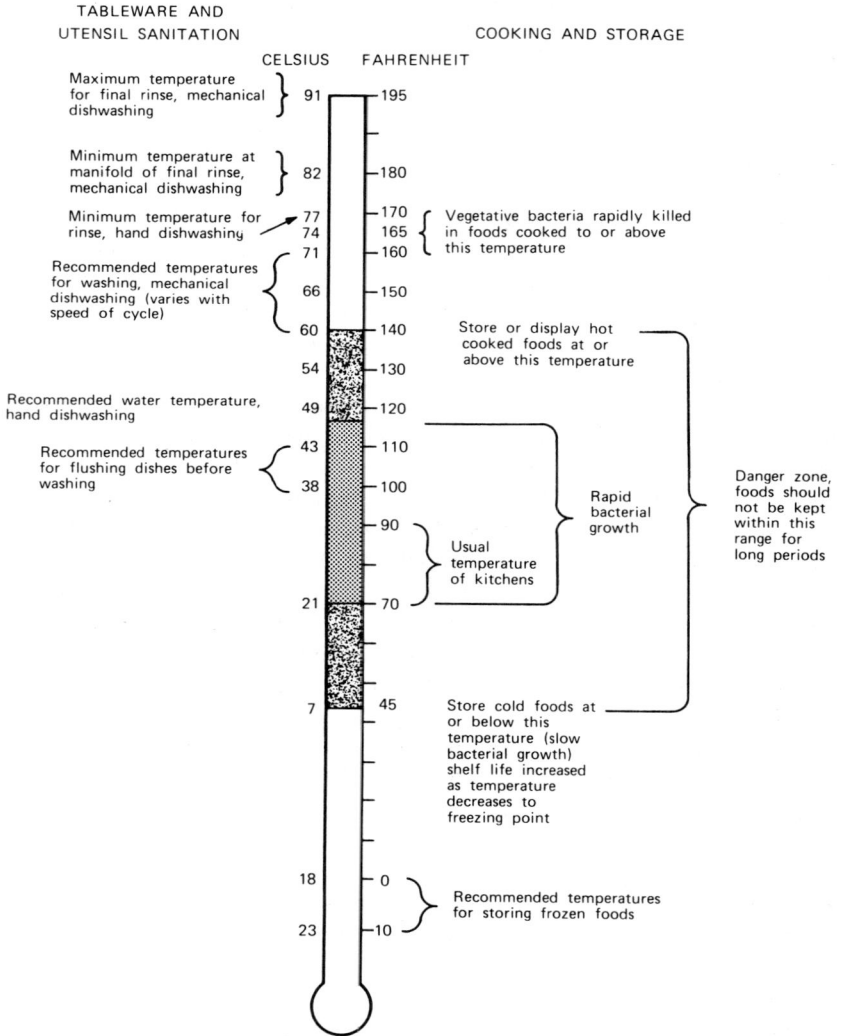

TABLEWARE AND
UTENSIL SANITATION   COOKING AND STORAGE

CELSIUS   FAHRENHEIT

Maximum temperature for final rinse, mechanical dishwashing   }   91   195

Minimum temperature at manifold of final rinse, mechanical dishwashing   }   82   180

Minimum temperature for rinse, hand dishwashing   77 — 170, 74 — 165, 71 — 160   { Vegetative bacteria rapidly killed in foods cooked to or above this temperature

Recommended temperatures for washing, mechanical dishwashing (varies with speed of cycle)   { 66 — 150, 60 — 140

Store or display hot cooked foods at or above this temperature

54 — 130

Recommended water temperature, hand dishwashing   49 — 120

Recommended temperatures for flushing dishes before washing   { 43 — 110, 38 — 100

— 90

Rapid bacterial growth

Danger zone, foods should not be kept within this range for long periods

Usual temperature of kitchens

21 — 70

— 45

7 — 45

Store cold foods at or below this temperature (slow bacterial growth) shelf life increased as temperature decreases to freezing point

18 — 0

Recommended temperatures for storing frozen foods

23 — 10

Fig. 29.1. Temperatures that affect the life or death of microorganisms.

the customer's esthetic values, but they seldom add enough organisms to cause illness. Nevertheless, food workers at the service level should not touch food with the bare hands but rather should use serving utensils or paper or wear disposable plastic gloves to grasp and dispense the product.

Foods on display should be protected against dirty surfaces, dust, insects, airborne droplets, rodents, and other animals, and hands of persons serving themselves. Such protection can be accomplished by storage under glass or similar cover, behind sneeze guards that baffle droplets expelled by customers, or by wrapping in plastic films or other packaging material.

High aerobic plate counts or indicator organism counts on foods from serving lines, or on foods ready to be served, suggest the failure of control measures; but only an inspection of each step of the operation can disclose possible sources of contamination and the likelihood that microorganisms have survived the cooking process, have reached the product later, or have multiplied.

IV. SUMMARY

Efforts should be made to prevent or minimize contamination of foods, particularly cooked foods, at every phase of preparation and service. Despite reasonable precautions, however, foods will sometimes come into kitchens contaminated or will become contaminated during preparation. When practicable, cooking and reheating should ensure that any contaminating vegetative organisms are killed. It is of particular importance to follow the adage "Keep foods hot, or keep foods cold, or don't keep foods long." In this context, "hot" is 60°C or higher, "cold" is 7°C or lower, and "long" is not more than 2 hr at room temperature for foods that permit rapid multiplication of pathogenic bacteria. Figure 29.1 shows temperatures that permit or encourage microbial growth and those that inhibit or kill microorganisms.

The procedures recommended in this chapter will increase shelf-life of foods and decrease the incidence of foodborne disease.

References

Abd-El-Malek, Y., and Gibson, T. (1948). Studies on the bacteriology of milk. II. The staphylococci and micrococci of milk. *J. Dairy Res.* **15**, 249–260.

Abdou, I. A., Abou-Zeid, A. A., El-Sherbeeny, M. R., and Abou-el-Gheat, Z. H. (1972). Antimicrobial activities of *Allium sativum, Allium cepa, Raphanus sativus, Capsicum frutescens, Eruca sativa, Allium kurrat* on bacteria. *Qual. Plant. Mater. Veg.* **22**(1), 29–35.

Abrahamson, A. E., Field, R., Buchbinder, L., and Catelli, A. V. (1952). A study of the control of the sanitary quality of custard-filled bakery products in a large city. *Food Res.* **17**, 268–277.

Adamcic, M., and Clark, D. S. (1970). Bacteria-induced biochemical changes in chicken skin stored at 5°C. *J. Food Sci.* **35**, 103–106.

Alderman, G. G., and Marth, E. H. (1974). Experimental production of aflatoxin on intact citrus fruit. *J. Milk Food Technol.* **37**, 451–456.

Alford, L. R., Holmes, N. E., Scott, W. J., and Vickery, J. R. (1950). Studies in the preservation of shell eggs. I. The nature of wastage in Australian export eggs. *Aust. J. Appl. Sci.* **1**, 208–214.

Allcroft, R., and Raymond, W. D. (1966). Toxic groundnut meal: biological and chemical assays of a large batch of a "reference" meal used for experimental work. *Vet. Rec.* **79**, 122–123.

Allen, L. A., Cooper, A. H., Cairns, A., and Maxwell, M. C. C. (1946). Microbiology of beet sugar manufacture. *Proc. Soc. Appl. Bacteriol.* **9**, 1–5.

Allen, L. A., Cairns, A., Eden, G. E., Wheatland, A. B., Wormwell, F., and Nurse, T. J. (1948a). Microbiological problems in the manufacture of sugar from beet. Part I. Corrosion in the diffusion battery and in the recirculation system. *J. Soc. Chem. Ind., London* **67**, 70–77.

Allen, L. A., Cooper, A. H., and Maxwell, M. C. C. (1948b). Microbiological problems in the manufacture of sugar from beet. II. Losses due to fermentation during the diffusion process. *J. Soc. Chem. Ind., London* **67**, 450–457.

Allen, V. D., and Stovall, W. D. (1960). Laboratory aspects of staphylococcal food poisoning from Colby cheese. *J. Milk Food Technol.* **23**, 271–274.

Allred, J. N., Walker, J. W., Beal, V. C., Jr., and Germaine, F. W. (1967). A survey to determine the *Salmonella* contamination rate in livestock and poultry feeds. *J. Am. Vet. Med. Assoc.* **151**, 1857–1860.

Almy, L. H., Macomber, H. I., and Hepburn, J. S. (1922). A study of methods of minimizing shrinkage in shell eggs during storage. *Ind. Eng. Chem.* **14,** 525–527.

American Association of Medical Milk Commissions (1976). "Methods and Standards for the Production of Certified Milk." Am. Assoc. Med. Milk Comm., Alphareta, Georgia.

American Bottlers of Carbonated Beverages (1957). "Manual for Testing Procedures." National Soft Drink Association (Formerly American Bottlers of Carbonated Beverages) Washington, D.C.

American Public Health Association (1970). "Recommended Procedures for the Examination of Sea Water and Shellfish," 4th ed. Am. Public Health Assoc., New York.

American Spice Trade Association (1971). "Cleanliness Specifications for Unprocessed Spices, Seeds and Herbs." Am. Spice Trade Assoc., Englewood Cliffs, New Jersey.

American Spice Trade Association (1976). "Official Microbiological Methods of the American Spice Trade Association." Am. Spice Trade Assoc., Englewood Cliffs, New Jersey.

Anderson, E. E., Esselen, W. B., Jr., and Fellers, C. R. (1949). Effect of acids, salt, sugar, and other food ingredients on thermal resistance of *Bacillus thermoacidurans*. *Food Res.* **14,** 499–510.

Anderson, E. E., Esselen, W. B., Jr., and Handleman, A. R. (1953). The effect of essential oils on the inhibition and thermal resistance of microorganisms in acid food products. *Food Res.* **18,** 40–47.

Anderson, P. H. R., and Stone, D. M. (1955). Staphylococcus food poisoning associated with spray dried milk. *J. Hyg.* **53,** 387–397.

Andrews, W. H., Wilson, C. R., Poelma, P. L., and Romero, A. (1977). Comparison of methods for the isolation of *Salmonella* from imported frog legs. *Appl. Environ. Microbiol.* **33,** 65–68.

Andrews, W. H., Wilson, C. R., Poelma, P. L., Romero, A., Rude, R. A., Duran, A. P., McClure, F. D., and Gentile, D. E. (1978). Usefulness of the stomacher in a microbiological regulatory laboratory. *Appl. Environ. Microbiol.* **35,** 89–93.

Andriyevs'kyi, I., and Bondarchuk, A. I. (1976). Dynamics of mycoflora development on cereal grains. *Ukr. Bot. Zh.* **33**(5), 498–501 [*Biol. Abstr.* **64,** 2017 (1977)].

Anellis, A., Lubas, J., and Rayman, M. M. (1954). Heat resistance in liquid eggs of some strains of the genus *Salmonella. Food Res.* **19,** 377–395.

Anema, P. J., and Michels, M. J. M. (1974). Microbiology of instant dry soup mixes. *Proc. Int. Symp. Food Microbiol.* **2,** 165–182. Fed. Assoc. Sci. Tec. (FAST), Milan. (In Ital., Engl. Abstr.)

Ang, Ö., Özek, Ö., Cetin, E. T., and Töreci, K. (1973). Salmonella serotypes isolated from tortoises and frogs in Istanbul. *J Hyg.* **71,** 85–88.

Angelotti, R., Foter, M. J., and Lewis, K. H. (1961a). Time–temperature effects on salmonellae and staphylococci in foods. I. Behavior in refrigerated foods. II. Behavior at warm holding temperatures. *Am. J. Public Health* **51,** 76–88.

Angelotti, R., Foter, M. J., and Lewis, K. H. (1961b). Time–temperature effects on salmonellae and staphylococci in foods. III. Thermal death time studies. *Appl. Microbiol.* **9,** 308–315.

Angelotti, R., Lewis, K. H., and Foter, M. J. (1963). Fecal streptococci in foods. Time temperature effects on. 1. Behavior in refrigerated foods and at warm-holding temperatures. *J. Milk Food Technol.* **26,** 296–301.

Anonymous (1958). A search for organisms of the *Salmonella* group in wheat and flour. *Mon. Bull. Minist. Health Public Health Lab. Ser.* **17**, 133–135.

Anonymous (1969). Know your spices. *Food Eng.* **41**(5), 88–91.

Anonymous (1978). FDA orders expansion of candy recall because of *Salmonella*. *Food Chem. News* **20**(2), 17–18.

Appert, N. (1810). "Le Livere de Tous les Ménages ou l'Art de Conserver, Pendant Plusieurs Années Toutes les Substances Animales et Vegetales." Chez Patris, Paris.

Appleman, M. D., Hess, E. P., and Rittenberg, S. C. (1949). An investigation of a mayonnaise spoilage. *Food Technol.* **3**, 201–203.

Arafa, A. S., and Chen, T. C. (1975). Effect of vacuum packaging on microorganisms on cut-up chickens and in chicken products. *J. Food Sci.* **40**, 50–52.

Arbuckle, W. S. (1972). "Ice Cream," 2nd ed. AVI, Westport, Connecticut.

Armijo, R., Henderson, D. A., Timothee, R., and Robinson, H. B. (1957). Food poisoning outbreaks associated with spray-dried milk. *Am. J. Public Health* **47**, 1093–1100.

Ashworth, L. J., Jr., McMeans, J. L., Pyle, J. L., Brown, C. M., Osgood, J. W., and Ponton, R. E. (1968). Aflatoxins in cotton seeds: Influence of weathering on toxin content of seeds and on a method for mechanically sorting seedlots. *Phytopathology* **58**, 102–107.

Association Internationale de l'Industrie des Bouillons et Potages, Commission Technique (AIIBP) (1977). Microbiological specifications for dry soups. *Alimenta* **16**, 191–193.

Atwell, B. D., and Crawford, G. H. (1867). Keeping eggs. U.S. Patent 65,988.

Auclair, J. E., and Berridge, N. J. (1953). The inhibition of microorganisms by raw milk, II. The separation and electrophoretic examination of two different inhibitory fractions. *J. Dairy Res.* **21**, 370–374.

Auclair, J. E., and Hirsch, A. (1953). The inhibition of microorganisms by raw milk, I. The occurrence of inhibitory and stimulatory phenomena. Methods of estimation. *J. Dairy Res.* **21**, 45–59.

Avens, J. S., and Miller, B. F. (1970a). Quantifying bacteria on poultry carcass skin. *Poult. Sci.* **49**, 1309–1315.

Avens, J. S., and Miller, B. F. (1970b). Optimum skin blending method for quantifying poultry carcass bacteria. *Appl. Microbiol.* **20**, 129–132.

Ayres, J. C. (1951). Some bacteriological aspects of spoilage of self-service meats. *Iowa State Coll. J. Sci.* **26**, 31–48.

Ayres, J. C. (1955). Microbiological implications in handling, slaughtering, and dressing meat animals. *Adv. Food Res.* **6**, 109–161.

Ayres, J. C. (1958). Methods for depleting glucose from egg albumen before drying. *Food Technol.* **12**, 186–189.

Ayres, J. C. (1959). Effect of sanitation, packaging and antibiotics on the microbial spoilage of commercially processed poultry. *Iowa State Coll. J. Sci.* **34**, 27–46.

Ayres, J. C. (1960a). Temperature relationships and some other characteristics of the microbial flora developing on refrigerated beef. *Food Res.* **25**, 1–18.

Ayres, J. C. (1960b). The relationship of organisms of the genus *Pseudomonas* to the spoilage of meat, poultry, and eggs. *J. Appl. Bacteriol.* **23**, 471–486.

Ayres, J. C. (1972). Mycotoxins. *Proc. Conf. Collaborators South. Agric. Exp. Stn.* ARS 72–97, pp. 17–24. Agric. Res. Serv., U.S. Dep. Agric., Washington, D.C.

Ayres, J. C., and Slosberg, H. M. (1949). Destruction of *Salmonella* in egg albumen. *Food Technol.* **3,** 180–183.

Ayres, J. C., and Stewart, G. F. (1947). Removal of sugar from raw egg white by yeast before drying. *Food Technol.* **1,** 519–526.

Ayres, J. C., and Taylor, B. (1956). Effect of temperature on microbial proliferation in shell eggs. *Appl. Microbiol.* **4,** 355–359.

Ayres, J. C., Ogilvy, W. S., and Stewart, G. F. (1950). Postmortem changes in stored meats. I. Microorganisms associated with the development of slime on eviscerated cut-up poultry. *Food Technol.* **4,** 199–205.

Ayres, J. C., Walker, H. W., Fanelli, M. J., King, A. W., and Thomas, F. (1956). Use of antibiotics in prolonging storage life of dressed chicken. *Food Technol.* **10,** 563–568.

Babel, F. J., and Hammer, B. W. (1944). Action of butter cultures in butter. A review. *J. Dairy Sci.* **27,** 79–141.

Bache-Wiig, C. (1903). Method of preserving eggs. U.S. Patent 739,137.

Baer, E. F., Gray, R. J. H., and Orth, D. S. (1976). Methods for the isolation and enumeration of *Staphylococcus aureus*. *In* "Compendium of Methods for the Microbiological Examination of Foods" (M. L. Speck, ed.), pp. 374–386. Am. Public Health Assoc., Washington, D.C.

Bailey, C. (1972). Spray washing of lamb carcasses. *Inst. Meat Bull.* No. 75, 3–12.

Bailey, C. (1975). Thawing methods for meat. *Proc. Eur. Symp. Food-Eng. Food Qual., 6th, Cambridge, Engl.* pp. 175–189.

Bailey, C., and Cox, R. P. (1976). The chilling of beef carcasses. *Proc. Inst. Refrig. (Session 1975–1976)* Mem. No. 737, pp. 1–11. Meat Res. Inst., Langford, Bristol, England.

Bailey, C., James, S. J., Kitchell, A. G., and Hudson, W. R. (1974). Air-, water-, and vacuum-thawing of frozen pork legs. *J. Sci. Food Agric.* **25,** 81–87.

Baker, D., Neustadt, M. H., and Zeleny, L. (1959). Relationships between fat acidity values and types of damage in grain. *Cereal Chem.* **36,** 308–311.

Baker, R. C. (1974). Microbiology of eggs. *J. Milk Food Technol.* **37,** 265–268.

Baldwin, R. R., Campbell, H. A., Thiessen, R., Jr., and Lorant, G. J. (1953). The use of glucose oxidase in the processing of foods with special emphasis on the desugaring of egg white. *Food Technol.* **7,** 275–282.

Bamburg, J. R., Strong, F. M., and Smalley, E. B. (1969). Toxin from moldy cereals. *J. Agric. Food Chem.* **17,** 443–450.

Bannerjee, D. (1967). Oxidative metabolism of non-protein nitrogen components by fish spoilage bacteria and their physiology of psychrotrophic growth during storage of fish (English sole). Ph.D. Thesis, Univ. of Washington, Seattle.

Banwart, G. J. (1964). Effect of sodium chloride and storage temperature on the growth of *Salmonella oranienburg* in egg yolk. *Poult. Sci.* **43,** 973–976.

Banwart, G. J., and Ayres, J. C. (1956). The effect of high temperature storage on the content of *Salmonella* and on the functional properties of dried egg white. *Food Technol.* **10,** 68–73.

Banwart, G. J., and Ayres, J. C. (1957). The effect of pH on the growth of *Salmonella* and functional properties of liquid egg white. *Food Technol.* **11,** 244–246.

Baran, W. L., and Stevenson, K. E. (1975). Survival of selected pathogens during processing of a fermented turkey sausage. *J. Food Sci.* **40,** 618–620.

Barber, F. W., and Fram, H. (1955). The problems of false coliform counts on fruit ice cream. *J. Milk Food Technol.* **18,** 88–90.

Barber, L. E., and Deibel, R. H. (1972). Effect of pH and oxygen tension on staphylococcal growth and enterotoxin formation in fermented sausage. *Appl. Microbiol.* **24,** 891–898.

Bard, J. C. (1973). Effect of sodium nitrite and sodium nitrate on botulinal toxin production and nitrosamine formation in wieners. *Proc. Meat Ind. Res. Conf.* pp. 61–68. Am. Meat Ind. Found., Chicago, Illinois.

Barker, W. H., Weaver, R. E., Morris, G. K., and Martin, W. T. (1974). Epidemiology of *Vibrio parahaemolyticus* infections in humans. *In* "Microbiology—1974" (D. Schlessinger, ed.), pp. 257–262. Am. Soc. Microbiol., Washington, D.C.

Barnes, E. M. (1960). The sources of the different psychrophilic organisms on chilled eviscerated poultry. *Proc. Int. Congr. Refrig., 10th, Copenhagen* **3,** 97–100.

Barnes, E. M. (1965). The effect of chlorinating chill tanks on the bacteriological condition of processed chicken. Suppl. Bull. Inst. Int. Froid, Commission 4, Karlsruhe, Annexe 1965–1, 219–225.

Barnes, E. M. (1974). Microbiological aspects of poultry chilling. *Proc. World Poult. Congr., 15th, New Orleans* pp. 549–552.

Barnes, E. M. (1975). The microbiological problems of sampling a poultry carcass. *Qual. Poult. Meat, Proc. Eur. Symp. Poult. Meat Qual., 2nd, Oosterbeck, Neth.* pp. (23)1–8.

Barnes, E. M. (1976). Microbiological problems of poultry at refrigerator temperatures—a review. *J. Sci. Food Agric.* **27,** 777–782.

Barnes, E. M., and Corry, J. E. L. (1969). Microbial flora of raw and pasteurized egg albumen. *J. Appl. Bacteriol.* **32,** 193–205.

Barnes, E. M., and Impey, C. S. (1968). Psychrophilic spoilage bacteria of poultry. *J. Appl. Bacteriol.* **31,** 97–107.

Barnes, E. M., and Shrimpton, D. H. (1968). The effect of processing and marketing procedures on the bacteriological condition and shelf life of eviscerated turkeys. *Br. Poult. Sci.* **9,** 243–251.

Barnes, E. M., Impey, C. S., and Parry, R. T. (1973). The sampling of chickens, turkeys, ducks, and game birds. *In* "Sampling—Microbiological Monitoring of Environments" (R. G. Board and D. W. Lovelock, eds.), pp. 63–75. Academic Press, New York.

Baross, J., and Liston, J. (1970). Occurrence of *Vibrio parahaemolyticus* and related hemolytic vibrios in marine environments of Washington State. *Appl. Microbiol.* **20,** 179–186.

Barrile, J. C., and Cone, J. F. (1970). Effect of added moisture on the heat resistance of *Salmonella anatum* in milk chocolate. *Appl. Microbiol.* **19,** 177–178.

Barrile, J. C., Cone, J. F., and Keeney, P. G. (1970). A study of salmonellae survival in milk chocolate. *Manuf. Confect.* **50**(9), 34–39.

Barrile, J. C., Ostovar, K., and Keeney, P. G. (1971). Microflora of cocoa beans before and after roasting at 150°C. *J. Milk Food Technol.* **34,** 369–371.

Bartlett, K. H., Trust, T. J., and Lior, H. (1977). Small pet aquarium frogs as a source of *Salmonella*. *Appl. Environ. Microbiol.* **33,** 1026–1029.

Bashford, T. E., Gillespy, T. J., and Tomlinson, A. J. H. (1960). "Report of the Fruit and Vegetable Canning and Quick Freezing Association." Chipping Camden, England.

Bauman, H. E. (1974). The HACCP concept and microbiological hazard categories. *Food Technol.* **28**(9), 30–34, 74.

Bean, G. A., Schillinger, J. A., and Klarman, W. L. (1972). Occurrence of aflatoxins and aflatoxin-producing strains of *Aspergillus* spp. in soybeans. *Appl. Microbiol.* **24,** 437–439.

Bean, P. G., and Salvi, A. (1970). The bacteriological quality of some raw materials used in the Italian canning industry. *Ind. Aliment.* (*Pinerolo, Italy*) **9**(4), 57–63. (In Ital.)

Beasley, J., Hopkins, G. B., McNab, D. J. N., Rickards, A. G., and King, G. J. G. (1967). Pet meat as a potential source of human salmonellosis. *Lancet* No. 7489, 560–563.

Beljaars, P. R., Schumans, J. C. H. M., and Koken, P. J. (1975). Quantitative fluoro-densitometric determination and survey of aflatoxins in nutmeg. *J. Assoc. Offic. Anal. Chem.* **58,** 263–271.

Bell, T. A., and Etchells, J. L. (1952). Sugar and acid tolerance of spoilage yeasts from sweet-cucumber pickles. *Food Technol.* **6,** 468–472.

Beloian, A., and Schlosser, G. C. (1963). Adequacy of cook procedures for the destruction of salmonellae. *Am. J. Public Health* **53,** 782–791.

Bem, Z., Hechelmann, H., Leistner, L., and Dresel, J. (1976). Mikrobiologie des DFD-Fleisches. *Fleischwirtschaft* **56,** 985–987.

Bennion, E. B. (1967). "Breadmaking. Its Principles and Practice," 4th ed., pp. 280–286. Oxford Univ. Press, London and New York.

Bergdoll, M. S., and Bennett, R. W. (1976). Staphylococcal enterotoxins. *In* "Compendium of Methods for the Microbiological Examination of Foods" (M. L. Speck, ed.), pp. 387–416. Am. Public Health Assoc., Washington, D.C.

Bergquist, D. H. (1966). Heat pasteurization of plain egg white and egg white powder. *In* "The Destruction of Salmonellae," ARS 74–37, pp. 65–67. Agric. Res. Serv., U.S. Dep. Agric., Albany, California.

Berlin, P. J. (1962). Kumiss. *Annu. Bull., Int. Dairy Fed.* Part IV, Sect. A, 4–16.

Berner, H., Kluberger, A., and Bresse, M. (1969). Investigations into a new method of chilling poultry. III. Testing aspects of hygiene for the new method. *Fleischwirtschaft* **49,** 1617–1620, 1623.

Bernstein, A. (1952). *Salmonella* infection of hen eggs. *Mon. Bull. Med. Res. Counc. G. B.*) **11**(3), 64–67 [*Biol. Abstr.* **26,** 25512 (1952)].

Berry, B. W., and Chen, A. A.-T. (1976). Bacterial, shelf-life, and consumer acceptance characteristics of chopped beef. *J. Milk Food Technol.* **39,** 405–407.

Beuchat, L. R. (1973). *Escherichia coli* on pecans: survival under various storage conditions and disinfection with propylene oxide. *J. Food Sci.* **38,** 1063–1066.

Beuchat, L. R. (1976). Sensitivity of *Vibrio parahaemolyticus* to spices and organic acids. *J. Food Sci.* **41,** 899–902.

Beuchat, L. R. (1978). Microbial alterations of grains, legumes, and oilseeds. *Food Technol.* **32**(5), 193–198.

Beuchat, L. R., and Heaton, E. K. (1975). *Salmonella* survival on pecans as influenced by processing and storage conditions. *Appl. Microbiol.* **29,** 795–801.

Bevan, D., and Bond, J. (1971). Micro-organisms in field and mill—a preliminary survey. *Proc. Conf. Queensl. Soc. Sugar Cane Technol., 38th* pp. 137–143.

Bierer, B. W., and Barnett, B. D. (1961). Effect of increasing wash water temperature on eggs contaminated with *Salmonella. Poult. Sci.* **40,** 1379.

Bierer, B. W., Valentine, H. D., Barnett, B. D., and Rhodes, W. H. (1961a). Germicidal efficiency of egg washing compounds on eggs artificially contaminated with *Salmonella typhimurium. Poult. Sci.* **40,** 148–152.

Bierer, B. W., Barnett, B. D., and Valentine, H. D. (1961b). Experimentally killing *Salmonella typhimurium* on egg shells by washing. *Poult. Sci.* **40,** 1009–1014.

Binsted, R., and Devey, J. D. (1970). "Soup Manufacture. Canning, Dehydration and Quick Freezing," 3rd ed. Food Trade Press, London.

Bitting, A. W. (1937). "Appertizing or the Art of Canning; Its History and Development." Trade Pressroom, San Francisco, California.

Blackburn, B. O., and Ellis, E. M. (1973). Lactose fermenting *Salmonella* from dried milk and milk drying plants. *Appl. Microbiol.* **26,** 672–674.

Blackwell, J. H., and Hyde, J. L. (1976). Effect of heat on foot and mouth disease virus (FMDV) in the components of milk from FMDV infected cows. *J. Hyg.* **77,** 77–83.

Blake, D. F., and Stumbo, C. R. (1970). Ethylene oxide resistance of microorganisms important in spoilage of acid and high-acid foods. *J. Food Sci.* **35,** 26–29.

Blake, P. A., Rosenberg, M. L., Florencia, J., Costa, J. B., Quintino, L. D. P., and Gangarosa, E. J. (1977). Cholera in Portugal, 1974 II. Transmission by bottled mineral water. *Am. J. Epidemiol.* **105**(4), 344–348.

Blaker, G. G., Newcomer, J. L., and Ramsey, E. (1961). Holding temperatures needed to serve hot foods hot. *J. Am. Diet. Assoc.* **38,** 455–457.

Blanchard, R. O., and Hanlin, R. T. (1973). Effect of propylene oxide treatment on the microflora of pecans. *Appl. Microbiol.* **26,** 768–772.

Blankenship, L. C., and Cox, N. A. (1976). Modified water rinse sampling for sensitive, non-adulterating salmonellae detection on eviscerated broiler carcasses. *J. Milk Food Technol.* **39,** 680–681.

Blankenship, L. C., Cox, N. A., Craven S. E., and Richardson, R. L. (1975). Total rinse method for microbiological sampling of the internal cavity of eviscerated broiler carcasses. *Appl. Microbiol.* **30,** 290–292.

Blount, W. P. (1961). Turkey "X" disease. *J. Br. Turkey Fed.* **9,** 52–61, 77.

Blum, H. B., and Fabian, F. W. (1943). Spice oils and their components for controlling microbial surface growth. *Fruit Prod. J.* **22,** 326–329, 347.

Board, R. G. (1964). The growth of gram-negative bacteria in the hen's egg. *J. Appl. Bacteriol.* **27,** 350–364.

Board, R. G. (1965a). Bacterial growth on and penetration of the shell membranes of the hen's egg. *J. Appl. Bacteriol.* **28,** 197–205.

Board, R. G. (1965b). The properties and classification of the predominant bacteria in rotten eggs. *J. Appl. Bacteriol.* **28,** 437–453.

Board, R. G. (1969). Microbiology of the hen's egg. *Adv. Appl. Microbiol.* **11,** 245–281.

Board, R. G., and Halls, N. A. (1973). The cuticle: a barrier to liquid and particle uptake by the shell of the hen's egg. *Br. Poult. Sci.* **14,** 69–97.

Board, R. G., Ayres, J. C., Kraft, A. A., and Forsythe, R. H. (1964). The microbiological contamination of egg shells and egg packing materials. *Poult. Sci.* **43,** 584–595.

Bolin, H. R., King, A. D., Jr., Stanley, W. L., and Jurd, L. (1972). Antimicrobial protection of moisturized Deglet Moor dates. *Appl. Microbiol.* **23,** 799–802.

Borden, G. (1856). Process and improvement for the concentration and preservation of milk. U.S. Patent 15,553.

Bothast, R. J., Lancaster, E. B., and Hesseltine, C. W. (1973). Ammonia kills spoilage molds in corn. *J. Dairy Sci.* **56**(2), 241–245.

Bothast, R. J., Adams, G. H., Hatfield, E. E., and Lancaster, E. B. (1975). Preservation of high-moisture corn: A microbiological evaluation. *J. Dairy Sci.* **58,** 386–391.

Botma, Y. (1973). The use of benzoic acid and its salts. A summary of the available regulations on this subject. *Naarden News* **24**, 8–9.

Botma, Y. (1974). Benelux Limonade Reglement. *Voedingsmiddelen Technol.* **7**(4), 18–21.

Brant, A. W., and Starr, P. B. (1962). Some physical factors related to egg spoilage. *Poult. Sci.* **41**, 1468–1473.

Brant, A. W., Patterson, G. W., and Walter, R. E. (1968). Batch pasteurization of liquid whole egg. I. Bacteriological and functional property evaluation. *Poult. Sci.* **47**, 878–885.

Breag, G. R., Coward, L. D. G., Nabney, J., and Robinson, F. V. (1972). Artificial drying of Jamaican pimento. *Proc. Conf. Spices* pp. 149–157. Trop. Prod. Inst., London.

Breed, A. F. (1928). Micrococci present in the normal cows udder. *N.Y. (Geneva) Agric. Exp. Stn., Tech. Bull.* No. 132.

Briggs, A., and Yazdany, S. (1970). Effect of sodium chloride on the heat and radiation resistance and on the recovery of heated or irradiated spores of the genus *Bacillus*. *J. Appl. Bacteriol.* **33**, 621–632.

Broadbent, J. A., and Oyeniran, J. O. (1968). A new look at mouldy cocoa. *In* "Bio-deterioration of Materials—Microbiological and Allied Aspects" (A. H. Walters and J. J. Elphick, eds.), pp. 693–702.

Brooks, J. (1960). Mechanism of the multiplication of *Pseudomonas* in the hen's egg. *J. Appl. Bacteriol.* **23**, 499–509.

Brooks, J., and Shrimpton, D. H. (1962). α-amylase in whole egg and its sensitivity to pasteurization temperatures. *J. Hyg.* **60**, 145–151.

Brooks, J., and Taylor, D. J. (1955). Eggs and egg products. *Food Invest. Organ. (London), Spec. Rep.* No. 60.

Brooks, M. A. (1969). General relationships between microorganisms and insects. *Proc. Symp. Biol. Contam. Grain Anim. Byprod.* pp. 17–20. Agric. Ext. Serv., Dep. Entomol. Fish., Wildl., Univ. Minnesota, Minneapolis.

Brown, B. I. (1972). Ginger storage in acidified sodium metabisulphite solutions. *J. Food Technol.* **7**, 153–162.

Brown, B. I., and Lloyd, A. C. (1972). Investigation of ginger storage in salt brine. *J. Food Technol.* **7**, 309–321.

Brown, H. J., and Gibbons, N. E. (1954). Air cell mold in oiled eggs. *Food Technol.* **8**, 307–311.

Brown, W. E., Baker, R. C., and Naylor, H. B. (1965). The role of the inner shell membrane in bacterial penetration of chicken eggs. *Poult. Sci.* **44**, 1323–1327.

Brown, W. E., Baker, R. C., and Naylor, H. B. (1966a). Egg spoilage as affected by the route of exposure. *Poult. Sci.* **45**, 279–283.

Brown, W. E., Baker, R. C., and Naylor, H. B. (1966b). The microbiology of cracked eggs. *Poult. Sci.* **45**, 284–287.

Brown, W. E., Baker, R. C., and Naylor, H. B. (1970). The effect of egg position in storage on susceptibility to bacterial spoilage. *Can. Inst. Food Technol. J.* **3**, 29–32.

Bruce, D., Zochowski, W., and Ferguson, I. R. (1977). *Campylobacter enteritis*. *Br. Med. J.* **iv**, 1219.

Bryan, C. S., Bryan, H. S., and Manson, K. (1946). Heat resistant bacteria from an unclean milking machine invade the udder of the cow. *Milk Plant Mon.* **35**(8), 30–32.

Bryan, F. L. (1968). What the sanitarian should know about staphylococci and salmonellae in non-dairy products. II. Salmonellae. *J. Milk Food Technol.* **31**, 131–140.

Bryan, F. L. (1973). Activities of the Center for Disease Control in public health problems related to the consumption of fish and fishery products. *In* "Microbial Safety of Fishery Products" (C. O. Chichester and H. D. Graham, eds.), pp. 273–302. Academic Press, New York.

Bryan, F. L. (1976). Public health aspects of cream-filled pastries. A review. *J. Milk Food Technol.* **39**, 289–296.

Bryan, F. L. (1977a). Diseases transmitted by foods contaminated by waste water. *J. Food Prot.* **40**, 45–56.

Bryan, F. L. (1977b). Plant and retail control. *Proc. Int. Symp. Salmonella Prospects Control* pp. 181–202. Univ. of Guelph, Guelph, Ontario.

Bryan, F. L. (1978a). Factors that contribute to outbreaks of foodborne disease. *J. Food Prot.* **41**, 816–827.

Bryan, F. L. (1978b). Impact of foodborne diseases and methods of evaluating control programs. *J. Environ. Health* **40**, 315–323.

Bryan, F. L., and Kilpatrick, E. G. (1971). *Clostridium perfringens* related to roast beef cooking, storage, and contamination in a fast food service restaurant. *Am. J. Public Health* **61**, 1869–1885.

Bryan, F. L., and McKinley, T. W. (1974). Prevention of foodborne illness by time–temperature control of thawing, cooking, chilling, and reheating turkeys in school lunch kitchens. *J. Milk Food Technol.* **37**, 420–429.

Bryan, F. L., and McKinley, T. W. (1979). Hazard analysis and control of roast beef preparation in foodservice establishments. *J. Food Prot.* **42**, 4–18.

Bryan, F. L., Ayres, J. C., and Kraft, A. A. (1968a). Contributory sources of salmonellae on turkey products. *Am. J. Epidemiol.* **87**, 578–591.

Bryan, F. L., Ayers, J. C., and Kraft, A. A. (1968b). Salmonellae associated with further-processed turkey products. *Appl. Microbiol.* **16**, 1–9.

Bryan, F. L., Ayres, J. C., and Kraft, A. A. (1968c). Destruction of salmonellae and indicator organisms during thermal processing of turkey rolls. *Poult. Sci.* **47**, 1966–1978.

Bryan, F. L., McKinley, T. W., and Mixon, B. (1971). Use of time-temperature evaluations in detecting the responsible vehicle and contributing factors of foodborne disease outbreaks. *J. Milk Food Technol.* **34**, 576–582.

Bryan, F. L., Seabolt, K. A., Peterson, R. W., and Roberts, L. M. (1978). Time–temperature observations of food and equipment in airline catering operations. *J. Food Prot.* **41**, 80–92.

Bryan, F. L., Fanelli, M. J., and Riemann, H. (1979). Salmonelloses. *In* "Foodborne Infections and Intoxications" (H. Riemann and F. L. Bryan, eds.), pp. 73–130. Academic Press, New York.

Buchanan, J. R., Sommer, N. F., and Fortlage, R. J. (1975). *Aspergillus flavus* infection and aflatoxin production in fig fruits. *Appl. Microbiol.* **30**, 238–241.

Buchanan, R. E., and Gibbons, N. E., eds. (1974). "Bergey's Manual of Determinative Bacteriology," 8th ed. Williams & Wilkins, Baltimore, Maryland.

Buchli, K., van Schothorst, M., and Kampelmacher, E. H. (1966). Untersuchungen über die hygienische Beschaffenheit von mit Wasser resp. Luft gekühltem Schlachtgeflügel. *Arch. Lebensmittelhyg.* **17**, 97–99.

Bugbee, W. M., Cole, D. F., and Nielsen, G. (1975). Microflora and invert sugars in juice from healthy tissue of stored sugar beets. *Appl. Microbiol.* **29**, 780–781.

Bulla, L. A., Jr., Kramer, K. J., and Speirs, R. D. (1978). Insects and microorganisms in stored grain and their control. *In* "Advances in Cereal Science and Technology" (Y. Pomeranz, ed.), Vol. 2, pp. 91–97. Am. Assoc. Cereal Chem., St. Paul, Minnesota.

Bullerman, L. B. (1974). Inhibition of aflatoxin production by cinnamon. *J. Food Sci.* **39**, 1163–1165.

Bullerman, L. B., and Hartung, T. E. (1973). Mycotoxin-producing potential of molds isolated from flour and bread. *Cereal Sci. Today* **18**, 346–347.

Bullerman, L. B., Baca, J. M., and Stott, W. T. (1975). An evaluation of potential mycotoxin-producing molds in corn meal. *Cereal Foods World* **20**, 248–250, 253.

Bullerman, L. B., Lieu, F. Y., and Seier, S. A. (1977). Inhibition of growth and aflatoxin production by cinnamon and clove oils, cinnamic aldehyde and eugenol. *J. Food Sci.* **42**, 1107–1109.

Burow, H. (1974). Untersuchungen zum *Clostridium perfringens*—Befall bei gekühlten Seefischen und Miesmuschheln in der Türkii. *Arch Lebensmittelhyg.* **25**, 39–42.

Burton, B. T. (1976). "Human Nutrition," 3rd ed. McGraw-Hill, New York.

Busta, F. F., and Speck, M. L. (1968). Antimicrobial effect of cocoa on salmonellae. *Appl. Microbiol.* **16**, 424–425.

Butler, E. E. (1960). Pathogenicity and taxonomy of *Geotrichum candidum*. *Phytopathology* **50**, 665–672.

Butler, R. W., and Josephson, H. E. (1962). Egg-containing cake-mixes as a source of *Salmonella*. *Can. J. Public Health* **53**, 478–482.

Buttiaux, R. (1959). Le contrôle bactériologique des eaux minérales *Rev. Hyg. Med. Soc.* **7**, 131–147.

Buttiaux, R., and Boudier, A. (1960). Comportement des bacteries autothropes dans les eaux minérales conservées en récipients hermetiquement clos. *Ann. Inst. Pasteur, Lille* **11**, 43–52.

Buttiaux, R., and Catsaras, M. (1966). Les bactéries psychrotrophes des viandes entreposées en chambre froid. Influence de la temperature et de l'humidité relative. *Ann. Inst. Pasteur, Lille* **17**, 107–116.

Cakebread, S. H. (1971). Chemistry of candy: factors in microbiological deterioration. *Manuf. Confect.* **51**(Apr.), 45–49.

Cameron, E. J., and Williams, C. C. (1928). The thermophilic flora of sugar in its relation to canning. *Zentralbl. Bakteriol., Parasitenkd. Infektionskr., Abt. 1* **76**, 28–37.

Campbell, J. R., and Marshall, R. T. (1975). "The Science of Providing Milk for Man." McGraw-Hill, New York.

Canale-Parola, E., and Ordal, Z. J. (1957). A survey of the bacteriological quality of frozen poultry pies. *Food Technol.* **11**, 578–582.

Cann, D. C. (1974). Bacteriological aspects of tropical shrimp. *In* "Fishery Products" (R. Kreuzer, ed.), pp. 338–344. Fishing News (Books), West Byfleet, Surrey, England.

Cann, D. C. (1977). Bacteriology of shellfish with reference to international trade. *In* "Handling, Processing and Marketing of Tropical Fish," pp. 377–394. Trop. Prod. Inst., London.

Capparelli, E., and Mata, L. (1975). Microflora of maize prepared as tortillas. *Appl. Microbiol.* **29**, 802–806.

Carlin, A. F., and Ayres, J. C. (1953) Effect of the removal of glucose by enzyme treatment on the whipping properties of dried albumen. *Food Technol.* **7,** 268–270.

Carlson, V. L., and Snoeyenbos, G. H. (1970). Effect of moisture on salmonellae populations in animal feeds. *Poult. Sci.* **49,** 717–725.

Carroll, B. J., and Ward, B. Q. (1967). Control of salmonellae in fish meal. *Fish. Ind. Res.* 4(1), 29–36.

Carruthers, A., and Oldfield, J. F. T. (1955). The activity of thermophilic bacteria in sugar-beet diffusion systems. *Annu. Tech. Conf. Br. Sugar Corp., 8th, Nottingham, Engl.* 47 pp.

Carruthers, A., Gallagher, P. J., and Oldfield, J. F. T. (1958). "Nitrate Reduction by Thermophilic Bacteria in Sugar Beet Diffusion Systems," Report Br. Sugar Corp., Nottingham, England.

Carse, W. A., and Locker, R. H. (1974). A survey of pH values at the surface of beef and lamb carcasses, stored in a chiller. *J. Sci. Food Agric.* **25,** 1529–1535.

Castell, C. H. (1944). Thermophilic bacteria in foods and in various ingredients entering into the manufacture of foods. *Food Res.* **9,** 410–414.

Castellani, A. G. (1953). Inhibiting effects of amino acids and related compounds upon the growth of enterotoxigenic micrococci in cream pastry. *Appl. Microbiol.* **1,** 195–199.

Castellani, A. G., Clarke, R. R., Gibson, M. I., and Meisner, D. F (1953). Roasting time and temperature required to kill food poisoning microorganisms introduced experimentally into stuffing in turkeys. *Food Res.* **18,** 131–138.

Castellani, A. G., Makowski, R., and Bradley, W. B. (1955). The inhibiting effect of serine upon the growth of indigenous flora of cream filling. *Appl. Microbiol.* **3,** 132–135.

Cathcart, W. H., and Merz, A. (1942). Staphylococci and *Salmonella* control in foods. III. Effect of chocolate and cocoa fillings on inhibiting growth of staphylococci. *Food Res.* **7,** 96–103.

Cathcart, W. H., Merz, A., and Ryberg, R. E. (1942). Staphylococci and *Salmonella* control in foods. IV. Effect of cooking bakery custards. *Food Res.* **7,** 100–103.

Cathcart, W. H., Godkin, W. J., and Barnett, G. (1947). Growth of *Staphylococcus aureus* in various pastry fillings. *Food Res.* **12,** 142–150.

Catsaras, M., Sampaio Ramos, M. H., and Buttiaux, R. (1961). Étude microbiologique des potages déshydratés ou concentrés du marché francais. *Ann. Inst. Pasteur, Lille* **12,** 163–174.

Cavett, J. J. (1962). The microbiology of vacuum packed sliced bacon. *J. Appl. Bacteriol.* **25,** 282–289.

Chang, P. K., Powrie, W. D., and Fennema, O. (1970). Sodium hexametaphosphate effect on the foam performance of heat-treated and yolk-contaminated albumen. *Food Technol.* **24,** 63–67.

Charlton, D. B., Nelson, F. E., and Werkman, C. H. (1934). Physiology of *Lactobacillus fructivorans* sp. nov. isolated from spoiled salad dressing. *Iowa State J. Sci.* **9,** 1–11.

Chatt, E. M. (1953). "Cocoa. Cultivation, Processing, Analysis." Wiley (Interscience), New York.

Chichester, C. O., and Graham, H. D. (1973). "Microbial Safety of Fishery Products." Academic Press, New York.

Chichester, D. F., and Tanner, F. W. (1972). Antimicrobial food additives. *In* "Handbook of Food Additives" (T. E. Furia, ed.), 2nd ed., pp. 115–184. CRC Press, Cleveland, Ohio.

Chipley, J. R., and Heaton, E. K. (1971). Microbial flora of pecan meat. *Appl. Microbiol.* **22**, 252–253.

Chohan, J. S., Dhanraj, K. S., Sunar, M. S., and Waraich, K. S. (1972). Relation of aeration to the growth of storage fungi and the resulting effects on grain quality. *Bull. Grain Technol.* **10**(4), 256–262 [*Biol. Abstr.* **57**, 60440 (1974)].

Christensen, C. M. (1946). The quantitative determination of molds in flour. *Cereal Chem.* **23**, 322–329.

Christensen, C. M. (1956). Deterioration of stored grains by molds. *Wallerstein Lab. Commun.* **19**, 31–48.

Christensen, C. M. (1965). Fungi in cereal grains and their products. *In* "Mycotoxins in Foodstuffs" (G. N. Wogan, ed.), pp. 9–14. MIT Press, Cambridge, Massachusetts.

Christensen, C. M. (1967). Increase in invasion by storage fungi and in fat acidity values of commercial lots of soybeans stored at moisture contents of 13.0–14.0%. *Phytopathology* **57**(6), 622–624.

Christensen, C. M. (1969). Influence of moisture content, temperature and time of storage upon invasion of rough rice by storage fungi. *Phytopathology* **59**, 145–148 [*Biol. Abstr.* **50**, 88846].

Christensen, C. M. (1971). Evaluating condition and storability of sunflower seeds. *Stored Prod. Res.* **7**(3), 163–169 [*Biol. Abstr.* **53**, 48814 (1972)].

Christensen, C. M. (1972). Moisture content of sunflower seeds in relation to invasion by storage fungi. *Plant Dis. Rep.* **56**(2), 173–175 [*Biol. Abstr.* **54**, 19400].

Christensen, C. M. (1978). Moisture and seed decay. *In* "Water Deficits and Plant Growth. Vol. 5: Water and Plant Diseases" (T. T. Kozlowski, ed.), pp. 199–219. Academic Press, New York.

Christensen, C. M., and Cohen, M. (1950). Numbers, kinds and sources of molds in flour. *Cereal Chem.* **27**, 178–185.

Christensen, C. M., and Gordon, D. R. (1948). Mold flora of stored wheat and corn and its relation to heating of moist grain. *Cereal Chem.* **25**, 40–51.

Christensen, C. M., and Kaufmann, H. H. (1964). "Questions and Answers Concerning Spoilage of Stored Grains by Storage Fungi." Agric. Ext. Serv., U.S. Dep. Agric., Univ. of Minnesota, St. Paul.

Christensen, C. M., and Kaufmann, H. H. (1974). Microflora. *In* "Storage of Cereal Grains and their Products," Monogr. Ser., Vol. 5 (revised), pp. 158–192. Am. Assoc. Cereal Chem., St. Paul, Minnesota.

Christensen, C. M., and Kaufmann, H. H. (1977a). Good grain storage. *Agric. Ext. Serv., Ext. Folder* No. 226 (revised), Univ. of Minn., St. Paul.

Christensen, C. M., and Kaufmann, H. H. (1977b). Spoilage, heating, binburning and fireburning: Their naure, cause and prevention in grain. *Feedstuffs* **49**(44), 39, 47.

Christensen, C. M., and Kaufmann, H. H. (1978). Spoilage of corn in ship transport. *Feedstuffs* **50**(7), 28, 32.

Christensen, C. M., and Kennedy, B. W. (1971). Filamentous fungi and bacteria in macaroni and spaghetti products. *Appl. Microbiol.* **21**, 144–146.

Christensen, C. M., Olafson, J. H., and Geddes, W. F. (1949). Grain storage studies VIII. Relation of molds in moist stored cottonseed to increased production of carbon dioxide, fatty acids and heat. *Cereal Chem.* **26**, 109–128.

Christensen, C. M., Fanse, H. A., Nelson, G. H., Bates, F., and Mirocha, C. J. (1967). Microflora of black and red pepper. *Appl. Microbiol.* **15**, 622–626.

Christensen, C. M., Nelson, G. H., Bates, F., and Meronuck, R. A. (1969). Potentially toxic fungi in cereal grains and seeds and their products. *Proc. Symp. Biol. Contam. Grain Anim. Byprod.* pp. 53–54. Agric. Ext. Serv., Dep. Entomol., Fish. Wildl., Univ. of Minnesota, Minneapolis.

Christiansen, L. N., Tomkin, R. B., Shaparis, A. B., Johnston, R. W., and Kautter, D. A. (1975). Effect of sodium nitrite and nitrate on *Clostridium botulinum* growth and toxin production in a summer style sausage. *J. Food Sci.* **40**, 488–490.

Christophersen, J., and Precht, H. (1952). Untersuchungen zum Problem der Hitzeresistenz. II. Untersuchungen an Hefezellen. *Biol. Zentralbl.* **71**, 585–601.

Chung, J. R. (1968). Post-mortem degradation of fish muscle proteins: the role of proteolytic *Pseudomonas* spp. and their mechanism of action. Ph.D. Thesis, Univ. of Washington, Seattle.

Chung, K. C., and Goepfert, J. M. (1970). Growth of Salmonella at low pH. *J. Food Sci.* **35**, 326–328.

Churchill, E. S., and Mallmann, W. L. (1950). Role of the airline hose of the milking machine in the contamination of milk. *J. Milk Food Technol.* **13**, 137–145.

Ciblis, E. (1970). Characterization of a bacteriophage of *Streptococcus thermophilus. Zentralbl. Bakteriol., Parasitenkd., Infektionskr. Hyg., Abt. 2* **125**, 541.

Clairemont, V. (1914). Preserving eggs. U.S. Patent 1,092,897 [*C.A.* **8**, 2010].

Clark, D. S. (1965a). Method of estimating the bacterial population on surfaces. *Can. J. Microbiol.* **11**, 407–413.

Clark, D. S. (1965b). Improvement of spray gun method of estimating bacterial populations on surfaces. *Can. J. Microbiol.* **11**, 1021–1022.

Clark, D. S. (1968). Growth of psychrotolerant pseudomonads and *Achromobacter* on chicken skin. *Poult. Sci.* **47**, 1575–1578.

Clark, D. S., and Lentz, C. P. (1969). Microbiological studies in poultry processing plants in Canada. *Can. Inst. Food Technol. J.* **2**, 33–36.

Clark, F. M., and Tanner, F. W. (1937). Thermophilic canned-food spoilage organisms in sugar and starch. *Food Res.* **2**, 27–39.

Clark, W. S., Jr., Reinbold, G. W., and Rambo, R. S. (1966). Enterococci and coliforms in dehydrated vegetables. *Food Technol.* **20**, 1353–1356.

Clise, J. D., and Swecker, E. E. (1965). Salmonellae from animal byproducts. *Public Health Rep.* **80**, 899–905.

Cliver, D. O. (1971). Transmission of viruses through foods. *Crit. Rev. Environ. Control* **1**, 551–579.

Cohen, M. L., and Blake, P. A. (1977). Trends in foodborne salmonellosis outbreaks: 1963–1975. *J. Food Prot.* **40**, 798–800.

Cole, D. F., and Bugbee, W. M. (1976). Changes in resident bacteria, pH, sucrose, and invert sugar levels in sugarbeet roots during storage. *Appl. Environ. Microbiol.* **31**, 754–757.

Collins, E. B. (1972). Biosynthesis of flavor components by microorganisms. *J. Dairy Sci.* **55**, 1022–1028.

Collins, R. N., Trager, M. D., Goldsby, J. B., Boring, J. R., III, Cohoon, D. B., and Barr, R. N. (1968). Interstate outbreak of *Salmonella newbrunswick* infection traced to powdered milk. *J. Am. Med. Assoc.* **203**, 838–844.

Collins-Thompson, D. L., Aris, B., and Hurst, A. (1973). Growth and enterotoxin B synthesis by *Staphylococcus aureus* S 6 in associative growth with *Pseudomonas aeruginosa. Can. J. Microbiol.* **19**, 1197–1201.

Collins-Thompson, D. L., Erdman, I. E., Milling, M. E., Burgener, D. M., Purvis, U. T., Loit, A., and Coulter, R. M. (1977). Microbiological standards for cheese: Survey and viewpoint of the Canadian Health Protection Branch. *J. Food Prot.* **40**, 411–414.

Collins-Thompson, D. L., Weiss, K. F., Riedel, G. W., and Charbonneau, S. (1978). Sampling plans and guidelines for domestic and imported cocoa from a Canadian national microbiological survey. *Can. Inst. Food Sci. Technol. J.* **11**, 177–179.

Colwell, R. R., and Liston, J. (1960). Microbiology of Shellfish: Bacteriological study of the natural flora of Pacific oysters (*Crassostrea gigas*). *Appl. Microbiol.* **8**, 104–109.

Commission of the European Communities (1976). "Evaluation of the Hygienic Problems Related to the Chilling of Poultry Carcasses," Information on Agric. No. 22. EEC, Brussels.

Comptroller General of the United States (1974). "Salmonella in Raw Meat and Poultry: An Assessment of the Problem," Report to Congress. U.S. Gen. Account. Off., Washington, D.C.

Consumers' Association of Canada (1976). *Salmonella* in poultry. *Can. Consumer* **6**, 33–35.

Cook, L. R. (1972). "Chocolate Production and Use." Books for Ind., New York.

Cords, B. R., and Tatini, S. R. (1973). Applicability of heat stable deoxyribonuclease assay for assessment of staphylococcal growth and the likely presence of enterotoxin in cheese. *J. Dairy Sci.* **56**, 1512–1519.

Coretti, K., and Inal, T. (1969). Rückstandsprobleme bei der Kaltenkeimung von Gewürzen mit T-Gas (Äthylenoxyd). *Fleischwirtschaft* **49**, 599–604.

Coretti, K., and Müggenburg, H. (1967). Keimgehalt von Trockensuppen und seine Beurteilung. *Feinkostwirtschaft* **4**, 76–78, 108–115.

Corlett, D. A., Jr. (1976). Canned foods—tests for cause of spoilage. In "Compendium of Methods for the Microbiological Examination of Foods" (M. Speck, ed.), pp. 632–673. Am. Public Health Assoc., Washington, D.C.

Corran, J. W., and Edgar, S. H. (1933). Preservative action of spices and related compounds against yeast fermentation. *J. Soc. Chem. Ind.* **52**, 149T–152T.

Corry, J. E. L., and Barnes, E. M. (1968). The heat resistance of salmonellae in egg albumen. *Br. Poult. Sci.* **9**, 253–260.

Costin, I. D., Voiculescu, D., and Gorcea, V. (1964). An outbreak of food poisoning in adults associated with *Escherichia coli* serotype 86:B7:H34. *Pathol. Microbiol.* **27**, 68–78.

Cotterill, O. J., and Glauert, J. (1969). Thermal resistance of salmonellae in egg yolk products containing sugar or salt. *Poult. Sci.* **48**, 1156–1166.

Cotterill, O. J., and Glauert, J. (1971). Thermal resistance of salmonellae in egg yolk containing 10% sugar or salt after storage at various temperatures. *Poult. Sci.* **50**, 109–115.

Cotterill, O. J., and Glauert, J. (1972). Destruction of *Salmonella oranienburg* in egg yolk containing various concentrations of salt at low temperatures. *Poult. Sci.* **51**, 1060–1061.

Cotterill, O. J., Glauert, J., and Krause, G. F. (1973). Thermal destruction curves for *Salmonella oranienburg* in egg products. *Poult. Sci.* **52**, 568–577.

Cotterill, O. J., Glauert, J., Steinhoff, S. E., and Baldwin, R. E. (1974). Hot-pack pasteurization of salted egg products. *Poult. Sci.* **53**, 636–645.

Cox, N. A., Mercuri, A. J., Thompson, J. E., and Chew, V. (1976). Swab and excised tissue sampling for total and Enterobacteriaceae counts of fresh and surface-frozen broiler skin. *Poult. Sci.* **55**, 2405–2408.

Cox, N. A., Mercuri, A. J., Tanner, D. A., Carson, M. O., Thompson, J. E., and Bailey, J. S. (1978). Effectiveness of sampling methods for *Salmonella* detection on processed broilers. *J. Food Prot.* **41**, 341–343.

Crane, F. M., Hansen, M., Yoder, R., Lepley, K., and Cox, P. (1972). Effect of processing feeds on molds, *Salmonella* and other harmful substances in feeds. *Feedstuffs* **44**(23), 26.

Craven, P. C., Mackel, D. C., Baine, W. B., Barker, W. H., Gangarosa, E. J., Goldfield, M., Rosenfeld, H., Altman, R., Lachapelle, G., Davies, J. W., and Swanson, R. C. (1975). International outbreak of *Salmonella eastbourne* infection traced to contaminated chocolate. *Lancet* No. 7910, 788–793.

Craven, S. E., and Lillard, H. S. (1974). Effect of microwave reheating of precooked chicken on *Clostridium perfringens*. *J. Food Sci.* **39**, 211–212.

Craven, S. E., Lillard, H. S., and Mercuri, A. J. (1975). Survival of *Clostridium perfringens* during preparation of precooked chicken parts. *J. Milk Food Technol.* **38**, 505–508.

Credit, C., Hedeman, R., Heywood, P., and Westhoff, D. (1972). Identification of bacteria isolated from pasteurized milk following refrigerated storage. *J. Milk Food Technol.* **35**, 708–709.

Cripps, M. H. (1972). Spice oleoresins: the process, the market and the future. *Proc. Conf. Spices* pp. 237–242. Trop. Prod. Inst., London.

Crisley, F. D., Angelotti, R., and Foter, M. J. (1964). Multiplication of *Staphylococcus aureus* in synthetic cream fillings and pies. *Public Health Rep.* **79**, 369–376.

Crossley, E. L. (1962). "Milk Hygiene—Hygiene in Milk Production, Processing and Distribution." FAO/WHO, Geneva.

Culkin, K. A., and Fung, D. Y. C. (1975). Destruction of *Escherichia coli* and *Salmonella typhimurium* in microwave-cooked soup. *J. Milk Food Technol.* **38**, 8–15.

Cunliffe, H. R., Blackwell, J. H., Dors, R., and Walker, J. S. (1979). Inactivation of milkborne foot and mouth disease virus at ultra-high temperatures. *J. Milk Food Technol.* **42**, 135–137.

Cunningham, F. E. (1966). Process for pasteurizing liquid egg white. *In* "'The Destruction of Salmonellae," ARS 74–37, pp. 61–65. Agric. Res. Serv., U.S. Dep. Agric., Albany, California.

Cunningham, F. E., and Lineweaver, H. (1965). Stabilization of egg-white proteins to pasteurizing temperatures above 60°C. *Food Technol.* **19**, 136–141.

Cunningham, H. M., and Lawrence, G. A. (1977). Effect of exposure of meat and poultry to chlorinated water on the retention of chlorinated compounds and water. *J. Food Sci.* **42**, 1504–1505.

Dabbah, R., Moats, W. A., and Edwards, V. M. (1971). Survivor curves of selected *Salmonella enteritidis* serotypes in liquid whole egg homogenates at 60°C. *Poult. Sci.* **50**, 1772–1776.

Dack, G. M. (1956). "Food Poisoning," 3rd ed. Univ. of Chicago Press, Chicago, Illinois.

Dack, G. M. (1961). Public health significance of flour bacteriology. *Cereal Sci. Today* **6**, 9–10.

Daftary, R. D., Pomeranz, Y., Hoseney, R. C., Shogren, M. D., and Finney, K. F.

(1970). Changes in wheat flours damaged by mold during storage: Effects in breadmaking. *J. Agric. Food Chem.* **18**(4), 617–619.

Dahlberg, A. C., and Kosikowski, F. V. (1948). The development of flavor in American cheddar cheese made from pasteurized milk with *Streptococcus faecalis*. *J. Dairy Sci.* **31**, 275–284.

Dahlberg, A. C., Adams, H. S., and Held, M. E. (1953). "Sanitary Milk Control and its Relation to the Sanitary, Nutritive and other Qualities of Milk," Publ. No. 250. Natl. Acad. Sci.–Natl. Res. Counc., Washington, D.C.

Dainty, R. H. (1971). The control and evaluation of spoilage. *J. Food Technol.* **6**, 209–224.

Dainty, R. H., Shaw, B. G., de Boer, K. A., and Scheps, E. S. J. (1975). Protein changes caused by bacterial growth on beef. *J. Appl. Bacteriol.* **39**, 72–81.

Dakin, J. C. (1962). Pasteurization of acetic acid preserves. *Recent Adv. Food Sci.* **2**, 128–141.

Dammers, J., Kampelmacher, E. H., Edel, W., and van Schothorst, M. (1966). Effect of decontamination of feed mixtures by heat treatment and gamma radiation on growth and feed conversion in fattening pigs. *In* "Food Irradiation," pp. 159–166. IAEA, Vienna.

Dam-Mikkelsen, H., Jessen Petersen, P., and Skovgaard, N. (1962). Tre tilfaelde af *Clostridium perfringens* levnedsmiddelforgiftning (Three cases of *Clostridium perfringens* food poisoning). *Nord. Veterinaermed.* **14**, 200–211.

Danziger, M. T., Steinberg, M. P., and Nelson, A. I. (1973). Effect of CO_2, moisture content and sorbate on safe storage for wet corn. *Trans. ASAE* **16**, 679–682.

Daoust, D. R., Read, R. B., Jr., and Litsky, W. (1959). Thermal inactivation studies on *Corynebacterium diphtheriae* and *Shigella paradysenteriae* in milk and milk products. *Bacteriol. Proc.* pp. 9–10.

D'Aoust, J. Y. (1977). *Salmonella* and the chocolate industry. A review. *J. Food Prot.* **40**, 718–727.

D'Aoust, J. Y., Aris, B. J., Thisdele, P., Durante, A., Brisson, N., Dragon, D., Lachapelle, G., Johnston, M., and Laidley, R. (1975). *Salmonella eastbourne* outbreak associated with chocolate. *Can. Inst. Food Sci. Technol. J.* **8**, 181–184.

Davidson, C. M., and Webb, G. (1973). The behaviour of salmonellae in vacuum-packaged cooked cured meat products. *Can. Inst. Food Sci. Technol. J.* **6**, 41–44.

Davies, D. G., and Harvey, R. W. S. (1972). Anthrax infection in bone meal from various countries of origin. *J. Hyg.* **70**, 455–457.

Davies, R. F., and Wahba, A. H. (1976). "*Salmonella* Infections of Charter Flight Passengers," Report on a visit to Spain (Canary Islands) 26 February–2 March, 1976 (Booklet). WHO, Reg. Off. Eur., Copenhagen.

Day, F. (1974). Status of the milling and baking industries in Latin America. *Cereal Sci. Today* **19**(4), 157–160.

Deibel, R. H., and Hartman, P. A. (1976). The enterococci. *In* "Compendium of Methods for the Microbiological Examination of Foods" (M. L. Speck, ed.), pp. 370–373. Am. Public Health Assoc., Washington, D.C.

Del Vecchio, H. W., Dayharsh, C. A., and Baselt, F. C. (1951). Thermal death time studies on beer spoilage organisms. I. *Proc. Am. Soc. Brew. Chem.* p. 45.

Denny, C. B., Goeke, D. J., Jr., and Steinberg, R. (1969). "Inoculation Tests of *Clostridium botulinum* in Canned Breads with Special Reference to Water Activity," Res. Rep. No. 4–69. Washington Res. Lab., Natl. Canners Assoc., Washington, D.C.

Department of National Health and Welfare (1977). "Proposed Amendments to the Food and Drugs Act and Regulations." Ottawa.

Department of National Health and Welfare (1978). "Ethylene Oxide," Item E.1 in Division 16, Table 8, pp. 67–25. Food Drug Regul., Ottawa.

Depew, F. M. (1968). The salmonellae—a current challenge. *Manuf. Confect.* **48**, 31–32.

Devillers, M. (1955). *Proc. C.I.T.S., 9th, Brussels* p. 88. (Cited in Moroz, 1963.)

Dhennin, L., and Labie, J. (1976). Thermorésistance du virus de la fièvre aphteuse dans le lait de vaches infecties. *Bull. Acad. Vet. Fr.* **49**, 243–249.

DiGirolamo, R. G. (1969). The uptake, elimination and effects of processing on the survival of poliovirus in West Coast oysters. Ph.D. Thesis, Univ. of Washington, Seattle.

DiGirolamo, R., Liston, J., and Matches, J. (1970). The effects of freezing on the survival of *Salmonella* and *E. coli* in Pacific oysters. *J. Food Sci.* **35**, 13–16.

Dixon, J. M. S., and Pooley, F. E. (1961). Salmonellae in a poultry processing plant. *Mon. Bull. Minist. Health Public Health Lab. Serv.* **20**, 30–33.

Dixon, J. M. S., and Pooley, F. E. (1962). Salmonellae in two turkey processing factories. *Mon. Bull. Minist. Health Public Health Lab. Serv.* **21**, 138–141.

Dockstader, W. B., and Groomes, R. J. (1971). Detection and survival of salmonellae in milk chocolate. *Bacteriol. Proc.* No. A36, p. 7.

Donald, W. W., and Mirocha, C. J. (1977). Chitin as a measure of fungal growth in stored corn and soybean seed. *Cereal Chem.* **54**(3), 466–474.

Douglas, J. S. (1973). Commercial *Scitamineae*. II. Production of good quality ginger. *Flavour Ind.* **4**, 340–342.

Duggan, D. E., Anderson, A. W., and Elliker, P. R. (1959). A frozen concentrate of *Lactobacillus acidolphilus* for preparation of a palatable acidophilus milk. *Food Technol.* **13**, 465–469.

Duitschaever, C. L. (1977). Incidence of *Salmonella* in retailed raw cut-up chicken. *J. Food Prot.* **40**, 191–192.

Duncan, C. L., and Colmer, A. R. (1964). Coliforms associated with sugarcane plants and juices. *Appl. Microbiol.* **12**, 173–177.

Dunkley, W. L., Hunter, G., Thornton, H. R., and Hood, E. G. (1942). Studies on surface taint butter, II; an odorous compound in skimmilk cultures of *Pseudomonas putrefaciens. Sci. Agric.* **22**, 347–355.

Dupont, H. I., Formal, S. B., Hornick, R. B., Snyder, M. J., Dibanti, J. P., Sheahan, D. G., LaBrec, E. H., and Kalas, J. P. (1971). Pathogenesis of *Escherichia coli* diarrhea. *N. Engl. J. Med.* **285**, 3–11.

Dupuy, P. (1959). Inhibition by sulphurous acid of the oxidation of ethanol by *Acetobacter rancens. Ann. Inst. Natl. Rech. Agron., Ser. E* **8**(III), 233–283; **8**(IV), 337–376.

Durkee (1977a). "Kaomel. Confectioner's Coating Butter," IND–217J S.C.M. Durkee Ind. Foods Group, Cleveland, Ohio.

Durkee (1977b). "Satina. Fractionated Lauric Hard Butters," IND–264. S.C.M. Durkee Ind. Foods Group, Cleveland, Ohio.

Dyett, E. J., and Shelley, D. (1966). The effects of sulphite preservative in British fresh sausages. *J. Appl. Bacteriol.* **29**, 439–446.

Eckert, J. W. (1975). Postharvest diseases of fresh fruits and vegetables—etiology and control. *In* "Postharvest Biology and Handling of Fruits and Vegetables" (N. F. Haard and D. K. Salunkhe, eds.), pp. 81–117. AVI, Westport, Connecticut.

Eckert, J. W., and Sommer, N. F. (1967). Control of diseases of fruits and vegetables by postharvest treatment. *Annu. Rev. Phytopathol.* **5**, 391–432.

Edel, W., Guinée, P. A. M., van Schothorst, M., and Kampelmacher, E. H. (1966). Studies about the occurrence of *Salmonella* in pigs fattened with pellets and non-pelleted meal. *Tijdschr. Diergeneeskd.* **91**(15), 962–966.

Eekhoff-Stork, N. M. (1976). "The World Atlas of Cheese" (A. Bailey, transl.). Paddington Press, London.

Egan, B. T. (1971). "Post-harvest Deterioration of Sugar Cane." Sugar Exp. Stn. Board, Brisbane, Australia.

Eisenberg, W. V., and Cichowicz, S. M. (1977). Machinery mold—indicator organism in food. *Food Technol.* **31**(2), 52–56.

Elliott, L. E., and Brant, A. W. (1957). Effect of saline and egg shell membrane on bacterial growth. *Food Res.* **22**, 241–250.

Elliott, R. P. (1954). Spoilage of shell eggs by pseudomonads. *Appl. Microbiol.* **2**, 158–164.

Elliott, R. P. (1958). Determination of pyoverdine, the fluorescent pigment of pseudomonads in frozen whole egg. *Appl. Microbiol.* **6**, 247–251.

Elliott, R. P. (1961). Quality vs. safety in frozen foods. *Frosted Food Field* **33**(6), 33–34.

Elliott, R. P. (1966). Microbial selection in frozen foods and chilled products. *Quick Frozen Foods* **29**(4), 66, 68–69.

Elliott, R. P. (1967). Bacteriological problems in the manufacture of oilseed proteins. *Conf. Eng. Unconventional Protein Prod., Am. Inst. Chem. Eng., Santa Barbara, Calif.* Unpublished.

Elliott, R. P., and Heiniger, P. K. (1965). Improved temperature–gradient incubator and the maximal growth temperature and heat resistance of *Salmonella*. *Appl. Microbiol.* **13**, 73–76.

Elliott, R. P., and Michener, H. D. (1961). Microbiological standards for chilled and frozen foods. A review. *Appl. Microbiol.* **9**, 452–468.

Elliott, R. P., and Michener, H. D. (1962). Growth of food poisoning and food spoilage microorganisms at refrigeration temperatures. *Bull. Int. Inst. Refrig. Annexe* **1962–1**, 455–465.

Elliott, R. P., and Michener, H. D. (1965). "Factors Affecting the Growth of Psychrophilic Microorganisms in Foods—A Review," Tech. Bull. No. 1320. U.S. Dep. Agric., Albany, California.

Elliott, R. P., and Straka, R. P. (1964). Rate of microbial deterioration of chicken meat at 2°C after freezing and thawing. *Poult. Sci.* **43**, 81–86.

Elliott, R. P., Straka, R. P., and Garibaldi, J. A. (1964). Polyphosphate inhibition of growth of pseudomonads from poultry meat. *Appl. Microbiol.* **12**, 517–522.

El Mossalami, E., and Sedik, M. F. (1973). Ice cooling of *Tilapia nilotica*. *Egypt. Vet. Med. J.* **22**, 149–171.

El Mossalami, E., and Youssef, A. (1965). Studies on bacterial contamination of spices used in meat products. *Zentralbl. Veterinaermed., Reihe B* **12**(2), 176–182.

Emmons, D. B., and Tuckey, S. L. (1967). "Cottage Cheese and Other Cultured Milk Products," Pfizer Cheese Monographs, Vol. III. Pfizer, New York.

Empey, W. A., and Scott, W. J. (1939). "Investigation on Chilled Beef. Part I. Microbial Contamination Acquired in the Meat Works," Bull. Counc. Sci. Ind. Res., No. 126. CSIRO, Melbourne.

Enright, J. B. (1961). The pasteurization of cream, chocolate milk and ice cream mixes containing the organism of Q fever. *J. Milk Food Technol.* **24,** 351–355.

Enright, J. B., Sadler, W. W., and Thomas, R. C. (1956). Observations on the thermal inactivation of the organism of Q fever. *J. Milk Food Technol.* **19,** 313–318.

Ercolani, G. L. (1976). Bacteriological quality assessment of fresh marketed lettuce and fennel. *Appl. Environ. Microbiol.* **31,** 847–852.

Erdman, I. E., and Thornton, H. R. (1951). Psychrophilic bacteria in Edmonton milk and cream. II. Kinds. *Can. J. Technol.* **29,** 238–242.

Escher, F. E., Koehler, P. E., and Ayres, J. C. (1973). Effect of roasting on aflatoxin content of artificially contaminated pecans. *J. Food Sci.* **38,** 889–892.

Eschmann, K. H. (1965). Gewürze—eine Quelle bakteriologischer Infektionen. *Alimenta* **4**(3), 83–87.

Eskin, N. A. M., Henderson, H. M., and Townsend, R. J., eds. (1971). "Biochemistry of Food," pp. 202–217. Academic Press, New York.

Essary, E. O., Moore, W. E. C., and Kramer, C. Y. (1958). Influence of scald temperatures, chill times, and holding temperatures on the bacterial flora and shelf-life of freshly chilled, tray-packed poultry. *Food Technol.* **12,** 684–687.

Esselen, W. B., Levine, A. S., and Brushway, M. J. (1956). Adequate roasting procedures for frozen stuffed poultry. *J. Am. Diet. Assoc.* **32,** 1162–1166.

Essex-Cater, A. J., Jones, D. M., and Swindell, F. (1963). An outbreak of human infection due to *Salmonella typhimurium,* phage type 4, associated with the use of unpasteurized liquid egg. *J. Hyg.* **61,** 323–330.

Etchells, J. L., and Bell, T. A. (1976). Pickle products. *In* "Compendium of Methods for the Microbiological Examination of Foods" (M. Speck, ed.), pp. 574–593. Am. Public Health Assoc., Washington, D.C.

Etchells, J. L., Bell, T. A., Monroe, R. J., Masley, P. M., and Demain, A. L. (1958). Populations and softening enzyme activity of filamentous fungi on flowers, ovaries, and fruit of pickling cucumbers. *Appl. Microbiol.* **6,** 427–440.

Etchells, J. L., Borg, A. F., and Bell, T. A. (1961). Influence of sorbic acid on populations and species of yeasts occurring in cucumber fermentations. *Appl. Microbiol.* **9,** 139–144.

Etchells, J. L., Costilow, R. N., Anderson, T. E., and Bell, T. A. (1964). Pure culture fermentation of brined cucumbers. *Appl. Microbiol.* **12,** 523–535.

Etchells, J. L., Borg, A. F., Kittel, I. D., Bell, T. A., and Fleming, H. P. (1966). Pure culture fermentation of green olives. *Appl. Microbiol.* **14,** 1027–1041.

Etchells, J. L., Borg, A. F., and Bell, T. A. (1968). Bloater formation by gas-forming lactic acid bacteria in cucumber fermentations. *Appl. Microbiol.* **16,** 1029–1035.

Etchells, J. L., Bell, T. A., Fleming, H. P., Kelling, R. E., and Thompson, R. L. (1973). Suggested procedure for the controlled fermentation of commercially brined pickling cucumbers—the use of starter cultures and reduction of carbon dioxide accumulation. *Pickle Pak Sci.* **3,** 4–14.

Fabian, F. W., and Wethington, M. C. (1950). Spoilage in salad and French dressing due to yeasts. *Food Res.* **15,** 135–137.

Fabian, F. W., Krehl, C. F., and Little, N. W. (1939). The role of spices in pickled-food spoilage. *Food Res.* **4,** 269–286.

Fahey, J. E. (1955). Some observations on "air sac" infection in chickens. *Poult. Sci.* **34,** 982–984.

Fallesen, K. B. (1976). The bacteriological quality of reconstituted soups, in relation to that of the dried soups from which they are prepared. *Dan. Veterinaertidsskr.* **59**(17), 714–717.

Fanelli, M. J., Peterson, A. C., and Gunderson, M. F. (1965). Microbiology of dehydrated soups. I. A survey. *Food Technol.* **19,** 83–86.

Fanse, H. A., and Christensen, C. M. (1966). Invasion by fungi of rice stored at moisture contents of 13.5–15.5%. *Phytopathology* **56,** 1162–1164.

Farkas, J. (1973). Radurization and radicidation of spices. *In* "Aspects of the Introduction of Food Irradiation in Developing Countries," IAEA PL-518/6, pp. 43–59. IAEA, Vienna.

Farmiloe, F. J., Cornford, S. J., Coppock, J. B. M., and Ingram, M. (1954). The survival of *Bacillus subtilis* spores in the baking of bread. *J. Sci. Food Agric.* **5,** 292–304.

Farrell, A. J., and Barnes, E. M. (1964). The bacteriology of chilling procedures used in poultry processing plants. *Br. Poult. Sci.* **5,** 89–95.

Feeney, R. E., MacDonnell, L. R., and Lorenz, F. W. (1953). High temperature treatment of shell eggs. *Poult. Sci.* **32,** 899.

Feeney, R. E., MacDonnell, L. R., and Lorenz, F. W. (1954). High temperature treatment of shell eggs. *Food Technol.* **8,** 242–245.

Feig, M. (1950). Diarrhea, dysentery, food poisoning, and gastroenteritis. *Am. J. Public Health* **40,** 1372–1394.

Felsenfeld O., Young, V. M., and Yoshimura, T. (1950). A survey of *Salmonella* organisms in market meat, eggs and milk. *J. Am. Vet. Med. Assoc.* **116,** 17–21.

Ferencik, M. (1970). Formation of histamine during bacterial decarboxylation of histidine in the flesh of some marine fishes. *J. Hyg., Epidemiol., Microbiol., Immunol.* **14,** 52–60.

Fievez, L., and Granville, A. (1965). Alterations de semiconserves de viande par des épices riches en spores bactériennes très résistantes à la chaleur. *Ann. Med. Vet.* **109,** 143–146.

Finlayson M. (1978). Salmonellae in Alberta poultry products and their significance in human infection. *Proc. Int. Symp. Salmonella Prospects Control, Univ. Guelph* pp. 156–180.

Firstenberg, R., Mannheim, C. H., and Cohen, A. (1974). Microbial quality of dehydrated onions. *J. Food Sci.* **39,** 685–688.

Fishbein, M., Mehlman, I. J., Chugg, L., and Olson, J. C., Jr. (1976). Coliforms, fecal coliforms, *E. coli* and enteropathogenic *E. coli*. *In* "Compendium of Methods for the Microbiological Examination of Foods" (M. L. Speck, ed.), pp. 277–300. Am. Public Health Assoc., Washington, D.C.

Flake, J. C., Dabbah, R., and Weddle, D. B. (1978). Reduction methods. *In* "Standard Methods for the Examination of Dairy Products" (E. H. Marth, ed.), 14th ed., pp. 187–195. Am: Public Health Assoc., Washington, D.C.

Flannigan, B., and Dickie, N. A. (1972). Distribution of microorganisms in fractions produced during pearling of barley. *Trans. Br. Mycol. Soc.* **59**(3), 377–391 [*Biol. Abstr.* **55,** 60289 (1973)].

Flannigan, B., and Hui, S. C. (1976). The occurrence of aflatoxin-producing strains of *Aspergillus flavus* in the mould floras of ground spices. *J. Appl. Bacteriol.* **41,** 411–418.

Fleming, H. P., Walter, W. M., Jr., and Etchells, J. L. (1973). Antimicrobial properties of oleuropein and products of its hydrolysis from green olives. *Appl. Microbiol.* **26,** 777–781.

Flippin, R. S., and Mickelson, M. N. (1960). Use of salmonellae antagonists in fermenting egg white. I. Microbial antagonists of salmonellae. *Appl. Microbiol.* **8,** 366–370.

Florian, M. L. E., and Trussell, P. C. (1957). Bacterial spoilage of shell eggs IV. Identification of spoilage organisms. *Food Technol.* **11,** 56–60.

Follstad, M. N. (1966). Microflora of nonprocessed raisins. *Phytopathology* **56,** 1413.

Foltz, V. D. (1966). Incidence of salmonellae in prepared and packaged foods. *Assoc. Food Drug Off. U.S., Q. Bull.* **30**(4), 181–185.

Food and Agriculture Organization (FAO) (1957). "Meat Hygiene," Agricultural Studies, No. 34. FAO, Rome.

Food and Agriculture Organization (FAO) (1976). "Proposed Draft Code of Hygienic Practice for Molluscan Shellfish," Codex Alimentarius, ALINORM 76/13A. FAO, Rome.

Food and Agriculture Organization/International Atomic Energy Agency/World Health Organization (FAO/IAEA/WHO) Expert Committee (1977). "Wholesomeness of Irradiated Food," Tech. Rep. Ser., No. 604. WHO, Geneva.

Food and Agriculture Organization/World Health Organization (FAO/WHO) (1974). Toxicological evaluation of certain food additives with a review of general principles and of specifications. Expert Committee on Food Additives, 17th Report, Geneva, 1973. *Tech. Rep. Ser.* No. 539; *FAO Nutr. Meet., Rep. Ser.* Nos. 53 and 53A.

Food and Agriculture Organization/World Health Organization (FAO/WHO) (1975). "Microbiological Specifications for Foods," Expert Consultation, EC/Microbial/75/Report 1. FAO, Rome.

Food and Agriculture Organization/World Health Organization (FAO/WHO) (1976a). "Recommended International Code of Hygienic Practice for Fresh Meat," Codex Alimentarius Commission. FAO, Rome.

Food and Agriculture Organization/World Health Organization (FAO/WHO) (1976b). "Recommended International Code of Hygienic Practice for Processed Meat Products," Codex Alimentarius Commission. FAO, Rome.

Food and Agriculture Organization/World Health Organization (FAO/WHO) (1976c). "Recommended International Code of Hygienic Practice for Poultry Processing," Codex Alimentarius Commission, CAC/RCP 14-1976. FAO, Rome.

Food and Agriculture Organization/World Health Organization (FAO/WHO) (1976d). "Report of the Third Session, Codex Committee on Edible Ices," Appendix II. FAO, Rome.

Food and Agriculture Organization/World Health Organization (FAO/WHO) (1977). "Food Standards Programme. Report of the 10th Session of the Coordinating Committee for Europe," Appendix II, Revised Draft European Regional Standard for Natural Mineral Waters, Sept 8, Alinorm 78/19. FAO, Rome.

Food and Agriculture Organization/World Health Organization (FAO/WHO) (1978). "Report of the 19th Session, Government Experts on Code of Principles Concerning Milk Products," pp. 1, 2, 61–63. FAO, Rome.

Forgacs, J., and Carll, W. T. (1962). Mycotoxicosis. *Adv. Vet. Sci.* **7,** 273–381.

Forsyth, W. G. C., and Quesnel, V. C. (1963). The mechanism of cacao curing. *Adv. Enzymol.* **25,** 457–492.

Forsythe, R. H. (1968). The science and technology of egg products manufacture in the United States. *In* "Egg Quality: A Study of the Hen's Egg" (T. C. Carter, ed.), pp. 262–304. Oliver & Boyd, Edinburgh.

Forsythe, R. H. (1970). Egg processing technology—Progress and sanitation programs. *J. Milk Food Technol.* **33,** 64–73.

Forsythe, R. H., Ayres, J. C., and Radlo, J. L. (1953). Factors affecting the microbiological populations of shell eggs. *Food Technol.* **7,** 49–56.

Foster, E. M. (1968). *Salmonella's* heat resistance examined. *Candy Ind. Confect. J.* **131**(10), 66–73.

Foster, E. M., Nelson, F. E., Speck, M. L., Doetsch, R. N., and Olson, J. C., Jr. (1957). "Dairy Microbiology." Prentice-Hall, Englewood Cliffs, New Jersey.

Foster, H. G., Lear, S. A., and Metzger, H. S. (1953). Time temperature studies on the inactivation of *Brucella abortus* strain 2308 in milk. *J. Milk Food Technol.* **16,** 116–120.

Fowler, H. W., and Fowler, F. G. (1964). "The Concise Oxford Dictionary," 5th ed. Oxford Univ. Press, London and New York.

Fowler, J. L., and Clark, W. S., Jr. (1975). Microbiology of delicatessen salads. *J. Milk Food Technol.* **38,** 146–149.

Frank, H. A. (1963). Factors affecting bacterial spoilage of animal products at elevated temperatures. *Food Technol.* **17,** 573–574, 576–578, 580.

Frank, H. K. (1966). Aflatoxine in Lebensmitteln. *Arch. Lebensmittelhyg.* **17,** 237–242.

Frank, H. K. (1972). Mykotoxine und ihre Produzenten in landwirtschaftlichen Produkten. *Ber. Landwirtsch.* **50,** 240–255.

Frank, J. F., and Marth, E. H. (1977a). Inhibition of enteropathogenic *Escherichia coli* by homofermentative lactic acid bacteria in skimmilk. I. Comparison of strains of *Escherichia coli. J. Food Prot.* **40,** 749–753.

Frank, J. F., and Marth, E. H. (1977b). Inhibition of enteropathogenic *Escherichia coli* by homofermentative lactic acid bacteria in skimmilk. II. Comparison of lactic acid bacteria and enumeration of methods. *J. Food Prot.* **40,** 754–759.

Frank, J. F., and Marth, E. H. (1978). Survey of soft and semisoft cheese for presence of fecal coliforms and serotypes of enteropathogenic *E. coli. J. Food Prot.* **41,** 198–200.

Frank, J. F., Torrey, G. S., Marth, E. H., Stuiber, D. A., Lindsey, R. C., and Lund, D. B. (1975). Control of aflatoxin production during fermentation of wild rice. *J. Milk Food Technol.* **38,** 73–75.

Frank, J. F., Marth, E. H., Lindsey, R. C., and Lund, D. B. (1976). Microorganisms and flavor development associated with the wild rice fermentation. *J. Milk Food Technol.* **39,** 606–613.

Frank, J. F., Marth, E. H., and Olson, N. F. (1977). Survival of enteropathogenic and nonpathogenic *Escherichia coli* during the manufacture of Camembert cheese. *J. Food Prot.* **40,** 835–842.

Frank, J. F., Marth, E. H., and Olson, N. F. (1978). Behavior of enteropathogenic *Escherichia coli* during manufacture and ripening of Brick Cheese. *J. Food Prot.* **41,** 111–115.

Frazier, W. C. (1967). Contamination, preservation and spoilage of cereals and cereal products. *In* "Food Microbiology," pp. 180–191. McGraw-Hill, New York.

Fromm, D. (1959). An evaluation of techniques commonly used to quantitatively determine the bacterial population of chicken carcasses. *Poult. Sci.* **38**, 887–893.

Fromm, D. (1963). Permeability of the hen's egg shell. *Poult. Sci.* **42**, 1271.

Fujino, T., Sakaguchi, G., Sakazaki, R., and Takeda, Y., eds. (1974). "International Symposium on *Vibrio parahaemolyticus*." Saikon, Tokyo.

Fulton, L. H., Gilpin, G. L., and Dawson, E. H. (1967). Turkeys roasted from frozen and thawed states. *J. Home Econ.* **59**, 728–731.

Funk, E. M. (1943). Stabilizing quality in shell eggs. *Mo. Agric. Exp. Stn., Res. Bull.* No. 362.

Funk, E. M. (1950). Maintenance of quality in shell eggs by thermostabilization. *Mo. Agric. Exp. Stn., Res. Bull.* No. 467.

Funk, E. M., Forward, J., and Lorah, M. (1954). Minimizing spoilage in shell eggs by thermostabilization. *Poult. Sci.* **33**, 532–538.

Gabis, D. A., and Silliker, J. H. (1974). Salmonella in natural animal casings. *Appl. Microbiol.* **27**, 66–71.

Gabis, D. A., Langlois, B. E., and Rudnick, A. W. (1970). Microbiological examination of cocoa powder. *Appl. Microbiol.* **20**, 644–645.

Gaines, S., Achavasmith, Y., Thareesarvat, M., and Dunagmani, C. (1964). Types and distribution of enteropathogenic *Escherichia coli* in Bangkok, Thailand. *Am. J. Hyg.* **80**, 388–394.

Gal, I. E. (1968). Über die antibakterielle Wirksamkeit von Gewürzpaprika. Aktivitätsprüfung von Capsicidin und Capsaicin, *Z. Lebensm.-Unters.-Forsch.* **138**(2), 86–92.

Gal, I. E. (1969). The bacteriostatic effect of capsaicine. (In Hung.) *Elelmiszervizsgalati Kozl.* **15**(2), 80–85 [*Food Sci. Tech. Abstr.* **1**, 11 T328].

Galbraith, N. S., Taylor, C. E. D., Cavanagh, P., Hagan, J. G., and Patton, J. L. (1962). Pet foods and garden fertilisers as sources of human salmonellosis. *Lancet* No. 7225, 372–374.

Galbraith, N. S., Taylor, C. E. D., Patton, J. L., and Hagan, J. G. (1964). *Salmonella* infection in poultry. *Med. Off.* **111**, 354–356.

Galesloot, T. E., Hassing, F., and Viringa, H. A. (1968). Symbiosis in yoghurt. I. Stimulation of *Lactobacillus bulgaricus* by a factor produced by *Streptococcus thermophilus*. *Neth. Milk Dairy J.* **22**, 50.

Galin, M., Wurch, I., and Linder, R. (1970). Mise en évidence d'un bacteriophage du *Streptococcus thermophilus* cause d'un accident de fermentation du yoghurt. *C. R. Acad. Sci., Ser. D* **270**, 424.

Galton, M. M., Harless, M., and Hardy, A. V. (1955a). *Salmonella* isolations from dehydrated dog meals. *J. Am. Vet. Med. Assoc.* **126**, 57–58.

Galton, M. M., Mackel, D. C., Lewis, A. L., Haire, W. C., and Hardy, A. V. (1955b). Salmonellosis in poultry and poultry processing plants in Florida. *Am. J. Vet. Res.* **16**, 132–137.

Galyean, R. D., Cotterill, O. J., and Cunningham, F. E. (1972). Yolk inhibition of lysozyme activity in egg white. *Poult. Sci.* **51**, 1346–1353.

Gardner, G. A., and Carson, A. W. (1967). Relationship between carbon dioxide production and growth of pure strains of bacteria on porcine muscle. *J. Appl. Bacteriol.* **30**, 500–510.

Gardner, G. A., and Patterson, R. L. S. (1975). A *Proteus inconstans* which produces "cabbage odour" in the spoilage of vacuum-packed sliced bacon. *J. Appl. Bacteriol.* **39**, 263–271.

Garibaldi, J. A. (1960). Factors in egg white which control growth of bacteria. *Food Res.* **25**, 337–344.

Garibaldi, J. A. (1966). Factors affecting the heat sensitivity of salmonellae. *In* "The Destruction of Salmonellae," ARS 74–37, pp. 34–37. Agric. Res. Serv., U.S. Dep. Agric., Albany, California.

Garibaldi, J. A. (1968). Acetic acid as a means of lowering the heat resistance of *Salmonella* in yolk products. *Food Technol.* **22**, 1031–1033.

Garibaldi, J. A. (1970). Role of microbial iron transport compounds in the bacterial spoilage of eggs. *Appl. Microbiol.* **20**, 558–560.

Garibaldi, J. A., and Bayne, H. G. (1960). The effect of iron on the *Pseudomonas* spoilage of experimentally infected shell eggs. *Poult. Sci.* **39**, 1517–1520.

Garibaldi, J. A., and Bayne, H. G. (1962). The effect of iron on the *Pseudomonas* spoilage of farm washed eggs. *Poult. Sci.* **41**, 850–853.

Garibaldi, J. A., and Stokes, J. L. (1958). Protective role of shell membranes in bacterial spoilage of eggs. *Food Res.* **23**, 283–290.

Garibaldi, J. A., Ijichi, K., and Bayne, H. G. (1969a). Effect of pH and chelating agents on the heat resistance and viability of *Salmonella typhimurium* Tm-1 and *Salmonella senftenberg* 775W in egg white. *Appl. Microbiol.* **18**, 318–322.

Garibaldi, J. A., Lineweaver, H., and Ijichi, K. (1969b). Number of salmonellae in commercially broken eggs before pasteurization. *Poult. Sci.* **17**, 491–496.

Garibaldi, J. A., Straka, R. P., and Ijichi, K. (1969c). Heat resistance of *Salmonella* in various egg products. *Appl. Microbiol.* **17**, 491–496.

Gayler, G. E., MacCready, R. A., Reardon, J. P., and McKernan, B. F. (1955). An outbreak of salmonellosis traced to watermelon. *Public Health Rep.* **70**, 311–313.

Gedek, B. (1973). Futtermittelverderb durch Bakterien und Pilze und seine nachteiligen Folgen. *Uebers. Tierernaehrg.* **1**, 45–46.

Geldreich, E. E. (1966). "Sanitary Significance of Fecal Coliforms in the Environment," Publ. WP–20–3. U.S. Dep. Inter., Cincinnati Water Res. Lab. (Robert A. Taft Sanit. Eng. Cent.), Cincinnati, Ohio.

Geldreich, E. E., and Bordner, R. H. (1971). Fecal contamination of fruits and vegetables—A review. *J. Milk Food Technol.* **34**, 184–195.

Gerhardt, U. (1969). Entkeimung von Gewürzen. *Gordian* **9**, 427–439.

Gibbons, N. E., and Moore, R. L. (1944). Dried whole egg powder. XII. The effect of drying, storage, and cooking on the *Salmonella* content. *Can. J. Res., Sect. F* **22**, 58–63.

Gibbons, N. E., Moore, R. L., and Fulton, C. O. (1944). Dried whole egg powder. XV. The growth of *Salmonella* and other organisms in liquid and reconstituted egg. *Can. J. Res., Sect. F* **22**, 169–173.

Gibbons, N. E., Fulton, C. O., and Reid, M. (1946). Dried whole egg powder. XXI. Pasteurization of liquid egg and its effect on quality of the powder. *Can. J. Res., Sect. F* **24**, 327–337.

Gibbs, P. A. (1973). The detection of *Clostridium welchii* in the differential reinforced clostridial medium technique. *J. Appl. Bacteriol.* **36**, 23–33.

Gibbs, P. A., Patterson, J. T., and Harvey, J. (1978a). Biochemical characteristics and enterotoxigenicity of *Staphylococcus aureus* strains isolated from poultry. *J. Appl. Bacteriol.* **44**, 57–74.

Gibbs, P. A., Patterson, J. T., and Thompson, J. K. (1978b). The distribution of *Staphylococcus aureus* in a poultry processing plant. *J. Appl. Bacteriol.* **44**, 401–410.

Gibson, B. (1973). The effect of high sugar concentrations on the heat resistance of vegetative micro-organisms. *J. Appl. Bacteriol.* **36**, 365–376.

Gilbert, G. L., Gambill, V. M., Spiner, D. R., Hoffman, R. K., and Phillips, C. R. (1964). Effect of moisture on ethylene oxide sterilization. *Appl. Microbiol.* **12**, 496–503.

Gilbert, R. J., and Taylor, A. J. (1976). *Bacillus cereus* food poisoning. *In* "Microbiology in Agriculture, Fisheries and Food" (F. A. Skinner and J. G. Carr, eds.), pp. 197–213. Academic Press, New York.

Gilbert, R. J., and Watson, H. M. (1971). Some laboratory experiments on various meat preparation surfaces with regard to surface contamination and cleaning. *J. Food Technol.* **6**, 163–170.

Gilbert, R. J., Stringer, M. F., and Peace, T. C. (1974). The survival and growth of *Bacillus cereus* in boiled and fried rice in relation to outbreaks of food poisoning. *J. Hyg.* **73**, 433–444.

Gilcreas, F. W., and Coleman, M. B. (1941). Studies of rebaking cream-filled pastries. *Am. J. Public Health* **31**, 956–958.

Gill, C. O. (1976). Substrate limitation of bacterial growth at meat surfaces. *J. Appl. Bacteriol.* **41**, 401–410.

Gill, C. O., and Newton, K. G. (1977). The development of aerobic spoilage flora on meat stored at chill temperatures. *J. Appl. Bacteriol.* **43**, 189–195.

Gill, C. O., Penney, N., and Nottingham, P. M. (1976). Effect of delayed evisceration on the microbial quality of meat. *Appl. Environ. Microbiol.* **31**, 465–468.

Gillespie, J. M., and Scott, W. J. (1950). Studies in the preservation of shell eggs. IV. Experiments in the mode of infection by bacteria. *Aust. J. Appl. Sci.* **1**, 514–530.

Gillespie, N. C., and Macrae, I. C. (1975). The bacterial flora of some Queensland fish and its ability to cause spoilage. *J. Appl. Bacteriol.* **39**, 91–100.

Gilliland, S. E., and Speck, M. L. (1977). Antagonistic action of *Lactobacillus acidophilus* toward intestinal and foodborne pathogens in associative cultures. *J. Food Prot.* **40**, 820–823.

Giolitti, G., Cantoni, C. A., Bianchi, M. A., and Renon, P. (1971). Microbiology and chemical change in raw hams of Italian type. *J. Appl. Bacteriol.* **34**, 51–61.

Girgis, A. N., El-Sherif, S., Rofael, N., and Nesheim, S. (1977). Aflatoxins in Egyptian foodstuffs. *J. Off. Anal. Chem.* **60**, 746–747.

Glass L. (1977). Botulism—Michigan. *Morbid. Mortal. Wkly. Rep.* **26**(14), 117.

Glezen, W. P., Hines, M. P., Kerbaugh, M., Green, M. E., and Koomen, J., Jr. (1966). *Salmonella* in two poultry processing plants. *J. Am. Vet. Med. Assoc.* **148**, 550–555.

Godkin, W. J., and Cathcart, W. H. (1953). The complementary action of subtilin and terramycin in preserving custard fillings. *Food Technol.* **7**, 282–285.

Goel, M. C., Gaddi, B. L., March, E. H., Stuiber, D. A., Lund, D. B., Lindsay, R. C., and Brickbauer, E. (1970). Microbiology of raw and processed wild rice. *J. Milk Food Technol.* **33**, 571–574.

Goel, M. C., Kulshrestha, D. C., Marth, E. H., Francis, D. W., Bradshaw, J. G., and Read, R. B., Jr. (1971). Fate of coliforms in yoghurt, buttermilk, sour cream and cottage cheese during refrigerated storage. *J. Milk Food Technol.* **34**, 54–58.

Goel, M. C., Marth, E. H., Stuiber, D. A., Lund, D. B., and Lindsay, R. C. (1972). Changes in the microflora of wild rice during curing by fermentation. *J. Milk Food Technol.* **35**, 385–391.

Goepfert, J. M. (1976). The aerobic plate count, coliform and *Escherichia coli* content of raw ground beef at the retail level. *J. Milk Food Technol.* **39**, 175–178.

Goepfert, J. M., and Biggie, R. A. (1968). Heat resistance of *Salmonella typhimurium* and *Salmonella senftenberg* 775W in milk chocolate. *Appl. Microbiol.* **16**, 1939–1940.

Goepfert, J. M., and Foster, E. M. (1968). Salmonellae in confections. *Proc. Salmonella Foods Conf., Pa. State Univ.* pp. 145–150.

Goepfert, J. M., and Hicks, R. (1969). Effect of volatile fatty acids on *Salmonella typhimurium. J. Bacteriol.* **97**, 956–958.

Goepfert, J. M., Olson, N. F., and Marth, E. H. (1968). Behavior of *Salmonella typhimurium* during manufacture and curing of Cheddar cheese. *Appl. Microbiol.* **16**, 862–866.

Goepfert, J. M., Iskander, I. K., and Amundson, C. H. (1970). Relation of the heat resistance of salmonellae to the water activity of the environment. *Appl. Microbiol.* **19**, 429–433.

Goepfert, J. M., Spira, W. M., and Kim, H. U. (1972). *Bacillus cereus:* Food poisoning organism. A review. *J. Milk Food Technol.* **35**, 213–227.

Goldblatt, L. A. (1969). "Aflatoxin: Scientific Background, Control, and Implications." Academic Press, New York.

Goldblith, S. A. (1971). A condensed history of the science and technology of thermal processing—Part 1 *Food Technol.* **25**, 1256–1262.

Goldblith, S. A. (1972). A condensed history of the science and technology of thermal processing—Part 2. *Food Technol.* **26**(1), 64–69.

Goldblith, S. A. (1976). Thermal processing in retrospect and prospect. *Food Technol.* **30**(6), 32–33, 46.

Golumbic, C. (1965). Fungal spoilage in stored food crops. *In* "Mycotoxins in Foodstuffs" (G. N. Wogan, ed.), pp. 49–67. MIT Press, Cambridge, Massachusetts.

Goo, V. Y. L., Ching, G. Q. L., and Gooch, J. M. (1973). Comparison of brilliant green agar and hektoen enteric media in the isolation of salmonellae from food products. *Appl. Microbiol.* **26**, 288–292.

Goodfellow, S. J., and Brown, W. L. (1978). Fate of *Salmonella* inoculated into beef for cooking. *J. Food Prot.* **41**, 598–605.

Goodliffe, J. P., and Heale, J. B. (1975). Incipient infections caused by *Botrytis cinerea* in carrots entering storage. *Ann. Appl. Biol.* **80**, 243–246.

Goresline, H. E. (1963). A discussion of the microbiology of various dehydrated foods. *In* "Microbiological Quality of Foods" (L. W. Slanetz, C. O. Chichester, A. R. Gaufin, and Z. J. Ordal, eds.), pp. 179–192. Academic Press, New York.

Goresline, H. E., Moser, R. E., and Hayes, K. M. (1950). A pilot scale study of shell egg thermostabilization. *Food Technol.* **4**, 426–430.

Goresline, H. E., Hayes, K. M., Moser, R. E., Howe, M. E., and Drewniak, E. E. (1951). Pasteurization of liquid egg under commercial conditions to eliminate *Salmonella. U.S. Dep. Agric., Circ.* No. 897.

Goto, A., Yamazaki, K., and Oka, M. (1971). Bacteriology of radiation sterilization of spices. *Food Irradiat.* **6**(1), 35–42.

Gottschalk, H. M. (1977). A review on spices. *In* "Food Irradiation Information" (W. T. Potter, P. S. Elias, and H. M. Gottschalk, eds.), Int. Proj. Field Food Irradiat., No. 7, pp. 7–30. Inst. Strahlentechnol., Karlsruhe.

Graikoski, J. T. (1973). Microbiology of cured and fermented fish. *In* "Microbial Safety of Fishery Products" (C. O. Chichester and H. D. Graham, eds.), pp. 97–102. Academic Press, New York.

Grau, F. H., and Smith, M. G. (1974). *Salmonella* contamination of sheep and mutton carcasses related to pre-slaughter holding conditions. *J. Appl. Bacteriol.* **37**, 111–116.

Grau, F. H., Brownlie, L. E., and Smith, M. G. (1969). Effects of food intake on numbers of salmonellae and *Escherichia coli* in rumen and faeces of sheep. *J. Appl. Bacteriol.* **32**, 112–117.

Graves, R. R., and Hesseltine, C. W. (1966). Fungi in flour and refrigerated dough products. *Mycopathol. Mycol. Appl.* **29**, 277–290.

Graves, R. R., Rogers, R. F., Lyons, A. J., Jr., and Hesseltine, C. W. (1967). Bacterial and actinomycete flora of Kansas–Nebraska and Pacific Northwest wheat and wheat flour. *Cereal Chem.* **44**, 288–299.

Gray, D. F., Harley, O. C., and Noble, J. L. (1960). The ecology and control of *Salmonella* contamination in bonemeal. *Aust. Vet. J.* **36**, 246–252.

Gray, J. T. (1887). Process of preserving eggs. U.S. Patent 358,656.

Greenberg, R. A. (1972). Nitrite in the control of *Clostridium botulinum*. *Proc. Meat Ind. Res. Conf., Am. Meat Inst. Found., Chicago, Ill.* pp. 25–34.

Grossklaus, D., and Levetzow, R. (1967). Die Kühlung des Schlachtgeflügels—ein hygienisches und lebensmittelrechtliches Problem. *Berl. Muench. Tieraetztl. Wochenschr.* **78**, 187–190.

Grosskopf, J. C., and Harper, W. J. (1969). Role of psychrotrophic sporeformers in long life milk. *J. Dairy Sci.* **52**, 897.

Grosskopf, J. C., and Harper, W. J. (1974). Isolation and identification of psychrotrophic sporeformers in milk. *Milchwissenshaft* **29**, 467–470.

Grünewald, T., and Münzer, R. (1972). Strahlenbehandlung von Kakaopulver. *Lebensm.-Wiss. Technol.* **5**(6), 203–206.

Guarino, P. A. (1972). Microbiology of spices, herbs and related materials. *Proc. Annu. Symp. Fungi Foods, 7th, West. N.Y. Sect., Inst. Food Technol., Rochester, N.Y.* pp. 16–18.

Guarino, P. A., and Peppler, H. J. (1976). Spices and condiments. *In* "Compendium of Methods for the Microbiological Examination of Foods" (M. L. Speck, ed.), pp. 568–573. Am. Public Health Assoc., Washington, D.C.

Guenther, E. (1948). "The Essential Oils," Vol. 1. Van Nostrand, New York.

Guenther, E. (1949). "The Essential Oils," Vol. 2. Van Nostrand, New York.

Guenther, E. (1950). "The Essential Oils," Vol. 4. Van Nostrand, New York.

Guerin, B., Guerin, M.-S., and Loilier, M. (1972). Emploi en sucrerie d'un nouvel inhibiteur de développements microbiens. *Sucr. Fr.* **113**, 203–211.

Gunderson, M. F., and Rose, K. D. (1948). Survival of bacteria in a precooked fresh-frozen food. *Food Res.* **13**, 254–263.

Hadlock, R. (1969). Schimmelpilzkontamination von Fleischerzeugnissen durch naturbelassene Gewürze. *Fleischwirtschaft* **49**, 1601–1609.

Hadlock, R. (1970). Aflatoxine bei Fleischprodukten und Untersuchungen über die Häufigkeit der Aflatoxinbildung durch *Aspergillus flavus*—Stämme. *Fleischwirtschaft* **50**, 1499–1502.

Haines, R. B. (1937). Microbiology in the preservation of animal tissues. *G.B. Dep. Sci. Ind. Res., Food Invest. Board, Spec. Rep.* **45**.

Haines, R. B. (1938). Observations on the bacterial flora of the hen's egg, with a description of new species of *Proteus* and *Pseudomonas* causing rots in eggs. *J. Hyg.* **38**, 338–355.

Haines, R. B. (1939). Microbiology in the preservation of the hen's egg. *G.B. Dep. Sci. Ind. Res. Food Invest. Board, Spec. Rep.* **47**.

Haines, R. B., and Elliott, E. M. (1944). Some bacteriological aspects of dehydrated foods. *J. Hyg.* **43,** 370.

Haines, R. B., and Moran, T. (1940). Porosity of, and bacterial invasion through the shell of the hen's egg. *J. Hyg.* **40,** 453–461.

Haines, R. B., Elliott, E. M. L., and Tomlinson, A. J. H. (1947). The bacteriology of dried egg. *Med. Res. Counc. (G.B.), Spec. Rep. Ser.,* No. 260, 8 pp.

Hall, C. W., and Hedrick, T. I. (1966). "Drying Milk Products." AVI, Westport, Connecticut.

Hall, H. E. (1971). The significance of *Escherichia coli* associated with nut meats. *Food Technol.* **25,** 230–232.

Hall, H. E., and Angelotti, R. (1965). *Clostridium perfringens* in meat and meat products. *Appl. Microbiol.* **13,** 352–357.

Hall, H. E., Brown, D. F., and Read, R. B., Jr. (1971). Effect of pasteurization on the direct microscopic count of eggs. *J. Milk Food Technol.* **34,** 209–211.

Hamblett, F. E., Hill, W. F., Jr., Akin, E. W., and Benton, W. H. (1969). Effect of turbidity on the rate of accumulation and elimination of Poliovirus type 1 by the Eastern Oyster (*Crassostrea virginica*). *Proc. Gulf Atl. Shellfish Sanit. Conf.* Public Health Serv., U.S. Dep. Health, Educ. Welfare, Washington, D.C.

Hamdy, M. K., Barton, N. D., and Brown, W. E. (1964). Source and portal of entry of bacteria found in bruised poultry tissue. *Appl. Microbiol.* **12,** 464–469.

Hammer, B. W., and Babel, F. J. (1957). "Dairy Bacteriology," 4th ed. Wiley, New York.

Hammer, B. W., and Hix, R. H. (1916). Studies on the numbers of bacteria present in milk which has undergone various changes. *Iowa Agric. Exp. Stn., Res. Bull.* No. 29.

Hammer, B. W., and Hussong, R. V. (1930). Bacteriology of butter, I: Influence of the distribution of the non-fatty constituents on the changes in bacterial content during holding. *Iowa Agric. Exp. Stn., Res. Bull.* No. 134.

Hankin, L., Dillman, W. F., and Stephens, G. R. (1977). Keeping quality of pasteurized milk for retail sale related to code date, storage temperature and microbial counts. *J. Food Prot.* **40,** 848–853.

Hanna, M. O., Zink, D. L., Carpenter, Z. L., and Vanderzant, C. (1976). *Yersinia enterocolitica*-like organisms from vacuum-packaged beef and lamb. *J. Food Sci.* **41,** 1254–1256.

Hansen, A. P. (1975). Understanding the microbiological deterioration of cacao. *Candy Snack Ind.* **140**(10), 44, 46, 47.

Hansen, A. P., and Keeney, P. G. (1970). Comparison of carbonyl compounds in moldy and nonmoldy cocoa beans. *J. Food Sci.* **35,** 37–40.

Hansen, A. P., and Welty, R. E. (1970). Microflora of raw cacao beans. *Mycopathol. Mycol. Appl.* **44,** 309–316.

Hardman, R. (1972). Spices and herbs: their families, secretory tissues and pharmaceutical aspects. *Proc. Conf. Spices* pp. 23–35. Trop. Prod. Inst., London.

Harein, P. K. (1969). The granary weevil as a carrier of fungi and pathogenic bacteria. *Proc. Symp. Biol. Contam. Grain Anim. Byprod.* pp. 89–91. Agric. Ext. Serv., Dep. Entomol., Fish., Wildl., Univ. of Minnesota, Minneapolis.

Harper, W. J., and Hall, C. W. (1976). "Dairy Technology and Engineering." AVI, Westport, Connecticut.

Harrington, R., Jr., and Karlson, A. G. (1965). Destruction of various kinds of mycobacteria in milk by pasteurization. *Appl. Microbiol.* **13,** 494–495.

Harrison, J. M., and Lee, J. S. (1969). Microbial evaluation of Pacific Shrimp processing. *Appl. Microbiol.* **18**, 188–192.

Harry, E. G. (1963). The relationship between egg spoilage and the environment of the egg when laid. *Br. Poult. Sci.* **4**, 91–100.

Hartung, T. E., and Stadelman, W. J. (1963). *Pseudomonas fluorescens* penetration of egg shell membranes as influenced by shell porosity, age of egg, and degree of bacterial challenge. *Poult. Sci.* **42**, 147–150.

Hawthorne, J. R. (1950). Dried albumen: Removal of sugar by yeast before drying. *J. Sci. Food Agric.* **1**, 199–201.

Hawthorne, J. R., and Brooks, J. (1944). Dried egg. VIII. Removal of the sugar of egg pulp before drying. A method of improving the storage life of spray-dried whole egg. *J. Soc. Chem. Ind.* **63**, 232–234.

Health and Welfare Canada (1976, 1978). "Food-borne Disease in Canada: Annual Summary 1973, 1974." Health Prot. Branch, Ottawa.

Health and Welfare, Canada (1979). Staphylococcal intoxication from Swiss type cheese—Quebec and Ontario. *Can. Dis. Wkly. Rep.* **5**, 110–112.

Heath, H. B. (1964). Bacteria-free spices in food processing. *Food Process. Packag.* April, 144–147.

Heath, H. B. (1967). The effects of microbiological condition of minor ingredients in foods. The effect of spices. *Soc. Chem. Ind., London.* Unpublished.

Heath, J. L., and Wallace, J. (1978). Dilute acid immersion as a method of cleaning shell eggs. *Poult. Sci.* **57**, 149–155.

Heins, H. G., and Ulmann, R. M. (1971). The decontamination of spice by ionizing rays. *Vleesdistrib. Vleestechnol.* **6**(10), 15–17. (In Dutch.)

Heller, C. L., Roberts, B. C., Amos, A. J., Smith, M. E., and Hobbs, B. C. (1962). The pasteurization of liquid whole egg and the evaluation of the baking properties of frozen whole egg. *J. Hyg.* **60**, 135–143.

Henderson, S. F. (1916). Art of preserving and sterilizing eggs. U.S. Patent 1,177,105.

Henderson, S. M., and Lorenz, F. W. (1951). Cooling and holding eggs on the ranch. *Calif. Agric. Exp. Stn., Circ.* No. 405. Univ. of California, Davis.

Hendricks, S. L., Belknap, R. A., and Hausler, W. J., Jr. (1959). Staphylococcal food intoxication due to Cheddar cheese. I. Epidemiology. *J. Milk Food Technol.* **22**, 313–317.

Hess, G. W., Moulthrop, J. I., and Norton, H. R., II (1970). New decontamination efforts and techniques for elimination of *Salmonella* from animal protein rendering plants. *J. Am. Vet. Med. Assoc.* **157**, 1975–1980.

Hesseltine, C. W. (1967). Aflatoxins and other mycotoxins. *Health Lab. Sci.* **4**, 222–228.

Hesseltine, C. W. (1968). Flour and wheat: Research on their microbiological flora. *Bakers Dig.* **42**(3), 40–42, 66.

Hesseltine, C. W. (1976). Conditions leading to mycotoxin contamination of foods and feeds. *In* "Mycotoxins and other Fungal Related Food Problems" (J. V. Rodricks, ed.), Adv. Chem. Ser., No. 149, pp. 1–22. Am. Chem. Soc., Washington, D.C.

Hesseltine, C. W., and Graves, R. R. (1966). Microbiology of flours. *Econ. Bot.* **20**, 156–168.

Hesseltine, C. W., Graves, R. R., Rogers, R., and Burmeister, H. R. (1969). Aerobic and facultative microflora of fresh and spoiled refrigerated dough products. *Appl. Microbiol.* **18**, 848–853.

Hettinga, D. H., and Reinbold, G. W. (1975). Split defect of Swiss cheese. II. Effect of low temperatures on the metabolic activity of *Propionibacterium*. *J. Milk Food Technol.* **38**, 31–35.

Hettinga, D. H., Reinbold, G. W., and Vedamuthu, E. R. (1974). Split defect of Swiss cheese. I. Effect of strain of *Propionibacterium* and wrapping material. *J. Milk Food Technol.* **37**, 322–328.

Higginbottom, C. (1953). The effect of storage at different relative humidities on the survival of microorganisms in milk powder and in pure cultures dried in milk. *J. Dairy Res.* **20**, 65.

Hilker, J. S. (1976). Confectionary products. *In* "Compendium of Methods for the Microbiological Examination of Foods" (M. L. Speck, ed.), pp. 608–613. Am. Public Health Assoc., Washington, D.C.

Hill, G. A. (1925). *Clostridium multifermentans* in chocolate cream candies. *J. Bacteriol.* **10**, 413–420.

Hobbs, B. C. (1971). Food poisoning from poultry. *In* "Poultry Diseases and World Economy" (R. F. Gordon and B. M. Freeman, eds.), pp. 65–80. Br. Poult. Sci., Edinburgh.

Hobbs, B. C. (1972). Food poisoning in England and Wales. *In* "The Microbiological Safety of Food" (B. C. Hobbs and J. H. B. Christian, eds.), pp. 129–142. Academic Press, New York.

Hobbs, B. C. (1974). Microbiological hazards of meat production. *Food Manuf.* **49**(10), 29–34, 54.

Hobbs, B. C., and Gilbert, R. J. (1978). The vehicle of infection. *In* "Food Poisoning and Food Hygiene," 4th ed., pp. 51–76. Arnold, London.

Hobbs, B. C., and Smith, M. E. (1954). The control of infection spread by synthetic cream. *J. Hyg.* **52**, 230–246.

Hobbs, B. C., and Wilson, J. G. (1959). Contamination of wholesale meat supplies with salmonellae and heat resistant *Clostridium welchii*. *Mon. Bull. Minist. Health Public Health Lab. Serv.* **18**, 198–206.

Hobbs, W. E. (1968). Salmonellae in cereals and grain products. *Salmonellae Foods Conf., Pa. State Univ.* pp. 151–159. University Park, Pennsylvania.

Hobbs, W. E., and Greene, V. W. (1976). Cereal and cereal products. *In* "Compendium of Methods for the Microbiological Examination of Foods" (M. L. Speck, ed.), pp. 599–607. Am. Public Health Assoc., Washington, D.C.

Hodge, B. E. (1960). Control of staphylococcal food poisoning. *Public Health Rep.* **75**, 355–361.

Hodges, T. O., Converse, H. H., and Sauer, D. B. (1971). Some effects of cooling rates on quality of high moisture corn. *Trans. ASAE* **14**, 649–655.

Hoecker, W. H., and Hammer, B. W. (1945). Bacteriology of butter. IX: Salt distribution in butter and its effect on bacterial growth. *Iowa Agric. Exp. Stn., Res. Bull.* No. 339.

Holden, W. S. (1970). "Water Treatment and Examination," 8th ed. Williams & Wilkins, Baltimore, Maryland.

Holtzapffel, D., and Mossel, D. A. A. (1968). The survival of pathogenic material in, and the microbial spoilage of, salads containing meat, fish and vegetables. *J. Food Technol.* **3**, 223–239.

Honda, T., Taga, S., Takeda, T., Hasibuan, M. A., Takeda, Y., and Miwatani, T. (1976a). Identification of lethal toxin with the thermostable direct haemolysin produced by *Vibrio parahaemolyticus*, and some physiochemical properties of the purified toxin. *Infect. Immun.* **13**, 133–139.

Honda, T., Goshima, K., Takeda, Y., Sugino, Y., and Miwatani, T. (1976b). Demonstration of the cardiotoxicity of the thermostable direct hemolysin (lethal toxin) produced by *Vibrio parahaemolyticus*. *Infect. Immun.* **13**, 163–171.

Hoover, W. J. (1974). Status of the milling and baking industries in Africa and Asia. *Cereal Sci. Today* **19**(4), 153–156.

Horie, Y., Yamazaki, M., Miyaki, K., and Udagawa, S. (1971). On the fungal contents of spices. *Shokuhin Eiseigaku Zasshi* **12**, 516–519.

Horwitz, W., ed. (1975). Thermophilic bacterial spores in sugars (9)—Official first action. *In* "Official Methods of Analysis of the Association of Official Analytical Chemists," pp. 920–921. A.O.A.C., Washington, D.C.

Houghtby, G., and Liston, J. (1965). Effect of heating on *Staphylococcus aureus* in frozen precooked seafoods. *Food Technol.* **19**, 874–877.

Houser, L. (1965). "National Shellfish Sanitation Program Manual of Operations: Part 1. Sanitation of Shellfish Growing Areas. U.S. Public Health Serv., Washington, D.C.

Howat, G. R. (1968). Chocolate manufacture. *Process Biochem.* **3**(10), 31–35.

Howell, G. L. H. (1948). Notes on the factors governing quality in cocoa and the fermentation of different qualities of cocoa. *Proc. Colon. Microbiol. Res. Inst., Off. Open., Trinidad* pp. 57–61.

Howie, J. W. (1968). Typhoid in Aberdeen, 1964. *J. Appl. Bacteriol.* **31**, 171–178.

Hoy, W. A., and Rowlands, A. (1948). Wet storage methods for the treatment of milking machine clusters. *Proc. Soc. Appl. Microbiol.* pp. 40–52.

Hruby, S., Maresova, P., and Pelech, L. (1960). Sanitary significance of liquid extracts of spices for the meat industry. *Cesk. Hyg.* **5**, 436–440. (In Czech.)

Hruby, S., Maresova, P., and Tvrznik, D. (1961). Sterilisation of spices by means of ethylene oxide. *Cesk. Hyg.* **6**, 310–314. (In Czech.)

Hsieh, F., Acott, K., and Labuza, T. P. (1976a). Death kinetics of pathogens in a pasta product. *J. Food Sci.* **41**, 516–519.

Hsieh, F., Acott, K., and Labuza, T. P. (1976b). Prediction of microbial death during drying of a macaroni product. *J. Milk Food Technol.* **39**, 619–623.

Huang, L. H., and Hanlin, R. T. (1975). Fungi occurring in freshly harvested and in-market pecans. *Mycologia* **67**, 689–700.

Hucker, G. J., and Pederson, C. S. (1942). A review of the microbiology of commercial sugar and related sweetening agents. *Food Res.* **7**, 459–480.

Hughes, J. R., Kent, J., Mann, S., Blum, E., Midura, T. F., and Werner, S. B. (1977). Type A botulism associated with commercial pot pie—California. *Morbid. Mortal. Wkly. Rep.* **26**, 186, 192.

Huhtanen, C. N., Naghski, J., Custer, C. S., and Russel, R. W. (1976). Growth and toxin production by *Clostridium botulinum* in moldy tomato juice. *Appl. Environ. Microbiol.* **32**, 711–715.

Hurst, A. (1972). Interaction of food starter cultures and food-borne pathogens: The antagonism between *Streptococcus lactis* and sporeforming microbes. *J. Milk Food Technol.* **35**, 418–423.

Huszka, T., Csefalvy, I., and Incze, K. (1973). Sterilization of powdered paprika by means of hydrochloric acid (in Hung.). *Konzerv. Paprikaip.* No. 6, 213–217 [*Food Sci. Technol. Abstr.* **7**(2) 2J246 (1975)].

Hyndman, J. B. (1963). Comparison of enterococci and coliform microorganisms in commercially produced pecan nut meats. *Appl. Microbiol.* **11**, 268–272.

I.C.P. Cocoa (1977). "Technical Data Sheet SC–COT 5590." I.C.P. Cocoa, Camden, New Jersey.

Iizuka, H. (1957). Studies on the microorganisms found in Thai rice and Burma rice. On the microflora of Thai rice. *J. Gen. Appl. Microbiol.* **3**(2), 146–161.

Iizuka, H. (1958). Studies on the microorganisms found in Thai rice and Burma rice. Part II. On the microflora of Burma rice. *J. Gen. Appl. Microbiol.* **4**(2), 108–119.

Iizuka, H., and Ito, H., (1968). Effect of Gamma-irradiation on the microflora of rice. *Cereal Chem.* **45**, 503–511.

Ijichi, K., Garibaldi, J. A., Kaufman, V. F., Hudson, C. A., and Lineweaver, H. (1973). Microbiology of a modified procedure for cooling pasteurized salt yolk. *J. Food Sci.* **38**, 1241–1243.

Imai, C. (1976). Some characteristics of psychrophilic bacteria isolated from green rotten eggs. *Poult. Sci.* **55**, 606–610.

Inal, T., Keskin, S., Tolgay, Z., and Tezcan, I. (1975). Gewürzsterilisation durch Anwendung von Gamma-Strahlen. *Fleischwirtschaft* **55**, 675–677.

Ingraham, J. L. (1958). Growth of psychrophilic bacteria. *J. Bacteriol.* **76**, 75–80.

Ingram, M. (1962). Microbiological principles in prepacking meats. *J. Appl. Bacteriol.* **25**, 259–281.

Ingram, M. (1966). Psychrophilic and psychrotrophic microorganisms. *Ann. Inst. Pasteur, Lille* **15**, 111–118.

Ingram, M. (1972a). Meat preservation—past, present and future. *R. Soc. Health J.* **92**, 121–131.

Ingram, M. (1972b). Meat chilling—the first reason why. *Proc. Symp. Meat Chill.— Why, How, 2nd* pp. 1.1–1.13. Agric. Res. Counc., Meat Res. Inst., Langford, Bristol, England.

Ingram, M., and Dainty, R. H. (1971). Changes caused by microbes in spoilage of meats. *J. Appl. Bacteriol.* **34**, 21–39.

Ingram, M., and Handford, P. M. (1957). The influence of moisture and temperature on the destruction of *Cl. botulinum* in acid bread. *J. Appl. Bacteriol.* **20**, 442–453.

Ingram, M., and Lüthi, H. (1961). Microbiology of fruit juices. *In* "Fruit and Vegetable Juice, Process Technology" (D. K. Tressler and M. A. Joslyn, eds.), pp. 117–163. AVI, Westport, Connecticut.

Ingram, M., and Roberts, T. A. (1976). The microbiology of the red meat carcass and the slaughterhouse. *R. Soc. Health J.* **96**, 270–276.

International Association of Milk, Food, and Environmental Sanitarians (IAMFES) (1976a). E-3-A sanitary standards for liquid egg products cooling and holding tanks. No. E-1300. *J. Milk Food Technol.* **39**, 568–575.

International Association of Milk, Food, and Environmental Sanitarians (IAMFES) (1976b). E-3-A sanitary standards for fillers and sealers of single service containers for liquid egg products. No. E-1700. *J. Milk Food Technol.* **39**, 576–579.

International Association of Milk, Food, and Environmental Sanitarians (IAMFES) (1976c). E-3-A sanitary standards for egg breaking and separating machines. No. E-0600. *J. Milk Food Technol.* **39**, 651–653.

International Association of Milk, Food, and Environmental Sanitarians (IAMFES) (1976d). E-3-A sanitary standards for shell egg washers. *J. Milk Food Technol.* **39**, 654–656.

International Atomic Energy Agency (IAEA) (1968). "Elimination of Harmful Organisms from Food and Feed by Irradiation," STI/PUB/200. IAEA, Vienna.

International Atomic Energy Agency (IAEA) (1973). "Factors Influencing the Economical Application of Food Irradiation," STI/PUB/331. IAEA, Vienna.

International Commission for Uniform Methods of Sugar Analysis (1954). Subject 24A. Determination of factors causing deterioration of raw sugars in storage. *Rep. Proc. 11th Sess.* pp. 95–101. Int. Comm. Uniform Methods Sugar Anal., Paris.

International Commission for Uniform Methods of Sugar Analysis (1966). Subject. 21. Microbiological tests. *Rep. Proc. 14th Sess.* pp. 120–122. Int. Comm. Uniform Methods Sugar Anal., Copenhagen.

International Commission on Microbiological Specifications for Foods (ICMSF) (1974). "Microorganisms in Foods. 2. Sampling for Microbiological Analysis: Principles and Specific Applications." Univ. of Toronto Press, Toronto.

International Commission on Microbiological Specifications for Foods (ICMSF) (1978). "Microorganisms in Foods. 1. Their Significance and Methods of Enumeration," 2nd ed. Univ. of Toronto Press, Toronto.

International Dairy Federation (1975). *Proc. Semin. Mastitis Control* Int. Dairy Fed., Brussels.

Iotov, I., Beev, K., and Trendafilova, Z. (1974). Comparative quality evaluation of oiled and nonoiled eggs during storage. *Khranit. Prom.* **23**, 22–23.

Iszard, M. S. (1927a). The value of lactic acid in the preservation of mayonnaise dressing and other dressings. *Cann. Age* **8**, 434–436.

Iszard, M. S. (1927b). The value of lactic acid in the preservation of mayonnaise dressing and other products. *J. Bacteriol.* **13**, 57–58.

Iszard, M. S. (1927c). Supplementary report on the use of lactic acid as a preservative in mayonnaise and allied products. *Spice Mill* **50**, 2426–2430.

Ito, H., Shibabe, S., and Iizuka, H. (1971). Effect of storage studies of microorganisms on Gamma-irradiated rice. *Cereal Chem.* **48**, 140–149.

Ito, H., Iizuka, H., and Sato, T. (1973). Identification of osmophilic *Aspergillus* isolated from rice and their radiosensitivity. *Agric. Biol. Chem.* **37**, 789–798.

Ito, K. A., Chen, J. K., Lerke, P. A., Seeger, M. L., and Unverferth, J. A. (1976). Effect of acid and salt concentration in fresh-pack pickles on the growth of *Clostridium botulinum* spores. *Appl. Environ. Microbiol.* **32**, 121–124.

Jackson, H. W., and Morgan, M. E. (1954). Identity and origin of the malty aroma substance from milk cultures of *Streptococcus lactis* var. *maltigenes. J. Dairy Sci.* **37**, 1316–1324.

Jacobs, J., Guinée, P. A. M., Kampelmacher, E. H., and van Keulen, A. (1963). Studies on the incidence of Salmonella in imported fish meal. *Zentralbl. Veterinaermed., Reihe B* **10**, 542–550.

Jacobs, M. B. (1959). "Manufacture and Analysis of Carbonated Beverages." Chem. Publ. Co., New York.

Jacobsen, N. P. (1910). Procedure for preservation of eggs. Dan. Patent 14,902.

Jacquot, R. (1961). Organic constituents of fish and other aquatic animal foods. *In* "Fish as Food" (G. Borgstrom, ed.), Vol. 1, pp. 145–209. Academic Press, New York.

Jakobsen, M., and Jensen, H. C. (1975). Combined effect of water activity and pH on the growth of butyric anaerobes in canned pears. *Lebensm.-Wiss. Technol.* **8**, 158–160.

James, S. J., and Rhodes, D. N. (1978). Cooking beef joints from the frozen or thawed state. *J. Sci. Food Agric.* **29**, 187–192.

James, S. J., Creed, P. G., and Roberts, T. A. (1977). Air thawing of beef quarters. *J. Sci. Food Agric.* **28**, 1109–1119.

Janssen, H. J. L., and Mulder, R. W. A. W. (1971). "Pasteurization of Whole-egg with Sugar, Prepared from Incubator-reject Eggs," Rep. No. 8171. Spelderholt Inst. Poult. Res., Beekberger, Netherlands. (In Dutch.)

Jarvis, B., and Morisetti, M. D. (1969). The use of antibiotics in food preservation. *Int. Biodeterior. Bull.* **5,** 39–51.

Jay, J. M. (1972). Mechanism and detection of microbial spoilage in meats at low temperatures: a status report. *J. Milk Food Technol.* **35,** 467–471.

Jaye, M., Kittaka, R. S., and Ordal, Z. J. (1962). The effect of temperature and packaging material on the storage life and bacterial flora of ground beef. *Food Technol.* **16**(4), 95–98.

Jensen, J. M., and Bortree, A. L. (1946). Storage and treatment of milking machine inflations. *J. Dairy Sci.* **29,** 849–859.

Jensen, J. P., Reinbold, G. W., Washam, C. J., and Vedamuthu, E. R. (1975). Role of enterococci in Cheddar cheese: Organoleptic considerations. *J. Milk Food Technol.* **38,** 142–145.

Jensen, L. B. (1954). "Microbiology of Meats," 3rd ed. Garrard, Champaign, Illinois.

Jensen, L. B., Wood, I. H., and Jansen, C. E. (1934). Swelling in canned chopped hams. *Ind. Eng. Chem.* **26,** 1118–1120.

Jezeski, J. J., Morris, H. A., Zottola, E. A., George, E., Jr., and Busta, F. F. (1961). The effect of penicillin in milk on the growth and survival of *Staphylococcus aureus* during the manufacture and curing of Cheddar and Colby cheese. *J. Dairy Sci.* **44,** 1160.

Johns, C. K., and Berard, H. L. (1945). Further bacteriological studies relating to egg drying. *Sci. Agric.* **25,** 551–565.

Johns, C. K., and Berard, H. L. (1946). Effect of certain methods of handling upon the bacterial content of dirty eggs. *Sci. Agric.* **26,** 11–15.

Johnson, H. C., and Liston, J. (1973). Sensitivity of *Vibrio parahaemolyticus* to cold in oysters, fish fillets and crabmeat. *J. Food Sci.* **38,** 437–441.

Jones, J. L., and Weimer, J. P. (1977). "Food Safety: Homemakers' Attitudes and Practices," Agric. Econ., Rep. No. 360. Econ. Res. Serv., U.S. Dep. Agric., Washington, D.C.

Jordan, F. T. W. (1956). The transmission of *Salmonella gallinarum* through the egg. *Poult. Sci.* **35,** 1019–1025.

Jordan, M. C., Powell, K. E., Corothers, T. E., and Murray, R. J. (1973). Salmonellosis among restaurant patrons. The incisive role of a meat slicer. *Am. J. Public Health* **63,** 982–985.

Julseth, R. M., and Deibel, R. H. (1974). Microbial profile of selected spices and herbs at import. *J. Milk Food Technol.* **37,** 414–419.

Juven, B. J., Kanner, J., and Weisslowicz, H. (1978). Influence of orange juice composition on the thermal resistance of spoilage yeasts. *J. Food Sci.* **43,** 1074–1076.

Kadis, S., Ciegler, A., and Ajl, S. J., eds. (1971–1972). "Microbial Toxins," Vols. VI–VIII. Academic Press, New York.

Kadis, V. W., Hill, D. A., and Pennifold, K. S. (1971). Bacterial content of gravy bases and gravies obtained in restaurants. *Can. Inst. Food Technol. J.* **4,** 130–132.

Kajs, T. M., Hagenmaier, R., Vanderzant, C., and Mattil, K. F. (1976). Microbiological evaluation of coconut and coconut products. *J. Food Sci.* **41,** 352–356.

896 **References**

Kameswara Rao, G., Malathi, M. A., and Vijayaraghavan, P. K. (1964). Preservation and packaging of Indian foods I. Preservation of chapaties. *Food Technol.* **18**, 108–110.

Kameswara Rao, G., Malathi, M. A., and Vijayaraghavan, P. K. (1966). Preservation and packaging of Indian foods. II. Storage studies on preserved chapaties. *Food Technol.* **20**, 1070–1073.

Kampelmacher, E. H., Guinee, P. A. M., van Schothorst, M., and Willems, H. M. C. C. (1965). Experimental studies to determine the temperature and duration of heat treatment required for decontamination of feed meals. *Zentralbl. Veterinaermed., Reihe B* **12**(1), 50–54.

Kampelmacher, E. H., Ingram, M., and Mossel, D. A. A., eds. (1969). "The Microbiology of Dried Foods." Grafische Ind., Haarlem, Neth.

Kaplan, A. M., Solowey, M., Osborne, W. W., and Tubiash, H. (1950). Resting cell fermentation of egg white by streptococci. *Food Technol.* **4**, 474–477.

Kaplan, H. S., and Melnick, J. L. (1952). Effect of milk and cream on the thermal inactivation of human poliomyelitis virus. *Am. J. Public Health* **42**, 525–534.

Kaplan, H. S., and Melnick, J. L. (1954). Effect of milk and other dairy products on the thermal inactivation of Coxsackie viruses. *Am. J. Public Health* **44**, 1174–1184.

Kapooria, R. G., and Genda, J. (1974). Studies on fungal flora developing on bread at Njala, Sierra Leone. *Ghana J. Sci.* **14**(1), 55–58 [*Biol. Abstr.* **60**, 48823 (1975)].

Karlson, K. E., and Gunderson, M. F. (1965). Microbiology of dehydrated soups. II. "Adding machine" approach. *Food Technol.* **19**(1), 86–90.

Karlsson, K.-A., Rutqvist, L., and Thal, E. (1963). Salmonella isolated from animals and animal feeds in Sweden during 1958–1962. *Nord. Veterinaermed.* **15**, 833–850.

Katoh, H. (1965). Studies on the growth rate of various food bacteria. III. The growth of *Vibrio parahaemolyticus* in raw fish meat. *Nippon Saikingaku Zasshi* **20**, 541–544.

Kaufman, V. F. (1969). Detection of leaks in the regeneration section of egg pasteurizers. *J. Milk Food Technol.* **32**, 94–98.

Kaufman, V. F. (1972). Locating leaks in egg pasteurizers. *J. Milk Food Technol.* **35**, 461–463.

Kaufman, V. F., Klose, A. A., Bayne, H. G., Pool, M. F., and Lineweaver, H. (1972). Plant processing of sub-atmospheric steam scalded poultry. *Poult. Sci.* **51**, 1188–1194.

Kavanagh, T. E., Reineccius, G. A., Keeney, P. G., and Weissberger, W. (1970). Mold induced changes in cacao lipids. *J. Am. Oil Chem. Soc.* **47**, 344–346.

Kawabata, T., Ishizaka, K., and Miura, T. (1955). Studies on the allergy-like food poisoning associated with putrefaction of marine products. *Jpn. J. Med. Sci. Biol.* **8**, 521–528.

Kelly, C. A., Dempster, J. F., and McLoughlin, A. J. (1974). The effects of washing lamb carcasses. *Proc. Eur. Meet. Meat Res. Workers, 20th, Dublin.*

Kelly, N. (1967). Sugar. *In* "Fundamentals of Food Processing Operations" (J. L. Heid and M. A. Joslyn, eds.), pp. 30–61. AVI, Westport, Connecticut.

Kemp, J. D., Langlois, B. E., Fox, J. D., and Varney, W. Y. (1975). Effects of curing ingredients and holding times and temperatures on organoleptic and microbiological properties of dry-cured sliced ham. *J. Food Sci.* **40**, 634–636.

Kempton, A. G., and Bobier, S. R. (1970). Bacterial growth in refrigerated, vacuum-packed luncheon meats. *Can. J. Microbiol.* **16**, 287–297.

Kent-Jones, D. W., and Amos, A. J. (1957). "Modern Cereal Chemistry," 5th ed. Northern Publ., Liverpool.

Keoseyan, S. A. (1971). Incidence of *Clostridium perfringens* in dehydrated soup, gravy, and spaghetti mixes. *J. Assoc. Off. Anal. Chem.* **54**, 106–108.

Khan, A. W. (1964). Changes in nonprotein nitrogenous constituents of chicken breast muscle stored at below freezing temperatures. *J. Agric. Food Chem.* **12**, 378–380.

Kielstein, P., Bathke, W., and Schimmel, D. (1970). Der Einfluss der Pelletierung auf den Salmonellagehalt im Futter. *Monatsh. Veterinaermed.* **26**, 12–17.

Kilara, A., and Shahani, K. M. (1973). Removal of glucose from eggs: A review. *J. Milk Food Technol.* **36**, 509–513.

Kim, H. U., and Goepfert, J. M. (1971). Occurrence of *Bacillus cereus* in selected dry food products. *J. Milk Food Technol.* **34**, 12–15.

Kimata, M. (1961). The histamine problem. *In* "Fish as Food" (C. Borgstrom, ed.), pp. 329–352. Academic Press, New York.

King, A. D., Jr., Michener, H. D., and Ito, K. A. (1969). Control of *Byssochlamys* and related heat-resistant fungi in grape products. *Appl. Microbiol.* **18**, 166–173.

King, A. D., Jr., Miller, M. J., and Eldridge, L. C. (1970). Almond harvesting, processing, and microbial flora. *Appl. Microbiol.* **20**, 208–214.

Kinner, J. A., Kotula, A. W., and Mercuri, A. J. (1968). Microbiological examination of ingredients of Eastern-type turkey rolls. *Poult. Sci.* **47**, 1442–1447.

Kinosita, R., and Shikata, T. (1965). On toxic moldy rice. *In* "Mycotoxins in Foodstuffs" (G. N. Wogan, ed.), pp. 111–132. MIT Press, Cambridge, Massachusetts.

Kintner, T. C., and Mangel, M. (1953a). Survival of staphylococci and salmonellae experimentally inoculated into salad dressing prepared with dried eggs. *Food Res.* **18**, 6–10.

Kintner, T. C., and Mangel, M. (1953b). Survival of staphylococci and salmonellae in puddings and custards prepared with experimentally inoculated dried egg. *Food Res.* **18**, 492–496.

Kishonti, E., and Sjöström, G. (1970). Influence of heat resistant lipases and proteases in psychrotrophic bacteria on product quality. *Proc. Int. Dairy Congr., 18th* IE, p. 501.

Kiss, I., Farkas, J., Ferenczi, S., Kalman, B., and Beczner, J. (1974). Effects of irradiation on the technological and hygienic qualities of several food products. *In* "Improvement of Food Quality by Irradiation," ST1/PUB/370, pp. 158–177. IAEA, Vienna.

Kissinger, J. C., and Zaika, L. L. (1978). Effect of major spices in Lebanon bologna on acid production by starter culture organisms. *J. Food Prot.* **41**, 429–431.

Kitchell, A. G., and Ingram, M. (1956). A comparison of bacterial growth on fresh meat and on frozen meat after thawing. *Ann. Inst. Pasteur, Lille* **8**, 121–131.

Kitchell, A. G., and Shaw, B. G. (1975). Lactic acid bacteria in fresh and cured meat. *In* "The Lactic Acid Bacteria in Beverages and Food" (J. G. Carr, C. V. Cutting, and G. C. Whiting, eds.), pp. 209–220. Academic Press, New York.

Klaushofer, H., and Parkkinen, E. (1966). Zur Frage der Bedeutung aerober und anaerober thermophiler Sporenbildner als Infektionsursache in Rübenzuckerfabriken. *Z. Zuckerind.* **16**(3), 125–130.

Klaushofer, H., Hollaus, F., and Pollach, G (1971). Microbiology of beet sugar manufacture. *Process Biochem.* **6**(6), 39–41.

Kleinert, J. (1972). Microbiologie et altération des denrées alimentaires. Importance de l'hygiene des main. *Rev. Int. Choc.* **27**(3), 60–68.

Kline, L., and Sonoda, T. T. (1951). Role of glucose in the storage deterioration of whole egg powder. I. Removal of glucose from whole egg melange by yeast fermentation before drying. *Food Technol.* **5**, 90–94.

Kline, L., and Sugihara, T. F. (1966). Effects of pasteurization on egg products. *Bakers Dig.* **40**(4), 40–42, 47–50.

Kline, L., and Sugihara, T. F. (1971). Microorganisms of the San Francisco sour dough bread process. II. Isolation and characterization of undescribed bacterial species responsible for the souring activity. *Appl. Microbiol.* **21**, 459–465.

Kline, L., Sugihara, T. F., Bean, M. L., and Ijichi, K. (1965). Heat pasteurization of raw liquid egg white. *Food Technol.* **19**, 1709–1718.

Kline, L., Sugihara, T. F., and Ijichi, K. (1966). Further studies on heat pasteurization of raw liquid egg white. *Food Technol.* **20**, 1604–1606.

Kline, L., Sugihara, T. F., and McCready, L. B. (1970). Nature of the San Francisco sour dough French bread process. I. Mechanics of the process. *Bakers Dig.* **44**(2), 48–50.

Klose, A. A., Lineweaver, H., and Palmer, H. H. (1968). Thawing turkeys at ambient air temperatures. *Food Technol.* **22**, 1310–1314.

Klose, A. A., Kaufman, V. F., and Pool, M. F. (1971). Scalding poultry by steam at subatmospheric pressures. *Poult. Sci.* **50**, 302–304.

Knapp, A. W. (1937). "Cacao Fermentation." Bale, Sons & Curnow, London.

Knight, R. A., and Menlove, E. M. (1961). Effect of the bread-baking process on destruction of certain mold spores. *J. Sci. Food Agric.* **12**, 653–656.

Knowles, N. R. (1953). Observations on the microbiology of raw and heat-treated liquid egg. *Proc. Soc. Appl. Bacteriol.* **16**, 107–118.

Knowles, N. R. (1956). The prevention of microbial spoilage in whole shell eggs by heat treatment methods. *J. Appl. Bacteriol.* **19**, 293–300.

Koburger, J. A., and Farhat, B. Y. (1975). Fungi in foods. VI. A comparison of media to enumerate yeasts and molds. *J. Milk Food Technol.* **38**, 466–468.

Koedam, A. (1977). Animikrobielle Wirksamkeit ätherischer Ölle. *Riechst., Aromen, Kosmet.* **27**(1), 6, 8–11; **27**(2), 36–41.

Koegh, B. P. (1966). Bacteriological problems in the condensing and drying of milk. *Food Technol. Aust.* **18**, 126.

Kohl, W. F. (1971). A new process for pasteurizing egg whites. *Food Technol.* **25**, 1176–1184.

Kokal, D. (1965). Viability of *Escherichia coli* on English walnut meats (*Juglans regia*). *J. Food Sci.* **30**, 325–332.

Kokal, D., and Thorpe, D. W. (1969). Occurrence of *Escherichia coli* in almonds of nonpareil variety. *Food Technol.* **23**, 227–232.

Komarik, S. L., Tressler, D. K., and Long, L. (1974). "Food Products Formulary Series. Vol. 1: Meats, Poultry, Fish, and Shellfish." AVI, Westport, Connecticut.

Konowalchuk, J., and Speirs, J. I. (1975). Survival of enteric viruses on fresh vegetables. *J. Milk Food Technol.* **38**, 469–472.

Konowalchuk, J., and Speirs, J. I. (1978a). Antiviral effect of commercial juices and beverages. *Appl. Environ. Microbiol.* **35**, 1219–1220.

Konowalchuk, J., and Speirs, J. I. (1978b). Antiviral effects of apple beverages. *Appl. Environ. Microbiol.* **36**, 798–801.

Kooiman, W. J. (1977). Microbiological aspects of soft drinks and fruit spreads. Unpublished data.

Kopeikovskii, V. M., and Kostenko, V. K. (1963). The effect of heat drying conditions on the mold microflora of sunflower seeds. *Izv. Vyssh. Uchebn. Zaved., Pishch. Tekhnol.* **33**(2), 26–28 [*Biol. Abstr.* **45**, 21075 (1964)].

Kopeloff, N. (1926). *"Lactobacillus acidophilus."* Williams & Wilkins, Baltimore, Maryland.

Kopeloff, N., and Kopeloff, L. (1919). The deterioration of cane sugar by fungi. *La. Agric. Exp. Stn., Bull.* No. 166. (Cited in Mayeux, 1960.)

Koppensteiner, G., and Windisch, S. (1971). Der osmotische Wert als begrenzender Faktor für das Wachstum und die Gärung von Hefen. *Arch. Mikrobiol.* **80**, 300–314.

Kosikowski, F. V. (1977). "Cheese and Fermented Milk Foods," 2nd ed. Edwards, Ann Arbor, Michigan.

Kosker, O., Esselen, W. B., Jr., and Fellers, C. R. (1949). Mustard as a preservative for fruit juices. *Glass Packer* **28**, 818–820, 836, 838, 841.

Kosker, O., Esselen, W. B., Jr., and Fellers, C. R. (1951). Effect of allylisothiocyanate and related substances on the thermal resistance of *Aspergillus niger, Saccahromyces ellipsoideus,* and *Bacillus thermoacidurans. Food Res.* **16**, 510–514.

Kotter, L., and Terplan, G. (1960). Zur Gewinnung und Verarbeitung von Blutplasma. *Monatsh. Veterinaermed.* **15**, 429–436.

Kotula, A. W., and Kinner, J. A. (1964). Airborne microorganisms in broiler processing plants. *Appl. Microbiol.* **12**, 179–184.

Kotula, A. W., Lusby, W. R., Crouse, J. D., and de Vries, B. (1974). Beef carcass washing to reduce bacterial contamination. *J. Anim. Sci.* **39**, 674–679.

Kraft, A. A., Ayres, J. C., Weiss, K. F., Marion, W. W., Balloun, S. L., and Forsythe, R. H. (1963). Effect of method of freezing on survival of microorganisms on turkey. *Poult. Sci.* **42**, 128–137.

Kraft, A. A., Torrey, G. S., Ayres, J. C., and Forsythe, R. H. (1967a). Factors influencing bacterial contamination of commercially produced liquid egg. *Poult. Sci.* **46**, 1204–1210.

Kraft, A. A., Ayres, J. C., Forsythe, R. H., and Schultz, J. R. (1967b). Keeping quality of pasteurized liquid egg yolk. *Poult. Sci.* **46**, 1282.

Kramer, A., and Farquhar, J. (1977). Fate of microorganisms during frozen storage of custard pies. *J. Food Sci.* **42**, 1138–1139.

Krishnaswamy, M. A., Patel, J. D., and Parthasarathy, N. (1971). Enumeration of micro-organisms in spices and spice mixtures. *J. Food Sci. Technol.* **8**(4), 191–194.

Krol, B., and Tinbergen, B. J., eds. (1974). *Proc. Int. Symp. Nitrite Meat Prod., Pudoc, Wageningen, Neth.*

Krugers Dagneaux, E. L., and Mossel, D. A. A. (1968). The microbiological condition of dried soups. *In* "The Microbiology of Dried Foods" (E. H. Kampelmacher, M. Ingram, and D. A. A. Mossel, eds.), pp. 411–425. Grafische Ind., Haarlem, Neth.

Kurtzman, C. P. (1976). Salad dressings. *In* "Compendium of Methods for The Microbiological Examination of Foods" (M. L. Speck, ed.), pp. 594–598. Am. Public Health Assoc., Washington, D.C.

Kurtzman, C. P., and Ciegler, A. (1970). Mycotoxin from a blue-eye mold of corn. *Appl. Microbiol.* **20**, 204–207.

Kurtzman, C. P., Rogers, R., and Hesseltine, C. W. (1971). Microbiological spoilage of mayonnaise and salad dressings. *Appl. Microbiol.* **21,** 870–874.

Labuza, T. P. (1976). Drying food: Technology improves on the sun. *Food Technol.* **30**(6), 37–38, 42, 44, 46.

Ladiges, W. C., and Foster, J. F. (1974). Bacterial pathogens in dry dog foods. *J. Am. Vet. Med. Assoc.* **165,** 181–182.

Lahellec, G., Meurier, C., and Bennejean, G. (1973). *Journ. Res. Avic. Cunic.* December. (Cited in Barnes, 1974.)

Laidley, R., Handzel, S., Severs, D., and Butler, R. (1974). *Salmonella weltevreden* outbreak associated with contaminated pepper. *Epidemiol. Bull. (Dep. Natl. Health Welfare, Ottawa)* **18**(4), 62.

Lamprech, E. D. (1968). Refrigerated dough. *Bull. Assoc. Food Drug Off. U.S.* **32,** 168–173.

Lancaster, M. C., Jenkins, F. P., Philp, J. M., Sargeant, K., Sheridan, A., O'Kelly, J., and Carnaghan, R. B. A. (1961). Toxicity associated with certain samples of ground nuts. *Nature (London)* **192,** 1095–1097.

Langsrud, T., and Reinbold, G. W. (1974). Flavor development and microbiology of Swiss cheese—a review. IV. Defects. *J. Milk Food Technol.* **37,** 26–41.

Lannuier, G. L., and Matz, S. A. (1967). Refrigerated dough products. *Cereal Sci. Today* **12,** 478–480.

Larkin, E. P. (1973). The public health significance of viral infections of food animals. *In* "Microbiological Safety of Food" (B. C. Hobbs and J. H. B. Christian, eds.), pp. 257–270. Academic Press, New York.

Last, F. T., and Price, D. (1969). Yeasts associated with living plants and their environs. *In* "The Yeasts" (A. H. Rose and J. S. Harrison, eds.), Vol. 1, pp. 183–218. Academic Press, New York.

Lategan, P. M., and Vaughn, R. H. (1964). The influence of chemical additives on the heat resistance of *Salmonella typhimurium* in liquid whole egg. *J. Food Sci.* **29,** 339–344.

Lawler, F. K. (1965). Thaws frozen eggs fast. *Food Eng.* **37**(8), 72, 75–76.

Lawrence, B. M. (1978). Recent progress in essential oils. *Perfum. Flavorist* **2**(12), 44–49.

Lawrie, R. A. (1975). Meat components and their variability. *In* "Meat" (D. J. A. Cole and R. A. Lawrie, eds.), pp. 249–268. Butterworth, London.

Laycock, R. A., and Regier, L. W. (1970). Pseudomonads and achromobacters in the spoilage of irradiated haddock of different preirradiation quality. *Appl. Microbiol.* **20,** 333–341.

Lee, J. S., and Pfeifer, D. K. (1973). Aerobic microbial flora of smoked salmon. *J. Milk Food Technol.* **36,** 143–145.

Lee, L. S., Cucullu, A. F., and Goldblatt, L. A. (1968). Appearance and aflatoxin content of raw and dry roasted peanut kernels. *Food Technol.* **22,** 1131–1134.

Lee, W. H., Staples, C. L., and Olson, J. C., Jr. (1975). *Staphylococcus aureus* growth and survival in macaroni dough and the persistence of enterotoxins in the dried products. *J. Food Sci.* **40,** 119–120.

Leete, C. S. (1925). Relation between the bacterial count of whole milk and that of cream and skimmilk separated from it. *J. Agric. Res.* **31,** 695–699.

Leistner, L. (1973). Sprüh-Luft Kühlung von Schlachthahnchen ein Alternativ-Verfahren zum Spinchiller. *Proc. Poult. Meat Symp., Roskilde, Den.* A13, pp. 1–8.

Leistner, L., Johantges, J., Deibel, R. H., and Niven, C. F., Jr. (1961). The occurrence and significance of salmonellae in meat animals and animal by-product feeds. *Am. Meat Inst. Found., Circ.* **64**, 9–20.

Leistner, L., Rossmanith, E., and Woltersdorf, W. (1972). Rationalisierung des Sprüh-Kühlverfahrens für Schlachthahnchen. *Fleischwirtschaft* **52**, 362–364.

Leistner, L., Hechelmann, H., Kashiwazaki, M., and Albertz, R. (1975). Nachweis von *Yersinia enterocolitica* in Faeces und Fleisch von Schweinen, Rindern, und Geflügel. *Fleischwirtschaft* **55**, 1599–1602.

Leitao, M. F., Delazari, I., and Mazzoni, H. (1973–1974). Microbiology of dehydrated foods (in Port.). *Coletanea Inst. Tecnol. Aliment.* **5**, 223–241 [*Food Sci. Technol. Abstr.* **7**, 9B72 (1975)].

Leonberger, J. M., and Bernard, D. (1977). "*C. botulinum* Studies," Ann. Rep. Natl. Canners Assoc., Washington, D.C.

Lerche, M., Rievel, H., and Goerttler, V. (1957). "Lehrbuch der tierärztlichen Lebensmittelüberwachung." Schaper, Hannover.

Levanon, Y., and Rossetini, S. M. O. (1965). A laboratory study of farm processing of cocoa beans for industrial use. *J. Food Sci.* **30**, 719–722.

Lewis, M. N., Weiser, H. H., and Winter, A. R. (1953). Bacterial growth in chicken salad. *J. Am. Diet. Assoc.* **29**, 1094–1099.

Licciardello, J. J., Nickerson, J. T. R., and Goldblith, S. A. (1965). Destruction of salmonellae in hard boiled eggs. *Am. J. Public Health* **55**, 1622–1628.

Lifshitz, A., and Baker, R. C. (1964). Some physical properties of the egg shell membranes in relation to their resistance to bacterial penetration. *Poult. Sci.* **43**, 527–528.

Lifshitz, A., Baker, R. C., and Naylor, H. B. (1964). The relative importance of chicken egg exterior structures in resisting bacterial penetration. *J. Food Sci.* **29**, 94–99.

Lillard, H. S. (1971). Occurrence of *Clostridium perfringens* in broiler processing and further processing operations. *J. Food Sci.* **36**, 1008–1010.

Lillard, H. S., Hanlin, R. T., and Lillard, D. A. (1970). Aflatoxigenic isolates of *Aspergillus flavus* from pecans. *Appl. Microbiol.* **19**, 128–130.

Lillard, H. S., Klose, A. A., Hegge, R. I., and Chew, V. (1973). Microbiological comparison of steam (at sub-atmospheric pressure) and immersion-scalded broilers. *J. Food Sci.* **38**, 903–904.

Lillehoj, E. B., Fennell, D. I., and Hara, S. (1975). Fungi and aflatoxin in a bin of stored white maize. *J. Stored Prod. Res.* **11**(1), 47–51 [*Biol. Abstr.* **60**, 31158].

Lindberg, A. M., and Nickels, C. (1976). Microbiological standards of some species and herbs. *Livsmedelstek.* No. 4, 160–164.

Lindley, P. (1972). Chocolate and sugar confectionery, jams and jellies. *In* "Quality Control in the Food Industry" (S. M. Herschdoerfer, ed.), Vol. 3, pp. 259–295. Academic Press, New York.

Lineweaver, H., Cunningham, H. E., Garibaldi, J. A., and Ijichi, K. (1967). "Heat Stability of Egg White Proteins under Minimal Conditions that Kill Salmonellae," ARS 74–39. Agric. Res. Serv., U.S. Dep. Agric., Albany, California.

Lineweaver, H., Palmer, H. H., Putnam, G. W., Garibaldi, J. A., and Kaufman, V. F. (1969). "Egg Pasteurization Manual," ARS 74–48. Agric. Res. Serv., U.S. Dep. Agric., Albany, California.

Ling, A. C., and Lund, D. B. (1978). Fouling of heat transfer surfaces by solutions of egg albumen. *J. Food Prot.* **41**, 187–194.

Liston, J. (1956). Quantitative variations in the bacterial flora of flatfish. *J. Gen. Microbiol.* **15**, 305–314.

Liston, J. (1974). Influence of seafood handling procedures on *Vibrio parahaemolyticus*. *In* "International Symposium on *Vibrio parahaemolyticus*" (T. Fujino, G. Sakaguchi, R. Sakazaki, and Y. Takeda, eds.), Saikon, Tokyo.

Liston, J. (1976). *Vibrio parahaemolyticus*. *In* "Food Microbiology: Public Health and Spoilage Aspects" (M. P. de Figueiredo and D. F. Splittstoesser, eds.), AVI, Westport, Connecticut.

Liston, J. (1978). Microbial spoilage of fish and seafood. *Proc. Int. Conf. Global Impacts Appl. Microbiol., 4th, 1973* pp. 645–660.

Liston, J., Stansby, M. E., and Olcott, H. S. (1976). Bacteriological and chemical basis for deteriorative changes. *In* "Industrial Fishery Technology" (M. E. Stansby, ed.), pp. 345–358. Krieger, Huntington, New York.

Lloyd, W. E., and Harriman, L. A. (1957). Method of treating egg whites. U.S. Patent 2,776,214.

Llwewllyn, G. C., Benevides, J., and Eadie, T. (1978). Differential production of aflatoxin on natural and heat-treated cocoa beans. *J. Food Prot.* **41,** 785–787.

Loeb, J. R., and Mayne, R. Y. (1952). Effect of moisture on the microflora and formation of free fatty acids in rice bran. *Cereal Chem.* **29,** 163–175.

Lofton, C. B., Morrison, S. M., and Leiby, R. D. (1962). The Enterobacteriaceae of some Colorado small mammals and birds and their possible role in gastroenteritis in man and domestic animals. *Zoonoses Res.* **1,** 277–293.

Loken, K. I., Culbert, K. H., Solee, R. E., and Pomeroy, B. S. (1968). Microbiological quality of protein feed supplements produced by rendering plants. *Appl. Microbiol.* **16,** 1002–1005.

Long, H. F., and Hammer, B. W. (1938). Examination of butter with the Burri Smear Culture Technic. *Iowa State Coll. J. Sci.* **12,** 441–450.

Longrée, K. (1964). Cooling fluid food under agitation. *J. Am. Diet. Assoc.* **44,** 477–479.

Longrée, K. (1972). "Quantity Food Sanitation," 2nd ed. Wiley, New York.

Longrée, K., and White, J. C. (1955). Cooling rates and bacterial growth in food prepared and stored in quantity. I. Broth and white sauce. *J. Am. Diet. Assoc.* **31,** 124–132.

Longrée, K., White, J. C., Cutlar, K. L., and Willman, A. R. (1959). Bacterial growth in potato and turkey salads. *J. Am. Diet. Assoc.* **35,** 38–44.

Longrée, K., Moragne, L., and White, J. C. (1960). Cooling starch-thickened food items with cold tube agitation. *J. Milk Food Technol.* **23,** 330–336.

Longrée, K., Moragne, L., and White, J. C. (1963). Cooling menu items by agitation under refrigeration. *J. Milk Food Technol.* **26,** 317–322.

López, L. C., and Christensen, C. M. (1967). Effect of moisture content and temperature on invasion of stored corn by *Aspergillus flavus*. *Phytopathology* **57,** 588–590.

Lorenz, F. W., and Starr, P. B. (1952). Spoilage of washed eggs. I. Effect of sprayed versus static water under different washing temperatures. *Poult. Sci.* **31,** 204–214.

Lorenz, F. W., Starr, P. B., Starr, M. P., and Ogasawara, F. X. (1952). The development of *Pseudomonas* spoilage in shell eggs. I. Penetration through the shell. *Food Res.* **17,** 351–360.

Lovelace, T. E., Tubiash, H., and Colwell, R. R. (1968). Quantitative and qualitative commensal bacterial flora of *Crassostrea virginica* in Chesapeake Bay. *Proc. Natl. Shellfish Assoc.* **58,** 82–87.

Lubieniecki-von Schelhorn, M. (1973). Influence of relative humidity conditions on the thermal resistance of several kinds of spores of moulds. *Acta Aliment. Acad. Sci. Hung.* **2**, 163–171.

Lüthi, H. (1959). Microorganisms in noncitrus juices. *Adv. Food Res.* **9**, 221–285.

Lund, B. M. (1971). Bacterial spoilage of vegetables and certain fruits. *J. Appl. Bacteriol.* **34**, 9–20.

Lund, B. M., and Nicholls, J. C. (1970). Factors influencing the soft-rotting of potato tubers by bacteria. *Potato Res.* **13**, 210–214.

Lundbeck, H. (1974). Prevention of *Salmonella* infections in the chicken industry. (Mimeo.) *Latin Am. Congr. Microbiol., 6th, Caracas.* Unpublished.

Lundbeck, H., Plazikowski, U., and Silverstolpe, L. (1955). The Swedish Salmonella outbreak of 1953. *J. Appl. Bacteriol.* **18**, 535–548.

Lutey, R. W., and Christensen, C. M. (1963). Influence of moisture content, temperature and length of storage upon survival of fungi in barley kernels. *Phytopathology* **53**, 713–717.

Lynch, H. (1977). Microbiological standards of refined sugar. *In* "Cane Sugar Handbook" (G. P. Meade and J. C. P. Chen, eds.), 10th ed., pp. 416–422. Wiley, New York.

Lynt, R. K., Kautter, D. A., and Read, R. B., Jr. (1975). Botulism in commercially canned foods. *J. Milk Food Technol.* **38**, 546–550.

Macara, T. J. R. (1943). Dried meat. II. The growth of moulds on dried meat. *J. Soc. Chem. Ind.* **62**, 104–106.

McBride, G. B., Brown, B., and Skura, B. J. (1978). Effect of bird type, growers and season on the incidence of salmonellae in turkeys. *J. Food Sci.* **43**, 323–326.

McCall, C. E., Collins, R. N., Jones, D. B., Kaufmann, A. F., and Brachman, P. S. (1966). An interstate outbreak of salmonellosis traced to a contaminated food supplement. *Am. J. Epidemiol.* **84**, 32–39.

McCleskey, C. S., Faville, L. W., and Barnett, R. O. (1947). Characteristics of *Leuconostoc mesenteroides* from cane juice. *J. Bacteriol.* **54**, 697–708.

McCoy, J. H. (1975). Trends in *Salmonella* food poisoning in England and Wales 1941–72. *J. Hyg.* **74**, 271–282.

McDivitt, M. E., and Hammer, M. L. (1958). Cooling rate and bacterial growth in cornstarch pudding. *J. Am. Diet. Assoc.* **34**, 1190–1194.

McDowall, F. H. (1953). "The Buttermakers Manual," Vol. I. New Zealand Univ. Press, Wellington.

McKinley, T. W., and Clarke, E. J., Jr. (1964). Imitation cream filling as a vehicle of staphylococcal food poisoning. *J. Milk Food Technol.* **27**, 302–304.

McMaster, L. (1975). Thermophilic bacteria associated with the cane sugar diffusion process M.S. Thesis, Univ. of Natal, Durban, South Africa.

McMaster, L., and Ravnö, A. B. (1977). The occurrence of lactic acid and associated micro-organisms in cane sugar processing. *Proc. Int. Soc. Sugar Cane Technol.* **16**, 1–15.

McMeekin, T. A. (1977). Spoilage association of chicken leg muscle. *Appl. Microbiol.* **33**, 1244–1246.

McMeekin, T. A., and Thomas, C. J. (1978). Retention of bacteria on chicken skin after immersion in bacterial suspensions. *J. Appl. Bacteriol.* **45**, 383–387.

McNally, E. H. (1952). Effect of drying after washing on the incidence of eggs infected by spoilage microorganisms. *Poult. Sci.* **31**, 1102–1104.

Majumder, S. K. (1974). "Control of Microflora and Related Production of Mycotoxins in Stored Sorghum, Rice and Groundnut." P. M. Kuruvilla, Wesley Press, Mysore, India.

Majumder, S. K., Narasimhan, K. S., and Parpia, H. A. B. (1965). Microecological factors of microbial spoilage and the occurrence of mycotoxins on stored grains. *In* "Mycotoxins in Foodstuffs" (G. N. Wogan, ed.), pp. 27–47. MIT Press, Cambridge, Massachusettts.

Mann, E. J. (1973). Digest of World Literature: Continuous buttermaking. *Dairy Ind.* **38,** 520–521.

Maravalhas, N. (1966). Mycological deterioration of cocoa beans during fermentation and storage. *In* "Microbiological Deterioration," Monograph, No. 23, pp. 98–104. Soc. Chem. Ind., London.

Marcus, K. A., and Amling, H. J. (1973). *Escherichia coli* field contamination of pecan nuts. *Appl. Microbiol.* **26,** 279–281.

Marier, R., Wells, J. G., Swanson, R. C., Callahan, W., and Mehlman, I. J. (1973). An outbreak of enteropathogenic *Escherichia coli* foodborne disease traced to imported French cheese. *Lancet ii,* 1376–1378.

Marshall, C. R., and Walkley, V. T. (1951a). Some aspects of microbiology applied to commercial apple juice production. Part I. Distribution of microorganisms on the fruit. *Food Res.* **16,** 448–456.

Marshall, C. R., and Walkley, V. T. (1951b). Some aspects of microbiology applied to commercial apple juice production. Part II. Microbiological control of processing. *Food Res.* **16,** 515–521.

Marshall, C. R., and Walkley, V. T. (1952a). Some aspects of microbiology applied to commercial apple juice production. Part III. Isolation and identification of apple juice spoilage organisms. *Food Res.* **17,** 123–131.

Marshall, C. R., and Walkley, V. T. (1952b). Some aspects of microbiology applied to commercial apple juice production. Part IV. Development characteristics and viability of spoilage organisms in apple juice. *Food Res.* **17,** 197–203.

Marshall, C. R., and Walkley, V. T. (1952c). Some aspects of microbiology applied to commercial apple juice production. Part V. Thermal death rates of spoilage organisms in apple juice. *Food Res.* **17,** 204–211.

Marshall, C. R., and Walkley, V. T. (1952d). Some aspects of microbiology applied to commercial apple juice production. VI. The significance of changes in the relative incidence of spoilage organisms during processing. *Food Res.* **17,** 307–314.

Marth, E. H. (1967). Aflatoxins and other mycotoxins in agricultural products. *J. Milk Food Technol.* **30,** 192–198.

Marth, E. H. (1969). Salmonella and salmonellosis associated with milk and milk products. A review. *J. Dairy Sci.* **52,** 283–315.

Marth, E. H., and Calanog, B. G. (1976). Toxigenic fungi. *In* "Food Microbiology: Public Health and Spoilage Aspects" (M. P. de Figueiredo and D. F. Splittstoesser, eds.), pp. 210–256. AVI, Westport, Connecticut.

Marthedal, H. E. (1973). The occurrence of salmonellosis in poultry in Denmark, 1935–1971 and the eradication programme established. *In* "*The* Microbiological Safety of Food" (B. C. Hobbs and J. H. B. Christian, eds.), pp. 211–227. Academic Press, New York.

Martin, C. R., and Sauer, D. B. (1976). Physical and biological characteristics of grain dust. *Trans. ASAE* **19,** 720–723.

Martin, J. H., and Blackwood, P. W. (1971). Effect of pasteurization conditions, type of bacteria, and storage temperature on the keeping quality of UHT-processed soft-serve frozen dessert mixes. *J. Milk Food Technol.* **34,** 256–259.

Matches, J. R., and Liston, J. (1968a). Growth of salmonellae on irradiated and non-irradiated seafoods. *J. Food Sci.* **33**, 406–410.

Matches, J. R., and Liston, J. (1968b). Low temperature growth of *Salmonella*. *J. Food Sci.* **33**, 641–645.

Matches, J. R., Liston, J., and Curran, D. (1974). *Clostridium perfringens* in the environment. *Appl. Microbiol.* **28**, 655–660.

Matsuyama, A., Thornley, M. J., and Ingram, M. (1964). The effect of freezing on the radiation sensitivity of vegetative bacteria. *J. Appl. Bacteriol.* **27**, 110–124.

Mattil, K. F. (1964). Butter and margarine. *In* "Bailey's Industrial Oil and Fat Products" (D. Swern, ed.), pp. 317–352. Wiley (Interscience), New York.

Matz, S. A., and Milner, M. (1951). Inhibition of respiration and preservation of damp wheat by means of organic chemicals. *Cereal Chem.* **28**, 196–207.

Maxcy, R. B. (1976). Fate of post-cooking microbial contaminants of some major menu items. *J. Food Sci.* **41**, 375–378.

May, K. N. (1962). Bacterial contamination during cutting and packaging chicken in processing plants and retail stores. *Food Technol.* **16**(8), 89–91.

May, K. N. (1974). Chilling of poultry meat. 3. Changes in microbiological numbers during final washing and chilling of commercially slaughtered broilers. *Poult. Sci.* **53**, 1282–1285.

May, K. N., Irby, J. D., and Carmon, J. L. (1962). Shelf life and bacterial counts of excised poultry tissue. *Food Technol.* **16**(2), 66–68.

Mayeux, P. A. (1960). Some studies on the microbial flora of sugar cane. M.S. Thesis, Louisiana State Univ., Baton Rouge.

Mayr, G. E., and Suhr, H. (1972). Preservation and sterilization of pure and mixed spices. *Proc. Conf. Spices* pp. 201–207. Trop. Prod. Inst., London.

Mead, G. C. (1974). Bacteriological control in the processing of poultry. *Vet. Rec.* **95**, 569–572.

Mead, G. C. (1975). Hygiene aspects of the chilling process. *Qual. Poult. Meat, Proc. Eur. Symp. Poult. Meat.* 2nd, *Oosterbeck, Neth.* pp. (35) 1–8.

Mead, G. C., and Impey, C. S. (1970). The distribution of clostridia in poultry processing plants. *Br. Poult. Sci.* **11**, 407–414.

Mead, G. C., and Thomas, N. L. (1973). Factors affecting the use of chlorine in the spin-chilling of eviscerated poultry. *Br. Poult. Sci.* **14**, 99–117.

Mead, G. C., Adams, B. W., and Parry, R. T. (1975). The effectiveness of in-plant chlorination in poultry processing. *Br. Poult. Sci.* **16**, 517–526.

Meara, P. J. (1973). Salmonellosis in slaughter animals as a source of human food poisoning. *J. S. Afr. Vet. Assoc.* **44**, 215–233.

Mehlman, I. J., Simon, N. T., Sanders, A. C., and Olson, J. C., Jr. (1974). Problems in the recovery and identification of enteropathogenic *Escherichia coli* from foods. *J. Milk Food Technol.* **37**, 350–356.

Mehlman, I. J., Fishbein, M., Gorbeck, S. L., Sanders, A. C., Eide, E. L., and Olson, J. C., Jr. (1976). Pathogenicity of *Escherichia coli* recovered from food. *J. Assoc. Off. Anal. Chem.* **59**, 67–80.

Meiering, A. G., Bakker-Arkema, F. W., and Bickert, W. G. (1966). Short time sealed storage of high moisture small grains. *Mich. Agric. Exp. Stn., Q. Bull.* **48**(3), 465–470 [*Biol. Abstr.* **47**, 66321].

Mellor, D. B., and Banwart, G. J. (1965). Recovery of *Salmonella derby* from inoculated egg shell surfaces following sanitizing. *Poult. Sci.* **44**, 1244–1248.

Mercuri, A. J., and Kotula, A. W. (1964). Relation of "breast swab" to "drip" bacterial counts in tray-packed chicken fryers, *J. Food Sci.* **29**, 854–858.

Mercuri, A. J., Thompson, J. E., Rowan, J. D., and Norris, K. H. (1957). Use of the automatic green-rot detector to improve the quality of liquid egg. *Food Technol.* **11**, 374–377.

Merory, J. (1968). "Food Flavorings. Composition, Manufacture and Use," 2nd ed. AVI, Westport, Connecticut.

Metchnikoff, E. (1907). "Quelques remarques sur le lait aigni" (Scientifically Soured Milk and its Influence in Arresting Intestinal Putrefaction). Putnam, New York.

Metchnikoff, E. (1908). "The Prolongation of Life." Putnam, New York.

Meursing, E. H. (1969). Quality control of cocoa powder: methods for analysing seven properties. *Manuf. Confect.* **49**(11), 43–47.

Meursing, E. H. (1976). "Cocoa Powders for Industrial Processing," 2nd ed. Cacaofabriek De Zaan, Koog aan de Zaan, Netherlands.

Meursing, E. H., and Slot, H. (1968). The microbiological condition of cocoa powder. *In* "The Microbiology of Dried Foods" (E. H. Kampelmacher, M. Ingram, and D. A. A. Mossel, eds.), pp. 433–445. Bilthoven, Netherlands.

Meyer, J., and Oxhøj, P. (1964). A salmonellosis epidemic. *Medlemsbl. Dan. Dyrlaegeforen.* **47**, 810–819. (In Dan.)

Meyer, M. T., and Vaughn, R. H. (1969). Incidence of *Escherichia coli* in black walnut meats. *Appl. Microbiol.* **18**, 925–931.

Meyrath, J. (1962). Problems in fruit juice pasteurization. *Recent Adv. Food Sci.* **2**, 117–127.

Michels, M. J., and Visser, F. M. W. (1976). Occurrence and thermoresistance of spores of psychrophilic and psychrotropic aerobic sporeformers in soil and foods. *J. Appl. Bacteriol.* **41**, 1–11.

Michener, H. D., and Elliott, R. P. (1964). Minimum growth temperatures for food-poisoning, fecal-indicator and psychrophilic microorganisms. *Adv. Food Res.* **13**, 349–396.

Michener, H. D., Boyle, F. P., Notter, G. K., and Guadagni, D. G. (1968). Microbiological deterioration of frozen parfried potatoes upon holding after thawing. *Appl. Microbiol.* **16**, 759–761.

Mickelson, M. N., and Flippen, R. S. (1960). Use of salmonellae antagonists in fermenting egg white. II. Microbiological methods for the elimination of salmonellae from egg white. *Appl. Microbiol.* **8**, 371–377.

Milanović, A., and Beganović, A. (1974). Microflora of fodder of animal origin. *Veterinaria (Sarajevo)* **23**, 467–475.

Miller, M. W., Fridley, R. B., and McKillop, A. A. (1963). The effects of mechanical harvesting on the quality of prunes. *Food Technol.* **17**, 1451–1453.

Miller, W. A. (1959). Dry cleaning slightly soiled eggs versus washing to prevent penetration of spoilage bacteria. *Poult. Sci.* **38**, 906–910.

Miller, W. A., and Crawford, L. B. (1953). Some factors influencing bacterial penetration of eggs. *Poult. Sci.* **32**, 303–309.

Miller, W. A., and Smull, M. L. (1955). Efficiency of cooling practices in preventing growth of micrococci. *J. Am. Diet. Assoc.* **31**, 469–473.

Milner, M., and Geddes, W. F. (1946). Grain storage studies. IV. Biological and chemical factors involved in the spontaneous heating of soybeans. *Cereal Chem.* **23**, 449–470.

Milner, M., Christensen, C. M., and Geddes, W. F. (1947). Grain storage studies. VI. Wheat respiration in relation to moisture content, mold growth, chemical deterioration, and heating. *Cereal Chem.* **24**, 182–189.

Milone, N. A., and Watson, J. A. (1968). "Thermal Inactivation of *Salmonella senftenberg* 775W in Poultry Meat." U.S. Army Med. Res. and Dev. Command, Off. Surgeon Gen., Washington, D.C.

Milward, Z. (1976). Further experiments to determine the toxicity of propionic acid to fungi infesting stored grain. *Trans. Br. Mycol. Soc.* **66**(2), 319–324 [*Biol. Abstr.* **62**, 25482].

Minifie, B. W. (1970). "Chocolate, Cocoa and Confectionery: Science and Technology." AVI, Westport, Connecticut.

Ministry of Health (1954). Salmonella in duck eggs. *Mon. Bull. Minist. Health Public Health Lab. Serv.* **13**, 38–43.

Minor, T. E., and Marth, E. H. (1972a). *Staphylococcus aureus* and enterotoxin A in cream and butter. *J. Dairy Sci.* **55**, 1410–1414.

Minor, T. E., and Marth, E. H. (1972b). *Staphylococcus aureus* and staphylococcal food intoxications. A review. III. Staphylococci in dairy foods. *J. Milk Food Technol.* **35**, 77–82. [See *J. Milk Food Technol.* **35**, 228–241, for references cited in this series of papers.]

Mirocha, C. J. (1969). Mycotoxins in feeds and foods. *Proc. Symp. Biol. Contam. Grain Anim. Byprod.* pp. 59–68. Agric. Ext. Serv., Dep. Entomol., Fish. Wildl., Univ. of Minnesota, Minneapolis.

Mislivec, P. B., and Bruce, V. R. (1977). Incidence of toxic and other mold species and genera in soybeans. *J. Milk Food Technol.* **40**, 309–312.

Mislivec, P. B., and Tuite, J. (1970). Temperature and relative humidity requirements of species of *Penicillium* isolated from yellow dent corn kernels. *Mycologia* **62**, 75–88.

Mislivec, P. B., Dieter, C. T., and Bruce, V. R. (1975). Mycotoxin-producing potential of mold flora in dried beans. *Appl. Microbiol.* **29**, 522–526.

Mitsuda, H., and Nakajima, K. (1977). Storage of cooked rice. *J. Food Sci.* **42**, 1439–1443.

Miyajima, T., and Kamoi, I. (1960). Studies on sugar-loss microbes in processing beet juices. Part I. *Seito Gijutsu Kenkyukaishi* **9**, 87–95. (In Jpn.)

Moats, W. A. (1978). Egg washing—A review. *J. Food Prot.* **41**, 919–925.

Mocquot, G., and Hurel, C. (1970). The selection and use of some microorganisms for the manufacture of fermented and acidified milk products. *J. Soc. Dairy Technol.* **23**, 130.

Moerman, P. C., Krol, B., Kampelmacher, E. H., and van Noorle Jansen, L. M. (1966). Invloed van natriumsulfiet op de groei van Salmonellae en *E. coli* in rauw gehakt. (Influence of sodium sulfite on the growth of salmonellae and *E. coli* in raw minced meat.) *Conserva* **14**(7), 172–175.

Moon, N. J., and Reinbold, G. W. (1976). Commensalism and competition in mixed cultures of *Lactobacillus bulgaricus* and *Streptococcus thermophilus*. *J. Milk Food Technol.* **39**, 337–341.

Moragne, L., Longrée, K., and White, J. C. (1960). The effect of some selected factors on the cooling of food under refrigeration. *J. Milk Food Technol.* **23**, 142–150.

Moragne, L., Longrée, K., and White, J. C. (1961). Cooling custards and puddings with cold tube agitation. *J. Milk Food Technol.* **24**, 207–210.

Moragne, L., Longrée, K., and White, J. C. (1963). Effect of a scraper-lifter agitator on cooling time of food. *J. Milk Food Technol.* **26**, 182–184.

Morehouse, L. G., and Wedman, E. E. (1961). *Salmonella* and other disease-producing organisms in animal by-products. A survey. *J. Am. Vet. Med. Assoc.* **139**, 989–995.

Morey, R. G. (1954). Cultured and fermented milks. *Milk Dealer* **43**(4), 65, 74–77.

Mori, K., Sawada, H., Nabetani, O., and Maruo, S. (1974). The effects of essential oils and spice extracts on the prevention of slimy spoilage of wieners. *Nippon Shokuhin Kogyo Gakkai-Shi* **21**, 285–287. (In Jpn.)

Morita, T. N., and Woodburn, M. J. (1977). Stimulation of *Bacillus cereus* growth by protein in cooked rice combinations. *J. Food Sci.* **42**, 1232–1235.

Moroz, R. (1963). Methods and procedures for the analyses of microorganisms in sugar. *In* "Principles of Sugar Technology" (P. Honig, ed.), Vol. 3, pp. 373–449. Elsevier, Amsterdam.

Morozumi, S. (1978). Isolation, purification and antibiotic activity of o-methoxycinnamaldehyde from cinnamon. *Appl. Environ. Microbiol.* **36**, 577–583.

Morris, E. O. (1975). Yeasts from the marine environment. *J. Appl. Bacteriol.* **38**, 211–223.

Morris, G. K., and Wells, J. G. (1970). *Salmonella* contamination in a poultry-processing plant. *Appl. Microbiol.* **19**, 795–799.

Morris, G. K., McMurray, B. L., Galton, M. M., and Wells, J. G. (1969). A study of the dissemination of salmonellosis in a commercial broiler chicken operation. *Am. J. Vet. Res.* **30**, 1413–1421.

Morris, G. K., Martin, W. T., Shelton, W. H., Wells, J. G., and Brachman, P. S. (1970). Salmonellae in fish meal plants: relative amounts of contamination at various stages of processing and a method of control. *Appl. Microbiol.* **19**, 401–408.

Morris, T. G., and Ayres, J. C. (1960). Incidence of *Salmonellae* on commercially processed poultry. *Poult. Sci.* **39**, 1131–1135.

Morse, E. V., Duncan, M. A., Estep, D. A., Riggs, W. A., and Blackburn, B. O. (1976). Canine salmonellosis: A review and report of dog to child transmission of *Salmonella enteritidis*. *Am. J. Public Health* **66**, 82–84.

Morse, R. E. (1977). Spice and flavoring. *In* "McGraw-Hill Encyclopedia of Food, Agriculture and Nutrition" (D. N. Lapedes, ed.), pp. 623–624. McGraw-Hill, New York.

Mossel, D. A. A. (1955). The importance of the bacteriological condition of ingredients used as minor components in some canned meat products. *Ann. Inst. Pasteur Lille* **7**, 171–178.

Mossel, D. A. A. (1963). La survie des salmonellae dans les différents produits alimentaires. *Ann. Inst. Pasteur, Paris* **104**, 551–569.

Mossel, D. A. A., and Ingram, M. (1955). The physiology of the microbial spoilage of foods. *J. Appl. Bacteriol.* **18**, 232–268.

Mossel, D. A. A., and Krugers Dagneaux, E. L. (1963). Die hygienisch-bakteriologische Beurteilung von Trockenkochsuppen. *Arch. Lebensmittel. hyg.* **14**, 108–111.

Mossel, D. A. A., and Ratto, M. A. (1973). Wholesomeness of some types of semi-preserved foods. *J. Food Technol.* **8**, 97–103.

Mossel, D. A. A., and Sand, F. E. M. J. (1968). Occurrence and prevention of microbial deterioration of confectionery products. *Conserva* **17**(2), 23–32.

Mossel, D. A. A., and Scholts, H. H. (1964). Diagnostic, pronostic et prévention des altérations microbiennes des boissons hygiéniques. *Ann. Inst. Pasteur Lille* **15**, 11–30.

Mossel, D. A. A., van Schothorst, M., and Kampelmacher, E. H. (1967). Comparative study on decontamination of mixed feeds by radicidation and by pelletisation. *J. Sci. Food Agric.* **18,** 362–367.

Mossel, D. A. A., van Schothorst, M., and Kampelmacher, E. H. (1968). "Prospects for *Salmonella* Radicidation of Some Foods and Feeds with Particular Reference to the Estimation of the Dose Required. Elimination of Harmful Organisms from Food and Feed by Irradiation," pp. 43–57. IAEA, Vienna.

Mossel, D. A. A., Krol, B., and Moerman, P. C. (1972). Bacteriological and quality perspectives of *Salmonella* radicidation of frozen boneless meats. *Alimenta* **11,** 51–60.

Mossel, D. A. A., Meursing, E. H., and Slot, H. (1974). An investigation on the numbers and types of aerobic spores in cocoa powder and whole milk. *Neth. Milk Dairy J.* **28,** 149–154.

Mossel, D. A. A., Dijkmann, K. E., and Snijders, J. M. A. (1975). Microbial problems in handling and storage of fresh meats. *In* "Meat" (D. J. A. Cole and R. A. Lawrie, eds.), pp. 223–246. Butterworth, London.

Mountney, G. J. (1976). "Poultry Products Technology," 2nd ed. AVI, Westport, Connecticut.

Mourges, R., and Auclair, J. E. (1973). Durée de conservation à 4°C et 8°C du lait pasteurisé conditionné aseptiquement. *Lait* **53,** 481.

Moyle, A. I. (1966). Salmonellae in rendering plant by-products. *J. Am. Vet. Med. Assoc.* **149,** 1172–1176.

Mrak, E. M., and McClung, L. S. (1940). Yeasts occurring on grapes and in grape products in California. *J. Bacteriol.* **40,** 395–407.

Müggenburg, H. (1956). Verderbnisgefahren durch Zusatzstoffe. *Fleischwirtschaft* **8,** 193–195.

Münzner, R. (1974). Untersuchungen zur Strahlenbehandlung von Tierfutter. *Zentralbl. Bakteriol., Parasitenkd., Infektionskr. Hyg., Abt. 1: Orig. Reihe B* **158,** 588–592.

Mukherjee, S. K., Albury, M. N., Pederson, C. S., van Veen, A. G., and Steinkraus, K. H. (1965). Role of *Leuconostoc mesenteroides* in leavening the batter of idli, a fermented food of India. *Appl. Microbiol.* **13,** 227–231.

Mulder, R. W. A. W. (1971). Hygienische aspecten van de koeling van geslacht pluimvee. *Vlees* **6,** 11–13.

Mulder, R. W. A. W. (1973). "Shelf Life of Thawed Poultry Meat," Rep. No. 9873. Spelderholt Inst. Poult. Res., Beekbergen, Netherlands.

Mulder, R. W. A. W. (1974). Microbiological shelf life of vacuum packed smoked chickens. *Arch. Lebensmittel hyg.* **25,** 252–254.

Mulder, R. W. A. W. (1975). Radiation-inactivation of *Salmonella panama* and *Escherichia coli* K 12 present on deepfrozen poultry carcasses. *Qual. Poult. Meat, Proc. Symp. Poult. Meat Qual., 2nd, Oosterbeck, Neth.* pp. (14) 1–7.

Mulder, R. W. A. W. (1976). Microbiological aspects of poultry processing. *Vleesdistrib. Vleestechnol.* **2,** 20–22. (In Dutch.)

Mulder, R. W. A. W., and Dorresteijn, L. W. J. (1975). Microbiological quality of mechanically deboned poultry meat. *Qual. Poult. Meat, Proc. Eur. Symp. Poult. Meat Qual., 2nd, Oosterbeck, Neth.* pp. (50) 1–7.

Mulder, R. W. A. W., and Dorresteijn, L. W. J. (1977). Hygiene beim brühen von Schlachtgeflügel. (Hygiene during the scalding of broilers.) *Fleischwirtschaft* **57,** 2220–2222.

Mulder, R. W. A. W., and Gerrits, A. R. (1974). The microbiological shelf life of vacuum packed broiled chickens. *Arch. Lebensmittelhyg.* **25,** 108–111.

Mulder, R. W. A. W., and van der Hulst, M. C. (1973). The microflora of liquid whole egg made from incubator reject eggs. *J. Appl. Bacteriol.* **36,** 157–163.

Mulder, R. W. A. W., and Veerkamp, C. H. (1974). Improvements in poultry slaughterhouse hygiene as a result of cleaning before cooling. *Poult. Sci.* **53,** 1690–1694.

Mulder, R. W. A. W., Dorresteijn, L. W. J., and van der Broek, J. (1977a). Cross-contamination during the scalding and plucking of broilers. *Br. Poult. Sci.* **19,** 61–70.

Mulder, R. W. A. W., Notermans, S., and Kampelmacher, E. H. (1977b). Inactivation of salmonellae on chilled and deep frozen broiler carcasses by irradiation. *J. Appl. Bacteriol.* **42,** 179–185.

Muller, L. L. (1952). A method for determining the moisture distribution in butter and a review of its applications. *Aust. J. Dairy Technol.* **7,** 44–51.

Munch-Peterson, E. (1961). Staphylococcal carriage in man: An attempt at a quantitative survey. *Bull. W.H.O.* **24,** 761–769.

Munch-Peterson, E. (1963). Staphylococci in food and food intoxication. A review and an appraisal of phage typing results. *J. Food Sci.* **28,** 692–710.

Mundt, J. O., and Hammer, J. L. (1968). Lactobacilli on plants. *Appl. Microbiol.* **16,** 1326–1330.

Mundt, J. O., and Kitchen, H. M. (1951). Taint in Southern country-style hams. *Food Res.* **16,** 233–238.

Mundt, J. O., Anandam, E. J., and McCarty, I. E. (1966). Streptococceae in the atmosphere of plants processing vegetables for freezing. *Health Lab. Sci.* **3,** 207–213.

Mundt, J. O., Graham, W. F., and McCarty, I. E. (1967). Spherical lactic acid-producing bacteria of southern-grown raw and processed vegetables. *Appl. Microbiol.* **15,** 1303–1308.

Murdock, C. R., Crossley, E. L., Robb, J., Smith, M. E., and Hobbs, B. C. (1960). The pasteurization of liquid whole egg. *Mon. Bull. Minist. Health Public Health Lab. Serv.* **19,** 134–152.

Murdock, D. I. (1967). Methods employed by the citrus concentrate industry for detecting diacetyl and acetylmethylcarbinol. *Food Technol.* **21,** 643–647.

Murdock, D. I. (1968). Diacetyl test as a quality control tool in processing frozen concentrated orange juice. *Food Technol.* **22,** 90–94.

Murdock, D. I., Troy, V. S., and Folinazzo, J. F. (1953). Thermal resistance of lactic acid bacteria and yeast in orange juice and concentrate. *Food Res.* **18,** 85–89.

Murphy, R. P. (1965). Thermophilic bacterial contamination in sugar manufacture and the effect of ultraviolet rays. *Int. Sugar J.* **67,** 82–83.

Murphy, R. P. (1973). Microbiological contamination of dried vegetables. *Process Biochem.* **8**(10), 17–19.

Murphy, T. W., and Sutton, W. S. (1947). "The Pasteurization of Shell Eggs to Prevent Storage Rot and Maintain Quality: A Progress Report of Experimental Work," Misc. Publ. No. 3317. Dep. Agric. New South Wales, Australia.

Muys, I. G. T. (1971). Microbial safety in emulsions. *Process Biochem.* **6**(6), 25–28.

Nagarajan, V. and Bhat, R. V. (1973). Aflatoxin production in peanut varieties by *Aspergillus flavus* link and *Aspergillus parasiticus* speare. *Appl. Microbiol.* **25,** 319–321.

Nakamura, M., and Kelly, K. D. (1968). *Clostridium perfringens* in dehydrated soups and sauces. *J. Food Sci.* **33,** 424–426.

Nape, W. F., and Murphy, C. (1971). Recovery of salmonellae in feed mills, using terminally heated and regularly processed animal protein. *J. Am. Vet. Med. Assoc.* **159**, 1569–1572.

Narayan, K. G., and Takács, J. (1966). Incidence of clostridia in emergency slaughtered cattle. *Acta Vet. Hung.* **16**, 345–350.

Natarajan, C. P., Chari, C. N., and Mark, E. M. (1948). Yeast population in figs during drying. *Fruit Prod. J.* **27**, 242–243, 267.

National Academy of Sciences/National Research Council (NAS/NRC) (1969). "An Evaluation of the Salmonella Problem," Publ. No. 1683. NAS, Washington, D.C.

National Academy of Sciences/National Research Council (NAS/NRC) (1975). "Prevention of Microbial and Parasitic Hazards Associated with Processed Foods." NAS, Washington, D.C.

National Academy of Sciences/National Research Council (NAS/NRC) (1978). "Postharvest Food Losses in Developing Countries." NAS, Washington, D.C.

National Food Processors Association (1972). "Bacterial Standards for Sugar," June, 21, 1972, revised. Natl. Canners Assoc., Washington, D.C.

National Mastitis Council (1978). "Current Concepts of Bovine Mastitis," 2nd ed. Hoards Dairyman, Ft. Atkinson, Wisconsin.

National Soft Drink Association (1975). "Quality Specifications and Test Procedures for 'Bottlers' Granulated and Liquid Sugar." Natl. Soft Drink Assoc., Washington, D.C.

Necker, D. P., Kumta, U. S., and Sreenivasan, A. (1978). Radiation processing for the control of *Salmonella* in frog legs. *Use Radiat. Radioisot. Stud. Anim., Proc. Radiat. Symp., 1975, Izatnagar, India.*

Nelson, K. E., and Richardson, H. B. (1967). Storage temperature and sulfur dioxide treatment in relation to decay and bleaching of stored table grapes. *Phytopathology* **57**, 950–955.

Newcomer, J. L., Ramsey, E. W., and Eaton, H. D. (1962). Effect of air flow in a refrigerator on cooling rate. *J. Am. Diet. Assoc.* **40**, 39–40.

Newell, K. W. (1955). Paratyphoid B fever possibly associated with Chinese frozen egg. *Med. Res. Counc. (G.B.), Mon. Bull.* **14**(9), 146–154 [*Biol. Abstr.* **30**, 11001 (1956)].

Newell, K. W., Hobbs, B. C., and Wallace, E. J. G. (1955). Paratyphoid fever associated with Chinese frozen whole egg. Outbreaks in two bakeries. *Br. Med. J. ii*, 1296–1298.

Newton, K. G., Harrison, J. C. L., and Smith, K. M. (1977a). The effect of storage in various gaseous atmospheres on the microflora of lamb chops held at −1°C. *J. Appl. Bacteriol.* **43**, 53–59.

Newton, K. G., Harrison, J. C. L., and Smith, K. M. (1977b). Coliforms from hides and meat. *Appl. Environ. Microbiol.* **33**, 199–200.

Ng, H. (1966). Heat sensitivity of 300 *Salmonella* isolates. *In* "The Destruction of Salmonellae," ARS 74–37, pp. 39–41. Agric. Res. Serv., U.S. Dep. Agric., Albany, California.

Nichols, A. A., Davies, P. A., King, K. P., Winter, E. J., and Blackwall, F. L. C. (1971). Contamination of lettuce irrigated with sewage effluent. *J. Hortic. Sci.* **46**, 425–433.

Nickelson, R., Wyatt, L. E., and Vanderzant, C. (1975). Reduction of *Salmonella* contamination in commercially processed frog legs. *J. Food Sci.* **40**, 1239–1241.

Nicol, D. J., Shaw, M. K., and Ledward, D. A. (1970). Hydrogen sulfide production by bacteria and sulfmyoglobin formation in prepacked chilled beef. *Appl. Microbiol.* **19**, 937–939.

Niewiarowicz, A., Trojan, M., and Zielinska, T. (1967). Removal of glucose from raw egg white with the aid of enzyme containing yeast extract. *Przem. Spozyw.* **21**, 15–17 [*C. A.* **70**, 36512 (1969)].

Nikolaeva, S. A. (1967). *Cl. perfringens* in foodstuffs and in semimanufactured products of the canning industry. *Gig. Sanit.* **32**(5), 30–34. (In Russ.)

Nofsinger, G. W., Bothast, R. J., Lancaster, E. B., and Bagley, E. B. (1977). Ammonia-supplemented ambient temperature drying of high-moisture corn. *Trans. ASAE* **20**(6), 1151–1154, 1159.

Norda (1974). "Recreated Spice Particles." Norda, New York.

Norrish, R. S. (1966). An equation for the activity coefficients and equilibrium relative humidities of water in confectionery syrups. *J. Food Technol.* **1**, 25–39.

Northolt, M. D., Wiegersma, N., and van Schothorst, M. (1978). Pasteurization of dried egg white by high temperature storage. *J. Food Technol.* **13**, 25–30.

Notermans, S., and Kampelmacher, E. H. (1974). Attachment of some bacterial strains to the skin of broiler chickens. *Br. Poult. Sci.* **15**, 573–585.

Notermans, S., and Kampelmacher, E. H. (1975). Further studies on the attachment of bacteria to skin. *Br. Poult. Sci.* **16**, 487–496.

Notermans, S., Jeunink, J., van Schothorst, M., and Kampelmacher, E. H. (1974). Vergelijkende onderzoekingen over kruiskontaminatie in een spinchiller en in een sproeiriniger. (Comparative studies on cross contamination in a spin chiller and a spray cleaner.) *Tijdschr. Diergeneesk.* **99**, 352–357.

Notermans, S., van Leusden, F. M., van Schothorst, M., and Kampelmacher, E. H. (1975a). Salmonella-contaminatie van Slachtkuikens Tijdens Het Slachtproces in Enkele Pluimveeslachterijen. (Contamination of broiler chickens by *Salmonella* during processing in a number of poultry-processing plants.) *Tijdschr. Diergeneesk.* **100**, 259–264.

Notermans, S., van Schothorst, M., van Leusden, F. M., and Kampelmacher, E. H. (1975b). Onderzoekingen over het Kwantitatief Voorkomen van Salmonellae bij Diepvrieskuikens. (Quantitative studies for the presence of salmonellae in deep frozen broiler chickens.) *Tijdschr. Diergeneeskd.* **100**, 648–653.

Notermans, S., Kampelmacher, E. H., and van Schothorst, M. (1975c). Studies on sampling methods used in the control of hygiene in poultry processing. *J. Appl. Bacteriol.* **39**, 55–61.

Notermans, S., van Schothorst, M., and Kampelmacher, E. H. (1975d). Der Einfluss des Keimgehaltes des Spinchciller-Wassers auf den Keimgehalt des Tauwassers von Gefrierhähnchen. (The influence of the bacterial content of spin-chiller water on the bacterial content of thaw water from frozen chickens.) *Fleischwirtschaft* **55**, 1087–1090.

Notermans, S., Hindle, V., and Kampelmacher, E. H. (1976). Comparison of cotton swab versus alginate swab sampling method in the bacteriological examination of broiler chickens. *J. Hyg.* **77**, 205–210.

Notermans, S., van Leusden, F. M., and van Schothorst, M. (1977). Suitability of different bacterial groups for determining faecal contamination during post scalding stages in the processing of broiler chickens. *J. Appl. Bacteriol.* **43**, 383–389.

Nottingham, P. M. (1974). Analysis of microbiological data as a guide to quality processing. *Proc. Meat Ind. Res. Conf., 16th* Publ. No. 394, pp. 41–48. Meat Ind. Res. Inst. New Zealand Hamilton.

Nottingham, P. M., and Wyborn, R. (1975). Microbiology of beef processing. II. Chilling and aging. *N. Z. J. Agric. Res.* **18**, 23–27.

Nottingham, P. M., Penney, N., and Wyborn, R. (1972). *Salmonella* infection in calves and other animals. III. Further studies with calves and pigs. *N. Z. J. Agric. Res.* **15**, 279–283.

Nottingham, P. M., Rushbrook, A. J., and Jury, K. E. (1975). The effect of plating technique and incubation temperature on bacterial counts. *J. Food Technol.* **10**, 273–279.

Novak, A. F. (1973). Microbiological considerations in the handling and processing of crustacean shellfish. *In* "Microbial Safety of Fishery Products" (C. O. Chichester and H. D. Graham, eds.), pp. 59–73. Academic Press, New York.

Nunez, W. J., and Colmer, A. R. (1968). Differentiation of *Aerobacter-Klebsiella* isolated from sugarcane. *Appl. Microbiol.* **16**, 1875–1878.

Nurmi, E. (1966). Effect of bacterial inoculations on characteristics and microbial flora of dry sausage. Thesis, Univ. of Helsinki, Helsinki.

Nurmi, E., and Rantala, M (1973). New aspects of *Salmonella* infection in broiler production. *Nature (London)* **241**, 210–211.

Nury, F. S., Miller, M. W., and Brekke, J. E. (1960). Preservative effect of some antimicrobial agents on high-moisture dried fruits. *Food Technol.* **14**, 113–115.

Oberhauser, M. (1975). Untersuchungen über den Salmonellenbefall des im Handel feilgehalten Geflügels. *Fleischwirtschaft* **55**, 1424.

Ogilvy, W. S., and Ayres, J. C. (1951). Post-mortem changes in stored meats. II. The effect of atmospheres containing carbon dioxide in prolonging the storage life of cut-up chicken. *Food Technol.* **5**, 97–102.

Okabe, S. (1974). Statistical review of food poisoning in Japan, especially that of *Vibrio parahaemolyticus*. *In* "International Symposium on *Vibrio parahaemolyticus*" (T. Fujino, G. Sakaguchi, R. Sakazaki, and Y. Takeda, eds.), Saikon, Tokyo.

Okwelogu, T. N. (1979). Moisture meters. *In* "Postharvest Grain Loss Assessment Methods" (K. L. Harris and C. J. Lindblad, eds.), pp. 167–186. Am. Assoc. Cereal Chem., St. Paul, Minnesota.

Oldfield, J. F. T., Dutton, J. V., and Teague, H. J. (1971). The significance of invert and gum formation in deteriorated beet. *Int. Sugar J.* **73**, 3–8, 35–40, 66–68.

Oldfield, J. F. T., Dutton, J. V., and Shore, M. (1974a). Effects of thermophilic activity in diffusion on sugar beet processing. Part I. *Int. Sugar J.* **76**, 260–263.

Oldfield, J. F. T., Dutton, J. V., and Shore, M. (1974b). Effects of thermophilic activity in diffusion on sugar beet processing. Part 2. *Int. Sugar J.* **76**, 301–305.

Olson, J. C., Jr., and Neilsen, A. J. (1955). Changes in bacterial counts and flavor of dry milks after recombination by various means and storage at low temperatures. *J. Dairy Sci.* **38**, 361–370.

Olson, J. C., Jr., Casman, E. P., Baer, E. F., and Stone, J. E. (1970). Enterotoxigenicity of *Staphylococcus aureus* cultures isolated from acute cases of bovine mastitis. *Appl. Microbiol.* **20**, 605–607.

Olson, N. F. (1969). "Ripened Semi-soft Cheese," Pfizer Cheese Monograph, Vol. IV. Pfizer, New York.

Omura, J., Price, R. J., and Olcott, H. S. (1978). Histamine-forming bacteria isolated from spoiled skipjack tuna and jack mackerel. *J. Food Sci.* **43**, 1779–1781.

Orillo, C. A., Sison, E. C., Luis, M., and Pederson, C. S. (1969). Fermentation of Philippine vegetable blends. *Appl. Microbiol.* **17,** 10–13.

Orthoefer, J. G., Schrieber, M., Nichols, J. B., and Schneider, N. (1968). Salmonella contamination in a rendering plant. *Avian Dis.* **12,** 303–311.

Ostergren, B. (1974). Practical biological testing in confectionery plants. *Manuf. Confect.* **54**(6), 62, 64, 66, 69.

Ostovar, K. (1971). Isolation and characterization of microorganisms involved in the fermentation of Trinidad's cacao beans. Ph.D. Thesis, Pennsylvania State Univ., University Park [*Diss. Abstr.* **33,** 344–B (1972)].

Ostovar, K. (1973). A study on survival of *Staphylococcus aureus* in dark and milk chocolate. *J. Food Sci.* **38,** 663–664.

Ostovar, K., and Keeney, P. G. (1973). Isolation and characterization of microorganisms involved in the fermentation of Trinidad's cacao beans. *J. Food Sci.* **38,** 611–617.

Ostovar, K., and Ward, K. (1976). Detection of *Staphylococcus aureus* from frozen and thawed convenience pasta products. *Lebensm.- Wiss. Technol.* **9**(4), 218–219 [*Biol. Abstr.* **63,** 20448 (1977)].

Owen, W. L. (1977). Microbiology of sugar manufacture and refining. *In* "Cane Sugar Handbook" (G. P. Meade and J. C. P. Chen, eds.), 10th ed., pp. 405–422. Wiley, New York.

Pace, P. J. (1975). Bacteriological quality of delicatessen foods: are standards needed? *J. Milk Food Technol.* **38,** 347–353.

Pace, P. J., Silver, K. J., and Wisniewski, H. J. (1977). *Salmonella* in commercially produced dried dog food: possible relationship to a human infection caused by *Salmonella enteritidis* serotype Havana. *J. Food Prot.* **40,** 317–321.

Pal, N., and Kundu, A. K. (1972). Studies on *Aspergillus* spp. from Indian spices in relation to aflatoxin production. *Sci. Cult.* **38,** 252–254.

Pallanch, M. J. (1971). Dried products. *In* "By Products from Milk" (B. H. Webb and E. D. Whittier, eds.), 2nd ed., Chap. 5. AVI, Westport, Connecticut.

Palumbo, S. A., Rivenburgh, A. I., Smith, J. L., and Kissinger, J. C. (1975). Identification of *Bacillus subtilis* from sausage products and spices. *J. Appl. Bacteriol.* **38,** 99–105.

Pantaléon, J., and Rosset, R. (1964). Sur la présence de *Salmonella* dans les grenouilles destinées à la consommation humaine. *Ann. Inst. Pasteur Lille* **15,** 225–227.

Papavassiliou, J., Tzannetis, S., Leka, H., and Michopoulos, G. (1967). Coli-aerogenes bacteria on plants. *J. Appl. Bacteriol.* **30,** 219–223.

Papavizas, G. C., and Christensen, C. M. (1957). Grain storage studies. XXV. Effect of invasion by storage fungi upon germination of wheat seed and upon development of sick wheat. *Cereal Chem.* **34**(5), 350–359.

Papavizas, G. C., and Christensen, C. M. (1958). Grain storage studies. XXVI. Fungus invasion and deterioration of wheats stored at low temperatures and moisture contents of 15 to 18 percent. *Cereal Chem.* **35,** 27–34.

Park, C., Rayman, K., Stankiewicz, Z., and Hauschild, A. (1976). Toxic effect of cocoa on *Salmonella* during the pre-enrichment period. *Proc. Annu. Meet. Can. Fed. Biol. Soc., 19th, Halifax, Nova Scotia.*

Park, C. E., El Derea, H. B., and Rayman, M. K. (1978). Evaluation of staphylococcal thermonuclease (TNA) assay as a means of screening foods for growth of staphylococci and possible enterotoxin production. *Can. J. Microbiol.* **24,** 1135–1139.

Park, C. E., Stankiewicz, Z. K., Rayman, M. K., and Hauschild, A. H. W. (1979). Inhibitory effect of cocoa powder on the growth of a variety of bacteria in different media. *Can. J. Microbiol.* **25,** 233–235.

Park, H. S., Marth, E. H., and Olson, N. F. (1973). Fate of enteropathogenic strains of *Escherichia coli* during the manufacture and ripening of Camembert cheese. *J. Milk Food Technol.* **36,** 543–546.

Park, S. E., Graham, M. J., Prucha, J. M., and Brennan, J. M. (1932). Pasteurization of milk artificially infected with two strains of *Brucella suis. J. Bacteriol.* **24,** 461–471.

Parkinson, T. L. (1966). The chemical composition of eggs. *J. Sci. Food Agric.* **17,** 101–111.

Parry, J. W. (1969a). "Spices. Vol. 1: The Story of Spices. The Spices Described." Chem. Publ. Co., New York.

Parry, J. W. (1969b). "Spices. Vol. 2: Morphology, Histology, Chemistry." Chem. Publ. Co., New York.

Parry, W. H. (1963). The problem of *Salmonella* food poisoning. *Med. Off.* **109,** 27–32.

Patashnik, M. (1953). A simplified procedure for thermal process evaluation. *Food Technol.* **7,** 1–6.

Patrick, T. E., Goodwin, T. L., Collins, J. A., Wyche, R. C., and Love, B. E. (1972). Steam versus hot-water scalding in reducing bacterial loads on the skin of commercially processed poultry. *Appl. Microbiol.* **23,** 796–798.

Patrick, T. E., Collins, J. A., and Goodwin, T. L. (1973). Isolation of *Salmonella* from carcasses of steam- and water-scalded poultry. *J. Milk Food Technol.* **36,** 34–36.

Patterson, J. T. (1967). The incidence of enteropathogenic *E. coli* and salmonellae in processed broilers and on New York dressed poultry. *North. Irel. Minist. Agric. Rec. Agric. Res.* **16,** 151–155.

Patterson, J. T. (1968). Bacterial flora of chicken carcasses treated with high concentrations of chlorine. *J. Appl. Bacteriol.* **31,** 544–550.

Patterson, J. T. (1969). Salmonellae in meat and poultry, poultry plant cooling waters and effluents, and animal feedingstuffs. *J. Appl. Bacteriol.* **32,** 329–337.

Patterson, J. T. (1971). Microbiological assessment of surfaces. *J. Food Technol.* **6,** 63–72.

Patterson, J. T. (1972a). Salmonellae in processed poultry. *North Irel. Minist. Agric. Rec. Agric. Res.* **20,** 1–6.

Patterson, J. T. (1972b). Microbiological sampling of poultry carcasses. *J. Appl. Bacteriol.* **35,** 569–575.

Patterson, J. T. (1972c). Salmonellae in animal feedingstuffs. *North Irel. Minist. Agric. Rec. Agric. Res.* **20,** 27–33.

Patterson, J. T. (1973). Airborne micro-organisms in poultry processing plants. *Br. Poult. Sci.* **14,** 161–165.

Patterson, J. T., and Gibbs, P. A. (1973). Observations on the microbiology of cooked chicken carcasses. *J. Appl. Bacteriol.* **36,** 689–697.

Patterson, J. T., and Gibbs, P. A. (1977). Incidence and sources of Enterobacteriaceae found on frozen broilers. *Proc. Eur. Symp. Poult. Meat Qual., 3rd* pp. 69–75.

Patterson, R. L. S., and Edwards, R. A. (1975). Volatile amine production in uncured pork during storage. *J. Sci. Food Agric.* **26,** 1371–1373.

Paul, P., Symonds, H., Varozza, A., and Stewart, G. F. (1957a). Effect of glucose removal on storage stability of egg yolk solids. *Food Technol.* **11**, 494–498.

Paul, P., Varozza, A., Stewart, G. F., and Bergquist, D. H. (1957b). Effects of several pasteurization schedules on performance of egg yolk solids in doughnuts. *Food Technol.* **11**, 508–509.

Payne, D. B., and Perigo, J. A. (1960). Observations on the microbiology of carbonated soft drinks in cans. *Ann. Inst. Pasteur Lille* **11**, 99–104.

Pederson, C. S. (1930). Bacterial spoilage of a Thousand Island dressing. *J. Bacteriol.* **20**, 99–106.

Pederson, C. S. (1938). A bacteriological study of raw cane sugar plants. *J. Bacteriol.* **35**, 74–75.

Pederson, C. S. (1971). "Microbiology of Food Fermentations." AVI, Westport, Connecticut.

Pederson, C. S., and Hucker, G. (1946). The significance of bacteria in sugar mills. *Proc. Meet. Asoc. Tech. Azucar Cuba, 20th* pp. 225–230.

Pederson, C. S., and Kelly, C. D. (1938). Development of pink color in sauerkraut. *Food Res.* **3**, 583–588.

Pederson, C. S., Niketić, G., and Albury, M. N. (1962). Fermentation of the Yugoslavian pickled cabbage. *Appl. Microbiol.* **10**, 86–89.

Peel, B. (1976). Occurrence of salmonellae in raw and pasteurized liquid whole egg. *Queensl. J. Agric. Anim. Sci.* **33**(1), 13–21 [*Biol. Abstr.* **64**, 38129 (1977)].

Pelhate, J. (1976). Microflora of moist maize: Determination of its development. *Bull. Organ. Eur. Mediterr. Prot. Plant.* **6**(2), 91–100.

Pendergrass, A., and Isenberg, F. M. R. (1974). The effect of relative humidity on the quality of stored cabbage. *Hortic. Sci.* **9**, 226–227.

Penniston, V. A., and Hedrick, L. R. (1947). The reduction of bacterial count in egg pulp by use of germicides in washing dirty eggs. *Food Technol.* **1**, 240–244.

Perquin, L. H. C. (1940). On the incidental occurrence of rod-shaped, dextran producing bacteria in a beet-sugar factory. *Antonie van Leeuwenhoek; J. Microbiol. Serol.* **6**, 227–249.

Petersen, P. J. (1964). An outbreak of food infection caused by mayonnaise with *Salmonella typhimurium*. *Medlemsbl. Dan. Dyrlaegeforen.* **47**, 284–287. (In Dan.)

Peterson, A. C., and Gunnerson, R. E. (1974). Microbiological critical control points in frozen foods. *Food Technol.* **28**(9), 37–44.

Peterson, G. T., Fox, J. F., and Martin, L. E. (1960). Problems in the preparation and handling of hot vended canned foods. *Food Technol.* **14**, 89–91.

Petrushina, L. I. (1976). Reproduction of staphylococci and the formation of type A enterotoxin in cooked macaroni products. *Vopr. Pitan.* **5**, 61–64 [*Biol. Abstr.* **64**, 38131 (1977)].

Pette, J. W., and Lolkema, H. (1950a). Growth stimulating factors for *Staphylococcus thermophilus*. *Neth. Milk Dairy J.* **4**, 209.

Pette, J. W., and Lolkema, H., (1950b). Acid production and aroma formation in yoghurt. *Neth. Milk Dairy J.* **4**, 261.

Pette, J. W., and Lolkema, H. (1950c). Symbiosis and antibiosis in mixed cultures of *Lactobacillus bulgaricus* and *Streptococcus thermophilus*. *Neth. Milk Dairy J.* **4**, 1970.

Pettit, R. E., and Taber, R. A. (1968). Factors influencing aflatoxin accumulation in peanut kernels and the associated mycoflora. *Appl. Microbiol.* **16**, 1230–1234.

Pettit, R. E., Taber, R. A., Schroeder, H. W., and Harrison, A. L. (1971). Influence of fungicides and irrigation practice on aflatoxin in peanuts before digging. *Appl. Microbiol.* **22,** 629–634.

Pfeiffer, E. H. (1972). Hygienic evaluation of packaging material for food (plastics or cardboard). *Zentralbl. Bakteriol., Parasitenkd., Infectionskr. Hyg., Abt. 1: Orig., Reihe B* **156**(4/5), 446–455.

Philbrook, F. R., MacCready, R. A., van Roekel, H., Anderson, E. S., Smyser, C. F., Sanen, F. J., and Groton, W. M. (1960). Salmonellosis spread by a dietary supplement of avian source. *N. Engl. J. Med.* **263,** 713–718.

Pier, A. C., Gray, D. M., and Fossatti, M. J. (1958). A newly recognized pathogen of the mastitis complex. *Am. J. Vet. Res.* **16,** 319.

Pirie, D. G., and Clayson, D. H. F. (1964). Some causes of unreliability of essential oils as microbial inhibitors in foods. *In* "Microbial Inhibitors in Foods" (N. Molin, ed.), pp. 145–150. Almqvist & Wiksell, Stockholm.

Pitt, J. I. (1974). Xerophilic fungi and the spoilage of foods of plant origin. *In* "Water Relations of Foods" (R. B. Duckworth, ed.), pp. 273–307. Academic Press, New York.

Pitt, J. I., and Christian, J. H. B. (1968). Water relations of xerophilic fungi isolated from prunes. *Appl. Microbiol.* **16,** 1853–1858.

Pivnick, H. (1974). *Salmonella* in chocolate—estimation of a potential risk. *Latin Am. Congr. Microbiol., 6th, Caracas.* Unpublished.

Pivnick, H. (1978). Canadian microbiological standards for foods. *Food Technol.* **32,** 58–60.

Pivnick, H., Rubin, L. J., Barnett, H. W., Nordin, H. R., Ferguson, P. A., and Perrin, C. H. (1967). Effect of sodium nitrite and temperature on toxinogenesis by *Clostridium botulinum* in perishable cooked meats vacuum-packed in air-impermeable plastic pouches. *Food Technol.* **21,** 204–206.

Pivnick, H., Erdman, I. E., Manzatiuk, S., and Pommier, E. (1968). Growth of food poisoning bacteria on barbecued chicken. *J. Milk Food Technol.* **31,** 198–201.

Pivnick, H., Erdman, I. E., Collins-Thompson, D., Roberts, G., Johnston, M. A., Conley, D. R., Lachapelle, G., Purvis, U. T., Foster, R., and Milling, M. (1976). Proposed microbiological standards for ground beef based on a Canadian survey. *J. Milk Food Technol.* **39,** 408–412.

Pivnick, H., Handzel, S., and Lior, H. (1978). Product contamination with *Salmonella. Int. Symp. Salmonella Prospects Control, Univ. Guelph.* Guelph, Canada.

Pohja, M. S. (1957). Vergleichende Untersuchungen über den Mikrobengehalt fester und flüssiger Gewürze. *Fleischwirtschaft* **9,** 547–548.

Pollard, A. (1959). Organic acids and amino acids in fruit juices. *Rep. Int. Fruit Juice Congr., 5th, Vienna.*

Pomeranz, Y. (1957). Effectiveness of sorbic acid in preservation of damp wheat. *Food Res.* **22,** 553–561.

Pomeranz, Y., Daftary, R. D., Shogren, M. D., Hoseney, R. C., and Finney, K. F. (1968). Changes in biochemical and bread-making properties of storage-damaged flour. *J. Agric. Food Chem.* **16**(1), 92–96.

Pomeroy, B. S., Solee, R. E., Mahesh, C. K., and Loken, K. I. (1969). The *Salmonella* problem in animal feeds. *Proc. Symp. Biol. Contam. Grain Anim. Byprod.* pp. 69–76. Agric. Ext. Serv., Univ. of Minnesota, Minneapolis.

Porter, D. M., and Garren, K. H. (1970). Endocarpic microorganisms of two types of windrow-dried peanut fruit (*Arachis hypogaea* L.). *Appl. Microbiol.* **20,** 133–138.

Portnoy, B. L., Goepfert, J. M., and Harmon, S. M. (1976). An outbreak of *Bacillus cereus* food poisoning resulting from contaminated vegetable sprouts. *Am. J. Epidemiol.* **103**, 589–594.

Post, F. J., Bliss, A. H., and O'Keefe, W. B. (1961). Studies on the ecology of selected food poisoning organisms in food. I. Growth of *Staphylococcus aureus* in cream and a cream product. *J. Food Sci.* **26**, 436–441.

Powers, E. M., Lawyer, R., and Masuoka, Y. (1975). Microbiology of processed spices. *J. Milk Food Technol.* **38**, 683–687.

Powers, E. M., Latt, T. G., and Brown, T. (1976). Incidence and levels of *Bacillus cereus* in processed spices. *J. Milk Food Technol.* **39**, 668–670.

Powers, J. J. (1976). Effect of acidification of canned tomatoes on quality and shelf life. *CRC Crit. Rev. Food Sci. Nutr.* **7**, 371–395.

Poynter, S. F. B., and Mead, G. C. (1964). Volatile organic liquids and slime production. *J. Appl. Bacteriol.* **27**, 182–195.

Pramanik, A. K., and Khanna, P. N. (1977). Incidence of *Salmonella* in slaughtered poultry and its public health significance. *Indian J. Public Health* **16**, 41–43.

Preonas, D. L., Nelson, A. I., Ordal, Z. J., Steinberg, M. P., and Wei, L. S. (1969). Growth of *Staphylococcus aureus* MF31 on the top and cut surfaces of southern custard pies. *Appl. Microbiol.* **18**, 68–75.

Presnell, M. W., Miescier, J. J., and Hill, W. F., Jr. (1967). *Clostridium botulinum* in marine sediments and in the oyster (*Crassostrea virginica*) from Mobile Bay. *Appl. Microbiol.* **15**, 668–669.

Presnell, M. W., Cummins, J. M., and Miescier, J. J. (1969). Influence of selected environmental factors on the elimination of bacteria by the Eastern oyster (*Crassostrea virginica*). *Proc. Gulf Atl. Shellfish Sanit. Conf.* Public Health Serv., U.S. Dep. Health, Educ. Welfare, Washington, D.C.

Public Health Laboratory Service (1973). *Bacillus cereus* food poisoning. *Br. Med. J. iii*, 647.

Punch, J. D., Olson, J. C., Jr., and Thomas, E. L. (1965). Psychrotrophic bacteria. III. Population levels associated with flavor or physical change in milk. *J. Dairy Sci.* **48**, 1179–1183.

Put, H. M. C., De Jong, J., Sand, F. E. M. J., and van Grinsven, A. M. (1976). Heat resistance studies on yeast spp. causing spoilage in soft drinks. *J. Appl. Bacteriol.* **50**, 135–152.

Pyler, E. J. (1973). "Baking Science and Technology," pp. 210–221. Siebel, Chicago, Illinois.

Quesnel, V. C. (1968). Oxygen consumption and heat production during the fermentation of cacao. *Turrialba* **18**(2), 110–114.

Quevedo, F. (1965). Les enterobacteriaceae dans la farine de poisson. *Ann. Inst. Pasteur Lille* **16**, 157–162.

Quevedo, F., and Carranza, N. (1966). Le rôle des mouches dans la contamination des aliments au Perou. *Ann. Inst. Pasteur Lille* **17**, 199–202.

Qvist, S. (1976). Microbiology of sliced vacuum-packed meat products. *Meat Res. Congr., 22nd, Malmö.*

Raj, H., and Liston, J. (1961). Survival of bacteria of public health significance in frozen seafoods. *Food Technol.* **15**, 429–434.

Ramadan, F. M., El-Zanfaly, R. T., Alian, A. M., and El-Wakeil, F. A. (1972). On the antibacterial effects of some essential oils. II. Studies on semi-solid agar phase. *Chem., Mikrobiol., Technol. Lebensm.* **2**, 96–102.

Randolph, H. E., Chakraborty, B. K., Hampton, O., and Bogart, D. L. (1973).

Microbial counts of individual producer and comingled Grade A raw milk. *J. Milk Food Technol.* **36,** 146–151.

Rantala, M. (1974). Cultivation of a bacterial flora able to prevent the colonization of *Salmonella infantis* in the intestines of broiler chickens, and its use. *Acta Pathol. Microbiol. Scand., Sect. B* **82,** 75–80.

Rao, N. M., Nandy, S. C., Joseph, K. T., and Santappa, M. (1978). Control of *Salmonella* in frog legs by chemical and physical methods. *Indian J. Exp. Biol.* **16,** 593–596.

Read, R. B., Jr., Schwartz, C., and Litsku, W. (1961). Studies on thermal destruction of *Escherichia coli* in milk and milk products. *Appl. Microbiol.* **9,** 415–418.

Recca, J., and Mrak, E. M. (1952). Yeasts occurring in citrus products. *Food Technol.* **6,** 450–454.

Reddy, S. G., Chen, M. L., and Patel, P. J. (1975). Influence of lactic cultures on the biochemical, bacterial, and organoleptic changes in beef. *J. Food Sci.* **40,** 314–318.

Rehm, H. J. (1959). Beitrag zur Wirkung von Konservierungsmittel-Kombinationen. I. Grundlagen und Überblick zur Kombinationswirkung chemischer Konservierungsmittel. *Z. Lebensm.-Unters.-Forsch.* **110,** 283–291.

Rehm, H. J., and Meyer, H. (1967). Mykotoxinbildung in Fruchtsäften. *Ind. Obst-Gemueseverwert.* **52,** 675–677.

Rehm, H. J., and Wittmann, H. (1962). Beitrag zur Kenntnis der antimicrobiellen Wirkung der schwefligen Säure. I. Übersicht über einflussnehmende Factoren auf die antimikrobielle Wirkung der schwefligen Säure. *Z. Lebensm.-Unters. -Forsch.* **118,** 413–429.

Reinbold, G. W. (1963). "Italian Cheese Varieties," Pfizer Cheese Monograph, Vol. I. Pfizer, New York.

Reinbold, G. W. (1972). "Swiss Cheese Varieties," Pfizer Cheese Monograph, Vol. V. Pfizer, New York.

Reinbold, G. W., and Reddy, M. S. (1973). Bacteriophage for *Streptococcus thermophilus. Dairy Ind.* **38,** 413.

Reinbold, G. W., and Reddy, M. S. (1974). Sensitivity or resistance of dairy starter and associated microorganisms to selected antibiotics. *J. Milk Food Technol.* **37,** 517–521.

Reinbold, G. W., and Wilson, H. L. (1965). "American Cheese Varieties," Pfizer Cheese Monograph, Vol. II. Pfizer, New York.

Reinke, W. C., and Baker, R. C. (1966). The effect of pasteurizing liquid whole egg on viscosity, α-amylase, and *Salmonella senftenberg. Poult. Sci.* **45,** 1321–1327.

Rettger, L. F., and Cheplin, H. A. (1921). "A Treatise on the Transformation of the Intestinal Flora with Special Reference to the Implantation of *Bacillus acidophilus.*" Yale Univ. Press, New Haven, Connecticut.

Rettger, L. F., Levy, M. N., Weinstein, L., and Weiss, J. E. (1935). "*Lactobacillus acidophilus* and its Therapeutic Application." Yale Univ. Press, New Haven, Connecticut.

Reusse, U., Meyer, H., and Tillack, J. (1976a). Zur Methodik des Salmonellen-Nachweises aus gefrorenem Geflügel. *Arch. Lebensmittelhyg.* **27,** 98–100.

Reusse, U., Hafke, A., and Geister, R. (1976b). Feststellung des Enterobacteriaceen-Gehaltes von Fischmehl zur Beurteilung der Salmonellen-Freiheit. *Zentralbl. Bakteriol., Parasitenkd., Infektionskr. Hyg., Abt. 1: Orig., Reihe B* **162,** 288–306.

Reuter, G. (1975). Classification problems, ecology and some biochemical activities of lactobacilli of meat products. *In* "The Lactic Acid Bacteria in Beverages and Food" (J. G. Carr, C. V. Cutting, and G. C. Whiting, eds.), pp. 221–229. Academic Press, New York.

Rey, C. R., Walker, H. W., and Rohrbaugh, P. L. (1975). The influence of temperature on growth, sporulation, and heat resistance of spores of six strains of *Clostridium perfringens. J. Milk Food Technol.* **38,** 461–465.

Rice, S. L., Eitenmiller, R. R., and Koehler, P. E. (1976). Biologically active amines in food. A review. *J. Milk Food Technol.* **39,** 353–358.

Rice, S. L., Beuchat, L. R., and Worthington, R. E. (1977). Patulin production by *Byssochlamys* spp. in fruit juices. *Appl. Environ. Microbiol.* **34,** 791–796.

Richard, J., and Auclair, J. E. (1967). Étude des contaminations microbiennes du lait à la production. *Rev. Lait. Fr.* **236,** 227–230.

Riemann, H., ed. (1969). "Food-borne Infections and Intoxications." Academic Press, New York.

Rieschel, H., and Schenkel, J. (1971). Das Verhalten von Mikroorganismen, speziell Salmonellen in Schokoladenwaren. *Alimenta* **10**(2), 57–66.

Riha, W. E., and Solberg, M. (1970). Microflora of fresh pork sausage casings. 2. Natural casings. *J. Food Sci.* **35,** 860–863.

Rizzuto, A. B., and Jacobson, G. R. (1964). Use of membrane filters for determining microbial counts in sugar products. *Am. Soft Drink J.* September, 30–38.

Rizzuto, A. B., Skole, R. D., and Borodkin, S. E. (1964). The significance of the lactate fermenting anaerobic bacteria in sugar refining. Part I. The isolation and identification of a lactate fermenting bacteria commonly found in raw sugar and commercial corn starch. *Proc. Tech. Sess. Cane Sugar Refin. Res., New Orleans* ARS 72–44, pp. 68–75. Agric. Res. Serv., U.S. Dep. Agric., New Orleans.

Roberts, A. C., and McWeeny, D. J. (1972). The uses of sulphur dioxide in the food industry. A review. *J. Food Technol.* **7,** 221–238.

Roberts, D. (1972). Observations on procedures for thawing and spit-roasting frozen dressed chickens and post-roasting care and storage: With particular reference to food-poisoning bacteria. *J. Hyg.* **70,** 565–588.

Roberts, D., Boag, K., Hall, M. L. M., and Shipp, C. R. (1975). The isolation of salmonellas from British pork sausages and sausage meat. *J. Hyg.* **75,** 173–184.

Roberts, T. A. (1976). Establishing microbiological guidelines. *Inst. Meat Bull.* No. 94, 24–27.

Roberts, T. A., and Smart, J. L. (1976). The occurrence and growth of *Clostridium* spp. in vacuum-packed bacon with particular reference to *Cl. perfringens* (welchii) and *Cl. botulinum. J. Food Technol.* **11,** 229–244.

Robertson, J. A., and Thomas, J. K. (1976). Chemical and microbial changes in dehulled confectionery sunflower kernels during storage under controlled conditions. *J. Milk Food Technol.* **39,** 18–23.

Robinson, D. S., Barnes, E. M., and Taylor, J. (1975). Occurrence of β-hydroxybutyric acid in incubator reject eggs. *J. Sci. Food Agric.* **26,** 91–98.

Robinson, R. J. (1967). Microbiological problems in baking. *Bakers Dig.* **41**(5), 80–83.

Robinson, R. J., Lord, T. H., Johnson, J. A., and Miller, B. S. (1958). The aerobic microbiological population of pre–ferments and the use of selected bacteria for flavor production. *Cereal Chem.* **35,** 295–305.

Rodricks, J. V., and Lovett, J. (1976). Toxigenic fungi. *In* "Compendium of Methods for the Microbiological Examination of Foods" (M. L. Speck, ed.), pp. 484–503. Am. Public Health Assoc., Washington, D.C.

Roelofsen, P. A. (1958). Fermentation, drying, and storage of cacao beans. *Adv. Food Res.* **8**, 225–296.

Rogers, A. B., Sebring, M., and Kline, R. W. (1966). Hydrogen peroxide pasteurization process for egg white. *In* "The Destruction of Salmonellae," ARS 74–37, pp. 68–72. Agric. Res. Serv., U.S. Dep. Agric., Albany, California.

Rogers, E. A., Maier, K., Aserkoff, B., Woodward, W., Young, L., Martin, W., and Wells, J. (1967). A large outbreak of salmonellosis following a turkey barbecue. *Salmonella Surveillance Rep.* (*U.S.*) **67**, 2–3.

Rogers, R. E., and Gunderson, M. F. (1958). Roasting of frozen stuffed turkeys. I. Survival of *Salmonella pullorum* in inoculated stuffing. *Food Res.* **23**, 87–95.

Rohan, T. A. (1958a). Processing of raw cocoa. I. Small-scale fermentation. *J. Sci. Food Agric.* **9**, 104–111.

Rohan, T. A. (1958b). Processing of raw cocoa. II. Uniformity in heap fermentation and development of methods for rapid fermentation of West African Amelonado cocoa. *J. Sci. Food Agric.* **9**, 542–551.

Rohan, T. A. (1963). Processing of raw cocoa for the market. *FAO Agric. Stud.* No. 60.

Rombouts, J. E. (1952). Observations on the microflora of fermenting cacao beans in Trinidad. *Proc. Soc. Appl. Bacteriol.* **15**(1), 103–111.

Rombouts, J. E. (1953). Contribution to the knowledge of the yeast flora of fermenting cacao 1. A critical review of the yeast species previously described from cacao. *Trop. Agric.* (*Trinidad*) **30**, 34–41.

Rosengarten, F. (1973). "The Book of Spices." Pyramid Publ., Moonachie, New Jersey.

Ross, K. D. (1975). Estimation of water activity in intermediate moisture foods. *Food Technol.* **29**(3), 26–34.

Rosser, F. T. (1942). Preservation of eggs. II. Surface contamination on egg shell in relation to spoilage. *Can. J. Res., Sect. D* **20**, 291–296.

Rowe, B. (1973). Salmonellosis in England and Wales. *In* "The Microbiological Safety of Food" (B. C. Hobbs and J. H. B. Christian, eds.), pp. 165–180. Academic Press, New York.

Rubinsten-Szturn, S., Courtieur, A. L., and Maka, G. (1964). A cheese contaminated with *Shigella sonnei* as a cause of food poisoning. *Bull. Acad. Natl. Med., Paris* **148**, 480–482.

Ruehle, G. L. A., and Kulp, U. L. (1915). Germ content of stable air and its effect upon the germ content of milk. *N.Y.* (*Geneva*) *Agric. Exp. Stn., Bull.* No. 409.

Rutherford, P. P., and Murray, W. W. (1963). The effect of selected polymers upon the albumen quality of eggs after storage for short periods. *Poult. Sci.* **42**, 499–505.

Rutqvist, L. (1961). Vorkommen von Salmonella in Futtermitteln Vegetabilischen Ursprunges. *Zentralbl. Veterinaermed.* **8**, 1016–1024.

Rutqvist, L., and Waxberg, G. (1962). Observations on the contamination of vegetable fodder with Salmonella. (In Swedish). *Proc. Nord. Vet. Congr., 9th, Copenhagen* Sect. B(16), pp. 1–7.

Ryberg, R. E., and Cathcart, W. H. (1942). Staphylococci and *Salmonella* control in foods. II. Effect of pure fruit fillings. *Food Res.* **7**, 10–15.

Rylander, J. A. (1902). Method of preserving eggs. U.S. Patent 696,495.

Saisithi, P., Kasemsarn, B., Liston, J., and Dollar, A. M. (1966). Microbiology and chemistry of fermented fish. *J. Food Sci.* **31,** 105–110.

Sakabe, Y. (1973). Studies on allergy-like food poisoning. 1. Histamine production by *Proteus morganii. Nara Igaku Zasshi* **24,** 248–256.

Sakazaki, R. (1969). Halophilic vibrio infections. *In* "Food Borne Infections and Intoxications" (H. Riemann, ed.), pp. 115–129. Academic Press, New York.

Salton, M. R. J., Scott, W. J., and Vickery, J. R. (1951). Studies in the preservation of shell eggs. VI. The effect of pasteurization on bacterial rotting. *Aust. J. Appl. Sci.* **2**(1), 205–222.

Salzer, U., Bröker, U., Klie, H., and Liepe, H. (1977). Wirkung von Pfeffer und Pfefferinhaltsstoffen auf die Mikroflora von Wurstwaren. *Fleischwirtschaft* **57,** 2011–2014, 2017–2021.

Samish, Z., and Etinger-Tulczynska, R. (1963). Distribution of bacteria within the tissue of healthy tomatoes. *Appl. Microbiol.* **11,** 7–10.

Samish, Z., Etinger-Tulczynska, R., and Bick, M. (1961). Microflora within healthy tomatoes. *Appl. Microbiol.* **9,** 20–25.

Samish, Z., Etinger-Tulczynska, R., and Bick, M. (1963). The microflora within the tissue of fruits and vegetables. *J. Food Sci.* **28,** 259–266.

Sand, F. E. M. J. (1969). An ecological survey of yeasts within a soft drinks plant. *Proc. Int. Symp. Yeasts, 4th, Hochsch. Bodenkult., Vienna* Part I, p. 263.

Sand, F. E. M. J. (1971). Zur Bakterien Flora von Erfrischungsgetränken. *Brauwelt* **111,** 252–264.

Sand, F. E. M. J. (1973). Recent investigation on the microbiology of fruit juice concentrates. *In* "Scientific-Technical Commission XIII. Technology of Fruit Juice Concentrates—Chemical Composition of Fruit Juices," Int. Fed. Fruit Juice Processors, Vienna.

Sanders, D. H. (1969). Fluorescent dye tracing of water entry and retention in chilling of broiler chicken carcasses. *Poult. Sci.* **48,** 2032–2037.

Sanders, D. H., and Blackshear, C. D. (1971). Effect of chlorination in the final washer on bacterial counts of broiler chicken carcasses. *Poult. Sci.* **50,** 215–219.

Sanders, E., D'Alessio, D., Bauman, M. L., Harvey, R. B., Aiken, J. F., Jr., Cross, W. D., Cook, L., and Lloyd, B. H. (1963). Food poisoning: Public health implications of an outbreak of *Salmonella typhimurium* gastroenteritis in Wichita. *J. Kans. Med. Soc.* **64,** 293–298, 302.

Sanders, G. P. (1953). Cheese varieties and descriptions. *U.S. Dep. Agric., Agric. Handb.* No. 54.

Sandine, W. E. (1979). "Lactic Starter Technology," Pfizer Monograph, Vol. VI. Pfizer, New York.

Sandine, W. E., Muralidhara, K. S., Elliker, P. R., and England, D.C. (1972). Lactic acid bacteria in food and health: A review with special reference to enteropathogenic *Escherichia coli* as well as certain enteric diseases and their treatment with antibiotics and lactobacilli. *J. Milk Food Technol.* **35,** 691–702.

Sauer, D. B., and Burroughs, R. (1974). Efficacy of various chemicals as grain mold inhibitors. *Trans. ASAE* **17,** 557–559.

Sauer, D. B., and Christensen, C. M. (1968). Germination percentage, storage fungi isolated from, and fat acidity values of export corn. *Phytopathology* **58,** 1356–1359.

Sauer, D. B., and Christensen, C. M. (1969). Some factors affecting increase in fat acidity values in corn. *Phytopathology* **59**(1), 108–110.

Saul, R. A., and Harris, K. L. (1979). Losses in grain due to respiration of grain and molds and other microorganisms. *In* "Postharvest Grain Loss Assessment Methods" (K. L. Harris and C. J. Lindblad, eds.), pp. 95–99. Am. Assoc. Cereal Chem., St. Paul, Minnesota.

Sauter, E. A., and Peterson, C. F. (1969). The effect of egg shell quality on penetration by *Pseudomonas fluorescens*. *Poult. Sci.* **48**, 1525–1528.

Sauter, E. A., and Peterson, C. F. (1974). The effect of egg shell quality on penetration by various salmonellae. *Poult. Sci.* **53**, 2159–2162.

Sauter, E. A., Harns, V., Stadelman, W. J., and McLaren, B. (1954). Effect of oil treating shell eggs on their functional properties after storage. *Food Technol.* **8**, 82–85.

Sauter, E. A., Peterson, C. F., and Lampman, C. E. (1962). The effectiveness of various sanitizing agents in the reduction of green rot spoilage in washed eggs. *Poult. Sci.* **41**, 468–473.

Scarr, M. P. (1949). Production of polysaccharides by rod-shaped bacteria in sugar factory and refinery processes. *Proc. Soc. Appl. Bacteriol.* **12**, 100–104.

Scarr, M. P. (1950). Production of polysaccharides by rod-shaped bacteria in sugar factory and refinery processes. *Int. Sugar J.* September, p. 300.

Scarr, M. P. (1953a). Studies on the taxonomy and physiology of osmophilic yeasts isolated from the sugar cane, the sugar beet, raw sugar and intermediate refinery products. Ph.D. Thesis, Univ. of London, London.

Scarr, M. P. (1953b). Observations on the multiplication of thermophilic bacteria in the preparation of canner's sugar. *Proc. Int. Congr. Microbiol., 6th, Rome (Sess. 19)* **7**, 57–60.

Scarr, M. P. (1963). Microbiological standards for sugar. *J. Sci. Food Agric.* **14**, 220–223.

Scarr, M. P. (1968). Symposium on growth of microorganisms at extremes of temperature. Thermophiles in sugar. *J. Appl. Bacteriol.* **31**, 66–74.

Scarr, M. P., and Rose, D. (1966). Study of osmophilic yeasts producing invertase. *J. Gen. Microbiol.* **45**, 9–16.

Schade, J. E., and King, A. D., Jr. (1977). Methyl bromide as a microbicidal fumigant for tree nuts. *Appl. Environ. Microbiol.* **33**, 1184–1191.

Schade, J. E., McGreevy, K., King, A. D., Jr., Mackey, B., and Fuller, G. (1975). Incidence of aflatoxin in California almonds. *Appl. Microbiol.* **29**, 48–53.

Schaffner, C. P., Mosbach, K., Bibit, V. C., and Watson, C. H. (1967). Coconut and *Salmonella* infection. *Appl. Microbiol.* **15**, 471–475.

Schalm, O. W., Carroll, E. J., and Jain, N. C. (1971). "Bovine Mastitis." Lea & Febiger, Philadelphia, Pennsylvania.

Schantz, E. J. (1973). Seafood toxicants. *In* "Toxicants Occurring Naturally in Foods," pp. 424–447. Natl. Acad. Sci., Washington, D.C.

Schenkel, J. (1972). Some microbiological aspects of confectionery manufacturing and the related question of plant hygiene. *Int. Rev. Sugar Confect.* **25**, 356–360. (In Ger.)

Schiemann, D. A. (1978). Occurrence of *Bacillus cereus* and the bacteriological quality of Chinese "take-out" foods. *J. Food Prot.* **41**, 450–454.

Schindler, A. F., and Eisenberg, W. V. (1968). Growth and production of aflatoxins by *Aspergillus flavus* on red pepper (*Capsicum frutescens* L.). *J. Ass. Off. Anal. Chem.* **51**, 911–912.

Schindler, A. F., Palmer, J. G., and Eisenberg, W. V. (1967). Aflatoxin production by *Aspergillus flavus* as related to various temperatures. *Appl. Microbiol.* **15**, 1006–1009.

924 References

Schmidhofer, T. (1969). Hygiene bei der Geflügelschlachtung. *Wien. Tieraerztl. Monatsschr.* **56**, 402–410.

Schmidt, C. F., Lechowich, R. V., and Folinazzo, J. F. (1961). Growth and toxin production by type E *Clostridium botulinum* below 40°F. *J. Food Sci.* **26**, 626–630.

Schmidt, E. W., Jr., Gould, W. A., and Weiser, H. H. (1969). Chemical preservatives to inhibit the growth of *Staphylococcus aureus* in synthetic cream pies acidified to pH 4.5 to 5.0. *Food Technol.* **23**, 1197–1199.

Schmidt-Lorenz, W. (1976). Microbiological characteristics of natural mineral waters. *Ann. Inst. Super. Sanita* **12**, 93–112.

Schultze, W. D., and Olson, J. C., Jr. (1960a). Studies on psychrophilic bacteria. I. Distribution in stored commercial dairy products. *J. Dairy Sci.* **43**, 346–350.

Schultze, W. D., and Olson, J. C., Jr. (1960b). Studies on psychrophilic bacteria. II. Psychrophilic coliform bacteria in stored commercial dairy products. *J. Dairy Sci.* **43**, 351–357.

Scott, D. (1953). Glucose conversion in the preparation of albumen solids by glucose oxidase-catalase system. *J. Sci. Food Agric.* **1**, 727–730.

Scott, J. C. (1953). "Health and Agriculture in China." Faber & Faber, London.

Scott, P. M. (1973). Modified method for the determination of mycotoxins in cocoa beans. *J. Assoc. Off. Anal. Chem.* **56**, 1028–1030.

Scott, P. M., and Kennedy, B. P. C. (1973). Analysis and survey of ground black, white and capsicum peppers for aflatoxins. *J. Assoc. Off. Anal. Chem.* **56**, 1452–1457.

Scott, P. M., and Kennedy, B. P. C. (1975). The analysis of spices and herbs for aflatoxins. *Can. Inst. Food Sci. Technol. J.* **8**, 124–125.

Scott, W. J. (1953). Water relations of *Staphylococcus aureus* at 30°C. *Aust. J. Biol. Sci.* **6**, 549–564.

Scott, W. J., and Vickery, J. R. (1954). Studies in the preservation of shell eggs. VII. The effect of pasteurization on the maintenance of physical quality. *Aust. J. Appl. Sci.* **5**, 89–102.

Sebald, M. (1970). Sur le botulisme en France de 1956 à 1970. *Bull. Acad. Natl. Med.* **154**, 703–707.

Seeder, W. A., Mossel, D. A. A., and van Zijl, F. H. (1969). About the growth of molds, especially of *Asp. flavus* on wheat flour with different water content. *Z. Lebensm.-Unters. -Forsch.* **140**(5), 276–278 [*Biol. Abstr.* **51**, 68699 (1970)].

Seeley, C. (1975). Universal use of Australian ginger in confectionery. *Confect. Prod.* May, pp. 243, 255.

Segalove, M., and Dack, G. M. (1941). Relation of time and temperature to growth and enterotoxin production of staphylococci. *Food Res.* **6**, 127–133.

Seiler, D. A. L. (1964). Factors affecting the use of mould inhibitors in bread and cake. *In* "Microbial Inhibitors in Food," pp. 211–220. Almqvist & Wiksell, Stockholm.

Seligmann, R., and Frank-Blum, H. (1974). Microbial quality of barbecued chickens from commercial rotisseries. *J. Milk Food Technol.* **37**, 473–476.

Seligmann, R., and Lapinsky, Z. (1970). *Salmonella* findings in poultry as related to conditions prevailing during transportation from the farm to the processing plant. *Ref. Vet.* **27**, 7–14.

Semeniuk, G. (1954). Microflora. *In* "Storage of Cereal Grains and their Products" (J. A. Anderson and A. E. Alcock, eds.), Monograph Series, Vol. 2, pp. 77–151. Am. Assoc. Cereal Chem., St. Paul, Minnesota.

Senser, F., and Rehm, H. J. (1965). Über das Vorkommen von Schimmelpilzen in Fruchtsäften. *Dtsch. Lebensm.-Rundsch.* **61,** 184–186.

Sera, H., and Ishida, Y. (1972). Bacterial flora in the digestive tract of marine fish. III. Classification of isolated bacteria. *Nippon Suisan Gakkaishi* **38**(8), 853–858.

Seviour, E. M., and Board, R. G. (1972). The behaviour of mixed bacterial infections in the shell membranes of the hen's egg. *Br. Poult. Sci.* **13,** 33–43.

Shafi, R., Cotterill, O. J., and Nichols, M. L. (1970). Microbial flora of commercially pasteurized egg products. *Poult. Sci.* **49,** 578–585.

Shapton, D. A., Lovelock, D. W., and Laurita-Longo, R. (1971). The evaluation of sterilization and pasteurization processes from temperature measurements in degrees Celsius (°C). *J. Appl. Bacteriol.* **34,** 491–500.

Sharf, J. M. (1960). The evolution of microbiological standards for bottled carbonated beverages in the U.S. *Ann. Inst. Pasteur Lille* **11,** 117–131.

Sharma, V. K., Kaura, Y. K., and Singh, I. P. (1974). Frogs as carriers of *Salmonella* and *Edwardsiella*. *Antonie van Leeuwenhoek; J. Microbiol. Serol.* **40,** 171–175.

Sharp, J. G. (1953). Dehydrated meat. *G.B. Dep. Sci. Ind. Res., Food Invest. Board, Spec. Rep.* No. 57.

Shehata, A. M. E. (1960). Yeasts isolated from sugar cane and its juice during the production of aguardente de cana. *Appl. Microbiol.* **8,** 73–75.

Shehata, T. E., and Collins, E. B. (1971). Isolation of psychrophilic species of *Bacillus* from milk. *Appl. Microbiol.* **21,** 466–469.

Shehata, T. E., and Collins, E. B. (1972). Sporulation and heat resistance of psychrophilic strains of *Bacillus*. *J. Dairy Sci.* **55,** 1405–1409.

Sheneman, J. (1973). Survey of aerobic mesophilic bacteria in dehydrated onion products. *J. Food Sci.* **38,** 206–209.

Shenstone, F. S., and Vickery, J. R. (1958).: "Studies in the Preservation of Shell Eggs. VIII. The Effects of the Treatment of Shell Eggs with Oil," Tech. Pap. No. 7. Div. Food Preserv. Transp., CSIRO, Melbourne.

Shewan, J. M. (1962). The bacteriology of fish and spoiling fish and some related chemical changes. *Recent Adv. Food Sci.* **1,** 167–193.

Shewan, J. M. (1971). The microbiology of fish and fishery products—a progress report. *J. Appl. Bacteriol.* **34,** 299–315.

Shewan, J. M. (1974). *In* "Industrial Aspects of Biochemistry" (B. Spencer, ed.), pp. 475–490. North Holland/Am. Elsevier, Amsterdam.

Shewan, J. M. (1977). The bacteriology of fresh and spoiling fish and the biochemical changes induced by bacterial action. *In* "Handling, Processing and Marketing of Tropical Fish," pp. 51–66. Tro. Prod. Inst., London.

Shewan, J. M., and Liston, J. (1955). A review of food poisoning caused by fish and fishery products. *J. Appl. Bacteriol.* **18,** 522–534.

Shibaski, I. (1969). Antimicrobial activity of alkyl esters of p-hydroxybenzoic acid. *Hakke Kogaku Zasshi* **47,** 167–177.

Shibasaki, I., and Tsuchido, T. (1973). Enhancing effect of chemicals on the thermal injury of microorganisms. *Acta Aliment.* **2,** 327–349.

Shiflett, M. A., Lee, J. S., and Sinnhuber, R. O. (1966). Microbial flora of irradiated Dungeness crabmeat and Pacific oysters. *Appl. Microbiol.* **14,** 411–415.

Shotwell, O. L., Hesseltine, C. W., Burmeister, H. R., Kwolek, W. F., Shannon, G. M., and Hall, H. H. (1969a). Survey of cereal grains and soybeans for the presence of aflatoxin. I. Wheat, grain sorghum, and oats. *Cereal Chem.* **46,** 446–454.

Shotwell, O. L., Hesseltine, C. W., Burmeister, H. R., Kwolek, W. F., Shannon, G. M., and Hall, H. H. (1969b). Survey of cereal grains and soybeans for the presence of aflatoxin. II. Corn and soybeans. *Cereal Chem.* **46**, 454–463.

Shotwell, O. L., Hesseltine, C. W., Goulden, M. L., and Vandergraft, E. E. (1970). Survey of corn for aflatoxin, zearalenone, and ochratoxin. *Cereal Chem.* **47**, 700–707.

Shotwell, O. L., Goulden, M. L., and Hesseltine, C. W. (1974). Aflatoxin: Distribution in contaminated corn. *Cereal Chem.* **51**, 492–499.

Shrimpton, D. H., and Barnes, E. M. (1960). A comparison of oxygen permeable and impermeable wrapping materials for the storage of chilled eviscerated poultry. *Chem. Ind. (London)* No. 49, 1492–1493.

Shrimpton, D. H., Monsey, J. B., Hobbs, B. C., and Smith, M. E. (1962). A laboratory determination of the destruction of α-amylase and salmonellae in whole egg by heat pasteurization. *J. Hyg.* **60**, 153–162.

Shrivastava, K. P. (1977). Isolation and identification of *Salmonella* present in frozen froglegs. *Indian J. Microbiol.* **17**, 54–57.

Siddiqui, S. M., and Reddy, C. V. (1974). Influence of delayed washing and oil dipping on the internal quality of dirty eggs stored at room temperature. *Indian J. Poult. Sci.* **9**, 197–203.

Siems, H., Hildebrandt, G., Inal, T., and Sinell, H. J. (1975). Erhefungen über das Vorkommen von Salmonellen bei gefrorenem Schlactgeflügel auf dem deutschen Markt. *Zentralbl. Bakteriol., Parasitenkd., Infektionskr. Hyg., Abt. 1: Orig., Reihe B* **160**, 84–97.

Silliker, J. H. (1963). Total counts as indexes of food quality. In "Microbiological Quality of Foods" (L. W. Slanetz, C. O. Chichester, A. R. Gaufin, and Z. J. Ordal, eds.), pp. 102–112. Academic Press, New York.

Silliker, J. H. (1967). The *Salmonella* problem at the candy manufacturer's level. *Manuf. Confect.* **47**(Oct.), 31–33.

Silliker, J. H. (1969). Some guidelines for the safe use of fillings, toppings, and icings. *Bakers Dig.* **43**(1), 51–54.

Silliker, J. H., and Gabis, D. A. (1973). I.C.M.S.F. methods studies. I. Comparison of analytical schemes for detection of *Salmonella* in dried foods. *Can. J. Microbiol.* **19**, 475–479.

Silliker, J. H., and Gabis, D. A. (1976). I.C.M.S.F. methods studies. VII. Indicator tests as substitutes for direct testing of dried foods and feeds for *Salmonella*. *Can. J. Microbiol.* **22**, 971–974.

Silliker, J. H., and McHugh, S. A. (1967). Factors influencing microbial stability of butter-cream type fillings. *Cereal Sci. Today* **12**(1), 63–5, 73–4.

Simidu, U., Kaneko, E., and Aiso, K. (1969). Microflora of fresh and stored flatfish *Kareius bicoloratus*. *Nippon Suisan Gakkaishi* **35**(1), 77–82.

Simmons, E. R., Ayres, J. C., and Kraft, A. A. (1970). Effect of moisture and temperature on ability of salmonellae to infect shell eggs. *Poult. Sci.* **49**, 761–768.

Simonsen, B. (1968). Aerobic spore-forming bacteria that "blow" canned meats. *Nord. Veterinaermed.* **20**, 121–127.

Simonsen, B. (1975). Microbiological aspects of poultry meat quality. *Qual. Poult. Meat, Proc. Eur. Symp. Poultry Meat Qual., 2nd, Oosterbeck, Neth.* **2**, 1–10.

Sinell, H.-J. (1973). Food infection communicated from animal to man. In "The Microbiological Safety of Food" (B. C. Hobbs and J. H. B. Christian, eds.), pp. 229–238. Academic Press, New York.

Singh, R. N., and Nelson, F. E. (1948). Coliform bacteria in butter. *J. Dairy Sci.* **31,** 726.

Sinha, R. N., and Wallace, H. A. H. (1977). Storage stability of farm-stored rapeseed and barley. *Can. J. Plant Sci.* **57,** 351–365.

Sinha, R. N., Wallace, H. A. H., and Chebib, F. S. (1969a). Principal-component analysis of interrelations among fungi, mites and insects in grain bulk ecosystems. *Ecology* **50**(4), 536–547.

Sinha, R. N., Wallace, H. A. H., and Chebib, F. S. (1969b). Canonical correlations of seed viability, seed-borne fungi and environment in bulk grain ecosystems. *Can. J. Bot.* **47**(1), 27–34.

Skole, R. D., and Newman, H. (1970). The effect of heat activation on the germination of aerobic thermophilic spores in granulated sugars. *Rep. Proc. 15th Sess., Int. Comm. Uniform Methods Sugar Anal., London.*

Skole, R. D., Newman, H., and Barnwell, J. L. (1967). The significance of the polysaccharide producing bacteria belonging to the genus *Bacillus,* in sugar refining. *Proc. 1966 Tech. Sess. Cane Sugar Refin., New Orleans* pp. 35–45. Cane Sugar Refin. Res. Proj., New Orleans, Louisiana.

Skole, R. D., Newman, H., and Barnwell, J. L. (1968). Occurrence and metabolic activity of the polysaccharide producing bacteria (Genus *Bacillus*) in sugar refining. *Publ. Tech. Pap. Proc. Annu. Meet. Sugar Ind. Technol., 27th, Montreal. Sugar Ind. Technol.* **27**(1), 69–79.

Skole, R. D., Newman, H., and Hogu, J. (1974). Evaluation of recent developments in rapid microbial estimations. *Proc. 16th Sess., Int. Comm. Uniform Methods Sugar Anal., Ankara.*

Skole, R. D., Hogu, J. N., and Rizzuto, A. B. (1977). Microbiology of sugar: a taxonomic study. *Tech. Sess. Cane Sugar Refin. Res., New Orleans.*

Skoll, S. L., and Dillenberg, H. O. (1963). *Salmonella thompson* in cake mix. *Can. J. Public Health* **54,** 325–329.

Skovgaard, N., and Nielsen, B. B. (1972). Salmonellas in pigs and animal feeding stuffs in England and Wales and in Denmark. *J. Hyg.* **70,** 127–140.

Slabyj, B. M., Dollar, A. M., and Liston, J. (1965). Postirradiational survival of *Staphylococcus aureus* in seafoods. *J. Food Sci.* **30,** 344–350.

Smart, J. L., Roberts, T. A., Stringer, M. F., and Shah, N. (1979). The incidence and serotypes of *Clostridium perfringens* on beef, pork and lamb carcasses. *J. Appl. Bacteriol.* **46,** 377–383.

Smith, F. R., and Arends, R. E. (1976). Nut meats. *In* "Compendium of Methods for the Microbiological Examination of Foods" (M. Speck, ed.), pp. 614–619. Am. Public Health Assoc., Washington, D.C.

Smith, L. D. S. (1963). *Clostridium perfringens* food poisoning. *In* "Microbiological Quality of Foods" (L. W. Slanetz, C. O. Chichester, A. R. Gaufin, and Z. J. Ordal, eds.), pp 7.7–83. Academic Press, New York.

Smith, M. A., and Niven, C. F., Jr. (1957). The occurrence of *Leuconostoc mesenteroides* in potato tubers and garlic cloves. *Appl. Microbiol.* **5,** 154–155.

Smith, R. S. (1919). Observations on the pasteurization and subsequent handling of milk in city milk plants. *J. Dairy Sci.* **2,** 487–503.

Smith, R. S. (1920). Bacterial control in milk plants. *J. Dairy Sci.* **3,** 540–551.

Smith, W. H. (1972). "Biscuits, Crackers and Cookies. I. Technology, Production, and Management," pp. 177–178. Appl. Sci. Publ., London.

Smith, W. L., Jr. (1962). Chemical treatments to reduce postharvest spoilage of fruits and vegetables. *Bot. Rev.* **28,** 411–445.

Smittle, R. B. (1977). Microbiology of mayonnaise and salad dressing: A review. *J. Food Prot.* **40**, 415–422.

Snoeyenbos, G. H., Morin, E. W., and Wetherbee, D. K. (1967). Naturally occurring *Salmonella in* "blackbirds" and gulls. *Avian Dis.* **11**, 642–646.

Snoeyenbos, G. H., Carlson, V. L., Smyster, C. F., and Olesiuk, O. M. (1969). Dynamics of *Salmonella* infection in chicks reared on litter. *Avian Dis.* **13**, 72–83.

Snow, D. (1949). The germination of mould spores at controlled humidities. *Ann. Appl. Biol.* **36**, 1–13 [*Biol. Abstr.* **23**, 21802 (1949)].

Snyder, I. S., Johnson, W., and Zottola, E. A. (1978). Significant pathogens in dairy products. *In* "Standard Methods for the Examination of Dairy Products" (E. H. Marth, ed.), pp. 11–32. Am. Public Health Assoc., Washington, D.C.

Sohn, J. Y., Ryeom, K., Kim, Y., Lee, M., On, O., and Ryu, J. K. (1973). A study on the distribution of *Clostridium welchii* in fish and shellfish in Korea. *Rep. Korean Natl. Inst. Health* **10**, 79–88.

Solowey, M., Spaulding E. H., and Goresline, H. E. (1946). An investigation of a source and mode of entry of Salmonella organisms in spray-dried whole-egg powder. *Food Res.* **11**, 380–390.

Solowey, M., McFarlane, V. H., Spaulding, E. H., and Chemerda, C. (1947). Microbiology of spray-dried whole egg. II. Incidence and types of *Salmonella*. *Am. J. Public Health* **37**, 971–982.

Speck, M. L. (1975). Contributions of microorganisms to foods and nutrition. *Nutr. News* **38**(4), 13–16.

Speck, M. L., ed. (1976a). "Compendium of Methods for the Microbiological Examination of Foods." Am. Public Health Assoc., Washington, D.C.

Speck, M. L. (1976b). Interactions among lactobacilli and man. *J. Dairy Sci.* **59**, 338–343.

Speck, M. L., and Tarver, F. R., Jr. (1967). Microbiological populations in blended eggs before and after commercial pasteurization. *Poult. Sci.* **46**, 1321.

Sperber, W. H., and Deibel, R. H. (1969). Accelerated procedure for *Salmonella* detection in dried foods and feeds involving only broth cultures and serological reactions. *Appl. Microbiol.* **17**, 533–539.

Spicher, G., and Weipfert, D. (1974). The behavior of the microflora of wheat related to the cleaning—and the flour mill—flow diagram: II. Communication: Investigations regarding the effects of the drying process to the microbe content on cereals. *Zentralbl. Bakteriol., Parasitenkd., Infektionskr. Hyg., Abt. 2* **129**(1/2), 102–114.

Splittstoesser, D. F. (1970). Predominant microorganisms on raw plant foods. *J. Milk Food Technol.* **33**, 500–505.

Splittstoesser, D. F. (1973). The microbiology of frozen vegetables. *Food Technol.* **27**, 54–56, 60.

Splittstoesser, D. F., and Segen, B. (1970). Examination of frozen vegetables for *Salmonellae. J. Milk Food Technol.* **33**, 111–113.

Splittstoesser, D. F., and Wettergreen, W. P. (1964). The significance of coliforms in frozen vegetables. *Food Technol.* **18**, 392–394.

Splittstoesser, D. F., Wettergreen, W. P., and Pederson, C. S. (1961). Control of microorganisms during preparation of vegetables for freezing. II. Peas and corn. *Food Technol.* **15**, 332–334.

Splittstoesser, D. F., Hervey, G. E. R., II, and Wettergreen, W. P. (1965). Contamination of frozen vegetables by coagulase-positive staphylococci. *J. Milk Food Technol.* **28,** 149–151.

Splittstoesser, D. F., Kuss, F. R., Harrison, W., and Prest, D. B. (1971). Incidence of heat-resistant molds in eastern orchards and vineyards. *Appl. Microbiol.* **21,** 335–337.

Splittstoesser, D. F., Groll, M., Downing, D. L., and Kaminski, J. (1977). Viable counts versus the incidence of machinery mold (*Geotrichum*) on processed fruits and vegetables. *J. Milk Food Technol.* **40,** 402–405.

Stadhouders, J., Cordes, M. M., and van Schouwenberg-van Foeken, A. W. J. (1978). The effect of manufacturing conditions on the development of staphylococci in cheese. Their inhibition by starter bacteria. *Neth. Milk Dairy J.* **32,** 193–203.

Stamer, J. R., Hrazdina, G., and Stoyla, B. O. (1973). Induction of red color formation in cabbage juice by *Lactobacillus brevis* and its relationship to pink sauerkraut. *Appl. Microbiol.* **26,** 161–166.

Staniforth, V. (1972). Spices or oleoresins: a choice? *Proc. Conf. Spices* pp. 193–197. Trop. Prod. Inst., London.

Stark, J. B., Goodban, A. E., and Owens, H. S. (1953). Beet sugar liquors. Determination and concentration of lactic acid in processing liquors. *J. Agric. Food Chem.* **1,** 564–566.

Statutory Instruments (1963). "The Liquid Egg (Pasteurisation) Regulations," Statutory Instrument, Food and Drugs Composition No. 1503. HM Stationery Off., London.

Steele, F. R., Jr., Vadehra, D. V., and Baker, R. C. (1967). Recovery of bacteria following pasteurization of liquid whole egg. *Poult. Sci.* **46,** 1322.

Steele, J. L. (1969). Deterioration of shelled corn as measured by carbon dioxide production. *Trans. ASAE* **12**(5), 685–689.

Stewart, D. J., and Patterson, J. T. (1962). Bacteriology of processed broilers. 2. Experiments in broiler processing. *North. Irel. Minist. Agric. Rec. Exp. Res.* **11,** Part 1, 65–71.

Stewart, G. F., and Kline, R. W. (1941). Dried egg albumen. I. Solubility and color denaturation. *Proc. Inst. Food Technol.* pp. 48–56.

Stokes, J. L., Osborne, W. W., and Bayne, H. G. (1956). Penetration and growth of *Salmonella* in shell eggs. *Food Res.* **21,** 510–518.

Stoloff, L., Trucksess, M., Anderson, P. W., Glabe, E. F., and Aldridge, J. G. (1978). Determination of the potential for mycotoxin contamination of pasta products. *J. Food Sci.* **43,** 228–230.

Stone, J. M. (1952). The action of lecithinase of *Bacillus cereus* on the globule membrane of milk fat. *J. Dairy Res.* **19,** 311–315.

Stott, J. A., Hodgson, J. E., and Chaney, J. C. (1975). Incidence of salmonellae in animal feed and the effect of pelleting on content of Enterobacteriaceae. *J. Appl. Bacteriol.* **39,** 41–46.

Straka, R. P., and Combes, F. M. (1951). The predominance of micrococci in the flora of experimental frozen turkey meat steaks. *Food Res.* **16,** 492–493.

Stringer, W. C., Bilskie, M. E., and Naumann, H. D. (1969). Microbial profiles of fresh beef. *Food Technol.* **23,** 97–102.

Stritar, J., Dack, G. M., and Jungewaelter, F. G. (1936). The control of staphylococci in custard-filled puffs and eclairs. *Food Res.* **1,** 237–246.

Strong, D. H., and Ripp, N. M. (1967). Effect of cookery and holding on hams and turkey rolls contaminated with *Clostridium perfringens*. *Appl. Microbiol.* **15,** 1172–1177.

Strong, D. H., Canada, J. C., and Griffiths, B. B. (1963). Incidence of *Clostridium perfringens* in American foods. *Appl. Microbiol.* **11**, 42–44.

Stuart, L. S., and Goresline, H. E. (1942a). Bacteriological studies on the "natural" fermentation process of preparing egg white for drying. *J. Bacteriol.* **44**, 541–549.

Stuart, L. S., and Goresline, H. E. (1942b). Studies of bacteria from fermenting egg white and the production of pure culture fermentations. *J. Bacteriol.* **44**, 625–632.

Sugihara, T. F. (1977). Non-traditional fermentations in the production of baked goods. *Bakers Dig.* **51**(5), 76, 78, 80, 142.

Sugihara, T. F. (1978a). Microbiology of the soda cracker process. I. Isolation and identification of microflora. *J. Food Prot.* **41**, 977–979.

Sugihara, T. F. (1978b). Microbiology of the soda cracker process. II. Pure culture fermentation studies. *J. Food Prot.* **41**, 980–982.

Sugihara, T. F., Kline, L., and McCready, L. B. (1970). Nature of the San Francisco sour dough French bread process. II. Microbiological aspects. *Bakers Dig.* **44**(2), 51–53, 56–57.

Sugihara, T. F., Kline, L., and Miller, M. W. (1971). Microorganisms of the San Francisco sour dough bread process. I. Yeasts responsible for the leavening action. *Appl. Microbiol.* **21**, 456–458.

Sullivan, R., Tierney, J. T., Larkin, E. P., Read, R. B., Jr., and Peeler, J. T. (1971). Thermal resistants of certain oncogenic viruses suspended in milk and milk products. *Appl. Microbiol.* **22**, 315–320.

Surkiewicz, B. F. (1966). Bacteriological survey of the frozen prepared foods industry. I. Frozen cream-type pies. *Appl. Microbiol.* **14**, 21–26.

Surkiewicz, B. F., Groomes, R. J., and Padron, A. P. (1967). Bacteriological survey of the frozen prepared foods industry. III. Potato products. *Appl. Microbiol.* **15**, 1324–1331.

Surkiewicz, B. F., Johnston, R. W., Moran, A. B., and Krumm, G. W. (1969). A bacteriological survey of chicken eviscerating plants. *Food Technol.* **23**, 1066–1069.

Surkiewicz, B. F., Johnston, R. W., Elliott, R. P., and Simmons, E. R. (1972). Bacteriological survey of fresh pork sausage produced at establishments under Federal inspection. *Appl. Microbiol.* **23**, 515–520.

Surkiewicz, B. F., Harris, M. E., Elliott, R. P., Macaluso, J. F., and Strand, M. M. (1975). Bacteriological survey of raw beef patties produced at establishments under federal inspection. *Appl. Microbiol.* **29**, 331–334.

Surkiewicz, B. F., Johnston, R. W., and Carosella, J. M. (1976). Bacteriological survey of frankfurters produced at establishments under federal inspection. *J. Milk Food Technol.* **39**, 7–9.

Sutherland, J. P., Patterson, J. T., and Murray, J. G. (1975). Changes in the microbiology of vacuum-packaged beef. *J. Appl. Bacteriol.* **39**, 227–237.

Suzuki, A., Kawanishi, T., Konuma, H., and Takayama, S. (1973). A survey on *Salmonella* contamination of imported poultry meats. *Eisei Shikenjo Hokoku* **403**, 88–95.

Suzuki, J. I., Dainius, B., and Kilbuck, J. H. (1973). A modified method for aflatoxin determination in spices. *J. Food Sci.* **38**, 949–950.

Swanson, M. H. (1959). Shell egg preservation in the midwest: Progress in shell treatments. *In* "Conference on Eggs and Poultry," ARS 74–12, pp. 41–42. Agric. Res. Serv., U.S. Dep. Agric., Albany, California.

Szabo, P. (1968). Reducing the bacteria count in paprika. *Konzerv- Paprikaip.* No. 4, 128–131. (In Hung.)

Takács, J. (1964). Hygienic and microbiological qualification of sausages. *Acta Vet. Hung.* **14,** 15–18.

Takács, J. (1969). Mikrobiologische Standards für Fleischerzeugnisse. *Fleischwirtschaft* **49,** 193–200.

Takács, J. (1977). Thermobacteriological calculations for the heat treatment of fully sterilized meat conserves. Doc. Vet. Sci. Thesis, Univ. of Budapest, Budapest.

Takács, J., and Zukál, E. (1962). Microbiological and chemical studies on ripening of sausage products conserved with salt and dehydration. *Acta Vet. Hung.* **12,** 257–261.

Takács, J., Jirkovszky, M., and Hegyi, Z. (1963). Studies on the microbiological and biochemical changes involved in the ripening process of Hungarian salami. *Acta Vet. Hung.* **13,** 119–135.

Tamminga, S. K., Beumer, R. R., Kampelmacher, E. H., and van Leusden, F. M. (1976). Survival of *Salmonella eastbourne* and *Salmonella typhimurium* in chocolate. *J. Hyg.* **76,** 41–47.

Tanaka, H., and Miller, M. W. (1963). Microbial spoilage of dried prunes. I. Yeasts and moulds associated with spoiled dried prunes. *Hilgardia* **34,** 167–170.

Taniguti, Z. (1971). Studies on *Clostridium perfringens* in seafoods. *Nagasaki Daigaku Suisangakubu Kenkyu Hokoku* No. 31, 1–67. (In Jpn., Engl. summ.)

Tanner, F. W. (1944). "The Microbiology of Foods," 2nd ed. Garrard, Champaign, Illinois.

Tanner, F. W., and Dubois, G. C. (1925). Some notes on the effect of heat on members of the colon-typhoid group in milk. *J. Dairy Sci.* **8,** 47–53.

Tatini, S. R., Jezeski, J. J., Olson, J. C., Jr., and Casman, E. P. (1971a). Factors influencing the production of staphylococcal enterotoxin A in milk. *J. Dairy Sci.* **54,** 312–320.

Tatini, S. R., Jezeski, J. J., Morris, H. A., Olson, J. C., Jr., and Casman, E. P. (1971b). Production of staphylococcal enterotoxin A in Cheddar and Colby cheeses. *J. Dairy Sci.* **54,** 815–825.

Tatini, S. R., Wesala, W. D., Jezeski, J. J., and Morris, H. A. (1973). Production of staphylococcal enterotoxin in blue, brick, mozzarella and Swiss cheese. *J. Dairy Sci.* **56,** 429–435.

Tatini, S. R., Cords, B. R., and Gramoli, J. (1976). Screening for staphylococcal enterotoxins in food. *Food Technol.* **30**(4), 64, 66, 70, 72, 73.

Taylor, M. H., Helbacka, N. V. L., and Bigbee, D. E. (1968). The effect of packaging giblets separately on quality of fresh broilers. *Poult. Sci.* **47,** 1963–1966.

Taylor, S. L., Guthertz, L. S., Leatherwood, M., and Lieber, E. R. (1979). Histamine production by *Klebsiella pneumoniae* and an incident of scombroid fish poisoning. *Appl. Environ. Microbiol.* **37,** 274–278.

Thatcher, F. S., and Loit, A. (1961). Comparative microflora of chlortetracycline-treated and nontreated poultry with special reference to public health aspects. *Appl. Microbiol.* **9,** 39–45.

Thatcher, F. S., and Montford, J. (1962). Egg-products as a source of salmonellae in processed foods. *Can. J. Public Health* **53,** 61–69.

Thatcher, F. S. Coutu, C., and Stevens, F. (1953). The sanitation of Canadian flour mills and its relationship to the microbial content of flour. *Cereal Chem.* **30,** 71–102.

Thatcher, F. S. Erdman, I. E., and Pontefract, R. D. (1967). Some laboratory and regulatory aspects of the control of *Cl. botulinum* in processed foods. *In* "Botulism 1966" (M. Ingram and T. A. Roberts, eds.), pp. 511–521. Chapman & Hall, London.

Thomas, S. B., and Druce, R. G. (1969). Psychrotrophic bacteria in refrigerated pasteurized milk. A review. *Dairy Ind.* **34**, 351–355, 430–433, 501–505.

Thomas, S. B., and Sekhar, C. V. (1946). Psychrophilic bacteria in raw and commercially pasteurized milk. *Proc. Soc. Appl. Bacteriol.* **1**, 47.

Thomas, S. B., Egdell, J. W., Clegg, L. F. L., and Cuthbert, W. A. (1950). Thermoduric organisms in milk: I. A review of the literature. *Proc. Soc. Appl. Bacteriol.* **13**(1), 27–64.

Thomas, S. B., Druce, R. G., Peters, G. J., and Griffiths, D. G. (1967). Incidence and significance of thermoduric bacteria in farm milk supplies: a reappraisal and review. *J. Appl. Bacteriol.* **30**, 265–298.

Thompson, J. D., and Johnson, D. (1963). Food temperature preferences of surgical patients. *J. Am. Diet. Assoc.* **43**, 209–211.

Thompson, S. S., Harmon, L. G., and Stine, C. M. (1978). Survival of selected organisms during spray drying of skimmilk and storage of nonfat dry milk. *J. Food Prot.* **41**, 16–19.

Thompson, T. L., Engelhard, W. E., and Pivnick, H. (1955). Pustule formation by *Lactobacilli* on fermented vegetables. *Appl. Microbiol.* **3**, 314–316.

Thomson, J. E. (1970). Microbial counts and rancidity of fresh fryer chickens as affected by packaging materials, storage, atmosphere, and temperature. *Poult. Sci.* **49**, 1104–1109.

Thomson, J. E., Whitehead, W. K., and Mercuri, A. J. (1974). Chilling poultry meat—A literature review. *Poult. Sci.* **53**, 1268–1281.

Thomson, J. E., Cox, N. A., Whitehead, W. K., Mercuri, A. J., and Juven, B. J. (1975). Bacterial counts and weight changes of broiler carcasses chilled commercially by water immersion and air-blast. *Poult. Sci.* **54**, 1452–1460.

Thomson, J. E., Cox, N. A., Bailey, J. S., Holladay, J. H., and Richardson, R. L. (1976). Bacteriological sampling of poultry carcasses by a template-swab method. *Poult. Sci.* **55**, 459–462.

Thornburg, H. E. (1915). Process of treating eggs. U.S. Patent 1,163,873.

Thornton, H. (1972). The public health danger of unsterilised pet foods. *Vet. Rec.* **91**, 430–432.

Tilbury, R. H. (1966). Microbiological deterioration of raw cane sugar in the tropics. *In* "Microbiological Deterioration in the Tropics," S.C.I. Monograph, No. 23, pp. 63–79. Soc. Chem. Ind., London.

Tilbury, R. H. (1967). Studies on the microbiological deterioration of raw cane sugar, with special reference to osmophilic yeasts and the preferential utilisation of laevulose in invert. M.S. Thesis, Univ. of Bristol, Bristol, England.

Tilbury, R. H. (1968). Biodeterioration of harvested sugar cane. *In* "Biodeterioration of Materials. Microbiological and Allied Aspects" (A. H. Walters and J. J. Elphick, eds.), pp. 717–730. Elsevier, Amsterdam.

Tilbury, R. H. (1969). The ecology of *Leuconostoc mesenteroides* and control of post-harvest biodeterioration of sugar cane in Jamaica. *Proc. Meet. West Indies Sugar Technol.* pp. 126–135.

Tilbury, R. H. (1970). Biodeterioration of harvested sugar cane in Jamaica. Ph.D. Thesis, Univ. of Aston, Birmingham, England.

Tilbury, R. H. (1972). Dextrans and dextranase. *Proc. Congr. Int. Soc. Sugar Cane Technol. 14th* pp. 1444–1458.

Tilbury, R. H. (1975). Occurrence and effects of lactic acid bacteria in the sugar industry. *In* "Lactic Acid Bacteria in Beverages and Food" (J. G. Carr, C. V. Cutting, and G. C. Whiting, eds.), pp. 177–191. Academic Press, New York.

Tilbury, R. H. (1976). The microbial stability of intermediate moisture foods with respect to yeasts. *In* "Intermediate Moisture Foods" (R. Davies, G. G. Birch, and K. J. Parker, eds.), pp. 138–165. Appl. Sci., London.

Tilbury, R. H., and French, S. (1974). Further studies on enzymic hydrolysis of dextran in mill juices by dextranase and fungal α amylase. *Proc. Congr. Int. Soc. Sugar Cane Technol., 15th* **3**, 1277–1287.

Tilbury, R. H., Orbell, C. J., Owen, J. W., and Hutchinson, M. (1976). Biodeterioration of sweetwaters in sugar refining. *Proc. Int. Biodegrad. Symp. 3rd* pp. 533–543. Appl. Sci., London.

Tilbury, R. H., Hollingsworth, B. S., Graham, S. D., and Pottage, P. (1978). Mill sanitation—a fresh approach to biocide evaluation. *Proc. Congr. Int. Soc. Sugar Cane Technol., 16th* **3**, 2749–2768.

Timoney, J. (1968). The sources and extent of *Salmonella* contamination in rendering plants. *Vet. Rec.* **83**, 541–543.

Tinbergen, B. J., and Krol, B., eds. (1977). *Proc. Int. Symp. Nitrite Meat Prod., 2nd, Pudoc, Wageningen, Neth.*

Tittiger, F. (1971). Studies on the contamination of products produced by rendering plants. *Can. J. Comp. Med.* **35**, 167–173.

Tittiger, F., and Alexander, D. C. (1971). Recontamination of products with salmonellae after rendering in Canadian plants. *Can. Vet. J.* **12**, 200–203.

Tjaberg, T. B., Underdal, B., and Lunde, G. (1972). The effect of ionizing radiation on the microbiological content and volatile constituents of spices. *J. Appl. Bacteriol.* **35**, 473–478.

Todd, E. C. D. (1978). Foodborne disease in six countries—A comparison. *J. Food Prot.* **41**, 559–565.

Tomiyasu, Y., and Zenitani, B. (1957). The spoilage of fish and its preservation by chemical agents. *Adv. Food Res.* **7**, 41–81.

Tompkin, R. B., and Kueper, T. V. (1973). Factors influencing detection of salmonellae in rendered animal by-products. *Appl. Microbiol.* **25**, 485–487.

Townsend, C. T. (1966). Spoilage in canned foods. *J. Milk Food Technol.* **29**, 91–94.

Tretsven, W. I. (1964). Bacteriological survey of filleting processes in the Greater Northwest. III. Bacterial and physical effects of pughing fish incorrectly. *J. Milk Food Technol.* **27**, 13–17.

Troller, J. A. (1973). The water relations of food-borne bacterial pathogens. A review. *J. Milk Food Technol.* **36**, 276–288.

Troller, J. A., and Christian, J. H. B. (1978). "Water Activity and Food." Academic Press, New York.

Tropical Products Institute (1972). Discussion. *Proc. Conf. Spices* pp. 231–232. Trop. Prod. Inst., London.

Tryhnew, L. J., Gunaratne, K. W. B., and Spencer, J. V. (1973). Effect of selected coating materials on the bacterial penetration of the avian egg shell. *J. Milk Food Technol.* **6**, 272–275.

Tucker, C. B., Arnold, W. M., Barrick, J. H., Fowinkle, E., and Daffron, D. B. (1968). An outbreak of salmonellosis—Shelby County, Tennessee. *Morbid. Mortal. Wkly. Rep.* **17**(45), 417–418.

Tucker, E. W. (1953). Infection of the bovine udder with *Mycobacterium* species. *Cornell Vet.* **43**, 576–599.

Tucker, J. F., and Gordon, R. F. (1968). The incidence of salmonellae in poultry packing stations. *Br. Vet. J.* **124,** 102–109.

Tucker, J. F., Brown, W. B., and Goodship, G. (1974). Fumigation with methyl bromide of poultry foods artificially contaminated with *Salmonella. Br. Poult. Sci.* **15,** 587–595.

Tullock, E. F., Jr., Ryan, K. J., Formal, S. B., adn Franklin, F. A. (1973). Invasive enteropathogenic *Escherichia coli* dysentery. *Ann. Intern. Med.* **79,** 13–17.

Tuomi, S., Matthews, M. E., and Marth, E. H. (1974). Temperature and microbial flora of refrigerated ground beef gravy subjected to holding and heating as might occur in a school foodservice operation. *J. Milk Food Technol.* **37,** 457–462.

Turnbull, P. C. B., and Snoeyenbos, G. H. (1973). The role of ammonia, water activity, and pH in the salmonellacidal effect of long-used poultry litter. *Avian Dis.* **17,** 72–86.

Tynecka, Z., and Gos, Z. (1973). The inhibitory action of garlic (*Allium sativum* L.) on growth and respiration of some microorganisms. *Acta Microbiol. Pol., Ser. B* **5**(22), No. 1, 51–62.

Ullmann, W. W., Brazis, A. R., Arledge, W. L., Schultze, W. D., and Lawton, W. C. (1978). Screening and confirmatory methods for the detection of abnormal milk. *In* "Standard Methods for the Examination of Dairy Products" (E. H. Marth, ed.), 14th ed., pp. 115–140. Am. Public Health Assoc., Washington, D.C.

Ulloa-Sosa, M., and Schroeder, H. W. (1969). Note on aflatoxin decomposition in the process of making tortillas from corn. *Cereal Chem.* **46,** 397–400.

U.S. Department of Agriculture (USDA) (1968). "Preventing Mycotoxins in Farm Commodities," ARS 20–16. U.S.D.A., Washington, D.C.

U.S. Department of Agriculture (USDA) (1975a). "Regulations Governing the Grading of Shell Eggs and United States Standards, Grades, and Weight Classes for Shell Eggs," 7 CFR Part 56. U.S. Gov. Print. Off., Washington, D.C.

U.S. Department of Agriculture (USDA) (1975b). "Regulation Governing the Insspection of Eggs and Egg Products," 7 CFR Part 59. U.S. Gov. Print. Off., Washington, D.C.

U.S. Department of Agriculture (USDA) (1976). Definition of meat and classes of meat, permitted uses, and labelling requirements. Animal and Plant Health Inspection Service. *Fed. Regist.* **41,** No. 82.

U.S. Department of Health, Education and Welfare (1966). "Eggs and Egg Products" (FDA) 21 CFR Part 42. U.S. Gov. Print. Off., Washin , .t, D.C.

U.S. Department of Health, Education and Welfare (1967–1979). "Foodborne and Waterborne Disease Outbreaks: Annual Summaries (1966–1978)." Cent. Dis. Control, Atlanta, Georgia.

U.S. Department of Health, Education and Welfare (1970a). "Botulism in the United States." Cent. Dis. Control, Atlanta, Georgia.

U.S. Department of Health, Education and Welfare (1970b). Staphylococcal food poisoning traced to butter—Alabama. *Morbid. Mortal. Wkly. Rep.* **19,** 271.

U.S. Department of Health, Education and Welfare (1971a). Staphylococcal gastroenteritis associated with salami. *Morbid. Mortal. Wkly. Rep.* **20,** 253, 258.

U.S. Department of Health, Education and Welfare (1971b). Gastroenteritis attributed to imported French cheese. *Morbid. Mortal. Wkly. Rep.* **20,** 427–428.

U.S. Department of Health, Education and Welfare (1971c). Follow-up on gastro-enteritis attributed to imported French cheese. *Morbid. Mortal. Wkly. Rep.* **20,** 445.

U.S. Department of Health, Education and Welfare (1974a). Type A botulism due to a commercial product—Georgia. *Morbid. Mortal. Wkly. Rep.* **23,** 417–418.

U.S. Department of Health, Education and Welfare (1974b). Human *Salmonella dublin* infections associated with consumption of certified milk. *Morbid. Mortal. Wkly. Rep.* **25,** 175–176.

U.S. Department of Health, Education and Welfare (1975a). "Dressings for Food. Mayonnaise," 21 CFR 25.1. U.S. Gov. Print. Off., Washington, D.C.

U.S. Department of Health, Education and Welfare (1975b). "Dressings for Food. Salad Dressing," 21 CFR 25.3. U.S. Gov. Print. Off., Washington, D.C.

U.S. Department of Health, Education and Welfare (1975c). Part 128C. Cacao products and confectionery. *Fed. Regist.* **40**(108), 24,170–24,172.

U.S. Department of Health, Education and Welfare (1975d). *Salmonella typhimurium* outbreak traced to a commercial apple cider—New Jersey. *Morbid. Mortal. Wkly. Rep.* **24,** 87–88.

U.S. Department of Health, Education and Welfare (1975e). *Salmonella newport* contamination of hamburger—Colorado and Maryland. *Morbid. Mortal. Wkly. Rep.* **24,** 438, 443.

U.S. Department of Health, Education and Welfare (1975f). Scrombroid poisoning—New York City. *Morbid. Mortal. Wkly. Rep.* **24,** 342, 347.

U.S. Department of Health, Education and Welfare (1976a). "Food-borne and Water-borne Disease Outbreaks: Annual Summary, 1975," HEW, Publ. No. (CDC) 76–8185. U.S. Gov. Print. Off., Washington, D.C.

U.S. Department of Health, Education and Welfare (1976b). "Smoked and Smoke-flavored Fish. Processes and Controls," 21 CFR 128a.7. U.S. Gov. Print. Off., Washington, D.C.

U.S. Department of Health, Education and Welfare (1976c). "Ammoniated Cotton-seed Meal," 21 CFR 121.319. U.S. Gov. Print. Off., Washington, D.C.

U.S. Department of Health, Education and Welfare (1976d). "Food Service Sanitation Manual," DHEW Publ. (FDA) 78–2081. U. S. Gov. Print. Off., Washington, D.C.

U.S. Department of Health, Education and Welfare (1976e). "Federal Food, Drug and Cosmetic Act, as Amended," Sect. 201, footnote. U.S. Gov. Print. Off., Washington, D.C.

U.S. Department of Health, Education and Welfare (1976f). "Processing and Bottling of Bottled Drinking Water," 21 CFR 128d. U.S. Gov. Print. Off., Washington, D.C.

U.S. Department of Health, Education and Welfare (1976g). "Thermally Processed Low-acid Foods Packaged in Hermetically Sealed Containers," 21 CFR 128b. U.S. Gov. Print. Off., Washington, D.C.

U.S. Department of Health, Education and Welfare (1977a). "Foodborne and Water-borne Disease Outbreaks: Annual Summary, 1976" HEW Publ. No (CDC) 78–8185. U.S. Gov. Print. Off., Washington, D.C.

U.S. Department of Health, Education and Welfare (1977b). Presumed staphylo-coccal food poisoning associated with whipped butter. *Morbid. Mortal. Wkly. Rep.* **26,** 268.

U.S. Department of Health, Education and Welfare (1977c). Compliance program evaluation. FY 74. Imported spice program (7309.05). *Food Drug Adm. By-Lines* **7**(5), 278–281.

U.S. Department of Health, Education and Welfare (1978a). "Ethylene Oxide," 21 CFR 193.200 and 21 CFR 193.380. U.S. Gov. Print. Off., Washington, D.C.

U.S. Department of Health, Education and Welfare (1978b). "Spices and other Natural Seasonings," 21 CFR 182.10. U.S. Gov. Print. Off., Washington, D.C.

U.S. Department of Health, Education and Welfare (1978c). 21 CFR 169.140 and 21 CFR 101.00. U.S. Gov. Print. Off., Washington, D.C.

U.S. Department of Health, Education and Welfare (1978d). "Salad Dressing," 21 CFR 169.150. U.S. Gov. Print. Off., Washington, D.C.

U.S. Department of Health, Education and Welfare (1978e). "Bacteriological Analytical Manual" (FDA). Assoc. Off. Anal. Chem., Washington, D.C.

U.S. Department of Health, Education and Welfare (1978f). "Milk and Cream," Code of Federal Regulations, Title 21, Chap. 1, Part 131. U.S. Gov. Print. Off., Washington, D.C.

U.S. Department of Health, Education and Welfare (1978g). "Cheeses and Related Cheese Products," Code of Federal Regulations, Title 21, Chap. 1, Part 133. U.S. Gov. Print. Off., Washington, D.C.

U.S. Department of Health, Education and Welfare (1979). Salmonellosis associated with consumption of nonfat powdered milk—Oregon. *Morbid. Mortal. Wkly. Rep.* **28**, 129–130.

Vadehra, D. V., and Baker, R. C. (1967). The effect of dry heat on internal quality and spoilage of eggs. *Poult. Sci.* **46**, 1330–1331.

Vadehra, D. V., and Baker, R. C. (1973). Effect of egg shell sweating on microbial spoilage of chicken eggs. *J. Milk Food Technol.* **36**, 321–322.

Vadehra, D. V., Baker, R. C., and Naylor, H. B. (1969a). *Salmonella* infection of cracked eggs. *Poult. Sci.* **48**, 954–957.

Vadehra, D. V., Steele, F. R., Jr., and Baker, R. C. (1969b). Shelf life and culinary properties of thawed frozen pasteurized whole egg. *J. Milk Food Technol.* **32**, 362–364.

Vadehra, D. V., Baker, R. C., and Naylor, H. B .(1970a). Role of cuticle in spoilage of chicken eggs. *J. Food Sci.* **35**, 5–6.

Vadehra, D. V., Baker, R. C., and Naylor, H. B. (1970b). Infection routes of bacteria into chicken eggs. *J. Food Sci.* **35**, 61–62.

Vadehra, D. V., Baker, R. C., and Naylor, H. B. (1972). Distribution of lysozyme activity in the exteriors of eggs from *Gallus gallus*. *Comp. Biochem. Physiol. B* **43**, 503–508.

Vajdi, M., and Pereira, R. R. (1973). Comparative effects of ethylene oxide, gamma irradiation and microwave treatments on selected spices. *J. Food Sci.* **38**, 893–895.

Vajdi, M., Pereira, R. R., and Gallop, R. A. (1973). How gamma irradiated and ethylene oxide treated spices affect the microbial quality, shelf life, and flavor of garlic sausage. *Food Prod. Dev.* **7**(7), 90–92.

Vakrilov, V., and Murgov, T. (1970). Modification of certain technological conditions in the production of chocolate desserts with the aim of reducing mold dissemination. *Khranit. Prom.* **19**(2), 24–25. (In Bulg.)

Vandegraft, E. E., Shotwell, O. L., Smith, M. L., and Hesseltine, C. W. (1973). Mycotoxin formation affected by fumigation of wheat. *Cereal Sci. Today* **18**(12,) 412–414.

van den Berg, L., and Lentz, C. P. (1966). Effect of temperature, relative humidity and atmospheric composition on changes in quality of carrots during storage. *Food Technol.* **20**, 954–957.

van der Meijs, C. C. J. M. (1970). Resultaten van een bacteriologisch en chemisch onderzoek van gemalen vlees. (Results from a bacteriological and chemical control of minced meat). *Tijdschr. Diergeneeskd.* **95**, 1180–1184.

van der Riet, W. B. (1976). Water sorption isotherms of beef biltong and their use in predicting critical moisture contents for biltong storage. *S. Afr. Food Rev.* **3**(6), 93–96.

Vanderzant, C., Mroz, E., and Nickelson, R. (1970). Microbial flora of Gulf of Mexico and pond shrimp. *J. Milk Food Technol.* **33**, 346–350.

Vanderzant, C., Nickelson, R., and Judkins, P. W. (1971). Microbial flora of pond reared brown shrimp (*Peneaus aztecus*). *Appl. Microbial.* **21**, 916–921.

Vanderzant, C., Thompson, C. A., Jr., and Ray, S. M. (1973). Microbial flora and level of *Vibrio parahaemolyticus* of oysters (*Crassostrea virginica*), water and sediment from Galveston Bay. *J. Milk Food Technol.* **36**, 447–452.

Van Houten (undated). "Product Specifications of Van Houten Cocoa Powder," H.D. 1125. Pine Brook, New Jersey.

van Schothorst, M., and Kampelmacher, E. H. (1967). Salmonella in meat imported from South American countries. *J. Hyg.* **65**, 321–325.

van Schothorst, M., and van Leusden, F. M. (1977). Microbiologische specificaties van eiprodukten. (Microbiological specifications for egg products.) *Voeding-smiddelen Technol.* **10**(22), 16–19.

van Schothorst, M., Guinee, P. A. M., Kampelmacher, E H., and van Keulen, A. (1965). Prevalence of salmonellae in poultry in the Netherlands. *Zentralbl. Veterinaermed., Reihe B* **12**, 422–428.

van Schothorst, M., Mossel, D. A. A., Kampelmacher, E. H., and Drion, E. F. (1966). The estimation of the hygienic quality of feed components using an Enterobacteriaceae enrichment test. *Zentralbl. Veterinaermed., Reihe B* **13**, 273–285.

van Schothorst, M., Edel, W., and Kampelmacher, E. H. (1970). Voortgezette onder-zoekingen over het voorkomen van Salmonella in gehakt, in de maand juli, 1965–1969. (Further studies about the occurrence of *Salmonella* in minced meat in July 1965–1969.) *Tijdschr. Diergeneeskd.* **95**, 279–282.

van Schothorst, M., Notermans, S., and Kampelmacher, E. H. (1972). Hygiene in poultry slaughter. *Fleischwirtschaft* **6**, 749–752.

van Schothorst, M., Northholt, M. D., Kampelmacher, E. H., and Notermans, S. (1976). Studies on the estimation of the hygienic condition of frozen broiler chickens. *J. Hyg.* **76**, 57–63.

van Schouwenberg-van Foeken, A. W. J., Stadhauders, J., and Jans, J. A. (1978). The thermonuclease test for assessment of growth of coagulase positive staphylo-cocci in Gouda cheese with a normal acidity development. *Neth. Milk Dairy J.* **32**, 217–231.

Vasconcelos, G. J., and Lee, J. S. (1972). Microbial flora of Pacific Oysters (*Crasso-stera gigas*) subjected to ultraviolet-irradiated seawater. *Appl. Microbial.* **23**, 11–16.

Vassiliadis, P., Papoutsakis, G., Avramidis, D., Trichopoulos, D., and Papadakis, J. (1976). Recherche de *Salmonella* sur des carcasses de poulets à Athènes en 1976. *Arch. Inst. Pasteur Hell.* **22**, 23–28.

Vaughn, R. H. (1951). The microbiology of dehydrated vegetables. *Food Res.* **16**, 429–438.

Vaughn, R. H. (1970). Incidence of various groups of bacteria in dehydrated onions and garlic. *Food Technol.* **24,** 189–191.

Vaughn, R. H., Won, W. D., Spencer, F. B., Pappagianus, D., Foda, I. O., and Krumperman, P. H. (1953). *Lactobacillus plantarum,* the cause of "yeast spots" on olives. *Appl. Microbiol.* **1,** 82–85.

Vaughn, R. H., Jakubczyk, T., MacMillan, J. D., Higgins, T. E., Davé, B. A., and Crampton, V. M. (1969a). Some pink yeasts associated with softening of olives. *Appl. Microbiol.* **18,** 771–775.

Vaughn, R. H., King, A. D., Jr., Nagel, C. W., Ng, H., Levin, R. E., MacMillan, J. D., and York, G. K., II (1969b). Gram negative bacteria associated with sloughing, a softening of California ripe olives. *J. Food Sci.* **34,** 224–227.

Vaughn, R. H., Stevenson, K. E., Davé, B. A., and Park, H. C. (1972). Fermenting yeasts associated with softening and gas-pocket formation in olives. *Appl. Microbiol.* **23,** 316–320.

Vedamuthu, E. R., Hankin, L., Ordal, Z. J., and Vanderzant, C. (1978). Thermoduric, thermophilic and psychrotrophic bacteria. *In* "Standard Methods for the Examination of Dairy Products" (E. H. Marth, ed.), 14th ed., pp. 107–113. Am. Public Health Assoc., Washington, D.C.

Veerkamp, C. H. (1974). The simultaneous scalding and plucking of broiler carcasses compared with an industrial method of processing. *Proc. World Poult. Congr., 15th, New Orleans* pp. 450–451.

Veerkamp, C. H., and Hofmans, G. J. P. (1973). New development in poultry processing, simultaneous scalding and plucking. *Poult. Int.* **12,** 16–18.

Velaudapillae, T. G., Niles, R., and Nagaratnum, W. (1969). Salmonellae, shigellae and enteropathogenic *Escherichia coli* in uncooked food. *J. Hyg.* **67,** 187–191.

Vernon, E. (1977). Food poisoning and *Salmonella* infections in England and Wales, 1973–75: An analysis of reports to the Public Health Laboratory Service. *Public Health* **91,** 225–235.

Vernon, E., and Tillett, H. E. (1974). Food poisoning and *Salmonella* infections in England and Wales, 1969–1972. *Public Health* **88,** 225–235.

Viringa, H. A., Galesloot, T. E., and Davelaar, H. (1968). Symbiosis in yoghurt. II. Isolation and identification of a growth factor for *Lactobacillus bulgaricus* produced by *Streptococcus thermophilus. Neth. Milk Dairy J.* **22,** 114.

Voeten, A. C., Brus, D. H. J., and Jaartsveld, F. H. J. (1974). Distribution of *Salmonella* within a closed integrated broiler flock. *Tijdschr. Diergeneeskd.* **99,** 1093–1109. (In Dutch.)

Voigt, M. N., and Eitenmiller, R. R. (1978). Role of histidine and tyrosine decarboxylases and mono- and diamine oxidases in amine buildup in cheese. *J. Food Prot.* **41,** 182–186.

Voigt, M. N., Eitenmiller, R. R., Koehler, P. E., and Hamdy, M. K. (1974). Tyramine, histamine and tryptamine content of cheese. *J. Milk Food Technol.* **37,** 377–381.

Vojnovich, C., and Pfeifer, V. F. (1967). Reducing the microbial population of flour during milling. *Cereal Sci. Today* **12,** 54–55, 58–60.

Vojnovich, C., Pfeifer, V. F., and Griffin, E. L., Jr. (1970). Reducing microbial populations in dry-milled corn products. *Cereal Sci. Today* **15** (12) 401–411.

Vojnovich, C., Anderson, R. A., and Ellis, J. J. (1972). Microbial reduction in stored and dry-milled corn infected with southern corn leaf blight. *Cereal Chem.* **49**(3), 346–353.

Vollum, R. L., and Taylor, J. (1947). *Salmonella oranienburg* outbreak in Woodstock. *In* "The Bacteriology of Spray-dried Eggs with Particular Reference to Food Poisoning," Med. Res. Counc., Spec. Rep. Ser. No. 260, HM Stationery Off., London.

von Schelhorn, M. (1951). Control of microorganisms causing spoilage in fruit and vegetable products. *Adv. Food Res.* **3**, 429–482.

Wadsworth, R. V., and Howat, G. R. (1954). Cocoa fermentation. *Nature (London)* **174**, 392–394.

Wagenaar, R. O. (1952). The bacteriology of surface-taint butter: A review. *J. Dairy Sci.* **35**, 403–423.

Wagenaar, R. O., and Dack, G. M. (1954). Further studies on the effect of experimentally inoculating canned bread with spores of *Clostridium botulinum. Food Res.* **19**, 521–529.

Wain, H., and Blackstone, P. A. (1956). Staphylococcal gastroenteritis: Report of a major outbreak. *Am. J. Dig. Dis.* **1**, 424–429.

Walker, H. W., and Ayres, J. C. (1956). Incidence and kinds of microorganisms associated with commercially dressed poultry. *Appl. Microbiol.* **4**, 345–349.

Walker, H. W., and Ayres, J. C. (1959). Microorganisms associated with commercially processed turkeys. *Poult. Sci.* **38**, 1351–1355.

Walker, H. W., and Ayres, J. C. (1970). Yeasts as spoilage organisms. *In* "The Yeasts" (A. H. Rose and J. S. Harrison, eds.), Vol. 3, pp. 463–527. Academic Press, New York.

Walker, P., Cann, D., and Shewan, J. M. (1970). The bacteriology of "scampi": (*Nephrops norvegicus*). I. Preliminary bacteriological, chemical, and sensory studies. *J. Food Technol.* **5**, 375–385.

Walkowiak, E., Aleksandrowska, I., Wityk, A., and Watychowicz, I. (1971). Sterilization of spices used in the meat industry by UV irradiation (in Pol.). *Med. Weter.* **27**(11), 694 [*Food Sci. Technol. Abstr.* **6**, 5S569].

Wallace, G. I. (1938). The survival of pathogenic microorganisms in ice cream. *J. Dairy Sci.* **21**, 35–36.

Wallace, G. I., and Park, S. E. (1933). The behavior of *Clostridium botulinum* in frozen fruits and vegetables. *J. Infect. Dis.* **52**, 150–156.

Wallace, H. A. H., and Sinha, R. N. (1962). Fungi associated with hot spots in farm stored grain. *Can. J. Plant Sci.* **42**, 130–141.

Wallace, H. A. H., and Sinha, R. N. (1975). Microflora of stored grain in international trade. *Mycopathologia* **57**(3), 171–176.

Wallace, H. A. H., Sinha, R. N., and Mills, J. T. (1976). Fungi associated with small wheat bulks during prolonged storage in Manitoba. *Can. J. Bot.* **54**(12), 1332–1343.

Wallhäuser, K. H., and Schmidt, H. (1967). "Sterilisation-Desinfektion-Konservierung-Chemotherapie." Thieme, Stuttgart.

Walsh, D. E., and Funke, B. R. (1975). The influence of spaghetti extruding, drying, and storage on survival of *Staphylococcus aureus. J. Food Sci.* **40**, 714–716.

Walsh, D. E., Funke, B. R., and Graalum, K. R. (1974). Influence of spaghetti extruding conditions, drying and storage on the survival of *Salmonella typhimurium J. Food Sci.* **39**, 1105–1106.

Waltking, A. E. (1971). Fate of aflatoxin during roasting and storage of contaminated peanut products. *J. Assoc. Off. Anal. Chem.* **54**, 533–539.

Walz, E. (1956). Ein eindrucksvolles Beispiel von Verkeimung (Bazillen und Schimmelpilzbehaftung) in der Fleischwirtschaft verwendeter Gewürze. *Arch. Lebensmittelhyg.* **7**(11/12), 138–143.

Warmbrod, F., and Fry, L. (1966). Coliform and total bacterial counts in spices, seasonings and condiments. *J. Assoc. Off. Anal. Chem.* **49**, 678–680.

Washam, C. J., Olson, H. C., and Vedamuthu, E. R. (1977). Heat resistant psychrotrophic bacteria isolated from pasteurized milk. *J. Food Prot.* **40**, 101–108.

Watanabe, K. (1965). Technological problems of handling and distribution of fresh fish in southern Brazil. *In* "The Technology of Fish Utilization" (R. Kreuzer, ed.), pp. 44–46. Fishing News, London.

Watson, A. J., and McFarlane, V. H. (1948). Microbiology of spray-dried whole egg. IV. Molds. *Food Technol.* **2**, 15–22.

Watson, S. A., and Yahl, K. R. (1971). Survey of aflatoxins in commercial supplies of corn and grain sorghum used for wet-milling. *Cereal Sci. Today* **16**, 153–155, 163.

Watt, B. K., and Merrill, A. L. (1963). "Composition of Foods," Agric. Handb. No. 8. Consumer Food Econ. Res. Div., Agric. Res. Serv., U.S. Dep. Agric., Washington, D.C.

Webb, B. H. (1970). Condensed products. *In* "By-Products from Milk" (B. H. Webb and E. O. Whittier, eds.), 2nd ed., AVI, Westport, Connecticut.

Webb, T. A., and Mundt, J. O. (1978). Molds on vegetables at the time of harvest. *Appl. Environ. Microbiol.* **35**, 655–658.

Weber, C. W. (1947). Safety factors of HTST pasteurizer. *J. Milk Food Technol.* **10**, 14–24.

Webster, R. C., and Esselen, W. B. (1956). Thermal resistance of food poisoning organisms in poultry stuffing. *J. Milk Food Technol.* **19**, 209–212.

Weckbach, L. S., and Langlois, B. E. (1977). Effect of heat treatments on survival and growth of a psychrotroph and on nitrogen fractions in milk. *J. Food Prot.* **40**, 857–862.

Weckel, K. G., Hawley, R., and McCoy, E. (1964). Translocation and equilibration of moisture in canned frozen bread. *Food Technol.* **18**, 1480–1482.

Weihrauch, J. L., Brignoli, C. A., Reeves, J. B., III, and Iverson, J. L. (1977). Fatty acid composition of margarines, processed fats, and oils: A new compilation of data for tables of food composition. *Food Technol.* **31**(2), 80–85, 91.

Weinzirl, J. (1922). The cause of explosion in chocolate candies. *J. Bacteriol.* **7**, 599–604.

Weinzirl, J. (1927). Sugar as a source of the anaerobes causing explosion of chocolate candies. *J. Bacteriol.* **13**, 203–207.

Weiser, H. H., Winter, A. R., and Lewis, M. N. (1954). The control of bacteria in chicken salad. I. *Micrococcus pyogenes* var. *aureus. Food Res.* **19**, 465–471.

Weissmann, M. A., and Carpenter, J. A. (1969). Incidence of salmonellae in meat and meat products. *Appl. Microbiol.* **17**, 899–902.

Wells, J. M., and Payne, J. A. (1975). Mycoflora of pecans treated with heat, low temperatures, or methyl bromide for control of the pecan weevil. *Phytopathology* **65**, 1393–1395.

Werner, H.-P., Klein, H.-J., and Rotter, M. (1970). Die Empfindlichkeit verschiedener Mikroorganismen gegen Äthylenoxyd. *Zentralbl. Bakteriol., Parasitenkd., Infektionskr. Hyg., Abt. 1: Orig.* **214**, 262–271.

Wesley, F., Rourke, B., and Darbishire, O. (1965). The formation of persistent toxic chlorohydrins in foodstuffs by fumigation with ethylene oxide and with propylene oxide. *J. Food Sci.* **30**, 1037–1042.

West, N. S., Gililland, J. R., and Vaughn, R. H. (1941). Characteristics of coliform bacteria from olives. *J. Bacteriol.* **41**, 341–353.

Westhoff, D. C., and Engler, T. (1973). The fate of *Salmonella typhimurium,* and *Staphylococcus aureus* in cottage cheese whey. *J. Milk Food Technol.* **36**, 19–22.

Westhoff, D. C., and Feldstein, F. (1976). Bacteriological analysis of ground beef. *J. Milk Food Technol.* **39**, 401–404.

Wethington, M. C., and Fabian, F. W. (1950). Viability of food-poisoning staphylococci and salmonellae in salad dressing and mayonnaise. *Food Res.* **15**, 125–134.

White, A., and White, H. R. (1962). Some aspects of the microbiology of frozen peas. *J. Appl. Bacteriol.* **25**, 62–71.

White, C. H., and Custer, E. W. (1976). Survival of *Salmonella* in cheddar cheese. *J. Milk Food Technol.* **39**, 328–331.

Wilder, A. N., and MacCready, R. A. (1966). Isolation of *Salmonella* from poultry. Poultry products and poultry processing plants in Massachusetts. *N. Engl. J. Med.* **274**, 1453–1460.

Wilkerson, W. B., Ayres, J. C., and Kraft, A. A. (1961). Occurrence of enterococci and coliform organisms on fresh and stored poultry. *Food Technol.* **15**, 286–292.

Wilkin, M., and Winter, A. R. (1947). Pasteurization of egg yolk and white. *Poult. Sci.* **26**, 136–142.

Wilkinson, R. J., Mallmann, W. L., Dawson, L. E., Irmiter, T. F., and Davidson, J. A. (1965). Effective heat processing for the destruction of pathogenic bacteria in turkey rolls. *Poult. Sci.* **44**, 131–136.

Williams, E. F., and Spencer, R. (1973). Abattoir practices and their effect on the incidence of *Salmonella* in meat. *In* "The Microbiological Safety of Food" (B. C. Hobbs and J. H. B. Christian, eds.), pp. 41–46. Academic Press, New York.

Williams, J. E. (1978). Paratyphoid infection. *In* "Diseases of Poultry" (M. S. Hofstad, B. W. Calnek, C. F. Helmboldt, W. M. Reid, and H. W. Yoder, Jr., eds.), 7th ed., Iowa State Univ. Press, Ames.

Williams, J. E., Dillard, L. H., and Hall, G. O. (1968). The penetration patterns of *Salmonella typhimurium* through the outer structures of chicken eggs. *Avian Dis.* **12**, 445–466.

Williams, L. P., and Hobbs, B. C. (1975). *Enterobacteriaceae* infections. *In* "Diseases Transmitted from Animals to Man" (W. T. Hubbert, W. F. McCulloch, and P. R. Schnurrenberger, eds.), pp. 33–109. Thomas, Springfield, Illinois.

Williams, M. L. B. (1967). A new method for evaluating surface contamination of raw meat. *J. Appl. Bacteriol.* **30**, 498–499.

Williams, O. B., and Mrak, E. M. (1949). An interesting outbreak of yeast spoilage in salad dressing. *Fruit Prod. J.* **28**, 141, 153.

Wilson, B. J. (1966). Fungal toxins. *In* "Toxicants Occurring Naturally in Foods," Publ. No. 1354, pp. 126–146. Natl. Acad. Sci./Natl. Res. Counc., Washington, D.C.

Wilson, B. J. (1978). Hazards of mycotoxins to public health. *J. Food Prot.* **41**, 375–384.

Wilson, C. R., and Andrews, W. H. (1976). Sulfite compounds as neutralizers of spice toxicity for *Salmonella. J. Milk Food Technol.* **39**, 464–466.

Wilson, D. M., and Nuovo, G. J. (1973). Patulin production in apples decayed by *Penicillium expansum. Appl. Microbiol.* **26**, 124–125.

Wilson, E., Paffenbarger, R. S., Jr., Foter, M. J., and Lewis, K. H. (1961). Prevalence of salmonellae in meat and poultry products. *J. Infect. Dis.* **109,** 166–171.

Winter, A. R. (1952). Production of pasteurized frozen egg products. *Food Technol.* **6,** 414–415.

Winter, A. R., Stewart, G. F., McFarlane, V. H., and Solowey, M. (1946). Pasteurization of liquid egg products III. Destruction of *Salmonella* in liquid whole egg. *Am. J. Public Health* **36,** 451–460.

Winter, A. R., Shield, P., MacDonald, L., and Prudent, I. (1954). Improving the keeping quality of eggs during marketing by hot oil treatment. *Food Technol.* **8,** 515–518.

Wirth, F., Takács, J., and Leistner, L. (1971). Hitzebehandlung und F-Werte für langfristig lagerfähige Fleischkonserven ("Vollkonserven"). *Fleischwirtschaft* **51,** 923–935.

Wiseblatt, L. (1967). Reduction of the microbial populations in flours incorporated into refrigerated foods. *Cereal Chem.* **44**(3), 269–280.

Wistreich, H. E., Thundiyil, G. J., and Juhn, H. (1975). Ethanol vapor sterilization of natural spices and other foods. U.S. Patent 3,908,031.

Witlin, B., and Smyth, R. D. (1957). "Soapiness" in "white" chocolate candies. *Am. J. Pharm.* **129,** 135–142.

Wogan, G. N., ed. (1965). "Mycotoxins in Foodstuffs." MIT Press, Cambridge, Massachusetts.

Wohl, A. (1974). Spices. *In* "Encyclopedia of Food Technology" (A. H. Johnson and M. S. Peterson, eds.), pp. 840–845. AVI, Westport, Connecticut.

Wolf, J. (1977). Codex changes in chocolate standards will focus on cocoa butter extender quality. *Candy Snack Ind.* **142**(4), 32, 34, 36, 38–40.

Woodard, W. G. (1968). The survival and growth of bacteria in turkey rolls during processing and storage. M.S. Thesis, Iowa State Univ., Ames.

Woodburn, M. (1964). Incidence of salmonellae in dressed broiler-fryer chickens. *Appl. Microbiol.* **12,** 492–495.

Woodburn, M., Bennion, M., and Vail, G. E. (1962). Destruction of salmonellae and staphylococci in precooked poultry products by heat treatment before freezing. *Food Technol.* **16**(6), 98–100.

Woodward and Dickerson (undated). "Standard Specifications." Woodward and Dickerson, Inc., Philadelphia, Pennsylvania.

World Health Organization (WHO) (1973). *Wkly. Epidemiol. Rec.* **48,** 377–384.

World Health Organization (WHO) (1974). "Fish and Shellfish Hygiene," Tech. Rep. Ser., No. 550. WHO, Geneva.

World Health Organization (WHO) (1976). "Recommended International Standard for Chocolate," CAC/RS 87-1976. Food Agric. Organ., Rome.

World Health Organization (WHO) (1977). "Aviation Catering, Report of a Working Group, Torremolinos, 1976," ICP/FSP 004. Reg. Off. Eur. WHO, Copenhagen.

World Health Organization (WHO) (1978). "Report of the Codex Committee on Cocoa Products and Chocolate," Codex Alimentarius Commission, 12th Session, Alinorm 78/10. Food Agric. Organ., Rome.

Worth, R. M. (1963). Health in rural China: from village to commune. *Am. J. Hyg.* **77,** 228–239.

Wright, W. J., Bice, C. W., and Fogelberg, J. M. (1954). The effect of spices on yeast fermentation. *Cereal Chem.* **31,** 100–112.

Wrinkle, C., Weiser, H. H., and Winter, A. R. (1950). Bacterial flora of frozen egg products. *Food Res.* **15,** 91–98.

Wu, C. M., Koehler, P. E., and Ayres, J. C. (1972). Isolation and identification of xanthotoxin (8-methoxypsoralen) and bergapten (5-methoxypsoralen) from celery infected with *Sclerotinia sclerotiorum. Appl. Microbiol.* **23,** 852–856.

Yesair, J., and Williams, O. B. (1942). Spice contamination and its control. *Food Res.* **7,** 118–126.

Yeterian, M., Chugg, L., Smith, W., and Coles, C. (1974). Are microbiological quality standards workable? *Food Technol.* **28**(10), 23–32.

York, L. R., and Dawson, L. E. (1973). Shelf life of pasteurized liquid whole egg. *Poult. Sci.* **52,** 1657–1658.

Young, C. K., and Nelson, F. E. (1978). Survival of *Lactobacillus acidophilus* in "Sweet Acidophilus Milk" during refrigerated storage. *J. Food Prot.* **41,** 248–250.

Zaika, L. L., Zell, T. E., Smith, J. L., Palumbo, S. A., and Kissinger, J. C. (1976). The role of nitrite and nitrate in Lebanon bologna, a fermented sausage. *J. Food Sci.* **41,** 1457–1460.

Zaika, L. L., Zell, T. E., Palumbo, S. A., and Smith, J. L. (1978). Effect of spices and salt on fermentation of Lebanon bologna-type sausage. *J. Food Sci.* **43,** 186–189.

Zak, D. L., Ostovar, K., and Keeney, P. G. (1972). Implication of *Bacillus subtilis* in the synthesis of tetramethylpyrazine during fermentation of cocoa beans. *J. Food. Sci.* **37,** 967–968.

Žakula, R., Popvić, P., and Nikolić, L. (1964). Physical properties and microflora of treated and untreated sheep casings used in the production of canned frankfurters. *Proc. Conf. Eur. Meat Res. Workers, 10th, Roskilde, Den.*

Zapatka, F. A., Varney, G. W., and Sinskey, A. J. (1977). Neutralization of the bactericidal effect of cocoa powder on salmonellae by casein. *J. Appl. Bacteriol.* **42,** 21–25.

Zecha, B. C., McCapes, R. H., Dungan, W. M., Holte, R. J., Worcester, W. W., and Williams, J. E. (1977). The Dillon Beach Project—A five-year epidemiological study of naturally occurring *Salmonella* infections in turkeys and their environment. *Avian Dis.* **21,** 141–159.

Zehren, V. L., and Zehren, V. F. (1968a). Examination of large quantities of cheese for staphylococcal enterotoxin A. *J. Dairy Sci.* **51,** 635–644.

Zehren, V. L., and Zehren, V. F. (1968b). Relation of acid development during cheese making to development of staphylococcal enterotoxin A. *J. Dairy Sci.* **51,** 645–649.

Ziegler, F., and Stadelman, W. J. (1955). Increasing shelf-life of fresh chicken meat by using chlorination. *Poult. Sci.* **34,** 1389–1391.

Zimmerman, W. J. (1971). Saltcure and drying-time and temperature effects on viability of *Trichinella spiralis* in drycured hams. *J. Food Sci.* **36,** 58–62.

Zottola, E. A., and Busta, F. F. (1971). Microbiological quality of further-processed turkey products. *J. Food Sci.* **36,** 1001–1004.

Zottola, E. A., and Jezeski, J. J. (1969). Comparisons of short-time holding procedures to determine thermal resistance of *Staphylococcus aureus. J. Dairy Sci.* **52,** 1855–1857.

Zottola, E. A., and Marth, E. H. (1966). Thermal inactivation of bacteriophages active against lactic streptococci. *J. Dairy Sci.* **49,** 1338.

Zottola, E. A., Al-Dulaimi, A. N., and Jezeski, J. J. (1965). Heat resistance of *Staphylococcus aureus* isolated from milk and cheese. *J. Dairy Sci.* **48,** 774.

Zottola, E. A., Jezeski, J. J., and Al-Dulaimi, A. N. (1969). Effect of short-time subpasteurization treatments on the destruction of *Staphylococcus aureus* in milk for cheese manufacture. *J. Dairy Sci.* **52,** 1707–1714.

Zottola, E. A., Schmeltz, D. L., and Jezeski, J. J. (1970). Isolation of salmonellae and other airborne microorganisms in turkey processing plants. *J. Milk Food Technol.* **33,** 395–399.

Appendix I

The International Commission on Microbiological Specifications for Foods: Its Purposes and Accomplishments

HISTORY AND PURPOSE

The ICMSF was formed in 1962 by the parent body, the International Association of Microbiological Societies (IAMS), in response to the need for internationally acceptable and authoritative decisions on microbiological limits for foods commensurate with public health safety and particularly for foods moving in international commerce. Its overall purpose is to appraise public health aspects of the microbiological content of foods. Through the IAMS, the Commission is linked to the International Union of Biological Societies and to the World Health Organization (WHO) and, hence, is a body of the United Nations.

The founding terms of reference are as follows: (a) to assemble, correlate, and evaluate evidence about the microbiological quality of foods; (b) to consider whether microbiological criteria are necessary for any particular food; (c) where necessary, to propose such criteria; and (d) to suggest appropriate methods of sampling and examination. More descriptively, the Commission seeks to aid in providing comparable standards of judgment in different countries, to foster safe movement of foods in international commerce, and to dissipate difficulties caused by disparate microbiological criteria and methods of analysis. Fulfillment of such objectives will be of great value to the food industry, to the expansion of international trade in foods, to national control agencies, to the international agencies more concerned with the humanitarian aspects of food distribution, and, eventually, to the health of the consuming public.

The ICMSF is essentially a scientific advisory body which provides basic information through extensive study and makes recommendations based on such information. The results of studies are published either as books or papers and thus are available to interested individuals, governments, and national and international organizations to use as desired. The group provides the facts without prejudices and thereby fills a useful role as an authoritative base. Primarily through cross-membership, close liaison is enjoyed with other organizations involved in international standards, such as the Codex Alimentarius Commission, the International Standards Organization, the International Dairy Federation, and the Association of Official Analytical Chemists.

At meetings, the ICMSF functions as a work party, not as a forum for the reading of papers. Much of the work is done by subcommittees during the interval between meetings, often with assistance of nonmember consultants. The general meetings are largely directed to assessing the work of the subcommittees, debating to achieve a consensus, editing of draft submissions, and planning. Meetings have been held in Montreal, Canada (1962, the founding meeting); Cambridge, England (1965); Moscow, USSR (1966); London, England (1967); Dubrovnik, Yugoslavia (1969); Mexico City, Mexico (1970); Opatija, Yugoslavia (1971); Langford, England (1972); Ottawa, Canada (1973); Caracas, Venezuela (1974); Alexandria, Egypt (1976) and Cairo, Egypt (1977 and 1978).

MEMBERSHIP AND SUBCOMMISSIONS

The membership consists of 21 food microbiologists from 14 countries (pp. 953–954), whose combined professional interests include research, public health, official food control, education, and industrial research and development. They are drawn from government laboratories in public health, agriculture, and food technology, from universities, and from the food industry. In addition, the Commission engages consultants from time to time to assist with specific aspects of its studies. All members and consultants are chosen on the basis of their expertise in areas of food microbiology, not as national delegates; all work voluntarily without fees or honoraria.

To promote similar activities among food microbiologists on a regional scale, subcommissions have been created in various areas of the world. To date, three have been established (see memberships, pp. 954–955): one in the Balkan-Danubian region (the Balkan-Danubian Subcommission, BDS), composed of eight members; one in Latin America (the Latin American Subcommission, LAS), composed of ten members; and one in the Middle East-North African region (the Middle East-North African Subcommission, MENAS), composed of eight members. Each is an autonomous body which conducts studies on problems of specific concern to its region.

ACCOMPLISHMENTS

Since all studies made by the Commission are published, a list of its publications is a record of its accomplishments:

Books

1. "Microorganisms in Foods 1: Their Significance and Methods of Enumeration" (1978). 2nd ed., 434 pp. Univ. of Toronto Press, Toronto. (Spanish transl. available for 1st ed., 1973; Editorial Ascribia, Zaragoza, Spain.)

2. "Microbiological Specifications and Testing Methods for Irradiated Foods" (1970). Compiled and edited in cooperation with the Food and Agriculture Organization and the International Atomic Energy Agency. Published as *Tech. Rep. Ser., I.A.E.A.* No. 104, 122 pp. (Available in Engl., Fr., Ger., Russ.)

3. "Microorganisms in Foods 2: Sampling for Microbiological Analysis: Principles and Specific Applications" (1974). 213 pp. Univ. of Toronto Press, Toronto. (Span. transl. in progress; Editorial Ascribia, Zaragoza, Spain.)

The book "Microorganisms in Foods 1: Their Significance and Methods of Enumeration" has been widely acclaimed. Over 4000 copies of three printings in English have been sold. It has proved invaluable to food microbiologists in government control agencies, industry, and in teaching and research institutions. Volume 2, "Sampling for Microbiological Analysis: Principles and Specific Applications," has also received excellent reviews, and is proving of the greatest value to all agencies and food companies involved in assessing the microbiological quality of foods.

Articles

1. Thatcher, F. S., (1963). The microbiology of specific frozen foods in relation to public health: report of an international committee. *J. Appl. Bacteriol.* **26,** 266–285.

2. Cominazzini, C. (1969). Il comitato internazionale per la defini-

zione delle caratteristiche microbiologiche degli alimenti ed il suo contributo per la tutela igienica dei medesimi. *Cron. Chim.* **25,** 16–22.

3. Mendoza, S., and Quevedo, F. (1971). Comisión internacional de Especificaciones Microbiológicas de los Alimentos. *Bol. Inst. Bacteriol. Chile* **13,** 45–48.

4. Thatcher, F. S. (1971). The International Committee on Microbiological Specifications for Foods: its purposes and accomplishments. *J. Assoc. Off. Anal. Chem.* **54,** 836–841.

5. Silliker, J. H., and Gabis, D. A. (1973). ICMSF methods studies: I. Comparison of analytical schemes for detection of *Salmonella* in dried foods. *Can. J. Microbiol.* **19,** 475–479.

6. Gabis, D. A., and Silliker, J. H. (1974). ICMSF methods studies: II. Comparison of analytical schemes for detection of *Salmonella* in high moisture foods. *Can. J. Microbiol.* **20,** 663–669.

7. Idziak, E., and Erdman, I. E. (1974). ICMSF methods studies: III. An appraisal of 16 contemporary methods for the detection of *Salmonella* in meringue powder. *Can. J. Microbiol.* **19,** 475–479.

8. Erdman, I. E. (1973). ICMSF methods studies: IV. International collaborative assay for the detection of *Salmonella* in raw meats. *Can. J. Microbiol.* **20,** 715–720.

9. Silliker, J. H., and Gabis, D. A. (1974). ICMSF methods studies: V. The influence of selective enrichment media and incubation temperatures on the detection of salmonellae in frozen meats. *Can. J. Microbiol.* **20,** 813–816.

10. Gabis, D. A., and Silliker, J. H. (1974). ICMSF methods studies: VI. The influence of selective enrichment media and incubation temperatures on the detection of salmonellae in dried foods and feeds. *Can. J. Microbiol.* **20,** 1509–1511.

11. Silliker, J. H., and Gabis, D. A. (1976). ICMSF methods studies: VII. Indicator tests as substitutes for direct testing of dried foods and feeds for *Salmonella. Can. J. Microbiol.* **22,** 971–974.

12. Hauschild, A. H. W., Gilbert, R. J., Harmon, S. M., O'Keeffe, M. F., and Vahlefeld, R. (1977). ICMSF methods studies: VIII. Comparative study for the enumeration of *Clostridium perfringens* in foods. *Can. J. Microbiol.* **23,** 884–892.

13. Gabis, D. A., and Silliker, J. H. (1977). ICMSF methods studies: IX. The influence of selective enrichment broths, differential plating media, and incubation temperatures on the detection of *Salmonella* in dried foods and feed ingredients. *Can. J. Microbiol.* **23,** 1225–1231.

14. Rayman, M. K., Devoyod, J. J., Purvis, U., Kusch, D., Lanier, J.,

Gilbert, R. J., Till, D. D., and Jarvis, G. A. (1978). ICMSF methods studies: X An international comparative study of four media for the enumeration of *Staphylococcus aureus* in foods. *Can. J. Microbiol.* **24,** 274–281.

15. Clark, D. S. (1978). The International Commission on Mricrobiological Specifications for Foods. *Food Technol.* **32,** 51–54, 57.

16. Silliker, J. H., Gabis, D. A., and May, A. (1979). ICMSF methods studies: XI. Collaborative/comparative studies on the determination of coliforms using the most probable number procedure. *J. Food Prot.* **42,** 638–644.

17. Hauschild, A. H. W., Desmarchelier, P., Gilbert, R. J., Harmon, S. M., and Vahlefeld, R. (1979). ICMSF methods studies: XII. Comparative study for the enumeration of *Clostridium perfringens* in feces. *Can. J. Microbiol.* **25,** 953–963.

18. Rayman, M. K., Jarvis, G. A., Davidson, C. M., Long, S., Allen, J. M., Tong, T., Dodsworth, P., McLaughlin, S., Greenberg, S., Shaw, B. G., Beckers, H. J., Qvist, S., Nottingham, P. M., and Stewart, B. J. (1979). ICMSF methods studies: XIII. An international comparative study of the MPN procedure and the Anderson -Baird-Parker direct plating method for the enumeration of *Escherichia coli* biotype I in raw meats. *Can. J. Microbiol.* **25,** 1321–1327.

Most of the papers deal with the results of the Commission's methods-testing program which is described in the section below. All studies have proved of substantial value, but perhaps the most significant are those described in ICMSF Methods Studies I and II, which report on compositing food samples for *Salmonella* analysis. Combining multiples of standard 25-g sample units into one composite for analysis gives the same assurance of detection as separate examination of each 25-g sample unit. This finding alone has made the testing program worthwhile, because statistically valid quality control sampling programs to demonstrate *Salmonella*-negative lots are now economically feasible. The cost of *Salmonella* testing can be reduced to a fraction of what it was with no loss of accuracy.

Methods-Testing Program

The overall objective is to determine by detailed comparative analysis, involving laboratories in various countries, the best methods for the enumeration and identification of indicator and food-poisoning bacteria. In "Microorganisms in Foods 1: Their Significance and Methods of Enumeration," the Commission describes several of the best-known methods for some of the microbial categories, because it could not distinguish which one, if any, is superior. International comparative testing is seen as the only way to determine the most accurate methods.

The studies are being coordinated in Silliker Laboratories, Chicago Heights, Illinois, in the United States (Drs. J. H. Silliker and D. A. Gabis, coordinators), and at the Health Protection Branch of Health and Welfare Canada, in Ottawa, Canada (Drs. H. Pivnick and K. Rayman, coordinators). All projects are planned by a subcommittee of ICMSF members and consultants and are approved by the Commission in plenary session. Studies completed to date or in progress include those on methods for *Salmonella,* coliforms, *Staphylococcus, Clostridium perfringens* and *Escherichia coli.*

FINANCING

The Commission raises funds for its activities from government agencies in several countries, from the World Health Organization (WHO), from the International Union of Biological Societies (IUBS) and from the food industry. Assistance from government agencies has come in the form of grants for specific projects: the United States Department of Agriculture has given two grants in support of the methods-testing program; the United States Department of Health, Education, and Welfare sponsored Public Law 480 grants to support two general meetings on sampling of foods and to support three others on the preparation of this book; the government of Kuwait also contributed financially to the preparation of this book; and Health and Welfare Canada supported one general meeting on the preparation of the 2nd edition of Book 1. WHO has contributed annually in support of the methods-testing program and general expenditures, and the IUBS has granted funds in support of administration costs and to help meet expenses for the publication of this book. Over 60 food companies and agencies in 8 countries contribute to the Commision's Sustaining Fund, mostly on a yearly basis (see Appendix II).

Appendix II

Contributors to the Sustaining Fund of ICMSF

Amatil Ltd., Box 145, GPO Sydney NSW 2001, Australia

American Can Co., America Lane, Greenwich CT, 06830, USA

Arnotts Biscuits Pty Ltd., 170 Kent St., Sydney NSW 2000, Australia

Atlantic Sugar, Ltd., P.O. Box 7, Montreal, Quèbec, Canada H3C 1C5

Bacon and Meat Manufacturers Association, 1–2 Castle Lane, London, England, SW1E 6DU

Beatrice Foods Co., 1526 South State St., Chicago, IL, 60605, USA

Beecham Group Ltd., Beecham House, Great West Rd., Brentford, Middlesex, England

Brooke-Bond Oxo Ltd., Trojan Way, Purley Way, Croydon, England, CR9 9EH

Brown and Polson Ltd., Clay Gate House, Littleworth Rd., Esher, Surrey, England

Burns Foods Ltd., P.O. Box 1300, Calgary, Alta., Canada, T2P 2L4

Cadbury Schweppes Foods Ltd., Bournville, Birmingham, England

Cadbury Schweppes Ltd., P.O. Box 88, St. Kilda West, Victoria 3182, Australia

Campbell Soup Co. Ltd., 60 Birmingham St., Toronto, Ont., Canada, M8V 2B8

Canada Packers Ltd., 2211 St. Clair Ave. W., Toronto 9, Ont., Canada, M69 1K4.

Carlo Erba Institute for Therapeutic Research, Milan 20159, Italy

Central Alberta Dairy Pool, 5302 Gaetz Ave., Red Deer, Alta., Canada

Centro Studi sull' Alimentazione, Gino Alfonso Sada, P.za Diaz 7-20123, Milan, Italy

Christie Brown and Co., Ltd., 2150 Lakeshore Blvd. W., Toronto, Ont., Canada, M8V 1A3

Coca-Cola Co., 310 North Ave. N.W., Atlanta, GA, 30301, USA

CPC International Inc., International Plaza, Englewood Cliffs, NJ, 07632, USA

CSR Ltd., Box 1630 G.P.O., Sydney NSW 2001, Australia

Del Monte Corporation, 205 North Wiget Lane, Walnut Creek, CA, 94598, USA

Difco Laboratories, Detroit, MI, 48232, USA

Distillers Co. Ltd., 21 St. James Square, London, S.W. 1, England

Findus Ltd., Bjuv, Sweden

Geinoca Food Services Ltd., 6205 Airport Road, Bldg. B, Suite 205, Mississauga, Ont., Canada L4V 1E3

General Foods Corporation, Technical Center, White Plains, NY, 10625, USA

General Foods Canada Ltd., Box 4019, Terminal A, Toronto, Ont., Canada

Gerber Products Co., 445 State St., Fremont, MI, 49412, USA

H. J. Heinz Co. Ltd., Hayes Park, Hayes, Middlesex, England

Home Juice Co. Ltd., 175 Fenmar Dr., Weston, Ont., Canada

Horne & Pitfield Foods Ltd., 14550 112th Ave., P.O. Box, 2266, Edmonton 15, Alta., Canada, T5J 2P6

Infant Formula Council, 64 Perimeter Center East, Atlanta, GA, 30346, USA

International Union of Biological Societies, 51 Bd. de Montmorency, 75016, Paris, France

ITT Continental Baking Co., P.O. Box 731, Rye, NY, 10580, USA

J. Sainsbury Ltd., Stamford House, Stamford St., London, England, SE1 9LC

Jannock Corporation Ltd., P.O. Box 7, Montreal 101, Quebec, Canada

John Labatt Ltd., 150 Simcoe St., P.O. Box 5050, London, Ont., Canada, N6A 4M3

Joseph Rank Ltd., Millcrat House, Eastcheap, London, E.C. 3, England

Kellogg/Salada Canada Ltd., 6700 Finch Ave. W., Rexdale, Ont., Canada M9W 5P2

Kraft Foods Ltd., Box 1673N, G.P.O., Melbourne, 3001, Australia

Kuwait Ministry of Health, Kuwait, Kuwait

Langnese-Iglo GmbH, Hauptverwaltung, Postfach 10 40 29, 2000 Hamburg 1, West Germany

Maizena Gesellschaft mbH, Postfach 560, Knorrstr. 1, 71 Heilbronn/Neckar, Germany

Maple Leaf Mills Ltd., P.O. Box 710, Station K, Toronto, Ont., Canada, M4P 2X5

Marks and Spencer Ltd., Michael House, Baker St., London, W.1, England

Mars Ltd., Dundee Rd., Trading Estate, Slough, Bucks., England

McCormick and Co., Inc., Baltimore, Md, 21202, USA

Milk Marketing Board, Thames Ditton, Surrey, England, KT7 OEL

National Fisheries Institute Inc. 1101 Connecticut Ave., N.W. Washington, DC, 20036, USA

New Zealand Food Manufacturers' Federation, Industry House, 38–44 Courtenay Place, Wellington, New Zealand

Plasmon SPA, Corso, Garibaldi 97.99, 20121 Milano, Italy

RHM Research Ltd., Lincoln Road, High Wycombe, Bucks., England, H12 30N

R.J.R. Foods Inc., 4th and Main Sts., Winston Salem, NC, 27102, USA

Reckitt and Colman Ltd., Carrow, Norwich, England, NR1 2DD

Ross Laboratories, 615 Cleveland Ave., Columbus, OH, 43216, USA

Spillers Ltd., Old Charge House, Cannon St., London E.C.4, England

Standard Brands Ltd., 1 Dundas St. W., Toronto, Ont., Canada M5B 2H1

Swift Canadian Co. Ltd., 2 Eva Rd., Etobicoke, Ont., Canada M9C 4V5

Tate and Lyle Refineries Ltd., 21 Mincing Lane, London, England

Terre de Crodo, via Cristoforo Gluck 35, Milano 20125, Italy

Tesco Stores Ltd., Tesco House, Delamere Rd., Cheshunt, Waltham Cross, Herts., England

The Borden Co. Ltd., 1275 Lawrence Ave. E., Don Mills (Toronto), Ont., Canada M3A 1C5

The J. Lyons Group of Companies, Cadby Hall, London, England W14 09A

The Pillsbury Co., 608 Avenue S., Minneapolis, MN, 55402, USA

The Quaker Oats Co., 617 West Main St., Barrington, IL, 60010, USA

The Quaker Oats Co. Canada Ltd., Quaker Park, Peterborough, Ont., Canada K9J 7B2

Thomas J. Lipton, Inc., 800 Sylvan Ave., Englewood Cliffs, NJ, 07632, USA

Thomas J. Lipton, Ltd., 2180 Yonge St., Toronto, Ontario, Canada M452C4

Unigate (Australia) Pty Ltd., P.O. Box 13, Dandenong, Victoria 3175, New Zealand

Unilever Ltd., Unilever House, Blackfriars, London, E.C. 4, England

World Health Organization, Geneva, Switzerland.

Appendix III

Members and Consultants of ICMSF and Its Subcommissions.

MEMBERS OF THE ICMSF

Dr. H. Lundbeck (Chairman) Director, The National Bacteriological Laboratory, S-105 21, Stockholm, Sweden.

Dr. D. S. Clark (Secretary-Treasurer) Director, Bureau of Microbial Hazards, Food Directorate, Health Protection Branch, Health and Welfare Canada, Tunney's Pasture, Ottawa, Ontario, Canada. K1A 0L2.

Dr. A. C. Baird-Parker Head, Microbiological Research, Unilever Research, Colworth Laboratory, Unilever Limited, Colworth House, Sharnbrook, Bedford, England. MK44 1LQ.

Dr. Frank L. Bryan Chief, Foodborne Disease Training, Instructional Services Division, Bureau of Training, Center for Disease Control, Public Health Service, Department of Health, Education and Welfare, Atlanta, Georgia 30333. U.S.A.

Dr. J. H. B. Christian Director, Division of Food Research, C.S.I.R.O., P.O. Box 52, North Ryde, N.S.W. 2113, Australia.

Professor C. Cominazzini Professor in Charge of Hygiene, Faculty of Medicine, State University of Turin, Via Monte Nero 46 - 28100 Novara, Italy.

Professor Otto Emberger Chief, Department of Microbiology and Associate Professor of Hygiene, Faculty of Medical Hygiene, Charles' University of Prague, Srobárova 48, Praha 10, Vinohrady, Czechoslovakia.

Dr. Betty C. Hobbs Microbiology Department, Christian Medical College and Brown Memorial Hospital, Ludhiana, Punjab, India. Also, 1000, High Road, Whetstone, London N.20 OQG, England. Formerly Director, Food Hygiene Reference Laboratory, Central Public Health Laboratory, Colindale Avenue, London NW9 5HT, England.

Dr. Keith H. Lewis Professor of Environmental Health, School of Public Health, University of Texas, Health Sciences Center at Houston, P.O. Box 20186, Houston, Texas 77025, U.S.A.

Dr. G. Mocquot Chargé de mission à l'I.N.R.A., Technologie laitière, CNRZ, 78.350 Jouy-en-Josas, France.

Dr. N. P. Nefedjeva Chief, Laboratory of Food Microbiology, Institute of Nutrition, AMS USSR, Ustinsky pr. 2/14, Moscow G-240, USSR.

Dr. C. F. Niven, Jr. Director of Research,

Del Monte Research Center, 205 North Wiget Lane, Walnut Creek, California 94598, U.S.A.

Dr. P. M. Nottingham Head, Science Division, The Meat Industry Research Institute of New Zealand (Inc.), P.O. Box 617, Hamilton, New Zealand.

Dr. J. C. Olson, Jr. Consultant in Food Microbiology, 4982 Sentinel Dr., #204 Bethesda, MD. 20016, U.S.A. Formerly Deputy Assistant to the Director, Bureau of Foods, Food and Drug Administration, US Department of Health, Education and Welfare, Washington, D.C. 20204.

Dr. H. Pivnick Bureau of Microbial Hazards, Food Directorate, Health Protection Branch, Health and Welfare Canada, Tunney's Pasture, Ottawa, Onartio, Canada. K1A 0L2.

Dr. Fernando Quevedo Head, Food Microbiology and Hygiene Unit, Pan American Zoonoses Centre, Casilla 3092, 1000 Buenos Aires, Argentina.

Dr. T. A. Roberts Head of Food Quality and Control Division, Agricultural Research Council, Meat Research Institute, Langford, Bristol, BS18 7DY, England.

Dr. J. H. Silliker President, Silliker Laboratories, 1139 Dominguez St., Suite I, Carson, California. 90746, U.S.A.

Mr. Bent Simonsen Chief Microbiologist, Danish Meat Products Laboratory, Ministry of Agriculture, Howitzvej 13, DK-2000 Copenhagen F, Denmark.

Professor H. J. Sinell Director, Institute of Food Hygiene, Free University of Berlin, Koserstr. 20, 1000 Berlin 33, Germany.

Dr. M. van Schothorst Head of the Microbiology Section, Central Control Laboratory, Nestle Products Technical Assistance Co., Ltd., Case Postale 88, CH-1814 la Tour-de-Peitz, Switzerland.

BALKAN AND DANUBIAN SUBCOMMISSION

Professor Dr. J. Takács (Chairman) Director, Institute of Food Hygiene, University of Veterinary Sciences, 1400 Budapest, P.O. Box 2, Hungary. (Deceased, 1979.)

Dr. Milica Kalember-Radosavljevic (Secretary) Food Bacteriologist, Military Medical Academy, Institute of Hygiene, 2 Patserova Avenue, Belgrade, Yugoslavia.

Dr. Vladimir Bartl Head, Hygiene Laboratories, Hygiene Station for Middle Czech Region, Safarikova 14, 120 00 Praha 2, Czechoslovakia.

Dr. Deac Cornel Institutul de Igiena, Si Sanătate Publică, R3400 Cluj, Napoca, R.S., România.

Professor Dr. O. Prandl Director, Institute of Meat Hygiene and Veterinary Food Technology, Vienna III/40, Linke Bahngasse 11, Austria.

Dr. S. Tzannetis Faculty of Medicine, Dept. of Microbiology, National University of Athens, Athens 609, P.O. Box 1540, Greece.

Dr. Fuat Yanc Sehir Hifzissihha Müessesesi, Sarachanebasi, Istanbul, Turkey.

Professor Dr. C. Zachariev Director, N.P.O. Veterinarno Delo, U.L. Rabotničeska Klasa No. 1 Sofia, Bulgaria.

LATIN AMERICAN SUBCOMMISSION

Professor Josefina Gomez-Ruiz (Chairwoman) Central University of Venezuela, Apartado 50259, Caracas 105, Venezuela.

Dra. Silvia Mendoza G. (Secretary-Treasurer)

Division of Biological Sciences, Department of Bioengineering, Simon Bolivar University, Apartado 80659, Caracas 108, Venezuela.

Professor Dra. Nenufar Sosa de Caruso Director, Dairy Institute, Veterinary Faculty, University of Uruguay, Casilla de Correo 753, Montevideo, Uruguay.

Dr. Fernando Quevedo Head, Food Microbiology and Hygiene Unit, Pan American Zoonoses Centre, Casilla 3092, 1000 Buenos Aires, Argentina.

Dr. Sebastiao Timo Iaria Instituto de Ciencias Biomedicas, Universidade de Sao Paulo, Av. Dr. Arnaldo 715-C.P. 8099, Sao Paulo, E. Sao Paulo, Brazil.

Dra. Ethel G. V. Amato de Lagarde Head of Division Bacteriologia Sanitaria, Instituto Nacional de Microbiologia "Carlos G. Malbrán", Avda Velez Sársfield 563, Buenos Aires, Argentina.

Dra. Elvira Regús de Valera Calle José Contreras No. 98 (Altos), Zona 7, Sto. Domingo, República Dominicana

Dr. Mauro Faber de Freitas Leitao Head, Department of Food Microbiology, Instituto de Tecnologia de Alimentos, Caixa Postal 139, 13.100 Campinas Sao Paulo, Brazil.

Dr. Hernán Puerta Cardona Chairman, Food Hygiene Section, Escuela Nacional de Salud Pública, Universidad de Antioquia, Apartado Aéreo 51922, Medellin, Colombia

Dra. María Alina Ratto Head, Centro Latinamericano de Ensenanza e Investigación de Bacteriología Alimentaria (CLEIBA), Universidad Nacional Mayor de San Marcos de Lima - PERU, Apartado 5653

MIDDLE EAST–NORTH AFRICAN SUBCOMMISSION

Professor Refat Hablas (Chairman) Bacteriological Department, Faculty of Medicine, Al-Azhar University, El Houssein Hospital, Eldarrasa, Cairo, Egypt.

Dr. Hassan Sidahmed (Secretary) Head, Department of Bacteriology, P.O. Box 287, National Health Laboratory, Khartoum, Sudan.

Professor A. Alaoui Director of Institute Pasteur Maroc, Professor of Microbiology, Casa School of Medicine, Casablanca, Morocco.

Dr. Abdul-Kareem Nasir Al-Dulaimi Head, Department of Food and Water Hygiene, Central Public Health Laboratories, Andulis Square, Alwia, Baghdad, Iraq.

Mr. I. Kashoulis Analyst, Government Laboratory, Ministry of Health, Nicosia, Cyprus.

Professor El-Sayed El-Mossalami Head, Meat Hygiene Department, Faculty of Veterinary Medicine, Cairo University, Giza, Egypt.

Mr. Yacoub Khalid Motawa Head of Food Control Laboratory, Microbiological Section, Preventive Health, Public Health Laboratory, Ministry of Health, Kuwait.

Dr. Neji Othman Chief, Laboratory of Food Microbiology, Ministry of Health, National Institute of Nutrition, Tunis, Tunisia.

CONSULTANTS FOR THIS BOOK

Mr. J. P. Accolas INRA Laboratoire de microbiologie laitière et de génie alimentaire, 78.350 Jouy-en-Josas, France

Dr. E. M. Barnes Agriculture Research Council, Food Research Institute, Colney Lane, Norwich, England NR4 7UA

Dr. H. J. Beckers National Institute of Public Health, P.O. Box 1, 3720 BA Bilthoven, The Netherlands

Dr. C. F. Böhme Kalle, Niederlassung der Hoechst AG, Rheingaustrasse 190, 6200 Wiesbaden-Biebrich, West Germany

Dr. A. W. M. Brooymans Quaker Europe, Avenue Henri Matisse 16, Bte 6, B-1140 Brussels, Belgium

Mr. M. H. Brown Unilever Research, Colworth Laboratory, Colworth House, Sharnbrook, Bedford, England MK44 1LQ.

Dr. J. Bruijn Sugar Milling Research Institute, University of Natal, King George V Avenue, Durban 4001, South Africa

Dr. D. A. Corlett, Jr. Del Monte Research Center, 205 North Wiget Lane, Walnut Creek, California 94598, U.S.A.

Dr. C. M. Davidson Silliker Laboratories of Canada, 2222 South Sheridan Way, Mississauga, Ontario, Canada L5J 2M4

Dr. A. B. Dickinson Henshelwood Terrace, Jesmond, Newcastle-upon-Tyne, England

Dr. J. C. de Man Nestlé Products Technical Assistance Co. Ltd. (nestec), Case Postale 88, CH-1814 La Tour-de-Peilz, Switzerland

Mr. R. Paul Elliott Consultant in Food Microbiology, 1095 Lariat Lane, Pebble Beach, California 93953, U.S.A.

Mr. R. M. Friesen Griffith Laboratories Limited, 757 Pharmacy Ave., Scarborough, Ontario, Canada M1L 3J8

Dr. J. M. Goepfert Research Centre, Canada Packers Ltd., 2211 St. Clair Ave. W., Toronto, Ontario, Canada, M69 1K4.

Mr. M. Greenall Overseas Egg Company, London, England

Dr. R. A. Greenberg Swift Research and Development Laboratory, 1919 Swift Dr., Oak Brook, IL 60521

Dr. A. Hurst Bureau of Microbial Hazards, Food Directorate, Health Protection Branch, Health and Welfare Canada, Tunney's Pasture, Ottawa, Ontario, Canada K1A 0L2

Professor M. Ingram (Deceased) Formerly Director, Meat Research Institute, Langford, Bristol BS18 7DY England

Dr. M. Jemmali INRA Laboratoire de technologie alimentaire, Service des mycotoxines, 16 Rue Nicolas Fortin, 75013 Paris, France

Dr. J. J. Jezeski Research and Development, H. B. Fuller Co., Monarch Chemicals Division, 3900 Jackson St. Northeast, Minneapolis, Minn. 55421 U.S,A.

Professor P. G. Keeney Department of Food Science, College of Agriculture, Penn. State University, University Park, Pennsylvania 16802 U.S.A.

Mr. A. H. Klopp Griffith Laboratories Limited, 757 Pharmacy Avenue, Scarborough, Ontario, Canada M1L 3J8

Dr. C. Knowles Egg Marketing Board, London, England

Dr. O. E. Kolari Armour Food Company Research Center, 15101 N. Scottsdale Road, Scottsdale, Arizona 85260 U.S.A.

Mr. W. J. Kooiman Unilever Research, Duiven, Helhoek 30, Groessen, Postbus 2, Zevenaar, The Netherlands

Professor J. Liston Director, Institute for Food Science and Technology, College of Fisheries, University of Washington, Seattle, Washington, 98105 U.S.A.

Ms. Helen Lynch Corn Products Company, Federal Street, Yonkers, New York, 10702 U.S.A.

Dr. J. H. McCoy Public Health Laboratory, Hull, U.K.

Miss L. McMaster Department of Microbiology and Plant Pathology, University of Natal, Pietermaritzburg, South Africa

Dr. A. J. Mercuri U.S. Dept. of Agriculture, Science and Education Administration, Richard B. Russell Agricultural Research Center, P.O. Box 5677, Athens, Georgia, 30604, U.S.A.

Dr. E. H. Meursing Cacaofabriek de Zaan b.v. Koog aan de Zaan, Holland

Dr. M. J. M. Michels Unilever Research, P.O. Box 7, Zevenaar, The Netherlands

Dr. R. W. A. W. Mulder Spelderholt Institute for Poultry Research, 7361 DA Beekbergen, The Netherlands

Dr. G. Tuynenburg Muys Unilever Research Vlaardingen/Duiven Olivier van-Noortlaan 120 Vlaardingen, postbus 114 Nederlandse Unilever Bedrijven B.V. stat. zetel Rotterdam, h.reg. Rotterdam nr 53802, The Netherlands

Dr. A. Neitzert Kalle, Niederlassung der Hoechst AG, Rheingaustrasse 190, 6200 Wiesbaden-Biebrich, West Germany

Dr. M. D. Northolt National Institute of Public Health, P.O. Box 1, 3720 BA Bilthoven, The Netherlands

Dr. I. S. H. W. Notermans National Institute of Public Health, P.O. Box 1, 3720 BA Bilthoven, The Netherlands

Dr. S. A. Palumbo Agricultural Research Service, Eastern Regional Research Center, 600 East Mermaid Lane, Philadelphia, Pennsylvania 19118 U.S.A.

Dr. J. T. Patterson Ministry of Agriculture, Agricultural and Food Bacteriology Research Division, Newforge Lane, Belfast, BT9 5PX Northern Ireland

Dr. A. B. Ravno Huletts Sugar Limited, Private Bag X04, Mount Edgecombe 4300, Natal, South Africa

Mr. R. A. Seward Department of Food Science and Technology, University of Wisconsin, Madison, Wisconsin 53706 U.S.A.

Mr. B. H. Siebers The Nestle Company, Inc., Fulton, New York, 13069, U.S.A.

Dr. R. D. Skole Research and Development Department, Amstar Corporation, 49 South Second Street, Brooklyn, New York 11211, U.S.A.

Dr. Richard B. Smittle Silliker Laboratories of New Jersey, Inc. 2353 Beryllium Road Scotch Plains, New Jersey 07076, U.S.A.

Dr. R. H. Tilbury Group Research and Development, Tate and Lyle, Limited, Phillip Lyle Memorial Research Laboratory, University of Reading, Whiteknights, Reading, Berks., England

Dr. R. B. Tompkin Swift Research and Development Laboratory, 1919 Swift Drive, Oak Brook, Illinois, 60521 U.S.A.

Dr. R O Wagenaar Technical Center, General Mills, Inc. 9000 Plymouth Ave. N. Minneapolis, Minnesota, 55427, U.S.A.

Appendix IV

CHOICE OF CASE

An attempt has been made to coordinate Volume II with the Commission's second book, *Microorganisms in Foods, 2: Sampling for Microbiological Analysis,* by a treatment, in each commodity chapter, of "Choice of Case." "Case" is a classification of sampling plans varying from 1 (the least stringent) to 15 (the most stringent). As shown in Table A.1 of this appendix, the choice of Case, and therefore of the sampling plan it represents, depends on two aspects: first, on the relative hazard to food quality or consumer health that the presence of specified microorganisms represent and, second, on the expectation of their destruction or their survival and multiplication during the anticipated normal handling the food undergoes from the time of sampling until consumption. For more details see *Microorganisms in Foods, 2* (ICMSF, 1974).

TABLE A.1

Sampling Plan Stringency (Case) in Relation to Degree of Health Hazard and Condition of Use [a]

Type of hazard	Conditions under which food is expected to be handled and consumed after sampling		
	Reduce degree of hazard (e.g., cook)	Cause no change in hazard (e.g., consume dry)	May increase hazard (e.g., delay after reconstitution, with no cooking)
No direct health hazard Utility (general contamination, reduced self-life, spoilage; e.g. aerobic plate count)	Case 1	Case 2	Case 3
Health hazard Low, indirect (indicators, e.g., coliform group)	Case 4	Case 5	Case 6
Moderate, direct, limited spread e.g., *Staphylococcus aureus*, *Bacillus cereus*, *Clostridium perfringens*)	Case 7	Case 8	Case 9
Moderate, direct, potentially extensive spread (e.g., *Salmonella typhimurium*, *Vibrio parahaemolyticus*, Enteropathogenic *Escherichia coli*)	Case 10	Case 11	Case 12
Severe, direct (e.g., *Clostridium botulinum*, *Salmonella typhi*, *Vibrio comma*)	Case 13	Case 14	Case 15

[a] From ICMSF (1974).

Index

A

Absidia
  in spice, 743
Abuse, *see* Preventing abuse of foods
Acetic acid
  as preservative
    for grain, 690
    for mayonnaise and salad dressing, 754
Acetobacter
  in fermentation
    of cocoa beans, 806, 808, 809, 810
  in fruit, 651
  growth limits
    pH, 808
    temperature, 808
Achromobacter, see Acinetobacter, Moraxella, and *Pseudomonas*
Acidophilus milk, 517–518
Acinetobacter
  on egg shells, fillers, flats, 525
  on fish and shellfish, 574
  in fermentation,
    of wild rice, 676
  in molluscs, 593
  resistance
    to pasteurization in liquid egg, 552
  in spoilage
    of cooked crustaceae, 599, 600
    of crustaceae, 590
    of eggs, 530, 537, 538, 556
    of fish, 584–586

of meat, 344, 345, 350–354, 362, 365–366, 370
of poultry meat, 433–437
of shrimp, 590
of sour cream, 769
on tree nuts, 640
Actinomyces
  in raw cane sugar juice, 782
Actinomycetes
  on grain in field, 672
  in stored grain, 695
Aerobic plate count
  baked goods, 696
  bone meal, 460
  breakfast cereals, 695
  cereal grain, 695
  cereal snacks, 695
  chilled dough, 696
  cocoa, 816
  crustaceae, 572
  dried blood plasma, 379
  dried cereal mixes, 697
  dried egg
    criteria in 19 countries, 560
    FAO criteria, 560
  fish, 572
  fish meal, 465
  flour, 695, 700, 702
  frozen egg
    criteria in 19 countries, 560
    FAO criteria, 560
  frozen vegetables, 614
  gelatin, 380

Index

Aerobic plate count (*cont.*)
  grain in field, 672
  Hungarian sausages, 389
  incubator-reject eggs, 543
  liquid egg, 540, 546
    effect of pasteurization, 543, 545,
      546, 552
  milk, 486
  molluscs, 572
  pasta, 697, 717
  poultry meat
    after deboning and cutting, 432
    after evisceration, 428
    after picking, 427
    after scalding, 425–426
    after spray wash, 428
    to measure cleanliness, 444
    to measure spoilage, 444
  raw chilled meat, 351, 353, 360, 361,
    362, 363, 365, 368, 370, 373
  raw frozen meat, 373
  raw milk, 477
  raw vegetables, 608
  roasted cocoa beans, 813
  salads, 825
  shrimp, 590
  soy protein, 696
  spice, 738, 740–741, 745, 750
  sugar beet flume water, 790
  sugar cane, 780
  tree nuts, 640, 641
  wild rice, 675, 676
Aerococcus
  in frozen vegetables, 613
Aeromonas
  on egg shells, fillers, flats, 525
  in egg spoilage, 537
  on meat, 344, 350
  in meat spoilage, 364, 370
Aerosol, *see* Air
Aflatoxin, *see* Mycotoxin
Air
  as source of microorganisms, 343, 416,
    475, 607
    in bakeries, 712, 716
    in pastry, 724, 728
Air sacs, 425
Alcaligenes
  on egg shells, fillers, flats, 525
  in egg spoilage, 530, 537, 538, 556

on meat carcasses, 344
  in natural mineral waters, 835
  resistance
    to pasteurization in liquid egg, 552
  in spoilage
    of fish, 584
    of milk, 476, 483
    of sour cream, 769
Alcaligenes faecalis
  spoilage,
    of eggs, 537
Alimentary toxic aleukia
  from moldy grain, 685
Allspice, 742, 744
Alternaria
  invader of grain in field, 671
  producing mycotoxin in grain, 685
  in spoilage
    of raw fruit, 625
    of raw vegetables, 610
  on vegetables, 607
Alteromonas
  on fish and shellfish, 574
  in spoilage,
    of fish, 584–585
Aluminum
  effect on egg spoilage, 530
Aluminum sulfate
  effect on pasteurization of egg albumen,
    551
Ammonia
  as preservative,
    for blood plasma, 379
    for grain, 690
Anasakis
  in fish, 578
Angiostrongylus cantonensis
  in crustaceae, 578
Animal by-product meal, 459–466, *see*
    also Bone meal
  pelletizing, 461
  as source of animal disease, 415, 459
  as source of human disease, 461
Antagonism
  among grain fungi, 689
Antibiotic
  effect on cheese fermentation, 504
Antimicrobial substance
  in egg albumen, 529–530
  in fruit, 649

from soil, 416
in spoilage, 433–437
resistance
to low-temperature rendering, 376–
377
on shrimp, 592
in soy protein, 696
in wild rice, 676
Pyoverdine
in eggs, 530
Pyrodinium phoneus
in molluscs, 578–579
Pythium
in spoilage
of raw vegetables, 610

R

Raw cane sugar
spoilage, 786–788
Raw foods
as source of microorganisms
in foodborne disease, 850–852, 858
Raw fruit, 623–626
damage
from chilling, 626
from handling, 623–624
pathogens, 624–625
spoilage, 624
Raw milk, 471–478
certified, 477, 478–479
definition, 471–472
initial microflora, 472–476
spoilage, 476
Raw vegetables
harvesting, transport, storage, 609
initial microflora, 607
pathogens, 608, 610–611
spoilage, 609
Red bread
from *Geotrichum aurantiacum,* 714
from *Serratia,* 714
Red meat, *see* Meat
Redox potential, *see* Oxidation-reduction
potential
References, 862–944
Refined sugar
microbiological criteria, 801–802
as source of microorganisms

flat sour bacteria, 801
hydrogen swell bacteria, 801
soft drink spoilage bacteria, 801
sulfide stinker bacteria, 801
Reheating left-over food, 859–860
Rehydrating foods, 850
Relative humidity
effect on spoilage
of dried meat, 380–381
of meat, 350, 352
of raw cane sugar, 788
of raw fruit, 624
of raw vegetables, 612
Rhizomes, 739
Rhizopus
in spice, 743
in spoilage
of meat pies, 831, 832
of raw fruit, 525
of raw vegetables, 609–610
Rhizopus nigricans
in spoilage
of bread, 713
Rhodotorula
in fruit, 650
in spoilage
of olives, 633
of soft drinks and fruit juices,
655
Riboflavin
in egg albumen, 529
Roasting
cocoa beans, 813–814
Rodents
as source of microorganisms
in fish, 581
in foodborne disease, 846
in spice, 738
as vectors of fungi
in dried fruit, 631
Rope, 713, 716–717
Rumen,
Salmonella in, 346

S

Saccharomyces
in beet sugar, 799